Advances in Intelligent Systems and Computing

Volume 360

Series editor

Janusz Kacprzyk, Polish Academy of Sciences, Warsaw, Poland
e-mail: kacprzyk@ibspan.waw.pl

About this Series

The series "Advances in Intelligent Systems and Computing" contains publications on theory, applications, and design methods of Intelligent Systems and Intelligent Computing. Virtually all disciplines such as engineering, natural sciences, computer and information science, ICT, economics, business, e-commerce, environment, healthcare, life science are covered. The list of topics spans all the areas of modern intelligent systems and computing.

The publications within "Advances in Intelligent Systems and Computing" are primarily textbooks and proceedings of important conferences, symposia and congresses. They cover significant recent developments in the field, both of a foundational and applicable character. An important characteristic feature of the series is the short publication time and world-wide distribution. This permits a rapid and broad dissemination of research results.

Advisory Board

Chairman

Nikhil R. Pal, Indian Statistical Institute, Kolkata, India
e-mail: nikhil@isical.ac.in

Members

Rafael Bello, Universidad Central "Marta Abreu" de Las Villas, Santa Clara, Cuba
e-mail: rbellop@uclv.edu.cu

Emilio S. Corchado, University of Salamanca, Salamanca, Spain
e-mail: escorchado@usal.es

Hani Hagras, University of Essex, Colchester, UK
e-mail: hani@essex.ac.uk

László T. Kóczy, Széchenyi István University, Győr, Hungary
e-mail: koczy@sze.hu

Vladik Kreinovich, University of Texas at El Paso, El Paso, USA
e-mail: vladik@utep.edu

Chin-Teng Lin, National Chiao Tung University, Hsinchu, Taiwan
e-mail: ctlin@mail.nctu.edu.tw

Jie Lu, University of Technology, Sydney, Australia
e-mail: Jie.Lu@uts.edu.au

Patricia Melin, Tijuana Institute of Technology, Tijuana, Mexico
e-mail: epmelin@hafsamx.org

Nadia Nedjah, State University of Rio de Janeiro, Rio de Janeiro, Brazil
e-mail: nadia@eng.uerj.br

Ngoc Thanh Nguyen, Wroclaw University of Technology, Wroclaw, Poland
e-mail: Ngoc-Thanh.Nguyen@pwr.edu.pl

Jun Wang, The Chinese University of Hong Kong, Shatin, Hong Kong
e-mail: jwang@mae.cuhk.edu.hk

More information about this series at http://www.springer.com/series/11156

Hoai An Le Thi · Tao Pham Dinh
Ngoc Thanh Nguyen
Editors

Modelling, Computation and Optimization in Information Systems and Management Sciences

Proceedings of the 3rd International Conference on Modelling, Computation and Optimization in Information Systems and Management Sciences – MCO 2015 – Part II

Springer

Editors
Hoai An Le Thi
Laboratory of Theoretical and
    Applied Computer Science,
University of Lorraine - Metz,
France

Ngoc Thanh Nguyen
Division of Knowledge Management
    Systems,
Wroclaw University of Technology,
Poland

Tao Pham Dinh
Laboratory of Mathematics,
National Institute for
    Applied Sciences - Rouen,
France

ISSN 2194-5357   ISSN 2194-5365   (electronic)
Advances in Intelligent Systems and Computing
ISBN 978-3-319-18166-0   ISBN 978-3-319-18167-7   (eBook)
DOI 10.1007/978-3-319-18167-7

Library of Congress Control Number: 2015937024

Springer Cham Heidelberg New York Dordrecht London

Printed on acid-free paper

Springer International Publishing AG Switzerland is part of Springer Science+Business Media
(www.springer.com)

Preface

This volume contains 86 selected full papers (from 181 submitted ones) presented at the MCO 2015 conference, held on May 11–13, 2015 at University of Lorraine, France.

MCO 2015 is the third event in the series of conferences on Modelling, Computation and Optimization in Information Systems and Management Sciences organized by LITA, the Laboratory of Theoretical and Applied Computer Science, University of Lorraine.

The first conference, MCO 2004, brought together 100 scientists from 21 countries and was a great success. It included 8 invited plenary speakers, 70 papers presented and published in the proceedings, "Modelling, Computation and Optimization in Information Systems and Management Sciences", edited by Thi Hoai An and Pham Dinh Tao, Hermes Sciences Publishing, June 2004, 668 pages, and 22 papers published in the European Journal of Operational Research and in the Journal of Global Optimization. The second conference, MCO 2008 was jointly organized by LITA and the Computer Science and Communications Research Unit, University of Luxembourg. MCO 2008 gathered 66 invited plenary speakers and more than 120 scientists from 27 countries. The scientific program consisted of 6 plenary lectures and of the oral presentation of 68 selected full papers as well as 34 selected abstracts covering all main topic areas. Its proceedings were edited by Le Thi Hoai An, Pascal Bouvry and Pham Dinh Tao in Communications in Computer and Information Science 14, Springer. Two special issues were published in Journal of Computational, Optimization & Application (editors: Le Thi Hoai An, Joaquim Judice) and Advance on Data Analysis and Classification (editors: Le Thi Hoai An, Pham Dinh Tao and Ritter Guntter).

MCO 2015 covered, traditionally, several fields of Management Science and Information Systems: Computer Sciences, Information Technology, Mathematical Programming, Optimization and Operations Research and related areas. It will allow researchers and practitioners to clarify the recent developments in models and solutions for decision making in Engineering and Information Systems and to interact and discuss how to reinforce the role of these fields in potential applications of great impact. It would be a timely occasion to celebrate the 30th birthday of DC programming and DCA, an efficient approach in Nonconvex programming framework.

Continuing the success of the first two conferences, MCO 2004 and MCO 2008, MCO 2015 will be attended by more than 130 scientists from 35 countries. The International

Scientific Committee consists of more than 80 members from about 30 countries all the world over. The scientific program includes 5 plenary lectures and the oral presentation of 86 selected full papers as well as several selected abstracts covering all main topic areas. MCO 2015's proceedings are edited by Le Thi Hoai An, Pham Dinh Tao and Nguyen Ngoc Thanh in Advances in Intelligent Systems and Computing (AISC), Springer. All submissions have been peer-reviewed and we have selected only those with highest quality to include in this book.

We would like to thank all those who contributed to the success of the conference and to this book of proceedings. In particular we would like to express our gratitude to the authors as well as the members of the International Scientific Committee and the referees for their efforts and cooperation. Finally, the interest of the sponsors in the meeting and their assistance are gratefully acknowledged, and we cordially thank Prof. Janusz Kacprzyk and Dr. Thomas Ditzinger from Springer for their supports.

We hope that MCO 2015 significantly contributes to the fulfilment of the academic excellence and leads to greater success of MCO events in the future.

March 2015   Hoai An Le Thi
                                                        Tao Pham Dinh
                                                    Ngoc Thanh Nguyen

DC Programming and DCA:
Thirty Years of Developments

The year 2015 marks the 30th birthday of DC (Difference of Convex functions) programming and DCA (DC Algorithm) which were introduced by Pham Dinh Tao in 1985 as a natural and logical extension of his previous works on convex maximization since 1974. They have been widely developed since 1994 by extensive joint works of Le Thi Hoai An and Pham Dinh Tao to become now classic and increasingly popular.

DC programming and DCA can be viewed as an elegant extension of Convex analysis/Convex programming, sufficiently broad to cover most real-world nonconvex programs, but no too in order to be able to use the powerful arsenal of modern Convex analysis/Convex programming. This philosophy leads to the nice and elegant concept of approximating a nonconvex (DC) program by a sequence of convex ones for the construction of DCA: each iteration of DCA requires solution of a convex program. It turns out that, with appropriate DC decompositions and suitably equivalent DC reformulations, DCA permits to recover most of standard methods in convex and nonconvex programming. These theoretical and algorithmic tools, constituting the backbone of Nonconvex programming and Global optimization, have been enriched from both a theoretical and an algorithmic point of view, thanks to a lot of their applications, by researchers and practitioners in the world, to model and solve nonconvex programs from many fields of Applied Sciences, including Data Mining-Machine Learning, Communication Systems, Finance, Information Security, Transport Logistics & Production Management, Network Optimization, Computational Biology, Image Processing, Robotics, Computer Vision, Petrochemicals, Optimal Control and Automatic, Energy Optimization, Mechanics, etc. As a continuous approach, DC programming and DCA were successfully applied to Combinatorial Optimization as well as many classes of hard nonconvex programs such as Variational Inequalities Problems, Mathematical Programming with Equilibrium Constraints, Multilevel/Multiobjective Programming.

DC programming and DCA were extensively developed during the last two decades. They were the subject of several hundred articles in the high ranked scientific journals and the high-level international conferences, as well as various international research projects, and were the methodological basis of more than 50 PhD theses. More than 90 invited symposia/sessions dedicated to DC programming & DCA were presented in numerous international conferences. The ever-growing number of works using DC programming and DCA proves their power and their key role in Nonconvex programming/Global optimization and many areas of applications.

In celebrating the 30th birthday of DC programming and DCA, we would like to thank the founder, Professor Pham Dinh Tao, for creating these valuable theoretical and algorithmic tools, which have such a wonderful scientific impact on many fields of Applied Sciences.

Le Thi Hoai An

General Chair of MCO 2015

Organization

MCO 2015 is organized by the Laboratory of Theoretical and Applied Computer Science, University of Lorraine, France.

Organizing Committee

Conference Chair

Hoai An Le Thi   University of Lorraine, France

Members

Lydia Boudjeloud   University of Lorraine, France
Conan-Guez Brieu   University of Lorraine, France
Alain Gély   University of Lorraine, France
Annie Hetet   University of Lorraine, France
Vinh Thanh Ho   University of Lorraine, France
Hoai Minh Le   University of Lorraine, France
Duy Nhat Phan   University of Lorraine, France
Minh Thuy Ta   University of Lorraine, France
Thi Thuy Tran   University of Lorraine, France
Xuan Thanh Vo   University of Lorraine, France
Ahmed Zidna   University of Lorraine, France

Program Committee

Program Co-Chairs

Hoai An Le Thi   University of Lorraine, France
Tao Pham Dinh   National Institute for Applied Sciences - Rouen, France

Members

El-Houssaine Aghezzaf	University of Gent, Belgium
Tiru Arthanari	University of Auckland, New Zealand
Adil Bagirov	University of Ballarat, Australia
Younès Bennani	University Paris 13-University Sorbonne Paris Cité, France
Lyes Benyoucef	University of Aix-Marseille, France
Lydia Boudjeloud	University of Lorraine, France
Raymond Bisdorff	University of Luxembourg, Luxembourg
Pascal Bouvry	University of Luxembourg, Luxembourg
Stéphane Canu	INSA–Rouen, France
Emilio Carrizosa	Universidad de Sevilla, Sevilla, Spain
Suphamit Chittayasothorn	King Mongkut's Institute of Technology Ladkraban, Thailand
John Clark	University of York, UK
Rafael Correa	Universidad de Chile, Santiago, Chile
Van Dat Cung	Grenoble INP, France
Frédéric Dambreville	DGA, France
Mats Danielson	Stockholm University, Sweden
Mohamed Djemai	University of Valenciennes, France
Sourour Elloumi	National School of Computer Science for Industry and Business, France
Alexandre Dolgui	Mines Saint-Etienne, France
Love Ekenberg	Stockholm University, Sweden
Ronan M.T. Fleming	University of Luxembourg, Luxembourg
Fabián Flores-Bazán	University of Concepción, Chile
Yann Guermeur	CNRS, France
Nicolas Hadjisavvas	University of the Aegean, Greece
Mounir Haddou	INSA-Rennes, France
Jin-Kao Hao	University of Angers, France
Duong-Tuan Hoang	University of Technology of Sydney, Australia
Van-Ngai Huynh	University of Quy Nhon, Vietnam
Fadili Jalal	University of Caen-IUF, France
J.Jung Jason	Chung-Ang University, South Korea
Joaquim Judice	University Coimbra, Portugal
Kang-Hyun Jo	University of Ulsan, South Korea
Pang Jong-Shi	University of Southern California, USA
Djamel Khadraoui	CRP H. Tudor, Luxembourg
Arnaud Lallouet	University of Caen, France
Aron Larson	Stockholm University, Sweden
Hoai Minh Le	University of Lorraine, France
Van Cuong Le	University Paris 1 Panthéon-Sorbonne, France
Duan Li	Chinese University of Hong Kong, Hong Kong
Yufeng Liu	University of North Carolina at Chapel Hill, USA

Philippe Mahey	ISIMA, France
Ali Ridha Mahjoub	University Paris-Dauphine, France
Francois Malgouyres	University of Paul Sabatier, Toulouse, France
Tobias Marschall	Saarland University, Germany
Shashi-Kant Mishra	Banaras Hindu University, Varanasi, India
Mehryar Mohri	Courant Institute of Mathematical Sciences, New York, USA
Ali Mohammad-Djafari	CNRS, Supélec, France
Kazuo Murota	University of Tokyo, Japan
Amedeo Napoli	CNRS, France
Zuhair Nashed	University of Central Florida, USA
Dong Yen Nguyen	Vietnam Academy of Science and Technology, Vietnam
Ngoc Thanh Nguyen	Wroclaw University of Technology, Poland
Tri Dung Nguyen	University of Southampton, UK
Van Thoai Nguyen	Trier University, Germany
Viet Hung Nguyen	University Pierre and Marie Curie, France
Stefan Nickel	Karlsruhe Institute of Technology, Germany
Yi-Shuai Niu	Paristech-SJTU and Shanghai Jiao Tong University, China
M. Panos Pardalos	University of Florida, USA
Vangelis Paschov	University Paris-Dauphine, CNRS and IUF, France
Gabriel Peyre	University of Paris Dauphine, France
Duc Truong Pham	University of Birmingham, UK
Duc Pham-Hi	ECE Paris, France
Hoang Pham	Rutgers University, USA
Huyen Pham	University Paris Diderot (Paris 7), France
Olivier Pietquin	University Lille 1 - IUF, France
Cedric Pradalier	Georgiatech Lorraine, France
Christian Prins	University of Technology of Troyes, France
Alain Quilliot	ISIMA, France
Nidhal Rezg	University of Lorraine, France
D. Yaroslav Sergeyev	University of Calabria, Italia
Marc Sevaux	University Bretagne-Sud, France
Patrick Siarry	University Paris 12, France
Charles Soussen	University of Lorraine, France
Defeng Sun	National University of Singapore
Jie Sun	Curtin University, Australia
Q. Bao Truong	Northern Michigan University, USA
Truyen Tran	Deakin University, Australia
A. Ismael Vaz	University of Minho, Portugal
Dieter Weichert	University of Technology (RWTH) Aachen, Germany
Gerhard-Wilhelm Weber	Middle East Technical University, Turkey

Adilson Elias Xavier	Federal University of Rio de Janeiro, Rio de Janeiro, Brazil
Jack Xin	University of California at Irvine, USA
Adnan Yassine	University of Le Havre, France
Daniela Zaharie	West University of Timisoara, Romania
Ahmed Zidna	University of Lorraine, France

Organizers of Special Sessions/Workshops

Combinatorial Optimization

Viet Hung Nguyen   University Pierre and Marie Curie, France

DC Programming and DCA: Thirty Years of Developments

Hoai An Le Thi   University of Lorraine, France

Dynamic Optimization

Patrick Siarry   University of Paris-Est Créteil, France

Global Optimization and Semi-infinite Optimization

Mohand Ouanes   University of Tizi-Ouzou, Algeria

Maintenance and Production Control Problems

| Nidhal Rezg, Hajej Zied | University of Lorraine, France |
| Ali Gharbi | ETS Montreal, Canada |

Modeling and Optimization in Computational Biology

Tobias Marschall   Saarland University, Germany

Modeling and Optimization in Financial Engineering

Duan Li   The Chinese University of Hong Kong

Numerical Optimization

Adnan Yassine   University of Le Havre, France

Optimization Applied to Surveillance and Threat Detection

Frédéric Dambreville   ENSTA Bretagne & DGA, France

Post Crises Banking and Eco-finance Modelling

Duc Pham-Hi   ECE Paris, France

Spline Approximation & Optimization

Ahmed Zidna   University of Lorraine, France

Technologies and Methods for Multi-stakeholder Decision Analysis in Public Settings

Aron Larsson, Love Ekenberg,   Stockholm University, Sweden
Mats Danielson

Variational Principles and Applications

Q. Bao Truong   Northern Michigan University, US
Christiane Tammer   Faculty of Natural Sciences II,
                                         Institute of Mathematics, Germany
Antoine Soubeyran   Aix-Marseille University, France

Sponsoring Institutions

University of Lorraine (UL), France
Laboratory of Theoretical and Applied Computer Science, UL
UFR Mathématique Informatique Mécanique Automatique, UL
Conseil Général de la Moselle, France
Conseil Régional de Lorraine, France
Springer
IEEE France section
Mairie de Metz, France
Metz Métropole, France

Contents

Part II: Heuristic / Meta Heuristic Methods for Operational Research Applications

Part III: Optimization Applied to Surveillance and Threat Detection

Part IV: Maintenance and Production Control Problems

Part V: Scheduling

Part VI: Post Crises Banking and Eco-finance Modelling

Part VII: Transportation

Part VIII: Technologies and Methods for Multi-stakeholder Decision Analysis in Public Settings

Part I

Data Analysis and Data Mining

A Multi-criteria Assessment for R&D Innovation with Fuzzy Computing with Words

Wen-Pai Wang[1,*] and Mei-Ching Tang[2]

[1] National Chin-Yi University of Technology,
57, Sec. 2, Zhongshan Rd., Taiping Dist., Taichung 41170, Taiwan, R.O.C.
[2] National Chung Hsing University,
250 Kuo Kuang Rd., Taichung 402, Taiwan R.O.C.
wangwp@ncut.edu.tw,
nacctweb@dragon.nchu.edu.tw

Abstract. The assessment of research and development (R&D) innovation is inherently a multiple criteria decision making (MCDM) problem and has become a fundamental concern for R&D managers in the last decades. Research in identifying the relative importance of criteria used to select a favorable project has relied on subjective lists of criteria being presented to R&D managers. The conventional methods for evaluating corresponding R&D merits are inadequate for dealing with suchlike imprecise, heterogeneity or uncertainty of linguistic assessment. Whereas most attributes and their weights are linguistic variables and not easily quantifiable, 2-tuple fuzzy linguistic representation and multigranular linguistic computing manner are applied to transform the heterogeneous information assessed by multiple experts into a common domain and style. It is advantageous to retain consistency of evaluations. The proposed linguistic computing approach integrates the heterogeneity and determines the overall quality level and the performance with respect to specific quality attributes of an R&D innovation.

Keywords: R&D innovation, Heterogeneity, 2-tuple fuzzy linguistic, Group decision-making, Fuzzy linguistic computing.

1 Introduction

Technology has been extensively recognized as one of the major factors determining the competitiveness of an industry. The key to continued competitiveness for many enterprises lies in their ability to develop and implement new products and processes. In the current business environment of increased global competition and rapid change, research and development (R&D) is an investment companies make in their future. Firms need tools that can help determine the optimum allocation of resources. R&D managers and senior executives are becoming increasingly aware of challenges and opportunities in order to be more efficient and effective in delivering products and

---

* Corresponding author.

© Springer International Publishing Switzerland 2015
H.A. Le Thi et al. (eds.), *Model. Comput. & Optim. in Inf. Syst. & Manage. Sci.*,
Advances in Intelligent Systems and Computing 360, DOI: 10.1007/978-3-319-18167-7_1

services to customers. At the same time, without effective performance evaluation, R&D organizations will find it hard to motivate their R&D scientists and engineers. Achieving the desired outcomes of R&D innovation remains the most critical but yet elusive agenda for all firms [1-4].

While the changing nature of competition has placed firms under heavy pressure to rapidly develop and commercialize new innovations to ensure their survival, the successful release of new products undoubtedly brings firms plenty of profits and revenue growth. Doubts often arise about what and who actually determine project success. The nature of the project may differ from project to project depending on its subject, but the success of a project is the common goal for any organization that performs the project. Performance measurement therefore plays an important role in ensuring the project success and its subsequent usefulness to the sponsoring organization [5-7].

In general, the concept of performance measurement implies identification of certain performance metrics and criteria for their computation [3, 8]. A number of qualitative and quantitative metrics have been conducted to measure the performance of R&D innovation at various levels during the selection phase. However, the explicit consideration of multiple, conflicting objectives in a decision model has made the area of multiple criteria decision-making (MCDM) very challenging. The goal of this paper therefore makes a point of developing an evaluation approach to assessing R&D innovation for achieving business strategies; and further , the business decision mechanism is usually composed of multiple experts who implement alternatives evaluation and decision analysis in the light of association rules and criteria. Experts devote to judge by their experiential cognition and subjective perception in decision-making process. However, there exist considerable extent of uncertainty, fuzziness and heterogeneity [9-11]. Consequently, the heterogeneous information that includes crisp values, interval values and linguistic expression is likely to happen under different criterion.

In addition, it is prone to information loss happen during the integration processes, and gives rise to the evaluation result of performance level may not be consistent with the expectation of evaluators. Hence, developing an intimate method to calculate the performance ratings while the processes of evaluation integration and appropriately to manipulate the operation of qualitative factors and evaluator judgment in the evaluation process could brook no delay. The purpose of this paper is to propose a 2-tuple fuzzy linguistic computing approach to evaluate the suppliers' performance. The proposed approach not only inherits the existing characters of fuzzy linguistic assessment but also overcomes the problems of information loss of other fuzzy linguistic approaches.

2   Literature Review

For many firms, especially those that depend on innovation to stay in business, the key to continued competitiveness lies in their ability to develop and implement new products and processes. For these organizations, R&D is an integral function within the strategic management framework. Even firms with excellent technical skills must work within the limits of available funding and resources. R&D project selection decisions, then, are critical if the organization is to stay in business. There are many mathematical decision-making approaches proposed for suchlike decisions. Accordingly, project performance is a crucial issue for the success of a project, which has attracted much attention. Most studies have attempted to analyze the factors that determine the project success; quality, time, and budget [5, 8, 12].

A number of R&D selection models and methods have been proposed in practitioner and academic literature [1-4, 13].Included in the articles reviewed in their papers is those that utilize criteria and methods such as NPV, scoring models, mathematical programming models, and multi attribute approaches. Even with the number of proposed models, the R&D selection problem remains problematic and few models have gained wide acceptance. Mathematical programming techniques such as linear and integer programming are not commonly used in industry, primarily because of the diversity of project types, resources, and criteria used. They also found that many managers do not believe that the available methods for project selection improve the quality of their decisions. Baker and Freeland [14] identified the weaknesses by are:

1. inadequate treatment of multiple, often interrelated, criteria;
2. inadequate treatment of project interrelationships with respect to both value contribution and resource usage;
3. inadequate treatment of risk and uncertainty;
4. inability to recognize and treat nonmonetary aspects;
5. perceptions held by the R&D managers that the models are unnecessarily difficult to understand and use.

The R&D project selection problem plays a critical function in many organizations. A review of literature reveals three major themes relating to R&D project selection: (1) the need to relate selection criteria to corporate strategies; (2) the need to consider qualitative benefits and risks of candidate projects; and (3) the need to reconcile and integrate the needs and desires of different stakeholders [13]. Kim and Oh [3] showed most of the R&D personnel prefer their compensation based on their performance, and indicate that lack of fair performance evaluation system could be the biggest obstacle towards implementing such a compensation scheme.

Mohanty et al. [4] indicated that R&D project selection is indeed a complex decision- making process. It involves a search of the environment of opportunities, the generation of project options, and the evaluation by different stakeholders of multiple attributes, both qualitative and quantitative. Qualitative attributes are often accompanied by certain ambiguities and vagueness because of the dissimilar perceptions of organizational goals among pluralistic stakeholders, bureaucracy and the functional specialization of organizational members.

Other factors complicate the decision process. Often, especially in portfolio selection situations, different projects with different impacts must be compared. There may also be overlaps, synergies, and other interactions of the projects that must be considered. R&D projects are often initiated and championed in a bottom-up manner, where engineers or scientists may advocate projects that have great technical merit. However, financial or strategic benefits of the technology should be considered simultaneously.

It is however difficult and laborious to measure R&D projects' performance using traditional crisp value directly as the process of R&D project performance measurement is possessed of many intangible or qualitative factors and items [2, 4, 15]. Linguistic variable representation is therefore favorable for experts to express and evaluate the ratings of R&D projects under such situation. The fundamentals of 2-tuple fuzzy linguistic approach are to apply linguistic variables to stand for the difference of degree and to carry out processes of computing with words easier and without information loss during the integration procedure [16]. That is to say, decision participators or experts can use linguistic variables to estimate measure items and obtain the final evaluation result with proper linguistic variable. It is an operative method to reduce the decision time and mistakes of information translation and avoid information loss through computing with words.

3   Preliminary Concepts

Many aspects of different activities in the real world cannot be assessed in a quantitative form, but rather in a qualitative one, i.e., with vague or imprecise knowledge. Whereas characteristics of the fuzziness and vagueness are inherent in various decision-making problems, a proper decision-making approach should be capable of dealing with vagueness or ambiguity.

A fuzzy set $\tilde{A}$ in a universe of discourse X is characterized by a membership function $\mu_{\tilde{A}}(x)$, which associates with each element x in X a real number in the interval $[0,1]$. The function value $\mu_{\tilde{A}}(x)$ is termed the grade of membership of x in $\tilde{A}$. A fuzzy number is a fuzzy subset in the universe of discourse X that is both convex and normal. (See Fig. 1)

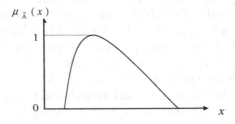

Fig. 1. Fuzzy number $\tilde{A}$

The fuzzy linguistic approach represents qualitative aspects as linguistic values by means of linguistic variables [17]. The concept of linguistic variable is very useful in dealing with situations which are too complex or too ill-defined to be reasonably described in conventional quantitative expressions. For easing the computation and identifying the diversity of each evaluation item, linguistic terms are often possessed of some characteristics like finite set, odd cardinality, semantic symmetric, ordinal level and compensative operation. At present, many aggregation operators have been developed to aggregate information. Herrera and Martinez [18] proposed a 2-tuple fuzzy linguistic representation model. The linguistic information with a pair of values is called 2-tuple that composed by a linguistic term and a number. This representation benefits to be continuous in its domain. It can express any counting of information in the universe of the discourse, and can be denoted by a symbol $L=(s_i, \alpha_i)$ where s_i denotes the central value of the ith linguistic term, and α_i indicates the distance to the central value of the ith linguistic term. (See Fig. 2)

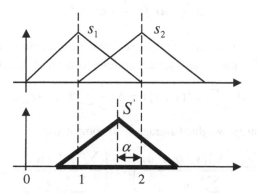

Fig. 2. Representation of 2-tuple linguistic

Suppose $L_1=(s_1, \alpha_1)$ and $L_2=(s_2, \alpha_2)$ are two linguistic variables represented by 2-tuples. The main algebraic operations are shown as follows:

$$L_1 \oplus L_2 = (s_1, \alpha_1) \oplus (s_2, \alpha_2) = (s_1 + s_2, \alpha_1 + \alpha_2)$$
$$L_1 \otimes L_2 = (s_1, \alpha_1) \otimes (s_2, \alpha_2) = (s_1 s_2, \alpha_1 \alpha_2)$$

where $\oplus$ and $\otimes$ symbolize the addition and multiplication operations of parameters, respectively.

Let (s_i, α_i) and (s_j, α_j) be two 2-tuples, with each one representing a counting of information as follows:

1. If $i > j$ then (s_i, α_i) is better than (s_j, α_j);
2. If $i = j$ and $\alpha_i > \alpha_j$ then (s_i, α_i) is better than (s_j, α_j);
3. If $i = j$ and $\alpha_i < \alpha_j$ then (s_i, α_i) is worse than (s_j, α_j);
4. If $i = j$ and $\alpha_i = \alpha_j$ then (s_i, α_i) is equal to (s_j, α_j), i.e. the same information.

The symbolic translation function Δ is presented to translate β into a 2-tuple linguistic variable [19]. Then, the symbolic translation process is applied to translate β ($\beta \in [0, 1]$) into a 2-tuple linguistic variable. The generalized translation function can be represented as [20]:

$$\Delta : [0,1] \rightarrow S \times [-\frac{1}{2g}, \frac{1}{2g}) \tag{1}$$

$$\Delta(\beta) = (s_i, \alpha_i) \tag{2}$$

where $i = round(\beta \times g)$, $\alpha_i = \beta - \dfrac{i}{g}$ and $\alpha_i \in [-\dfrac{1}{2g}, \dfrac{1}{2g})$.

A reverse function Δ^{-1} is defined to return an equivalent numerical value β from 2-tuple linguistic information $(s_i,\ \alpha_i)$. According to the symbolic translation, an equivalent numerical value β is obtained as follow [20].

$$\Delta^{-1}(s_i,\alpha_i) = \frac{i}{g} + \alpha_i = \beta \tag{3}$$

Let $S=\{(s_1,\ \alpha_1),\ldots,(s_n,\ \alpha_n)\}$ be a 2-tuple linguistic variable set and $W=\{w_1,\ldots,w_n\}$ be the weight set of linguistic terms, their arithmetic mean $\overline{S}$ is calculated as [16]

$$\overline{S} = \Delta[\frac{1}{n}\sum_{i=1}^{n}\Delta^{-1}(s_i,\alpha_i)] = \Delta(\frac{1}{n}\sum_{i=1}^{n}\beta_i) = (s_m,\alpha_m) \tag{4}$$

The 2-tuple linguistic weighted average $\overline{S}^W$ is computed as

$$\overline{S}^w = \Delta\left(\frac{\sum_{i=1}^{n}\Delta^{-1}(s_i,\alpha_i)\cdot w_i}{\sum_{i=1}^{n}w_i}\right) = \Delta\left(\frac{\sum_{i=1}^{n}\beta_i\cdot w_i}{\sum_{i=1}^{n}w_i}\right) = (s^w,\alpha^w) \tag{5}$$

Moreover, let $W=\{(w_1,\ \alpha_{w1}),\ldots,(w_n,\ \alpha_{wn})\}$ be the linguistic weight set of linguistic terms. The linguistic weighted average $\overline{S}^{LW}$ can be computed as

$$\overline{S}^{LW} = \Delta\left(\frac{\sum_{i=1}^{n}\beta_i\times\beta_{w_i}}{\sum_{i=1}^{n}\beta_{w_i}}\right) = (s^{LW},\alpha^{LW}) \tag{6}$$

with $\beta_i = \Delta^{-1}(s_i,\alpha_i)$ and $\beta_{w_i} = \Delta^{-1}(w_i,\alpha_{w_i})$

Table 1. Different types of linguistic variables

Type	No. of linguistic	Linguistic variable
A	3	Poor(s_0^3), average(s_1^3), good(s_2^3)
B	5	Very poor(s_0^5), poor(s_1^5), average(s_2^5), good(s_3^5), very good(s_4^5)
C	7	Extremely Poor(s_0^7), Poor(s_1^7), Medium Poor(s_2^7), Fair(s_3^7), Medium Good(s_4^7), Good(s_5^7), Extremely Good(s_6^7)
D	9	Extremely poor(s_0^9), very poor(s_1^9), poor(s_2^9), fair(s_3^9), average(s_4^9), good(s_5^9), very good(s_6^9), extremely good(s_7^9), excellent(s_8^9)

In general, decision makers will use the different types of 2-tuple linguistic variables based on their knowledge or experiences to express their opinions [21]. For example, the different types of linguistic variables are shown as Table 1. Each 2-tuple linguistic variable can be represented as a triangle fuzzy number. In order to aggregate the evaluation ratings of all decision-makers, a transformation function is needed to transfer these 2-tuple linguistic variables from different linguistic sets to a standard linguistic set at unique domain. According to the method of Herrera and Martinez [19], the domain of the linguistic variables will increase as the number of linguistic variable is increased. To overcome this drawback, a new translation function is applied to transfer a crisp number or 2-tuple linguistic variable to a standard linguistic term at the unique domain [20]. Suppose that the interval [0, 1] is the unique domain. The linguistic variable sets with different semantics (or types) will be defined by partitioning the interval [0, 1].

Furthermore, transforming a crisp number β ($\beta \in$ [0, 1]) into i^{th} linguistic term (s_i, α_i) ($s_i^{n(t)}, \alpha_i^{n(t)}$) of type t as

$$\Delta_t(\beta) = (s_i^{n(t)}, \alpha_i^{n(t)}) \tag{7}$$

where $i = round\,(\beta \times g_t)$, $\alpha_i^{n(t)} = \beta - \dfrac{i}{g_t}$, $g_t = n(t)\text{-}1$ and $n(t)$ is the number of linguistic variable of type t.

Transforming i^{th} linguistic term of type t into a crisp number β ($\beta \in$ [0, 1]) as

$$\Delta_t^{-1}(s_i^{n(t)}, \alpha_i^{n(t)}) = \frac{i}{g_t} + \alpha_i^{n(t)} = \beta \tag{8}$$

where $g_t = n(t)\text{-}1$ and $\alpha_i^{n(t)} \in [-\dfrac{1}{2g_t}, \dfrac{1}{2g_t})$.

Therefore, the transformation from i^{th} linguistic term $(s_i^{n(t)}, \alpha_i^{n(t)})$ of type t to k^{th} linguistic term $(s_k^{n(t+1)}, \alpha_k^{n(t+1)})$ of type $t+1$ at interval [0, 1] can be expressed as

$$\Delta_{t+1}(\Delta_t^{-1}(s_i^{n(t)}, \alpha_i^{n(t)})) = (s_k^{n(t+1)}, \alpha_k^{n(t+1)}) \tag{9}$$

where $g_{t+1} = n(t+1)\text{-}1$ and $\alpha_k^{n(t+1)} \in [-\dfrac{1}{2g_{t+1}}, \dfrac{1}{2g_{t+1}})$.

4 Procedure for Fuzzy Linguistic Evaluation

R&D innovation evaluation/selection is full of uncertainty and can be viewed as a multiple-criteria decision that is normally made by a review committee. Afterward the following questions are expected to be answered via the proposed approach:

> ➤ Which innovation ideas should be selected ultimately? (It means perfect ideas).
> ➤ Which innovation ideas must be supported via evaluation programs? (It means favorable ideas).
> ➤ Which innovation ideas no longer should be considered in any level? (It means bad or unsuitable ideas).

Based on the profound discussions with decision-makers (DMs) the most prevalent and meaningful concerns and sub-criteria for this exemplification can be concluded as follows.

C1. _Technical_: factors related to the project itself and the technology being investigated. Specific measures include:
  C11: Existence of project champion (EPC);
  C12: Probability of technical success (PTS);
  C13: Existence of required competence (ERC);

C_{14}: Availability of available resources (AAR);
C_{15}: Time to market (TM).

C2. *Market*: factors related to the success of the technology and its associated products as related to commercial and marketing. Specific measures include:

C_{21}: Probability of market success of product (PMS);
C_{22}: Number and strength of competitors (NSC);
C_{23}: Potential size of market (PSM);
C_{24}: Net present value (NPV);

C3. *Organizational*: includes internal and external cultural and political factors that might influence the decision.

C_{31}: Strategic fit (SF);
C_{32}: External regulations (ER);
C_{33}: Workplace safety (WS);

The algorithm procedure of the proposed evaluation approach is organized sequentially into six steps (see Fig. 3), and explained as follows:

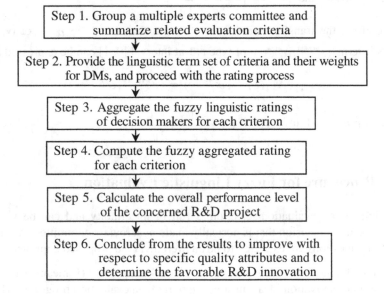

Fig. 3. Procedure of algorithm on proposing approach

Step 1.   Group the committee who are familiar with marketing, business strategies, selling and the relationship of supply and demand. Identify and divide the evaluation criteria into positive criteria (the higher the rating, the greater the preference) and negative criteria (the lower the rating, the greater the preference). For example, DMs choose one kind of linguistic variables respectively to determine the importance of each criterion and the performance of each factor with respect to each criterion. Afterward the rating outcome is shown in Tables 2 and 3.

Table 2. Linguistic evaluations of each decision maker for each criterion and elements

Criteria	Decision-makers			
C1. Technical:	D_1	D_2	D_3	D_4
C_{11}: Existence of project champion (EPC)	G	VG	A	VG
C_{12}: Probability of technical success (PTS)	VG	A	VG	G
C_{13}: Existence of required competence (ERC)	G	VG	G	G
C_{14}: Availability of available resources (AAR)	VG	G	A	VG
C_{15}: Time to market (TM)	A	A	VG	VG
C2. Market:				
C_{21}: Probability of market success of product (PMS)	VG	A	VG	A
C_{22}: Number and strength of competitors (NSC)	VG	G	A	G
C_{23}: Potential size of market (PSM)	A	G	VG	G
C_{24}: Net present value (NPV)	G	VG	VG	A
C3. Organizational:				
C_{31}: Strategic fit (SF)	VG	G	G	VG
C_{32}: External regulations (ER)	A	VG	A	VG
C_{33}: Workplace safety (WS)	G	VG	VG	G

Table 3. Linguistic evaluations of importance of each criterion and corresponding elements

Criteria	Decision-makers			
C1. Technical:	D_1	D_2	D_3	D_4
C_{11}: Existence of project champion (EPC)	VI	VI	VI	I
C_{12}: Probability of technical success (PTS)	VI	VI	VI	I
C_{13}: Existence of required competence (ERC)	A	VI	A	A
C_{14}: Availability of available resources (AAR)	I	I	A	A
C_{15}: Time to market (TM)	I	I	VI	I
C2. Market:	VI	VI	VI	VI
C_{21}: Probability of market success of product (PMS)	I	I	VI	I
C_{22}: Number and strength of competitors (NSC)	I	VI	VI	VI
C_{23}: Potential size of market (PSM)	I	A	VI	VI
C_{24}: Net present value (NPV)	A	I	A	I
C3. Organizational:				
C_{31}: Strategic fit (SF)	I	VI	VI	I
C_{32}: External regulations (ER)	I	I	I	I
C_{33}: Workplace safety (WS)	A	VI	A	I

Step 2. Selectable categories of linguistic terms in Table 1 for experts are prepared when they apply the linguistic importance variables to express the weight of each criterion and employ the linguistic rating variables to evaluate the performance of sub-criteria with respect to each criterion.

Step 3. Aggregate the fuzzy linguistic assessments of the N experts for each criterion by Eqs. (4), (5) and (6); In the meanwhile transform crisp numbers and linguistic terms mutually by Eqs. (7) and (8), and Transform different types of linguistic terms mutually by Eq. (9);

Step 4. Compute the fuzzy aggregated rating of $C_i(\overline{S}_i)$ through Eq. (5);

Table 4. Aggregation results

Criteria	Mean rating	Mean weighting	Weighted rating	Aggregated weighting
C1. Technical:				
C_{11}: Existence of project champion (EPC)	$(S_3, 0.0625)$	$(S_2, 0.125)$		
C_{12}: Probability of technical success (PTS)	$(S_3, 0.0625)$	$(S_2, 0.125)$		
C_{13}: Existence of required competence (ERC)	$(S_3, 0.0625)$	$(S_3, 0.0625)$	$(S_3, 0.0494)$	$(S_4, -0.0625)$
C_{14}: Availability of available resources (AAR)	$(S_3, 0.0625)$	$(S_4, 0)$		
C_{15}: Time to market (TM)	$(S_3, 0)$	$(S_3, 0.0625)$		
C2. Market:				
C_{21}: Probability of market success of product (PMS)	$(S_3, 0)$	$(S_3, 0.0625)$		
C_{22}: Number and strength of competitors (NSC)	$(S_3, 0)$	$(S_2, 0.125)$	$(S_3, 0.015)$	$(S_4, -0.0625)$
C_{23}: Potential size of market (PSM)	$(S_3, 0)$	$(S_4, -0.0625)$		
C_{24}: Net present value (NPV)	$(S_3, 0.0625)$	$(S_3, 0)$		
C3. Organizational:				
C_{31}: Strategic fit (SF)	$(S_3, 0.125)$	$(S_3, 0)$		
C_{32}: External regulations (ER)	$(S_3, 0)$	$(S_3, -0.0625)$	$(S_3, 0.0846)$	$(S_3, 0.125)$
C_{33}: Workplace safety (WS)	$(S_3, 0.125)$	$(S_3, -0.0625)$		

Step 5.   Compute the overall performance level (*OPL*) of a concerned R&D project, the linguistic term s_T, can be applied to represent the control and management performance level of projects as well as being the improvement index directly.

$$OPL = \Delta\left(\frac{\sum_{i=1}^{n}\beta_i \cdot \beta_{w_i}}{\sum_{i=1}^{n}\beta_{w_i}}\right) = (s_T, \alpha_T) \quad \text{with } \beta_i$$

$$= \Delta^{-1}(r_i, \alpha_i) \text{ and } \beta_{w_i} = \Delta^{-1}(w_i, \alpha_{w_i}) \tag{10}$$

According to the above-mentioned values, the *OPL* can be computed as:

$$OPL = \Delta\frac{\left(\begin{array}{l}\Delta^{-1}(s_3, 0.0494)\cdot\Delta^{-1}(s_4, -0.0625)+\Delta^{-1}(s_3, 0.015)\cdot\Delta^{-1}(s_4, -0.0625)\\+\Delta^{-1}(s_3, 0.0846)\cdot\Delta^{-1}(s_3, 0.125)\end{array}\right)}{[\Delta^{-1}(s_4, -0.0625)+\Delta^{-1}(s_4, -0.0625)+\Delta^{-1}(s_3, 0.125)]}$$

$$= \Delta\left(\frac{0.7994\cdot 0.9375 + 0.765\cdot 0.9375 + 0.8346\cdot 0.875}{0.9375 + 0.9375 + 0.875}\right)$$

$$= \Delta(0.7988) = (s_3, 0.0488)$$

Step 6.   Conclude from the results to improve with respect to specific quality attributes and to determine the favorable R&D innovation further. For example, in contrast with linguistic term set *S*, the obtained overall performance level (*OPL*) is 2-tuple fuzzy linguistic information. The transformed value $(S_3, 0.0488)$ represents slightly better than "Good".

5   Conclusions

In relation to the scenario of global competition of nowadays and rapid technology change, R&D innovation assessment is full of subjectivity, uncertainty and budget limitation, and can be therefore viewed as a group decision making problem in a multiple criteria environment. This paper applies 2-tuple fuzzy linguistic representation and multigranular linguistic computing manner to transform the heterogeneous information assessed by multiple experts into a common domain and style. It is advantageous to retain consistency of evaluations. The proposed linguistic computing approach integrates the heterogeneity and determines the overall quality level and the performance with respect to specific quality attributes of an R&D innovation. It is capable of providing the comprehension of R&D features for companies through the proposed measurement index and related dimensions of management and technology. In addition it offers the reference of regulating the market strategies for competitiveness through the proposed critical factors.

Acknowledgments. The research in this paper was partially supported by the National Science Council of Taiwan under grant NSC104-2914-I-167-003-A1.

References

1. Eilat, H., Golany, B., Shtub, A.: R&D Project Evaluation: An Integrated DEA and Balanced Scorecard Approach. Omega-Int. J. Manage. S. 36(5), 895–912 (2008)
2. Huang, C.C., Chu, P.Y., Chiang, Y.H.: A Fuzzy AHP Application in Government- Sponsored R&D Project Selection. Omega-Int. J. Manage. S. 36(6), 1038–1052 (2008)
3. Kim, B., Oh, H.: An Effective R&D Performance Measurement System: Survey of Korean R&D Researchers. Omega-Int. J. Manage. S. 30(1), 19–31 (2002)
4. Mohanty, R.P., Agarwal, R., Choudhury, A.K., Tiwari, M.K.: A fuzzy ANP-based approach to R&D project selection: a case study. Int. J. Prod. Res. 43(24), 5199–5216 (2005)
5. Hélène Sicotte, H., Langley, A.: Integration Mechanisms and R&D Project Performance. J. Eng. Technol. Manage. 17(1), 1–37 (2000)
6. Huang, Y.A., Chung, H.J., Lin, C.: R&D Sourcing Strategies: Determinants and Consequences. Technovation 29(3), 155–169 (2009)
7. Pillai, A.S., Joshi, A., Rao, K.S.: Performance Measurement of R&D Projects in a Multi-Project, Concurrent Engineering Environment. Int. J. Proj. Manag. 20(2), 165–177 (2002)
8. Klapka, J., Piňos, P.: Decision Support System for Multicriterial R&D and Information Systems Projects Selection. Eur. J. Oper. Res. 140(2), 434–446 (2002)
9. Chen, S.J., Hwang, C.L.: Fuzzy Multiple Attribute Decision Making–Methods and Applications. Springer, New York (1992)
10. Hwang, C.L., Yoon, K.: Multiple Attributes Decision Making Methods and Applications. Springer, New York (1981)
11. Wang, W.P.: Toward Developing Agility Evaluation of Mass Customization Systems Using 2-Tuple Linguistic Computing. Expert Syst. Appl. 36(2), 3439–3447 (2009)
12. Wi, H., Jung, M.: Modeling and Analysis of Project Performance Factors in an Extended Project-Oriented Virtual Organization (EProVO). Expert Syst. Appl. 37(2), 1143–1151 (2010)

13. Meade, L.M., Presley, A.: R&D Project Selection Using the Analytic Network Process. IEEE T. Eng. Manage. 49(1), 59–66 (2002)
14. Baker, N., Freeland, J.: Recent Advances in R&D Benefit Measurement and Project Selection Methods. Manage. Sci. 21(10), 1164–1175 (1975)
15. Hwang, H.S., Yu, J.C.: R&D Project Evaluation Model Based on Fuzzy Set Priority. Comput. Ind. Eng. 35(3-4), 567–570 (1998)
16. Herrera-Viedma, E., Herrera, F., Martinez, L., Herrera, J.C., Lopez, A.G.: Incorporating Filtering Techniques in a Fuzzy Linguistic Multi-Agent Model for Information Gathering on the Web. Fuzzy Set Syst. 148(1), 61–83 (2004)
17. Zadeh, L.A.: The Concept of a Linguistic Variable and its Application to Approximate Reasoning. Inform. Sciences 8(3, 4), 199–249(pt. I), 301–357(pt. II) (1975)
18. Herrera, F., Martinez, L.: An Approach for Combining Linguistic and Numerical Information Based on 2-Tuple Fuzzy Representation Model in Decision-Making. Int. J. Uncertain. Fuzz. 8(5), 539–562 (2000)
19. Herrera, F., Martinez, L.: A Model Based on Linguistic 2-Tuples for Dealing with Multigranular Hierarchical Linguistic Contexts in Multi-Expert Decision-Making. IEEE T. Syst. Man Cy. B 31(2), 227–234 (2001)
20. Tai, W.S., Chen, C.T.: A new evaluation model for intellectual capital based on computing with linguistic variable. Expert Syst. Appl. 36(2), 3483–3488 (2009)
21. Herrera, F., Martinez, L., Sanchez, J.: Managing Non-Homogeneous Information in Group Decision Making. Eur. J. Oper. Res. 166, 115–132 (2005)

DEA for Heterogeneous Samples

Fuad Aleskerov and Vsevolod Petrushchenko*

National Research University Higher School of Economics,
International Laboratory on Decision Choice and Analysis,
Myasnitskaya 20, 101000 Moscow, Russia
alesk@hse.ru, goroddt@yandex.ru

Abstract. Data Envelopment Analysis is not very well applicable when a sample consists of firms operating under drastically different conditions. We offer a new method of efficiency estimation on heterogeneous samples based on a sequential exclusion of alternatives and standard DEA approach. We show a connection between efficiency scores obtained via standard DEA model and the ones obtained via our algorithm. We also illustrate our model by evaluating 28 Russian universities and compare the results obtained by two techniques.

Keywords: Data Envelopment Analysis, Heterogeneity, Sequential Exclusion of Alternatives, Universities.

1 Introduction

There are a lot of papers dealing with the problem of adapting DEA to heterogeneous samples. Some classic references here are [2], [3], [4], [5], [9]. The short description of all above techniques can be found in [7].

Somewhat new idea is to combine the power of clustering models with DEA (see, e.g., [16], [19], [11], [13], [15], [18], [14], [17]).

Another approach introduced by [8] deals with the problem of extreme units in DEA.

The so-called context dependent DEA introduced in [6] is another model aimed at heterogeneity problem in efficiency assessment. Authors state that the model measures the relative attractiveness of evaluated items on a specific performance level against items exhibiting poorer performance.

Our model differs from the above mentioned approaches. We suggest to move the efficiency frontier in a special way, using the barycenter of the sample.

The next section presents our model. In the last section we illustrate the proposed method via evaluating the efficiency scores of 28 Russian universities and comparing them with standard DEA.

---

* We thank Decision Choice and Analysis Laboratory and Scientific Fund of National Research University Higher School of Economics for partial financial support.

© Springer International Publishing Switzerland 2015   15
H.A. Le Thi et al. (eds.), *Model. Comput. & Optim. in Inf. Syst. & Manage. Sci.*,
Advances in Intelligent Systems and Computing 360, DOI: 10.1007/978-3-319-18167-7_2

2   The Model

Suppose that all evaluated firms are characterized by N input and M output variables. We construct a sequence of linear programs which allow us to take into account heterogeneity of the sample. Besides, we want our model to have the following properties

1) Convergence of the procedure for any sample;
2) Efficient firms are exactly those which are efficient according to standard CRS (constant return to scale) model;
3) Clear connection between efficiency scores obtained via standard CRS model and our algorithm.

Let $x_i = (x_{1i}, \ldots, x_{Ni})$ and $q_i = (q_{1i}, \ldots, q_{Mi})$ be the vectors of inputs and outputs for i-th firm. We define the input and output parts of the barycenter as

$$b_x = (\overline{x}_1, \ldots, \overline{x}_N).$$

and

$$b_q = (\overline{q}_1, \ldots, \overline{q}_M),$$

where, as usual, the bar means the average value of a particular parameter.

We will also need the heterogeneity index $\mu \in [0, 1]$ which measures to what extent the sample is heterogeneous. The higher heterogeneity the higher the value of μ.

Let the whole sample be defined by the set of indices $I = \{1, \ldots, L\}$. Recall that there are N input and M output variables. We denote the group of 100% efficient (relatively to the standard CRS DEA model) companies as a subset $I_e = \{i_1, \ldots, i_S\} \subset I$. Now, let X_e be the $N \times S$ matrix of input parameters of all efficient firms, and Q_e be the $M \times S$ matrix of outputs for the same firms. We also define the following matrix

$$B_x^i = \big\| \underbrace{b_x^T, \ldots, b_x^T}_{S}, x_i^T \big\|,$$

where b_x^T is the input part of the barycenter written as column, x_i is the input vector for some inefficient firm, i.e., $i \in I \setminus I_e$. Similarly we define the $M \times (S+1)$ matrix

$$B_q^i = \big\| \underbrace{b_q^T, \ldots, b_q^T}_{S}, q_i^T \big\|,$$

where b_q^T is the output part of the barycenter written as column, and q_i is the vector of outputs for the i-th company, $i \in I \setminus I_e$. Let X_e^i and Q_e^i be the matrices X_e and Q_e with the one added column — x_i^T and q_i^T, respectively. Since the core idea is to move the frontier towards the barycenter, we can form the matrices

$$X_i = \mu B_x^i + (1 - \mu) X_e^i \text{ and } Q_i = \mu B_q^i + (1 - \mu) Q_e^i, \tag{1}$$

where the product of a matrix by a scalar is defined in the usual componentwise way.

Note that matrices (1) are defined only for inefficient companies, i.e. $i \in I \setminus I_e$. Also, the last column of X_i and Q_i are x_i^T and q_i^T, respectively.

Now we introduce the general form of the procedure. The first step is to solve the following linear program for every inefficient firm $i \in I \setminus I_e$.

$$\min_{\lambda, \theta_i^*} \theta_i^* \tag{2}$$

subject to

$$\begin{cases} -q_i + Q_i\lambda \geq 0; \\ \theta_i^* x_i - X_i\lambda \geq 0; \\ \lambda \geq 0, \end{cases} \tag{3}$$

where X_i and Q_i are defined in (1), λ is $(S+1) \times 1$ vector of constants, x_i and q_i are input and output vectors for the i-th inefficient firm. Finally, θ_i^* is the corrected efficiency score of the i-th inefficient company.

The algorithm works as follows. Let $Z_1 = I \setminus I_e$, i.e., Z_1 is the set of all firms inefficient according to the standard CRS model. The description of k-th step is as follows

1. Calculation of the barycenter of firms included in the set $Z_k \cup I_e$;
2. Calculation of new matrices X_i and Q_i for all companies i in Z_k;
3. Calculation of $\theta_i^* < 1$ for all companies i in Z_k;
4. All those companies i which get $\theta_i^* < 1$ are excluded from the sample, i.e. $Z_{k+1} = Z_k \setminus \{i | \theta_i^* < 1\}$;
5. If Z_{k+1} is empty then stop, if not – begin $(k+1)$-th stage of the algorithm.

We did not take into account only one case, when matrices (1) are organized in such a way that for all inefficient companies $\theta^* = 1$ holds, i.e., $Z_{k+1} = Z_k$ at some stage k. It means that the original frontier is moved too much. Therefore we have to decrease the value of μ and begin the procedure from the beginning. For instance, we can take the new value of the heterogeneity index as μ^2.

To conclude we make two remarks regarding the algorithm. The convergence is guaranteed by construction. The set of efficient firms is preserved as well. Since we use standard DEA CRS model, we can calculate a projection of every inefficient firm on the temporary frontier defined by (2)-(3) programm (see [7] for details). Then it holds that

$$E_F^{CRS} = E_P^{CRS} \cdot E_F^{New}, \tag{4}$$

where E_F^{CRS} is the standard efficiency score of the firm F, E_P^{CRS} is the efficiency of the projection P of the firm F on the new frontier defined by (2)-(3). Finally, E_F^{New} is the efficiency score of the firm F according to our procedure.

3   An Application to Russian Universities

We apply now our model to evaluate efficiency scores for 28 Russian universities, and compare the results with the standard DEA outcome. The detailed description of related research is presented in [1]. To apply DEA we choose three input parameters, which reflect main universities' resources – the level of state financing, quality of professorial and teaching staff and a quality of entrants.

1. Funding from federal budget (denoted as I_1);
2. The number of employees with a degree of Doctor of Science (denoted as I_2);
3. The quality of university entrants, to estimate this parameter we use a mean value of Universal State Exam (USE), which is mandatory for admission (denoted as I_3).

and two output parameters

1. The number of students who do not pay tuition (denoted as Q_1);
2. The number of published articles in refereed journals (denoted as Q_2).

The first output indicates the attractiveness of a university for the applicants and the second one is a proxy for the level of research in a university.

First, we calculate efficiency scores according to standard DEA model. After that we apply our technique taking three distinct values of heterogeneity index μ, namely, 0.2, 0.5 and 0.8. We would like to point out that these values are chosen only to test the developed model, we did not compute the index of heterogeneity of Russian universities in a fair way. Appendix 1 contains the detailed list of efficiency scores for all four cases.

We compare the results obtained via different models in two ways. First, we rank all universities according to their efficiency scores in each of four cases and compare different orderings via Kendall's distance, see [12].

Recall that the value of Kendall's distance lie between -1 and 1, where 1 means that two orderings are the same and -1 means that the two rankings are inverse.

Table 1 shows the Kendall's distance between all four types of efficiency evaluation models.

Table 1. Kendall's distances

	DEA	$\mu = 0.2$	$\mu = 0.5$	$\mu = 0.8$
DEA	1	-	-	-
$\mu = 0.2$	0.9086	1	-	-
$\mu = 0.5$	0.7849	0.8441	1	-
$\mu = 0.8$	0.7258	0.7097	0.6720	1

Note that the distance between ranks obtained via the technique is small. Moreover the nearer the value of μ to 1 the higher the bias of new efficiency scores from the ones obtained via standard DEA.

4   Conclusion

We have introduced a new algorithm of efficiency evaluation in the case when the sample is heterogeneous. One of the main assumptions used is that the geometric barycenter of a sample represents the average situation of the evaluated sector of economy. Taking it into account, the core idea of our technique is to move the efficiency frontier towards the barycenter. It allows to evaluate all inefficient firms more mildly.

Our algorithm has three important properties. First, the convergence for any sample is guaranteed. Second, the set of firms that are efficient according to the standard DEA model is preserved. Finally, there is the simple connection between efficiency scores obtained via DEA and our algorithm.

We tested our model on the real data set containing the information on five parameters on 28 Russian universities in 2008. The developed technique shows consistent results, i.e., our model does not crucially change the structure of a ranking by efficiency, however, the efficiency scores grow when the heterogeneity of the sample is increasing.

References

1. Abankina, I.V., Aleskerov, F.T., Belousova, V.Y., Bonch-Osmolovskaya, A.A., Petrushchenko, S., Ogorodniychuk, D.L., Yakuba, V.I., Zin'kovsky, K.V.: University Efficiency Evaluation with Using its Reputational Component. Lecture Notes in Management Science, vol. 4, pp. 244–253. Tadbir Operational Research Group, Ltd. (2012)
2. Banker, R.D., Morey, R.C.: Efficiency Analysis for Exogenously Fixed Inputs and Outputs. Operations Research 34, 513–521 (1986a)
3. Banker, R.D., Morey, R.C.: The Use of Categorical Variables in Data Envelopment Analysis. Management Science 32, 1613–1627 (1986b)
4. Bessent, A.M., Bessent, E.W.: Comparing the Comparative Efficiency of Schools through Data Envelopment Analysis. Educational Administration Quarterly 16, 57–75 (1980)
5. Charnes, A., Cooper, W.W., Rhodes, E.: Evaluating Program and Managerial Efficiency: An Application of Data Envelopment Analysis to Program Follow Through. Management Science 27, 668–697 (1981)
6. Chen, Y., Morita, H., Zhu, J.: Context-dependent DEA with an application to Tokyo public libraries. International Journal of Information Technology and Decision Making 4, 385–394 (2005)
7. Coelli, T.J., Rao, D.S.P., O'Donnell, C.J., Battese, G.E.: An Introduction to Efficiency and Productivity Analysis, 2nd edn. Springer, N.Y. (2005)
8. Despotis, D.K., Smirlis, Y.G.: Relaxing the impact of extreme units in Data Envelopment Analysis. International Journal of Information Technology and Decision Making 11, 893–907 (2012)
9. Ferrier, G.D., Lovell, C.A.K.: Measuring Cost Efficiency in Banking: Econometric and Linear Programming Evidence. Journal of Econometrics 46, 229–245 (1990)
10. Fried, H.O., Schmidt, S.S., Yaisawamg, S.: Incorporating the Operating Environment into a Nonparametric Measure of Technical Efficiency. Journal of Productivity Analysis 12, 249–267 (1999)

11. Hirschberg, J.G., Lye, J.N.: Clustering in a Data Envelopment Analysis Using Bootstrapped Efficiency Scores. Department of Economics – Working Papers Series 800, The University of Melbourne (2001)
12. Kendall, M.A.: New Measure of Rank Correlation. Biometrika 30, 81–89 (1938)
13. Lemos, C.A.A., Lima, M.P., Ebecken, N.F.F.: DEA Implementation and Clustering Analysis using the K-means algorithm. In: Data Mining VI – Data Mining, Text Mining and Their Business Applications, vol. 1, pp. 321–329. Skiathos (2005)
14. Marroquin, M., Pena, M., Castro, C., Castro, J., Cabrera-Rios, M.: Use of data envelopment analysis and clustering in multiple criteria optimization. Intelligent Data Analysis 12, 89–101 (2008)
15. Meimand, M., Cavana, R.Y., Laking, R.: Using DEA and survival analysis for measuring performance of branches in New Zealand's Accident Compensation Corporation. Journal of the Operational Research Society 53(3), 303–313 (2002)
16. Samoilenko, S., Osei-Bryson, K.M.: Determining sources of relative inefficiency in heterogeneous samples: Methodology using Cluster Analysis, DEA and Neural Networks. European Journal of Operational Research 206, 479–487 (2010)
17. Schreyögg, J., von Reitzenstein, C.: Strategic groups and performance differences among academic medical centers. Health Care Management Review 33(3), 225–233 (2008)
18. Sharma, M.J., Yu, S.J.: Performance based stratification and clustering for benchmarking of container terminals. Expert Systems with Applications 36(3), 5016–5022 (2009)
19. Shin, H.W., Sohn, S.Y.: Multi-attribute scoring method for mobile telecommunication subscribers. Expert Systems with Applications 26(3), 363–368 (2004)

Appendix: Efficiency Scores of Russian Universities

Efficiency scores (in percents) for different models

	DEA ($\mu = 0$)	$\mu = 0.2$	$\mu = 0.5$	$\mu = 0.8$
1	100	100	100	100
2	54	60	74	98
3	100	100	100	100
4	35	40	77	91
5	42	59	90	78
6	53	61	96	93
7	99	99	99	99
8	15	35	38	32
9	24	27	40	71
10	100	100	100	100
11	23	26	33	58
12	31	39	73	81
13	53	59	71	90
14	70	98	99	85
15	100	100	100	100
16	100	100	100	100
17	100	100	100	100
18	73	84	98	90
19	19	22	26	33
20	100	100	100	100
21	45	52	57	74
22	52	58	84	81
23	45	52	79	88
24	59	80	95	89
25	49	55	67	93
26	92	98	94	97
27	33	47	87	97
28	48	68	82	97
Mean	61.60	68.94	81.09	86.56

Efficient Optimization of Multi-class Support Vector Machines with MSVMpack

Emmanuel Didiot and Fabien Lauer

LORIA, Université de Lorraine, CNRS, Inria,
Nancy, France

Abstract. In the field of machine learning, multi-class support vector machines (M-SVMs) are state-of-the-art classifiers with training algorithms that amount to convex quadratic programs. However, solving these quadratic programs in practice is a complex task that typically cannot be assigned to a general purpose solver. The paper describes the main features of an efficient solver for M-SVMs, as implemented in the MSVMpack software. The latest additions to this software are also highlighted and a few numerical experiments are presented to assess its efficiency.

Keywords: Quadratic programming, classification, support vector machines, parallel computing.

1 Introduction

The support vector machine (SVM) [2,14] is a state-of-the-art tool for binary classification. While there is mostly one main algorithm for binary SVMs, their multi-class counterparts have followed various development paths. As a result, four main types of multi-class support vector machines (M-SVMs) can be found in the literature [15,4,11,8] while others are still developed.

From the optimization point of view, training an M-SVM amounts to solving a convex quadratic program with particular difficulties that make the task unsuitable for a general purpose solver. The paper describes how to deal with such difficulties and how to obtain an efficient implementation, in particular regarding parallel computing issues.

The described implementation was first released in 2011 as an open-source software, MSVMpack (available at http://www.loria/~lauer/MSVMpack/) [10], which remains, to the best of our knowledge, the only parallel software for M-SVMs. We here present the inner details of this implementation and its latest improvements.

Paper Organization. M-SVMs are introduced in Section 2 with their training algorithm. Then, Section 3 details their efficient optimization, while the latest additions to MSVMpack are exposed in Section 4. Finally, numerical experiments are reported in Section 5 and Section 6 highlights possible directions for further improvements.

© Springer International Publishing Switzerland 2015   23
H.A. Le Thi et al. (eds.), *Model. Comput. & Optim. in Inf. Syst. & Manage. Sci.,*
Advances in Intelligent Systems and Computing 360, DOI: 10.1007/978-3-319-18167-7_3

2   Multi-class Support Vector Machines

We consider Q-category classification problems, where labels $Y \in \{1, \ldots, Q\}$ are assigned to feature vectors $X \in \mathbb{R}^d$ via a relationship assumed to be of probabilistic nature, i.e., X and Y and random variables with an unknown joint probability distribution. Given a training set, $\{(x_i, y_i)\}_{i=1}^m$, considered as a realization of m independent copies of the pair (X, Y), the aim is to learn a classifier $f : \mathbb{R}^d \to \{1, \ldots, Q\}$ minimizing the risk defined as the expectation of the 0–1 loss returning 1 for misclassifications (when $f(X) \neq Y$) and 0 for correct classifications (when $f(X) = Y$).

Given a feature vector $x \in \mathbb{R}^d$, the output of an M-SVM classifier is computed as

$$f(x) = \arg \max_{k \in \{1, \ldots, Q\}} h_k(x) + b_k,$$

where the $b_k \in \mathbb{R}$ are bias terms and all component functions h_k belong to some reproducing kernel Hilbert space with the positive-definite function $K : \mathbb{R}^d \times \mathbb{R}^d \to \mathbb{R}$ as reproducing kernel [1]. As an example, the kernel function is typically chosen as the Gaussian RBF kernel function $K(x, x') = \exp\left(-\|x - x'\|^2/2\sigma^2\right)$.

The training algorithm for an M-SVM depends on its type, considered here as one of the four main types available in the literature, but can always be formulated as a convex quadratic program. More precisely, the dual form of the quadratic program is usually considered:

$$\max_{\alpha \in \mathcal{C}} J_d(\alpha) = -\frac{1}{2}\alpha^T H \alpha + c^T \alpha, \tag{1}$$

where $\alpha \in \mathbb{R}^{Qm}$ is the vector of dual variables, H and c depend on the type of M-SVM and the training data, and $\mathcal{C}$ is a convex set accounting for all linear constraints that also depend on the M-SVM type and the training parameters. Given the solution α^* to this quadratic program, the functions h_k and the bias terms b_k become available.

The four main types of M-SVMs are the ones of Weston and Watkins [15] (hereafter denoted WW), Crammer and Singer [4] (CS), Lee, Lin and Wahba [11] (LLW) and the M-SVM^2 of Guermeur and Monfrini [8] (MSVM2). We refer to these works for a detailed discussion of their features and the derivation of the corresponding dual programs (see also [7] for a generic M-SVM model unifying the four formulations and [9] for a comparison of a subset of them).

3   Efficient Optimization

The main algorithm used in MSVMpack for solving (1) is the Frank-Wolfe algorithm [6]. In this algorithm, each step is obtained via the solution of a linear program (LP), as depicted in the following procedure minimizing $-J_d(\alpha)$.

1. Initialize $t = 0, \alpha(0) = 0$.
2. Compute the gradient vector $g(t) = -\nabla J_d(\alpha(t))$.

3. Compute a feasible direction of descent as

$$u = \arg\max_{v \in \mathcal{C}} \ g(t)^T v. \qquad\qquad (LP)$$

4. Compute the step length $\lambda = \min\left\{1, \dfrac{g(t)^T (u - \alpha(t))}{(u - \alpha(t))^T H (u - \alpha(t))}\right\}$.
5. Update $\alpha(t+1) = \alpha(t) + \lambda(u - \alpha(t))$.
6. $t \leftarrow t + 1$.
7. Compute $U(t)$ and repeat from Step 2 until $J_d(\alpha(t))/U(t) > 1 - \epsilon$.

Checking the Kuhn-Tucker (KKT) optimality conditions during training may be too much time consuming for large data sets (large m). Thus, we measure the quality of $\alpha(t)$ thanks to the computation of an upper bound $U(t)$ on the optimum $J_d(\alpha^*)$. Given an $U(t)$ that converges towards this optimum, the stopping criterion is satisfied when the ratio $J_d(\alpha)/U(t)$ reaches a value close to 1. In MSVMpack, such a bound is obtained by solving the primal problem with the primal variables h_k fixed to values computed using the current $\alpha(t)$ instead of α^* in the formulas giving the optimal h_k's. This partial optimization requires little computation for all M-SVMs except the M-SVM^2, for which another simple quadratic program has to be solved.

3.1   Decomposition Method

Despite the quadratic programming form of (1), solving large-scale instances of such problems requires some care in practice. In particular, a major issue concerns the memory requirements: the Qm-by-Qm matrix H is dense as it is computed from the typically dense m-by-m kernel matrix K, and thus it simply cannot be stored in the memory of most computers. This basic limitation prevents any subsequent call to a general purpose solver in many cases.

Dedicated optimization algorithms have been proposed to train SVMs (see, e.g., [3]) and they all use decomposition techniques, i.e., block-coordinate descent or chunking, such as sequential minimal optimization (SMO) [12,13]. The basic idea is to optimize at each iteration only with respect to a subset of variables, here corresponding to a subset $\{(x_i, y_i) : i \in S\}$ of the data. With s the size of the subset $S = \{S_1, \ldots, S_s\}$ kept small, this makes training possible by requiring at each step only access to an s-by-m submatrix K_S of K with entries given by

$$(K_S)_{ij} = K(x_{S_i}, x_j), \quad i = 1, \ldots, s, \ j = 1, \ldots, m. \qquad (2)$$

From K_S, all the information required to perform a training step can be computed with little effort.

Working Set Selection. Most implementations of SVM learning algorithms include a dedicated working set selection procedure that determines the subset S and the variables that will be optimized in the next training step. Such procedures are of particular importance in order to reduce the numbers of iterations and of kernel evaluations required by the algorithm. In addition, these methods typically lead to convergence without having to optimize over all variables thanks to the sparsity of SVM solutions (given an initialization of α with zeros). In such cases, only a small subset of the kernel matrix need to be computed, thus also limiting the amount of memory required for kernel cache (see Sect. 3.2).

A working set selection strategy was proposed in [4] in order to choose the chunk of data considered in the next training step of the CS type of M-SVM. This strategy is based on the KKT conditions of optimality applied to the dual problem. The data points with maximal violation of the KKT conditions are selected. This violation can be measured for each data point (x_i, y_i) by

$$\psi_i = \max_{k \in \{j \ : \ \alpha_{ij}(t) > 0\}} g_{ik}(t) - \min_{k \in \{1, \ldots, Q\}} g_{ik}(t), \tag{3}$$

where $g_{ik}(t)$ is the partial derivative of $J_d(\alpha(t))$ with respect to α_{ik}, the $(iQ+k)$th dual variable which is associated to the ith data pair (x_i, y_i) and the kth category.

Unfortunately, there is no simple measure of the violation of the KKT conditions for the other types of M-SVMs. Thus, for these, we use a random selection procedure for the determination of S. [1]

In the following, the subscript $_S$ is used to denote subvectors (or submatrices) with all entries (or rows) associated to the working set, e.g., $\alpha_S(t)$ contains all dual variables α_{ik}, with $i \in S$ and $1 \leq k \leq Q$.

3.2   Kernel Cache

Kernel function evaluations are the most demanding computations in the training of an M-SVM. Indeed, every evaluation of $K(x_i, x_j)$ typically involves at least $O(d)$ flops and computing the s-by-m submatrix $\boldsymbol{K}_S$ of $\boldsymbol{K}$ at each training step requires $O(smd)$ flops.

Many of these computations can be avoided by storing previously computed values in memory. More precisely, the kernel cache stores a number of rows of $\boldsymbol{K}$ that depends on the available memory. When the solver needs to access a row of $\boldsymbol{K}$, either the row is directly available or it is computed and stored in the kernel cache. If the cache memory size is not sufficient to store the entire kernel matrix, rows are dropped from the cache when new ones need to be stored. While most binary SVM solvers use a so-called "last recently used" (LRU) cache to decide which kernel matrix row should be dropped, MSVMpack does not apply a particular rule. The reason is that, for binary SVMs, efficient working set selection methods often lead to many iterations based on the same rows, while the random selection of MSVMpack does not imply such a behavior.

---

[1] The implementation of Weston and Watkins model as proposed in [9] for the BSVM software does include a working set selection strategy. However, this strategy relies on a modification of the optimization problem in which the bias terms, b_k, are also regularized. Here, we stick to the original form of the problem as proposed in [15] and thus cannot use a similar strategy.

3.3   Parallel Computations

MSVMpack offers a parallel implementation of the training algorithms of the four M-SVMs (WW, CS, LLW, MSVM2). This parallelization takes place at the level of a single computer to make use of multiple CPUs (or cores) rather than a distributed architecture across a cluster. The approach is based on a simple design with a rather coarse granularity, in which kernel computations and gradient updates are parallelized during training.

Precomputing the Kernel Submatrix. As previously emphasized, the kernel computations, i.e., the evaluation of $K(x_i, x_j)$, are the most intensive ones compared to training steps and updates of the model (updates of α). In addition, they only depend on the training data, which is constant throughout training. Thus, it is possible to compute multiple kernel function values at the same time or while a model update is performed. Technically, the following procedure is executed on N_{CPU} CPUs (the steps truly running in parallel are shown in italic).

1. *Select a working set $S = \{S_1, \ldots, S_s\} \subset [\![1, m]\!]$ of s data indexes.*
2. *Compute the kernel submatrix $\boldsymbol{K}_S$ as in* (2).
3. Wait for the M-SVM model to become available (in case another CPU is currently taking a step and updating the model).
4. Take a training step and update the model, i.e., compute u_S, λ and $\alpha_S(t + 1) = \alpha_S(t) + \lambda(u - \alpha_S(t))$ as in the Frank-Wolfe algorithm depicted in Sect. 3.
5. Release the model (to unblock waiting CPUs).
6. Loop from step 1 until the termination criterion (discussed in Sect. 3) is satisfied.

Note that, for the CS model type, the working set selection is not random and actually makes use of the model (see section 3.1). In this case, it cannot be easily parallelized and the scheme above is modified to select the working set for the next training step between steps 4 and 5, while step 6 loops only from step 2.

With this procedure, a CPU can be blocked waiting in step 5 for at most $(N_{CPU} - 1)$ times the time of a training step. In most cases, this quantity is rather small compared to the time required to compute the kernel submatrix $\boldsymbol{K}_S$. In practice, this results in a high working/idle time ratio for all CPUs.

Computing the Gradient Update. The next most intensive computation after the kernel function evaluations concerns the gradient vector $g(t)$ that is required at each step.

First, note that a naive implementation computing the required subset of the gradient vector $g(t)$ before each step with $g_S(t) = \boldsymbol{H}_S\alpha(t) + c_S$, where $\boldsymbol{H}_S$ contains a subset of s rows of $\boldsymbol{H}$, is ineffective as it involves all the columns of $\boldsymbol{H}_S$. In addition, at every evaluation of the termination criterion, we require the entire gradient vector, which would involve the entire matrix $\boldsymbol{H}$.

A better strategy is to update the gradient at the end of each step with

$$g(t + 1) = g(t) + \boldsymbol{H}(\alpha(t + 1) - \alpha(t)),$$

where
$$\forall i \notin S, \ k \in [\![1, Q]\!], \ \alpha_{ik}(t+1) - \alpha_{ik}(t) = 0,$$
resulting in only $O(mQs)$ operations of the form:

$$\forall i \in [\![1, m]\!], \ k \in [\![1, Q]\!], \quad g_{ik}(t+1) = g_{ik}(t) - \sum_{j \in S} \sum_{1 \leq l \leq Q} H_{jl}(\alpha_{jl}(t+1) - \alpha_{jl}(t)).$$

This computation can be parallelized by splitting the gradient vector $g \in \mathbb{R}^{Qm}$ in $N_g \leq N_{CPU}$ parts, each one of which is updated by a separate CPU.

MSVMpack automatically balance the number N_K of CPUs precomputing kernel submatrices and the number N_g of those used for the gradient update along the training process. Indeed, at the beginning, the main work load is on the kernel matrix ($N_K = N_{CPU}$), but as the kernel cache grows, less and less computations are required to obtain a kernel submatrix. Thus, CPUs can be redistributed to the still demanding task of updating the gradient (N_g increases and $N_K = N_{CPU} - N_g + 1$). If the entire kernel matrix fits in memory, all CPUs eventually work for the gradient update.

Cross Validation. K-fold cross validation is a standard method to evaluate a classifier performance. After dividing the training set into N_{folds} subsets, the method trains a model on ($N_{folds} - 1$) subsets, test it on the remaining subset and iterates this for all test subsets. Thus, the cross validation procedure requires training of N_{folds} models and can also be parallelized with two possible scenarios. In case the number of CPUs is larger than the number of models to be trained ($N_{CPU} > N_{folds}$), each one of the N_{folds} models is trained in parallel with N_{CPU}/N_{folds} CPUs applying the scheme described above to compute the kernel function values and gradient updates. Otherwise, the first N_{CPU} models are trained in parallel with a single CPU each and the next one starts training as soon as the first one is done, and so on. When all models are trained, the N_{folds} test subsets are classified sequentially, but nonetheless efficiently thanks to the parallelization of the classification described next.

The parallelized cross validation also benefits from the kernel cache: many kernel computations can be saved as the N_{folds} models are trained in parallel from overlapping subsets of data. More precisely, MSVMpack implements the following scheme. A *master* kernel cache stores complete rows of the global kernel matrix computed from the entire data set. For each one of the N_{folds} models, a *local* kernel cache stores rows of a kernel submatrix computed with respect to the corresponding training subset. When a model requests a row that is not currently in the *local* kernel cache, the request is forwarded to the *master* cache which returns the complete row with respect to the entire data set. Then, the elements of this row are mapped to a row of the *local* cache. This mapping discards the columns corresponding to test data for the considered model and takes care of the data permutation used to generate the training subsets. In this scheme, every kernel computation performed to train a particular model benefits to other models as well. In the case where the entire kernel matrix fits in memory, all values are computed only once to train the N_{folds} models.

Making Predictions on Large Test Sets. Finally, the classification of large test sets can also benefit from parallelization via a straightforward approach. The test set is cut into N_{CPU} subsets and the classifier output is computed on each subset by a different CPU.

3.4 Vectorized Kernel Functions

In many cases, data sets processed by machine learning algorithms contain real numbers with no more than 7 significant digits (which is often the default when writing real numbers to a text file) or even integer values. On the other hand, the typical machine learning software loads and processes these values as double precision floating-point data, which is a waste of both memory and computing time. MSVMpack can handle data in different formats and use a dedicated kernel function for each format. This may be used to increase the speed of kernel computations on single precision floats or integers.[2]

In particular, proper data alignment in memory allows MSVMpack to use vectorized implementations of kernel functions. These implementations make use of the Single Instruction Multiple Data (SIMD) instruction sets available on modern processors to process multiple components of large data vectors simultaneously. With this technique, the smallest the data format is (such as with single-precision floats or short integers), the faster the kernel function is.

4 New Features in Latest MSVMpack

Since its first version briefly described in [10], MSVMpack was extended with a few features. Notably, it is now available for Windows and offers a Matlab interface. It also better handles unbalanced data sets by allowing the use of a different value of the regularization parameter for each category.

The parallelized cross validation described in Section 3.3 is an important new feature (available since MSVMpack 1.4) that improves the computational efficiency in many practical cases. Also, the software defaults are now to use as much memory and as many processors as available (early releases required the user to set these values) and the dynamical balancing between CPUs assigned to the kernel matrix and those assigned to the gradient updates was also improved.

5 Numerical Experiments

An experimental comparison of MSVMpack with other implementations of M-SVMs is provided here for illustrative purposes. The aim is not to conclude on what type of M-SVM model is better, but rather to give an idea of the efficiency of the different implementations. The following implementations are considered:

---

[2] Note that, in a kernel evaluation $v = K(x, z)$, only the format of x and z changes and the resulting value v is always in double precision floating-point format. All other computations in the training algorithms also remain double-precision.

- J. Weston's own implementation of his M-SVM model (WW) in Matlab included in the Spider[3],
- the BSVM[4] implementation in C++ of a modified version of the same M-SVM model (obtained with the -s 1 option of the BSVM software),
- K. Crammer's own implementation in C of his M-SVM model (CS) named MCSVM[5],
- and Lee's implementation in R of hers (LLW) named SMSVM[6].

Note that both the Spider and SMSVM are mostly based on non-compiled code. In addition, these two implementations require to store the kernel matrix in memory, which makes them inapplicable to large data sets. BSVM constitutes an efficient alternative for the WW model type. However, to the best of our knowledge, SMSVM is the only implementation (beside MSVMpack) of the LLW M-SVM model described in [11].

The characteristics of the data sets used in the experiments are given in Table 1. The ImageSeg data set is taken from the UCI repository[7]. The USPS_500 data set is a subset of the USPS data provided with the MCSVM software. The Bio data set is provided with MSVMpack. The CB513_01 data set corresponds to the first split of the 5-fold cross validation procedure on the entire CB513 data set [5]. MSVMpack uses the original MNIST data[8] with integer values in the range $[0, 255]$, while other implementations use scaled data downloaded from the LIBSVM homepage[9]. BSVM scaling tool is applied to the Letter data set downloaded from the UCI repository[10] to obtain a normalized data set in floating-point data format.

All methods use the same kernel functions and hyperparameters as summarized in Table 1. In particular, for MCSVM, we used $\beta = 1/C$. When unspecified, the kernel hyperparameters are set to MSVMpack defaults. Also, SMSVM requires $\lambda = \log_2(1/C)$ with larger values of C to reach a similar training error, so $C = 100000$ is used. Other low level parameters (such as the size of the working set) are kept to each software defaults. These experiments are performed on a computer with 2 Xeon processors (with 4 cores each) at 2.53 GHz and 32 GB of memory running Linux (64-bit).

The results are shown in Tables 2–3. Though no definitive conclusion can be drawn from this small set of experiments, the following is often observed in practice:

---

[3] http://people.kyb.tuebingen.mpg.de/spider/
[4] http://www.csie.ntu.edu.tw/~cjlin/bsvm/
[5] http://www.cis.upenn.edu/~crammer/code-index.html
[6] http://www.stat.osu.edu/~yklee/software.html
[7] http://archive.ics.uci.edu/ml/datasets/Image+Segmentation
[8] MNIST data sets are available at http://yann.lecun.com/exdb/mnist/. The data are originally stored as bytes, but the current implementation of the RBF kernel function in MSVMpack is faster with short integers than with bytes, so short integers are used in this experiment.
[9] http://www.csie.ntu.edu.tw/~cjlin/libsvmtools/datasets/
[10] http://archive.ics.uci.edu/ml/datasets/Letter+Recognition

- *Test error rates of different implementations of the same M-SVM model are comparable.* However, slight differences can be observed due to different choices for the stopping criterion or the default tolerance on the accuracy of the optimization.
- *Training is slower with* MSVMpack *than with other implementations for small data sets (with 15000 data or fewer).* This can be explained by better working set selection and shrinking procedures found in other implementations. For small data sets, these techniques allow other implementations to limit the computational burden related to the kernel matrix. Training with these implementations often converges with only few kernel function evaluations.
- *Training is faster with* MSVMpack *than with other implementations for large data sets (with 60000 data or more).* For large data sets, the true benefits of parallel computing, large kernel cache and vectorized kernel functions as implemented in MSVMpack become apparent. In particular, kernel function evaluations are not the limiting factor for MSVMpack in these experiments with this hardware configuration.
- *Embedded cross validation is fast.* Table 3 shows that a 5-fold cross validation can be faster with the parallel implementation that saves many kernel computations by sharing the kernel function values across all folds.
- MSVMpack *training is less efficient on data sets with a large number of categories (such as* Letter *with $Q = 26$).* However, for data sets with up to 10 categories (such as MNIST), MSVMpack compares favorably with other implementations.
- *Predicting the labels of a data set in test is faster with* MSVMpack *than with other implementations (always true in these experiments).* This is mostly due to the parallel implementation of the prediction. For particular data sets, data format-specific and vectorized kernel functions also speed up the testing phase.

Table 1. Characteristics of the data sets

Data set	#classes	#training examples	#test examples	data dim.	format	Training parameters
ImageSeg	7	210	2100	19	double	$C = 10$, RBF kernel, $\sigma = 1$ data normalized
USPS_500	10	500	500	256	float	$C = 10$, RBF kernel, $\sigma = 1$
Bio	4	12471	12471	68	bit	$C = 0.2$, linear kernel
CB513_01	3	65210	18909	260	short	$C = 0.4$, RBF kernel
CB513	3	84119	5-fold CV	260	short	$C = 0.4$, RBF kernel
MNIST	10	60000	10000	784	short double	$C = 1$, RBF kernel, $\sigma = 1000$ data normalized, $\sigma = 4.08$
Letter	26	16000	4000	16	float	$C = 1$, RBF kernel, $\sigma = 1$ data normalized

Table 2. Relative performance of different M-SVM implementations.

Data set	M-SVM model	Software	#SV	Training error (%)	Test error (%)	Training time	Testing time
ImageSeg	WW	Spider	113	5.24	11.05	11s	3.3s
		BSVM	98	2.38	9.81	0.1s	0.1s
		MSVMpack	192	0	10.33	1.2s	0.1s
	CS	MCSVM	97	2.86	8.86	0.1s	0.1s
		MSVMpack	141	0	9.24	3.3s	0.1s
	LLW	SMSVM	210	0.48	7.67	15s	0.1s
		MSVMpack	182	0	10.57	23s	0.1s
	MSVM^2	MSVMpack	199	0	10.48	4.5s	0.1s
USPS_500	WW	Spider	303	0.40	10.20	4m 19s	0.2s
		BSVM	170	0	10.00	0.2s	0.1s
		MSVMpack	385	0	10.40	2.5s	0.1s
	CS	MCSVM	284	0	9.80	0.5s	0.3s
		MSVMpack	292	0	9.80	30s	0.1s
	LLW	SMSVM	500	0	12.00	5m 58s	0.1s
		MSVMpack	494	0	11.40	1m 22s	0.1s
	MSVM^2	MSVMpack	500	0	12.00	22s	0.1s
Bio	WW	Spider	out	of	memory		
		BSVM	2200	6.23	6.23	11s	4.6s
		MSVMpack	4301	6.23	6.23	4m 00s	0.5s
	CS	MCSVM	1727	6.23	6.23	9s	4.2s
		MSVMpack	3647	6.23	6.23	8s	0.5s
	LLW	SMSVM	out	of	memory		
		MSVMpack	12467	6.23	6.23	2m 05s	1.1s
	MSVM^2	MSVMpack	12471	6.23	6.23	11m 44s	0.9s
CB513_01	WW	Spider	out	of	memory		
		BSVM	39183	19.46	25.70	1h 41m 52s	9m 02s
		MSVMpack	42277	16.94	25.55	10m 31s	26s
	CS	MCSVM	41401	17.07	25.45	4h 12m 35s	27m 11s
		MSVMpack	40198	16.93	25.41	4m 04s	32s
	LLW	SMSVM	out	of	memory		
		MSVMpack	54313	22.14	27.65	11m 46s	32s
	MSVM^2	MSVMpack	62027	14.31	25.32	1h 08m 54s	47s
MNIST	WW	Spider	out	of	memory		
		BSVM	13572	0.078	1.46	4h 03m 29s	2m 39s
		MSVMpack	14771	0.015	1.41	2h 50m 29s	15s
	CS	MCSVM	13805	0.038	2.76	1h 54m 26s	4m 09s
		MSVMpack	14408	0.032	1.44	49m 04s	15s
	LLW	SMSVM	out	of	memory		
		MSVMpack	47906	0.282	1.57	4h 36m 45s	45s
	MSVM^2	MSVMpack	53773	0.027	1.51	10h 55m 10s	55s
Letter	WW	Spider	out	of	memory		
		BSVM	7460	4.30	5.85	6m 20s	5s
		MSVMpack	7725	3.14	4.75	24m 38s	3s
	CS	MCSVM	6310	2.03	3.90	2m 38s	4s
		MSVMpack	7566	4.72	6.92	6m 54s	3s
	LLW	SMSVM	out	of	memory		
		MSVMpack	16000	16.90	18.80	3h 56m 08s	4s
	MSVM^2	MSVMpack	16000	5.08	7.28	*48h 00m 00s	3s

*: manually stopped before reaching the default optimization accuracy.

— MSVMpack *leads to models with more support vectors than other implementations*. This might be explained by the fact that no tolerance on the value of a parameter α_{ik} is used to determine if it is zero or not (and if the corresponding example x_i is counted as a support vector).

The ability of MSVMpack to compensate for the lack of cache memory by extra computations in parallel is illustrated by Table 4 for the WW model type (which uses a random working set selection and thus typically needs to compute the entire kernel matrix). In particular, for the Bio data set, changing the size of the cache memory has little influence on the training time. Note that this is also due to the nature of the data which can be processed very efficiently by the linear kernel function for bit data format. For the CB513_01 data set, very large cache sizes allow the training time to be divided by 2 or 3. However, for smaller cache sizes (below 8 GB) the actual cache size has little influence on the training time.

Table 3. Results of a 5-fold cross validation on the CB513 data set. The times for training and testing on a single fold are also recalled to emphasize the benefit of the parallel cross validation.

M-SVM model	Cross validation error	time	Single fold training time	test time
WW	23.63 %	21m 07s	10m 31s	26s
CS	23.55 %	12m 05s	4m 04s	32s
LLW	25.52 %	30m 29s	11m 46s	32s
MSVM²	23.36 %	2h 55m 05s	1h 08m 54s	47s

Table 4. Effect of the cache memory size on the training time of MSVMpack for the WW model type

Bio

Cache memory size in MB	10	60	120	300	600	1200
and in % of the kernel matrix	< 1%	5%	10%	25%	50%	100%
Training time	5m 55s	6m 00s	6m 13s	4m 29s	3m 56s	4m 00s

CB513_01

Cache memory size in MB	1750	3500	8200	16500	24500	30000
and in % of the kernel matrix	5%	10%	25%	50%	75%	92%
Training time	31m 49s	30m 40s	26m 06s	23m 00s	15m 52s	10m 31s

6   Conclusions

The paper discussed the main features of an efficient solver for training M-SVMs as implemented in MSVMpack.

Future work will focus on implementing methods to ease the tuning of hyperparameters, such as the computation of regularization paths. Extending the software to deal with cost-sensitive learning will also be considered. Another possible direction of research concerns the parallelization in a distributed environment in order to benefit from computing clusters in addition to multi-cores architectures.

References

1. Berlinet, A., Thomas-Agnan, C.: Reproducing Kernel Hilbert Spaces in Probability and Statistics. Kluwer Academic Publishers, Boston (2004)
2. Boser, B., Guyon, I., Vapnik, V.: A training algorithm for optimal margin classifiers. In: Proc. of the 5th Annual Workshop on Computational Learning Theory, pp. 144–152 (1992)
3. Bottou, L., Chapelle, O., DeCoste, D., Weston, J. (eds.): Large-Scale Kernel Machines. The MIT Press, Cambridge (2007)
4. Crammer, K., Singer, Y.: On the algorithmic implementation of multiclass kernel-based vector machines. Journal of Machine Learning Research 2, 265–292 (2001)
5. Cuff, J.A., Barton, G.J.: Evaluation and improvement of multiple sequence methods for protein secondary structure prediction. Proteins 34(4), 508–519 (1999)
6. Frank, M., Wolfe, P.: An algorithm for quadratic programming. Naval Research Logistics Quarterly 3(1-2), 95–110 (1956)
7. Guermeur, Y.: A generic model of multi-class support vector machine. International Journal of Intelligent Information and Database Systems 6(6), 555–577 (2012)
8. Guermeur, Y., Monfrini, E.: A quadratic loss multi-class SVM for which a radius-margin bound applies. Informatica 22(1), 73–96 (2011)
9. Hsu, C.W., Lin, C.J.: A comparison of methods for multi-class support vector machines. IEEE Transactions on Neural Networks 13(2), 415–425 (2002)
10. Lauer, F., Guermeur, Y.: MSVMpack: a multi-class support vector machine package. Journal of Machine Learning Research 12, 2269–2272 (2011), http://www.loria.fr/~lauer/MSVMpack
11. Lee, Y., Lin, Y., Wahba, G.: Multicategory support vector machines: Theory and application to the classification of microarray data and satellite radiance data. Journal of the American Statistical Association 99(465), 67–81 (2004)
12. Platt, J.: Fast training of support vector machines using sequential minimal optimization. In: Schölkopf, B., Burges, C., Smola, A. (eds.) Advances in Kernel Methods: Support Vector Learning, pp. 185–208. MIT Press (1999)
13. Shevade, S., Keerthi, S., Bhattacharyya, C., Murthy, K.: Improvements to the SMO algorithm for SVM regression. IEEE Transactions on Neural Networks 11(5), 1188–1193 (2000)
14. Vapnik, V.N.: The Nature of Statistical Learning Theory. Springer-Verlag New York, Inc. (1995)
15. Weston, J., Watkins, C.: Multi-class support vector machines. Tech. Rep. CSD-TR-98-04, Royal Holloway, University of London (1998)

Fuzzy Activation and Clustering of Nodes in a Hybrid Fibre Network Roll-Out

Joris-Jan Kraak and Frank Phillipson

TNO, PO Box 96800, 2509 JE The Hague, The Netherlands
frank.phillipson@tno.nl

Abstract. To design a Hybrid Fibre network, a selection of nodes is provided with active equipment and connected with fibre. If there is a need for a ring structure for high reliability, the activated nodes need to be clustered. In this paper a fuzzy method is proposed for this activation and clustering problem and is compared to a simple Activate First, Cluster Second (AFCS) method. The proposed method is shown to outperform the AFCS method in average.

Keywords: Hybrid Fibre, Network Design, Telecommunication, Cabinet, Planning.

1 Introduction

Nowadays on-line services that stream music and video have become immensely popular. Customers also make use of these services on an increasing number of devices and want to view video in higher resolutions. This pushes internet access providers to improve their network, enabling higher internet speeds. This could be achieved by providing Full Fibre to the Home (Full FttH) but also by introducing intermediate topological steps while moving from Full Copper to Full FttH, like Fibre to the Cabinet (FttCab) or Curb (FttCurb). We will focus in this paper on the design of an FttCab network. The translation to the FttCurb problem can be found in [1].

To design a FttCab network, a selection of cabinets is provided with active equipment, making it an *Activated Cabinet* (AC) and connected with fibre. The cabinets that are not activated are placed in cascade with an AC. If the copper distance from a house to the first AC is within a certain distance, e.g. 1 km, it is still able to receive the benefits of FttCab. We propose the following list of requirements for the FttCab network:

- At least a certain percentage of customers should be within a fixed distance of an AC.
- Each AC should be connected to the CO via two independent routes of fibre optics.
- There is an upper bound on the number of customers that can be connected to an AC, directly or via a cascade.
- The number of cascades on an AC is bounded by a maximum.

© Springer International Publishing Switzerland 2015
H.A. Le Thi et al. (eds.), *Model. Comput. & Optim. in Inf. Syst. & Manage. Sci.*,
Advances in Intelligent Systems and Computing 360, DOI: 10.1007/978-3-319-18167-7_4

— There also is a maximum to the amount of cabinet that may be served by a single fibre ring.
— The total cost needs to be minimized.

This problem is also addressed by [2]. There a three step approach for this planning problem is presented. The three sub-problems are:

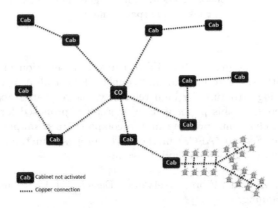

Fig. 1. Starting point

1. Activation: Which cabinets should be activated in order to reach the desired percentage of households at minimal costs? Fig. 1 shows the starting point for a very small example. All cabinets (Cab) are connected through copper with the CO. Several residences are connected to the cabinet; this is only shown for one cabinet in the illustration. Now a subset of the cabinets needs to be activated in order to reach the intended number of households over copper from an activated cabinet within the set distance which is shown in Fig. 2. In this picture the cabinets are shown that are to be activated, the fibre connection will follow in later steps.
2. Clustering: Which cabinet is served by which fibre ring? The cabinets now have to be divided into groups in order to determine which cabinets will be jointly connected by one fibre ring as depicted in Fig. 3.
3. Reliable routing: How will each fibre ring run? The physical route of the fibre rings needs to determined. What order will they be connected in, and how does this route run taking into account that one ring cannot use the same fibre or duct twice and the possibility to reuse existing infrastructure? An example is shown in Fig. 4.

All these subproblems are NP-hard problems, as was shown in [2], for which it proposes heuristic procedures, which work fast and are easily interpretable. The downside, however, is that the solutions provided are often not optimal and sometimes infeasible in the sense that not all fibre rings are edge disjoint. The

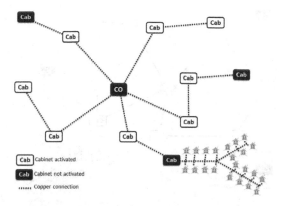

Fig. 2. Which cabinets must be activated?

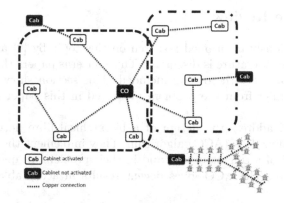

Fig. 3. Which cabinets are together on one ring?

combination of Clustering and (Reliable) Routing is already addressed by [3]. There a somewhat slower heuristic is proposed that guarantees more feasible and, in general, better solutions to the overall problem by solving the clustering and routing problem simultaneously.

The goal of this paper is to find an efficient way of solving the Activation and Clustering problem simultaneously. Similar to the goal of [2], the computation times should at most be in the order of a few minutes, and preferably in the order of seconds. In the remainder of this paper first research that has been done on similar problems will be discussed. Next the proposed heuristic is described in detail in section 3. In Section 4 the results in terms of incurred costs and computation times are presented. The last section is devoted to some conclusions and ideas for further research.

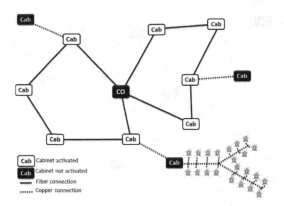

Fig. 4. Physical route of the fibre ring

2 Literature Review

Besides the previously mentioned research on this topic by [2] and [3], in this section some other literature is discussed. This concerns papers that address the full planning problem with routing included. This section shows in what way these problems differ from the problem presented in this paper and how they were solved.

Zhao et al. [4] address the design of a hierarchical fibre network using an Ant Colony Optimization (ACO) algorithm. They implement the use of a mesh structure instead of a ring structure and in their problem all nodes need to be connected to fibre. The ACO gives decent results in acceptable computation times.

Kalsch et al. [5] consider a fibre ring network where all ACs are known and edge capacity is not taken into account. Somewhat similar to Zhao et al., instead of a ring structure passing one CO, also a route starting from one CO and ending in another is feasible. Lastly, since the ACs are fixed, only connection costs are considered. Kalsch et al. decompose the problem by looking at each route cluster separately, similar to [2]. Then they use CPLEX to solve the NP-hard routing problem for each cluster. In total this takes up to 5 hours for a decently large network (135 rings). We would like to find solutions within minutes, therefore using a similar problem formulation and solving it with CPLEX is not a viable option for us. Unfortunately, we can also not make use of their clustering method, since it is only vaguely described in their article.

Caserta et al. [6] introduce a heuristic for finding a fibre ring network similar to the one Kalsch et al. designed. However, they do take into account activation costs and they set the capacity of the number of nodes per ring depending on the length of the ring. Due to the latter complexity they first have to enumerate a large amount of feasible rings and then pick rings that cover all ACs in a cost-efficient manner. Unfortunately, Caserta et al. only test their heuristic on one, large problem (784 rings). Furthermore, they do not compare their method to any other heuristics. The running time (they abort the heuristic after one week), although for a large test case, implies that their method is not suitable for our case.

In [7], Mateus et al. present an algorithm that addresses a hierarchical tree networking problem. In a tree networking problem each node only has to be connected to the CO through one route. Similar to in Zhao et al. [4], only the branch nodes need to be allocated, the leaf nodes are already fixed. The tree structure of the network is also different from our problem. Mateus et al. first locate the branch nodes on a centroid based method and then use simple heuristics to connect the leaf nodes to their branch node and the branch nodes to the CO. This method, due to its tree structure base, can compute fast results.

3 Fuzzy Activation and Clustering Heuristic

In this section, a heuristic is proposed that tries to simulate simultaneous decision-making on the activation and clustering problem. The heuristic initially implements fuzzy decision-making for both the activation and the clustering problems and is hence named Fuzzy Activation and Clustering Heuristic (FACH).

The key principal of FACH is that, in its first iterations, it only decides with a certain probability which cascades should be made and similarly which ACs should be clustered together. These probabilities grow to either 0 or 1 as the number of iterations increase, making the decisions more certain and less fuzzy. The diagram in Fig. 5 describes FACH on the highest level of abstractness. Most of the procedure comprises of one chain of consecutive methods that is performed a fixed number of times. This chain is called the main iteration. The main iteration consists of two important steps. These steps, the activation step and the clustering step, are fully described in Sections 3.2 and 3.1. We start by discussing the clustering step as it will give a better understanding for the procedures in the activation step. Finally, in Section 3.3, we discuss the merger step that tries to merge the clusters containing only one AC with other clusters.

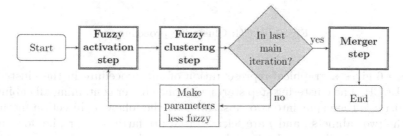

Fig. 5. FACH: Main procedure

3.1 Clustering Step

As mentioned in the problem description, [2] constructs clusters of ACs, based on a max-constrained K-Means Clustering Problem (KMCP). A commonly used algorithm to find a solution to this problem is Lloyd's algorithm [8]. This algorithm starts with a random allocation of centroids of a predefined number of clusters. Then, it assigns the memberships of the data points (ACs) to these clusters, which, in turn, results in updated locations of the centroids. Lloyd's algorithm continues updating the memberships and centroids iteratively, until the locations of the centroids do not change any more. This final situation is guaranteed to be a local optimum to the problem. In [2] this algorithm is extended to assure that clusters are not too large and to achieve a higher probability of finding the global optimum.

In this paper, fuzzy memberships are used, which means that the membership of cabinets is distributed among multiple clusters. In [9] the concept of fuzzy memberships to clusters is introduced. These fuzzy memberships are used to minimize an objective function that is similar to that of the KMCP, but has a fuzzy extension. This method is referred to as the Fuzzy C-Means Clustering Problem (FCMCP).

The main characteristic of the problem of this paper that makes the straight implementation of [9] impossible, is the absence of a fixed number of clusters C. Here, C is not an input parameter but a variable that depends on the maximum size of clusters. To adjust for this, FACH keeps increasing C until a feasible set of clusters is found. Another requirement of FCM is that the dissimilarities, or distances, between points are N-dimensional Euclidean. This can be accomplished by using multi-dimensional scaling (MDS) algorithms [10].

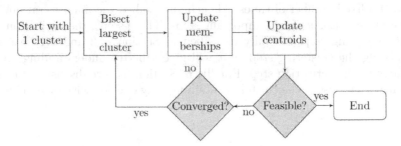

Fig. 6. FACH: Clustering procedure

Figure 6 gives a graphical representation of the procedure in the clustering step of FACH. The clustering step starts with one cluster containing all cabinets. This cluster is then split into two clusters using the following bisection method. First, the two cabinets i and j are selected that are farthest apart, i.e. for which the shortest path distance $d_{i,j}$ is highest. Next, each cabinet in the cluster will be grouped with the nearest selected cabinet. Of each of these two groups the location of their centroid is calculated. This is done by simply taking the average of all coordinates in a group. These two centroids now form a basis for two new clusters. Each cabinet is assigned a certain fuzzy membership to a cluster. Now that the memberships have been updated, the locations of two new centroids can be calculated by taking the weighted average of the coordinates of all members. This average is weighted by the active membership of cabinets. Active membership $ActiveMember_i^k$ is determined by the product of i's activeness $x_{i,i}$ and its membership to the corresponding cluster k. The activeness of a cabinet i is the only variable that is used in the clustering step of FACH and that is determined in the activation step. This variable is also fuzzy.

With the new locations of the centroids, the memberships can again be updated which implies new locations for centroids and so forth. This is continued until any of the following situations is reached:

- All clusters satisfy the capacity constraints.
- The locations of the centroids stop changing.
- The maximum number of iterations is reached.

In the finalization of the clustering step, FACH checks if the final main iteration has been reached. In that case, the active main memberships will be subjected to the merger step discussed in Section 3.3. Otherwise, the main iteration increases by one, all parameters are updated and FACH continues to another activation step.

3.2   Activation Step

Here the activation step of FACH, as presented in Fig.7 is described. It starts by activating all cabinets. Next, it searches for the cascade that has the highest cascade score. FACH will try to plan this cascade with the amount of its fractional score. This amount should be perceived as a probability of the cascade happening in the final solution and will remain the same value until the next activation step.

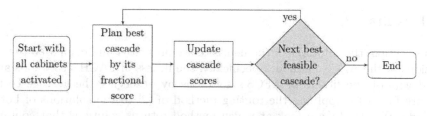

Fig. 7. FACH: Activation procedure

After each planning of a cascade, all influenced cascade scores, which will be defined later on, and fractional scores are updated. Next, FACH searches for the next cascade of any cabinet i on any cabinet j with the highest score. It keeps planning cascades on these until the number of reached clients is equal to the minimum necessary amount. At that point, it only searches for cascades that do not result in the loss of any clients. A cascade has no loss of clients when the AC is still in range of all the clients that are connected to the cascaded cabinet. Once there are no feasible cascades with a positive cascade score left, FACH ends the activation step and continues to the clustering step.

In its search it does not include cascades where either of the two involved cabinets i and j are fully cascaded ($x_{i,i} = 0$ or $x_{j,j} = 0$). Nor does it include cascades that are already performed in the incumbent activation step or cascades on ACs that are already at full capacity. Finally, FACH also avoids cascades of cabinet i on j that imply cascade costs that are higher than the costs of activating cabinet i.

How the cascade score is calculated is explained in detail in [11]. The cascade score is equal to zero for cascades that are not feasible or desirable. In other cases it depends on:

- The expected fibre costs and savings;
- The probability that two cabinets will end in the same cluster;
- The availability of a cabinet to receive cascades;
- The loss of clients;
- The expected number of clusters and the size of the clusters.

3.3   Merger Step

By implementing the cluster score, FACH avoids smaller sized clusters. However, it can still happen that, after the final main iteration, there are clusters containing only a few ACs. This is justifiable when this cluster is actually secluded or when its cabinets do not fit in an other nearby cluster. Therefore, a way was found to check for both of these possibilities, if neither is true then we merge the cluster with a larger nearby cluster. In our method we only check for possible mergers of clusters containing one AC. Although it might be beneficial to also try to do this for larger clusters, it would be harder to assure that these mergers are truly beneficial.

4   Results

In this section, the solutions of the heuristic are presented and compared to the solutions of the AFCS method. Specifically, the fitness of FACH's solutions are tested against the fitness of AFCS's solutions by observing the average costs that are found by applying the routing method of [2] to the solutions of both methods. We also take a look at which method returns solutions that contain more edge-disjoint fibre rings. For these comparisons, we use the test dataset containing 30 cities. The total dataset contained 45 different randomly picked cities, of which 15 were used as training set for the parameters of the method. The values of some parameters in FACH can not be determined by logical reasoning. For those parameters we performed some analysis to determine their optimal value. Actually, each city case was expanded into 5 separate cases by setting the desired penetration level (the percentage of clients reached) at 5 different values: 100%, 95% 90%, 85% and 80%. For the test cases, we also observe 3 different values for the maximum number of ACs that is allowed per cluster: 6, 8 and 10. For the training cases we only use $MaxACPerClus = 8$ to reduce the computation time necessary in the parameter analysis.

4.1 Fitness Solutions

Figure 8 shows for the 150 observed instances with $MaxACPerClus = 8$ the costs of applying both FACH and AFCS. The costs are set out against the size of the instances in terms of the number of cabinets present in the city of that instance. Also two second degree polynomials are shown that are fitted to the costs of both methods. Both curves increase almost linearly with respect to the number of cabinets. Since both curves start at almost the same value for the smallest city, it seems that FACH becomes increasingly profitable compared to AFCS when the city sizes increase.

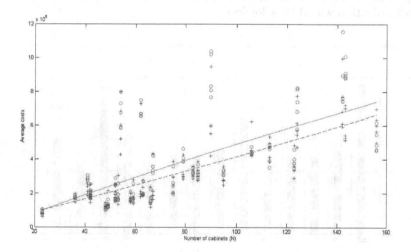

Fig. 8. Costs of FACH (pluses) and a second degree polynomial (dashed curve). Costs of AFCS (circles) and a second degree polynomial (solid curve).

To give a better insight in the performance, we grouped the instances on three of their properties. First, we put the instances into 5 different penetration level (β) groups. Second, we split the instances according to the size of their city into three equally large groups. These groups ('S', 'M' and 'L') contain the instances of 10 cities each. Group 'S' contains the cities with sizes $N = [23, 53]$, group 'M' those for which $N = (54, 83]$ holds and, finally, group 'L' represents all cities with sizes $N = (83, 156]$. Finally, we examine three different values of $MaxACPerClus$. This leaves us with a total of 11 groups, which are all represented in Fig. 9. This figure shows for each group two comparisons of FACH to AFCS. The first comparison (dark bar) is the average of the percentual difference per instance, whereas the second comparison (light bar) is the percentual difference of the average costs. In practice, this means that the second comparison is weighted by the size of the instances, while the first comparison is not. The error bars in Fig. 9 illustrate the 95 percent confidence intervals of the differences. Apparently, in all but one of the eight groups, FACH outperforms AFCS by a significant percentage. This percentage is especially high if we observe the weighted differences. This is a result from the fact that FACH performs increasingly better than AFCS for larger city sizes. The only group that does not have a conclusive preference, is the group containing instances for which $\beta = 100\%$, that is, for which we are not allowed to lose any clients. It seems that the further we decrease β, the higher FACH's profitability becomes. Finally, we take a look at the three pairs of bars at the far right of Fig. 9. These tell us that, although FACH results in significantly higher activation and cascade costs, the reduction in routing costs is much greater, thereby resulting in an overall reduction of costs. Of course, this is highly dependent on the sample that we take. If, for example, we would have only observed penetration levels of a 100% and 95%, the cost reduction would be a lot less.

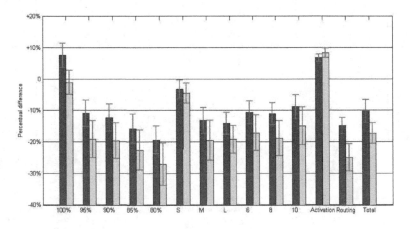

Fig. 9. Average cost difference of FACH compared to AFCS

4.2 Computation Times

One of the requirements of the heuristic was that it could be executed within the order of minutes. The heuristic was implemented in computing software MATLAB. The instances were solved by a PC with a 64-bits AMD 3.6 GHz Quad-Core Processor and with 4GB of RAM. Figure 10 displays the computation times of FACH as pluses for all 450 cases. The case with the longest computation time (396 seconds) still fulfils the requirement that was set. When we compare the computation times of FACH to those of AFCS (the circles), we see that FACH takes significantly longer to solve for almost all instances. Again exponential functions to both sets of computation times were fitted. We can conclude that the computation times increase exponentially in relation to number of cabinets

N in the city of the instance. However, since the curve for AFCS starts so low, it would never pose a problem for any realistically sized instance. FACH, on the other hand, could run into trouble when larger instances are given as input. For example, if we extrapolate the exponential function of FACH, then an instance of a city containing as much as 400 cabinets would cost about 66 hours to solve. Whereas AFCS would likely solve it within 14 minutes.

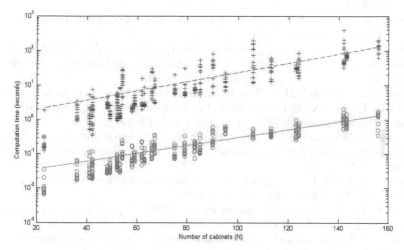

Fig. 10. The computation times in seconds for all the 450 cases in the test dataset

5   Conclusion and Discussion

In principle, we can say that we succeeded in finding a method that outperforms the AFCS approach. At least, on average this was the case for the instances that were examined. The computation time that was necessary for FACH to produce solutions was also within the limits of the set goal.

A point of discussion is the amount of parameters that is used. Since each parameter's optimal value is estimated with some uncertainty, the final parameter setting is likely to be deviated from the optimal setting. There is enough room for further exploration of FACH. Before a total conclusive answer on the effectiveness of FACH can be given, it should also be examined how well it performs against Daamen's simultaneous clustering and routing method. That is, FACH followed by the routing method of [2] should be compared against the activation method of [2] followed by the combined clustering and routing method of [3].

Finally, we argue that FACH could be adapted to be applied to other (capacitated) Location Routing Problems (LRPs). For example, the specific requirement that the rings are edge-disjoint, is not something that is addressed by FACH. This is something that should be solved by the specific routing method. Obviously, other LRPs would require different parameter settings and probably also some adaptation to the heuristic itself.

References

1. Phillipson, F.: Planning of fibre to the curb using g.fast in multiple rollout scenarios. In: Proceedings of the 3rd International Conference on Information Communication and Management (ICICM 2013), Paris, France (2013)
2. Phillipson, F.: Roll-out of reliable fiber to the cabinet: An interactive planning approach. IEEE/OSA Journal of Optical Communications and Networking 6(8), 705–717 (2014)
3. Daamen, R.C.J.: Heuristic methods for the design of edge disjoint circuits. Master's thesis, Tilburg University (2013)
4. Zhao, R., Liu, H., Lehnert, R.: Topology design of hierarchical hybrid fiber-vdsl access networks with aco. In: Fourth Advanced International Conference on Telecommunications, AICT 2008, pp. 232–237. IEEE (2008)
5. Kalsch, M.T., Koerkel, M.F., Nitsch, R.: Embedding ring structures in large fiber networks. In: 2012 XVth International Telecommunications Network Strategy and Planning Symposium (NETWORKS), pp. 1–6. IEEE (2012)
6. Caserta, M., Schwarze, S., Voß, S.: Developing a ring-based optical network structure with glass-through nodes. Journal of Telecommunications & Information Technology 2012(2) (2012)
7. Mateus, G., Cruz, F.R.B., Luna, H.P.L.: An algorithm for hierarchical network design. Location Science 2(3), 149–164 (1994)
8. Lloyd, S.: Least squares quantization in pcm. IEEE Transactions on Information Theory 28(2), 129–137 (1982)
9. Dunn, J.C.: Well-separated clusters and optimal fuzzy partitions. Journal of Cybernetics 4(1), 95–104 (1974)
10. Borg, I., Groenen, P.J.: Modern multidimensional scaling: Theory and applications. Springer (2005)
11. Kraak, J.J.: A heuristic method for the activation and clustering of nodes in a fibre to the cabinet/curb roll-out. Master's thesis, Erasmus University Rotterdam (2014)

Heuristic Ranking Classification Method for Complex Large-Scale Survival Data

Nasser Fard and Keivan Sadeghzadeh

Department of Mechanical and Industrial Engineering, Northeastern University,
Boston, MA, USA
{n.fard,k.sadeghzadeh}@neu.edu

Abstract. Unlike traditional datasets with a few explanatory variables, analysis
of datasets with high number of explanatory variables requires different approaches. Determining effective explanatory variables, specifically in a complex
and large-scale data provides an excellent opportunity to increase efficiency and
reduce costs. In a large-scale data with many variables, a variable selection
technique could be used to specify a subset of explanatory variables that are significantly more valuable to analyze specially in the survival data analysis. A
heuristic variable selection method through ranking classification to analyze
large-scale survival data which reduces redundant information and facilitates
practical decision-making by evaluating variable efficiency (the correlation of
variable and survival time) is presented. A numerical simulation experiment is
developed to investigate the performance and validation of the proposed method.

Keywords: ranking classification, decision-making, variable selection, large-
scale data, survival data.

1 Introduction

There are more than one trillion connected devices in the world and over 15 petabytes
$(15 \times 10^{15}$ bytes) of new data are generated each day [1]. As of 2012, overall created
data in a single day were approximated at 2.5 etabytes $(2.5 \times 10^{18}$ bytes) [2,3]. The
world's technological per-capita capacity to store information has approximately doubled every 40 months [2], [4] while annually worldwide information volume is growing at a minimum rate of 59 percent [5]. In 2012 total software, social media, and IT
services spending related to big data and analytics reported over $28 billion world-
wide by the IT research firm, Gartner, Inc. Also it has been predicted that IT organizations will spend $232 billion on hardware, software and services related to Big Data
through 2016.

Massive amount of data in high-dimensions are increasingly accessible from various sources such as transaction-based information, information-sensing mobile devices, remote sensing technologies, cameras, microphones, RFIDs, wireless sensor
networks as well as internet search and web analytics. It has become more difficult to
process the streaming high dimensional data in traditional application approaches and

© Springer International Publishing Switzerland 2015   47
H.A. Le Thi et al. (eds.), *Model. Comput. & Optim. in Inf. Syst. & Manage. Sci.*,
Advances in Intelligent Systems and Computing 360, DOI: 10.1007/978-3-319-18167-7_5

when and if a big data is compiled and processed correctly, it can enable any type of profession such as communication and network industry to make improved decisions [6,7,8,9,10,11].

Decision-making is increasingly based on the type and size of data, and also analytic methods rather than on experience and intuition. As stated in a broad survey [12], based on responses from organizations for techniques and tool types using in big data analysis, advanced analytics are among the most popular ones in terms of potential growth and commitment. As an analytical approach, decision-making is the process of finding the best option from all feasible alternatives [13,14]

The outcome of many scientific investigations, experiments and surveys are set of generated number of time points, representing failure or survival time, called survival data. To analyze survival data, when size and dimension are large, advanced analytics are advantageous for variable selection and variable reduction where different approaches could be used to determine a subset of variables that are significantly more valuable [6,7,8,9,10,11], [15]. Although recent methods may operate faster, involve different estimation methods and assumptions, such as Cox proportional hazard model [16], accelerated failure time [17], Buckley-James estimator [18], random survival forest [19], additive risk models [20], weighted least squares [21] and classification and regression tree (CART) [22], their robustness is not consistent [27, 28].

This study is motivated by the importance of the variable selection issue in the risk analysis. The objective of this study is to propose a methodology for variable selection in large-scale survival data. The aim is to reduce the volume of the explanatory variables and identify a set of most influential variables on the survival time. This paper provides heuristic algorithm for implementing variable selection through determining variable efficiency recognition procedure based on defined coefficient score in right-censored large-scale survival data. In this paper, the concept of survival data analysis, and accelerated failure time model are reviewed in Section 2. The methodology, including analytic model for the transformation of the explanatory variable dataset as well as method and algorithm is proposed in Section 3. The simulation experiment, result and analysis of variable selection decision-making algorithms are presented in Section 4. Finally, summary including the advantages of the proposed algorithm is discussed in Section 5. The computer package used in this research is the MATLAB® R2011b programming environment.

In this paper, a_{ij} denotes the element in the i'th row and the j'th column of r-by-q matrix $A = [a_{rq}]$ and column vector a_j denotes the j'th column of matrix A.

2   Definitions

In this section, applied introductions to survival data analysis and accelerated failure time model are presented.

2.1   Survival Data Analysis

Survival data analysis methods consider the time until the occurrence of an event. This analysis generally is defined as a set of methods for analyzing such data where subjects are usually followed over a specified time period. Survival data analysis has been extensively studied in reliability engineering and risk analysis. In survival data analysis, information on an event status and follow up time is used to estimate a survival function $S(t)$, which is defined as the probability that an object survives at least until time t where $F(t)$ is the cumulative distribution function

$$S(t) = 1 - P(T \leq t) = 1 - F(t) \tag{1}$$

Accordingly survival function is calculated by probability density function as:

$$S(t) = \int_t^\infty f(u)du \tag{2}$$

Methods to analyze survival data are parametric; which are based on survival function distributions such as exponential, semi-parametric; which do not assume knowledge of absolute risk and estimates relative risk, and nonparametric; which are useful when the underlying distribution of the problem is unknown. For moderate- to large-scale covariates, it is difficult to apply parametric or semi-parametric methods [21]. Nonparametric methods are used to describe survivorship in a population or for comparison of two or more populations. The Kaplan-Meier (KM) is the most commonly used nonparametric method for the survival function and has clear advantages [23,24]. As an applied method, accelerated failure time model is presented next.

2.2   Accelerated Failure Time Model

In survival data analysis, when distribution of the event time as a response variable is unknown, estimations and inferences become challenging and the traditional statistical methods are not practical. Cox proportional hazards model (PHM) [16] and the accelerated failure time (AFT) [17] are frequently used in these circumstances. For a comprehensive comparison see [25]. Similar to the multiple linear regression model, AFT assumes a direct relation between the logarithm of the survival time and the explanatory variables.

The event time in survival data may not always be observable. This is known as censored data. The random variables Y and C represent the time-to-event and censoring time, respectively. In this situation survival data is represented by (t, δ, U) where $t_i = \min(y_i, c_i)$ and the censoring indicator $\delta_i = 1$ if the event occurs, otherwise $\delta_i = 0$. The observed covariate U represents a set of variables. Let denote any observations by (t_i, δ_i, u_{ij}), $i = 1 \dots n$, $j = 1 \dots p$.

Defining random variables X as time-independent explanatory variable, x_j, $j = 1 \dots p$ corresponding i'th observation effects multiplicatively on y_i or additively on t_i where the relation between the explanatory variables and the response variable is considered linear.

The corresponding log-linear form of the AFT model with respect to time t is given by

$$\ln(t) = X\beta + \varepsilon \qquad (3)$$

The coefficient vector β is of length-p. Nonparametric maximum likelihood estimation does not work properly in the AFT model [26], where in many studies least squares method has been used to estimate parameters. To account for censoring, weighted least squares method is commonly used, as ordinary least squares does not work for censored data in the AFT model. In addition, regardless of the values of the coefficients and explanatory variables, the log form assures that the predicted values of t are positive [25].

Next section presents methodology which includes the proposed analytic model for transformation of the explanatory variable dataset of a survival data, as well as heuristic randomized methods.

3   Methodology

There are many advantages of using discrete values over continuous; as discrete variables are easy to understand and utilize, more compact, and more accurate. In this study, for a survival dataset including discrete values, a normalization analytical model is defined. The random variables T represent the survival time and the covariate U represents a set of variables. It is assumed that the hazard at time t only depends on the survivals at time t [17]. As a simple and well-known rescaling method, unity-based normalization scales in $[0, 1]$ range is used. For any n-by-p dataset in matrix $U = [u_{ij}]$, there are n independent observations and p variables. The original explanatory variables dataset may include any type of explanatory data as binary, continuous, categorical, or ordinal. To normalize the original explanatory variables dataset, we define minimum and maximum value of each explanatory variable value in a vector u_j for $j = 1 \dots p$ as:

$$umin_j = \min\{u_{ij}\}, \; umax_j = \max\{u_{ij}\} \quad i = 1 \dots n \qquad (4)$$

For any array u_{ij}, we allocate a normalized substituting array v_{ij} as:

$$v_{ij} = \frac{u_{ij} - umin_j}{umax_j - umin_j} \quad i = 1 \dots n, \; j = 1 \dots p \qquad (5)$$

The result of the transformation is an n-by-p dataset matrix $V = [v_{np}]$ will be used in the designed heuristic methods and algorithms. Also, we define survival time vector $t = [t_n]$ including all observed event times. Application of a robustness validation method proposed by [27,28], indicates that the change of correlation between variables before and after transformation is not significant. An approach is to develop models including combinations of the k variables, called selective, where $1 \leq k < p$. For a comprehensive discussion for evaluation a desired number of variables in any selected subset, k, see [27]. The following method utilizes random subset selection concepts to evaluate coefficient score of each variable through AFT model by applying weighed least square technique.

The preliminary step for the highest efficiency in proposed methods is to cluster the variables based on the correlation coefficient matrix of the original dataset $M = [m_{ij}]$, and choose a representative variable from each highly correlated cluster, then eliminate the other variables from the dataset. Let m_{ij} denote covariance of variables i and j. Pearson product-moment correlation coefficient matrix is defined as

$$m_{ij} = \frac{1}{n-1}\sum_{k=1}^{n}(v_{ik} - \bar{v}_i)(v_{jk} - \bar{v}_j) \tag{6}$$

v_{ik} and $\bar{v}_i$ in order represent value of variable i in observation k and mean of variable i, and similarly second parenthesis is defined for variable j. For instance, for any given dataset if three variables are highly correlated in M, only one of them is selected randomly and the other two are eliminated from the dataset. The outcome of this process assures that the remaining variables for applying methods are not highly correlated.

To obtain variable coefficients estimation, let w_{in} be the KM weights. Nonparametric estimate of survival function from the KM estimator for distinct ordered event times t_1 to t_n is:

$$\hat{S}(t) = \prod_{i=1}^{t}\left(1 - \frac{d_i}{n_i}\right) \tag{7}$$

At each event time t_j there are n_j subjects at risk and d_j is the number of subjects which experienced the event. For $i = 1 \ldots n$ these weights can be calculated using the expression

$$w_{in} = \hat{S}(\ln t_i) - \hat{S}(\ln t_{i-1}) \tag{8}$$

Applying (7) in (8), w_{in} is obtained [29] by

$$w_{1n} = \frac{\delta_1}{n}, \quad w_{in} = \frac{\delta_i}{n-i+1}\prod_{j=1}^{i-1}\left(\frac{n-j}{n-j+1}\right)^{\delta_i} \quad i = 2 \ldots n \tag{9}$$

Recall AFT model, solving (3) under the assumption that vector ε consists of random disturbances with zero mean, $E[\varepsilon \mid X] = 0$, then the estimator of β is calculated by least square method as

$$\hat{\beta} = (X^T W X)^{-1}X^T W \ln t \tag{10}$$

where X^T is transpose of X and W is a diagonal matrix with the KM weights. Equation (10) is used to estimate variable coefficients in heuristic decision-making methods where $X \equiv V$. As $k \leq n$, then k-by-p sub-dataset matrix $D = [d_{kp}]$ is defined as a selected subset of V where matrix V is defined in (5). Also vector $B = [b_k]$ will be utilized in following methods as a coefficient estimator vector corresponding to the explanatory variables for sub-dataset D.

3.1   Ranking Classification

The concept of proposed method is inspired by the semi-greedy heuristic [30,31]. The objective of this method is to introduce efficient variables in terms of correlation with survival time for an event time through randomly splitting approach by selecting the best local subset using a coefficient score. This method is a ranking classification using semi-greedy randomized procedure to split all explanatory variables into subset of predefined size k [27,28] using equal-width binning technique. The number of bins is equal to

$$c = \lceil p \, / \, k \rceil \tag{11}$$

Note that if p is not divisible by k, then one bin contains less than k variables. At each of l trials where l is a defined large reasonable number, a local coefficient score is calculated for any variable cluster at any bin. For i'th cluster, this coefficient score is given by

$$\tilde{\beta}^i = \left(\sum_{j=1}^{k} (\hat{\beta}_j)^2 \right)^{1/2} \tag{12}$$

$\hat{\beta}_j$ is obtained by weighted least squares estimation using (10) where $X \equiv D$. As a semi-greedy approach, the largest calculated $\tilde{\beta}^i$ in each trial gives the selected cluster including k variables and the coefficient of these variables will be counted into a cumulative coefficient score for each of the cluster variables respectively. To calculate this score summation for each variable over all trials, a randomization dataset matrix $\varXi = [\xi_{ck}]$ is set where each row as a bin is filled by k randomly selected variables' identification numbers in each trial. Ranking classification will be applied at the end of l trials based on accumulation of each variable score by vector $\boldsymbol{\Theta} = [\theta_p]$. Heuristic algorithm for the proposed method is following.

3.2   Heuristic Algorithm

Let $\boldsymbol{\Theta} = [\theta_p]$ be the variable efficiency rank vector where initially each array as the cumulative contribution score corresponding to a variable is zero

for $i = 1$ to l **do**

    Randomly split the dataset $\boldsymbol{V}$ into equally sized c subsets

    Compose $\varXi$ by c subsets

    for $j = 1$ to c **do**

        Compose the sub-dataset $\boldsymbol{D}^j$ for variable subset j in $\varXi$ including variables ξ_{jq} for $q = 1$ to k

        Calculate $\boldsymbol{W}^j$ over the sub-dataset $\boldsymbol{D}^j$

        Calculate coefficient vector $\boldsymbol{B}^j$ over the sub-dataset $\boldsymbol{D}^j$ using (10)

        Calculate coefficient score $\tilde{\beta}^j$ corresponding $\boldsymbol{B}^j$ over the sub-dataset $\boldsymbol{D}^j$ using (12)

    end for

Select a subset with the highest $\tilde{\beta}^j$, call this new subset $\boldsymbol{D}^*$

for $j = 1$ to k **do**

    Add the value of b_j^i corresponding the variable ξ_{ij} to the cumulative coefficient score θ_p of the variable q based on its identification number $= \xi_{ij}$

end for

end for

Normalize $\boldsymbol{\Theta} = [\theta_p]$

Return $\boldsymbol{\Theta}$ as variable efficiency cumulative coefficient score for ranking

The simulation experiment, results and analysis for abovementioned methods are following in next section.

4   Simulation Experiment, Results and Analysis

In this section, proposed method is verified by simulation experiment where experimental result and analysis are also discussed.

To evaluate the performance and investigate the accuracy of the proposed methods, simulation experiments are developed. We set $n = 1000$ observations, $p = 25$ variables. Four different simulation model are generated by pseudorandom algorithms for $i = 1 \ldots p$ as

$$u_i^1 \sim U(-1,1), \quad u_i^2 \sim N(0,1), \quad u_i^3 \sim Exp(1), \quad u_i^4 \sim Weib(2,1) \tag{13}$$

by uniform, normal, exponential, and Weibull distributions respectively, where $U^j, j = 1 \ldots 4$ is explanatory variable matrix including p variable vectors represent the original dataset $\boldsymbol{U}$ for j'th model. In each simulation, the coefficient components used in (11) are set as

$$\beta_{ii} = 10, \quad \beta_{jj} = 1, \quad \beta_{kk} = 0.1, \quad ii = 1 \ldots 5, \quad jj = 6 \ldots 15, \quad kk = 16 \ldots 25 \tag{14}$$

Simulated event time vector $\boldsymbol{Y}$ is obtained from (3) where $\boldsymbol{X} \equiv \boldsymbol{Y}$ and the censoring indicator is generated using a uniform distribution. To improve accuracy in calculating survival time vector $\boldsymbol{t}$, an error vector is considered as $E \sim N(0,0.1)$ in each simulation. In order to obtain k as an estimation of desired number of variables in any selected subset, principal component analysis (PCA) is used [27], which gives $k = 3$.

All Four simulation models are applied in the proposed algorithm. A sample of numerical results of simulation experiment when data is generated by the second model U^2 and applied in the proposed algorithm is depicted in Fig. 1. Variables with larger coefficient scores $\tilde{\beta}$ are more efficient and variables with smaller scores are ideal candidate to be eliminated from the dataset if it is desired. Each number represents a specific identity number of a variable

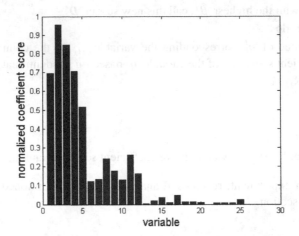

Fig. 1. Normalized coefficient score results for the simulation dataset for 100 trials

4.1   Numerical Experiment

A set of numerical simulation experiments are conducted to evaluate performance of the proposed method. Four defined simulation models in (13) are examined, and the result are presented in Table 1. These experiments are also designed for n=1000 observations and $p = 25$ variables following the same data generating algorithm. Number of significant efficient variables designed in each simulation model is set $m^* = \{5\}$ for each experiment. Each simulation is repeated 200 times independently. For each method, m represents average number of determined and selected efficient variables as an output of the proposed heuristic algorithm and $\tilde{p}$ denotes performance of the method, for 200 trials, where

$$\tilde{p} = m_{method}/m_{definition} = m/m^* \tag{15}$$

and $\bar{p}$ is the grand average of performance of each method. The adjusted censoring rates of δ_i for all numerical simulation experiments are $30 \pm 10\%$.

Table 1. Performance of the proposed methods based on four numerical simulation experiments with 200 replications

	Simulation Models								
	#1		#2		#3		#4		Ave
Method	m	$\tilde{p}$	m	$\tilde{p}$	m	$\tilde{p}$	m	$\tilde{p}$	$\bar{p}$
Proposed	4.2	0.84	4.4	0.88	4.1	0.82	3.9	0.78	0.83
Definition	5		5		5		5		

The robustness of this method is verified by replicated simulation experiments leading to similar results. As an outcome of proposed methods, to reduce the number of variables in the dataset for further analysis, it is recommended to eliminate those variables with lower rank. In addition, to concentrate on a reduced number of variables, a category of highest ranked variables is suggested.

5   Summary

Variable selection procedures are an integral part of data analysis. The information revolution brings larger datasets with more variables. In many areas, there are great interests in time and causes of events. The information including event time, censoring indicator and explanatory variables which are collected over a period of time is called survival data. The proposed method is beneficial to explanatory variable selection in decision-making process to obtain an appropriate subset of variables from among a large-scale survival data. By using such a novel method and reducing the number of variables, data analysis will be faster, simpler and more accurate. This procedure potentially is an applicable solution for many problems in a vast area of science and technology.

References

1. IBM: Information Integration and Governance (2011), http://www.ibm.com
2. McAfee, A., Brynjolfsson, E.: Big data: the management revolution. Harvard Business Review 90, 60–66 (2012)
3. IBM: What is Big Data? Bringing Big Data to the Enterprise (2013), http://www.ibm.com
4. Hilbert, M., Lopez, P.: The World's Technological Capacity to Store, Communicate, and Compute Information. Science 332(6025), 60–65 (2011)
5. Gartner (2011), http://www.gartner.com/newsroom/id/1731916
6. Hellerstein, J.: Parallel Programming in the Age of Big Data (2008), https://gigaom.com/2008/11/09/mapreduce-leads-the-way-for-parallel-programming
7. Segaran, T., Hammerbacher, J.: Beautiful Data: The Stories Behind Elegant Data Solutions. O'Reilly Media, Inc. (2009)
8. Feldman, D., Schmidt, M., Sohler, C.: Turning big data into tiny data: Constant-size coresets for k-means, pca and projective clustering. In: Proceedings of the Twenty-Fourth Annual ACM-SIAM Symposium on Discrete Algorithms, pp. 1434–1453. SIAM (2013)
9. Manyika, J., et al.: Big Data: The Next Frontier for Innovation, Competition, and Productivity. McKinsey Global Institute (2011)
10. Moran, J.: Is Big Data a Big Problem for Manufacturers? (2013), http://www.sikich.com/blog/post/Is-Big-Data-a-Big-Problem-for-Manufacturers#.VPswcU_F_BM
11. Brown, B., Chui, M., Manyika, J.: Are you ready for the era of 'big data'. McKinsey Quarterly 4, 24–35 (2011)
12. Russom, P.: Big Data Analytics. TDWI Best Practices Report, Fourth Quarter (2011)
13. Sadeghzadeh, K., Salehi, M.B.: Mathematical Analysis of Fuel Cell Strategic Technologies Development Solutions in the Automotive Industry by the TOPSIS Multi-Criteria Decision Making Method. International Journal of Hydrogen Energy 36(20), 13272–13280 (2010)

14. Chai, J., Liu, J.N., Ngai, E.W.: Application of decision-making techniques in supplier selection: A systematic review of literature. Expert Systems with Applications 40(10), 3872–3885 (2013)
15. Yao, F.: Functional Principal Component Analysis for Longitudinal and Survival Data. Statistica Sinica 17(3), 965 (2007)
16. Cox, D.R.: Regression Models and Life-Tables. Journal of the Royal Statistical Society 34(2), 187–220 (1972)
17. Kalbfleisch, J.D., Prentice, R.L.: The Statistical Analysis of Failure Time Data, vol. 360. John Wiley & Sons (2011)
18. Buckley, J., James, I.: Linear Regression with Censored Data. Biometrika 66(3), 429–436 (1979)
19. Ishwaran, H., Kogalur, U.B., Blackstone, E.H., Lauer, M.S.: Random Survival Forests. The Annals of Applied Statistics, 841–860 (2008)
20. Ma, S., Kosorok, M.R., Fine, J.P.: Additive Risk Models for Survival Data with High-Dimensional Covariates. Biometrics 62(1), 202–210 (2006)
21. Huang, J., Ma, S., Xie, H.: Regularized Estimation in the Accelerated Failure Time Model with High-Dimensional Covariates. Biometrics 62(3), 813–820 (2006)
22. Breiman, L., Friedman, J., Stone, C.J., Olshen, R.A.: Classification and Regression Trees. CRC Press (1984)
23. Lee, E.T., Wang, J.: Statistical Methods for Survival Data Analysis, vol. 476. John Wiley & Sons (2003)
24. Holford, T.R.: Multivariate Methods in Epidemiology. Oxford University Press (2002)
25. Mendes, A.C., Fard, N.: Accelerated Failure Time Models Comparison to the Proportional Hazard Model for Time-Dependent Covariates with Recurring Events. International Journal of Reliability, Quality and Safety Engineering 21(2) (2014)
26. Zeng, D., Lin, D.Y.: Efficient Estimation for the Accelerated Failure Time Model. Journal of the American Statistical Association 102(480), 1387–1396 (2007)
27. Sadeghzadeh, K., Fard, N.: Nonparametric Data Reduction Approach for Large-Scale Survival Data Analysis. IEEE (2015)
28. Sadeghzadeh, K., Fard, N.: Multidisciplinary Decision-Making Approach to High-Dimensional Event History Analysis through Variable Reduction. European Journal of Economics and Management 1(2), 76–89 (2014)
29. Stute, W., Wang, J.L.: The Strong Law under Random Censorship. The Annals of Statistics, 1591–1607 (1993)
30. Feo, T.A., Resende, M.G.: Greedy Randomized Adaptive Search Procedures. Journal of Global Optimization 6(2), 109–133 (1995)
31. Hart, J.P., Shogan, A.W.: Semi-Greedy Heuristics: An Empirical Study. Operations Research Letters 6(3), 107–114 (1987)

Mining the Adoption Intention of Location-Based Services Based on Dominance-Based Rough Set Approaches

Yang-Chieh Chin

Department of International Business, Asia University, 500, Lioufeng Rd.,
Wufeng, Taichung 41354, Taiwan
chin@asia.edu.tw

Abstract. Faced with the excessive opportunity, mobile service providers have unconstrained a variability of services, such as mobile instant messaging, mobile shopping and location-based services (LBS). Location-based service is a social media tool that allows users to write short text messages to public and private networks. This research focuses specifically on the mobile services on LBS. The main purposes of this study are to investigate and compare what recommendation sources influence the intention to use LBS and to combine gender, daily internet hour usage and past use experience to infer the usage of LBSs decision rules using a dominace-based rough-set approach. Data for this study were collected from 398 users and potential users. The analysis is grounded in the taxonomy of induction-related activities using a domiance-based rough set approach to infer the usage of LBSs decision rules. Finally, the study of the nature of LBS reflects essential practical and academic value.

Keywords: Location-base services, Dominance-based rough set approach, Recommendation source, Adoption intention.

1   Introduction

Compared to other mobile services, the unique advantage of location-based services (LBS) is locatability, which means that LBS can current the optimal information and services to users based on their location [1](Petrova & Wang, 2011).With the growth of users on the mobile services, the biggest benefit of mobile services is its ability to generate platform revenues by means of advertisements [2] and other applications. Xu et al. [3] also noted that multimedia location-based advertisements have a significant effect on users' purchase intention. Thus, how to stimulate the LBS adoption intention becomes a critical issue to platform marketers.

The contextual information and services revealed by LBS will decrease users' effort spent on information retrieval and improve their experience. However, LBS need to gather and utilize users' location information. This could arouse their privacy concern and produce their perceived privacy risk [4]. Therefore, even though LBS provides conveniences and benefits, some users are concerned about the use of LBS as

H.A. Le Thi et al. (eds.), *Model. Comput. & Optim. in Inf. Syst. & Manage. Sci.*,
Advances in Intelligent Systems and Computing 360, DOI: 10.1007/978-3-319-18167-7_6

another form of background check and that their privacy could be lost in cyberspace [5]. However, such concerns can be addressed by better and more accurate recommendations because people are influenced by others' recommendations when making decisions [6]. These recommendations can be classified as interpersonal sources, impersonal sources [7] and neutral sources [8]. Researchers have shed some light on the importance of recommendation sources in the context of product purchases [9], but little has been done on the relevance of these recommendation sources in the context of LBS usage. Thus, our primary goal in this study is to fill that gap by increasing our understanding of how the three primary categories of recommendation sources— interpersonal recommendations (e.g., word-of-mouth recommendations), impersonal recommendations (e.g., advertising recommendations), and neutral recommendations (e.g., expert recommendations)—influence users intention to adopt LBS.

The Dominance-based Rough Set Approach (DRSA) proposed by Greco et al. [10], effectively derived a set of decision rules based on preference-ordered data [11] and substituted the indiscernibility relation in classical rough set theory with a dominance relation that is reflexive and transitive [12]. In addition, the DRSA approach was motivated by representing preference models for multiple criteria decision analysis (MCDA) problems, where preference orderings on domains of attributes are quite typical in exemplary based decision-making. For instance, "if antecedent, then consequence" decision rules are derived from the options of users and express the relationships between attributes values and the intention to use LBSs. Therefore, another purpose of this study is to combine control variables (gender, daily internet hour usage, and past use experience), grounded in the taxonomy of induction-related activities using the DRSA, to infer the LBS-related decision rules.

2   Literature Review

2.1   LBS Adoption Intention

Adoption is a widely researched process that is often used to investigate the spread of information technology [13]. According to the literature on information technology adoption, adoption intention is an individual's intention to use, acquire, or accept a technology innovation [13]. LBS adoption intention has also received attention from past works. These information technology adoption theories such as the technology acceptance model (TAM), task technology fit (TTF) and the unified theory of acceptance and use of technology (UTAUT) as the theoretical bases [4].

2.2   Recommendations Source

Word-of-mouth (WOM) is the process of conveying information from person to person. WOM communication functions based on social networking and trust since people rely on families, friends, and others in their social network to provide reliable recommendations [14]. Prior studies have suggested that peer communications may be considered the most trustworthy type of recommendation source in making decisions [15]. Advertising recommendations, such as recommendations from site-sponsored advertisements, may be also regarded as a credibility cue [16]. Marketers can apply advertising techniques to social community websites for the express purpose of delivering information messages to recruit customers and users. Previous research has demonstrated that the perceived level of expertise positively impacts acceptance of source recommendations [17]. These recommendations may be also considered a credibility cue when making decisions [30].

3   Methodology

3.1   The Dominance-Based Rough Set Approach

The original rough set theory (RST) was proposed by Pawlak [18] as an effective mathematical approach for discovering hidden deterministic rules and associative patterns in quantitative or qualitative data and for handling unknown data distributions and information uncertainty. However, the main restriction for the use RST is that the domain of attributes is preference ordered. To help fill the gap, Greco et al. [10,11] proposed an extension of the rough set theory based on the dominance principle to incorporate the ordinal nature of the preference data into the classification problem—what is called dominance-based rough set approach (DRSA). In this study, to understand the influence of recommendation sources on the adoption intention of LBS, we begin by using the DRSA approach to extract the knowledge what we need.

3.2   Basic Concepts of Dominance-Based Rough Set

DRSA uses an ordered information table where in each row represents an object, which is defined a respondent to our survey, and each column represents an attribute, including preference-ordered domain and regular (no preference-ordered domain). Thus, the entries of the table are attribute values. Formally, an information system can be represented by the quintuple $IS = (U, Q, V, f)$, where U is a finite and non-empty set of objects, called the universe, that is composed of a certain set of objects, and Q is a non-empty finite set of attributes describing objects in the universe. An information system can be also divided into two subsets, C and D, which denote the finite set of condition attributes and the set of decision attributes, respectively. $V = \bigcup_{a \in Q} V_a$, in which V_a is a domain of the attribute a, and $f : U \times Q \rightarrow V$ is an information function such that $f(x, a) \in V_a$ for every $a \in Q$ and $x \in U$ [19].

3.3   Approximation of the Dominance Relation

In DRSA, which is based on there being at least one conditional attribute and classes' being preference-ordered, the approximation is a collection of upward and downward unions of classes. DRSA uses a dominance relation instead of indiscernibility relation

[19]. A rough set offers a means to describe vague classes through these lower and upper approximations.

According to Greco et al. [10, 11], first, let $\succeq_a$ be an outranking relation on U with respect to criterion $a \in Q$, such that $x \succeq_a y$ means "x is at least good as y with respect to criterion a." Suppose that $\succeq_a$ is a complete preorder. Furthermore, let $Cl = \{Cl_t, t \in T\}$, $T = \{1, ..., n\}$, be a set of decision classes of U that each $x \in U$ belongs to one and only one class $Cl_t = Cl$. Assume that, for all $r, s \in T$ such that $r \succ s$, the elements of Cl_r are preferred to the elements of Cl_s. Given the set of decision class Cl, it is possible to define upward and downward unions of classes, respectively,

$$Cl_t^{\geq} = \bigcup_{s \geq t} Cl_s, \quad Cl_t^{\leq} = \bigcup_{s \leq t} Cl_s, \quad t = 1, ..., n.$$

For example, $x \in Cl_t^{\geq}$ means "x belongs at least class Cl_t," while $x \in Cl_t^{\leq}$ means "x belongs to, at most, class Cl_t."

In dominance-based approaches, we say that x dominates y with respect to $P \subseteq C$ if $x \succeq_a y$ for all $a \in P$. Given $P \subseteq C$ and $x \subseteq U$, let $D_p^+(x) = \{y \in U : y \succeq x\}$ represent a set of objects dominating x, called a P-dominating set, and $D_p^-(x) = \{y \in U : x \succeq y\}$ represent a set of objects dominated by x, called a P-dominated set. We can adopt $D_p^+(x)$ and $D_p^-(x)$ to approximate a collection of upward and downward unions of decision classes.

The P-lower approximation of $\underline{P}(Cl_t^{\geq})$ of the unions of class $Cl_t^{\geq}$, $t \in \{2, 3, ..., n\}$, with respect to $P \subseteq C$ contains all objects x in the universe U, such that objects y that have at least the same evaluations for all the considered ordered attributes from P also belong to class Cl_t or better, as

$$\underline{P}(Cl_t^{\geq}) = \left\{ x \in U : D_p^+(x) \subseteq Cl_t^{\geq} \right\}$$

Similarly, the P-upper approximation of $\overline{P}(Cl_t^{\geq})$ is composed of all objects x in the universe U, whose evaluations on the criteria from P are not worse than the evaluations of at least one object y belonging to class Cl_t or better, as

$$\overline{P}(Cl_t^{\geq}) = \left\{ x \in U : D_p^-(x) \cap Cl_t^{\geq} \neq \varnothing \right\}$$

Analogously, the P-lower and P-upper approximations of $\underline{P}(Cl_t^\leq)$ and $\underline{P}(Cl_t^\leq)$, respectively, of the unions of class $Cl_t^\geq$, $t \in \{2,3,...,n\}$, with respect to $P \subseteq C$ are defined as

$$\underline{P}(Cl_t^\leq) = \left\{ x \in U : D_P^-(x) \subseteq Cl_t^\leq \right\}$$

$$\overline{P}(Cl_t^\leq) = \left\{ x \in U : D_P^+(x) \cap Cl_t^\leq \neq \varnothing \right\}$$

The P-boundaries (P-doubtable regions) of $Cl_t^\geq$ and $Cl_t^\leq$ are defined as

$$Bn_P(Cl_t^\geq) = \overline{P}(Cl_t^\geq) - \underline{P}(Cl_t^\geq)$$

$$Bn_P(Cl_t^\leq) = \overline{P}(Cl_t^\leq) - \underline{P}(Cl_t^\leq)$$

With each set $P \subseteq U$ we can estimate the accuracy of approximation of $Cl_t^\geq$ and $Cl_t^\leq$ using

$$\alpha_P(Cl_t^\geq) = \left| \frac{\underline{P}(Cl_t^\geq)}{\overline{P}(Cl_t^\geq)} \right| \qquad \alpha_P(Cl_t^\leq) = \left| \frac{\underline{P}(Cl_t^\leq)}{\overline{P}(Cl_t^\leq)} \right|$$

and the ratio

$$\gamma_P(\boldsymbol{Cl}) = \left| \frac{U - \left(\bigcup\limits_{t \in \{2,...,n\}} Bn_P\left(Cl_t^\geq \right) \right)}{U} \right| = \left| \frac{U - \left(\bigcup\limits_{t \in \{1,...,n-1\}} Bn_P\left(Cl_t^\leq \right) \right)}{U} \right|$$

The ratio, $\gamma_P(\boldsymbol{Cl})$, is called the quality of approximation of classification Cl by the set of attributes P or, in short, the quality of classification. It indicates the ratio of all P – correctly classified objects to all objects in the system.

3.4   Extraction of Decision Rules

A decision rule can be expressed as a logical manner of the *if* (antecedent) *then* (consequence) type of decision. The procedure of capturing decision rules from a set of initial data is known as induction [18]. For a given upward union of classes, $Cl_t^\geq$, the decision rules included under the hypothesis that all objects belonging to $\underline{P}(Cl_t^\geq)$ are positive and the others are negative. There are two types of decision rules as follows:

(1) $D_\geq$ decision rules ("at least" decision rules)

If $f(x, a_1) \geq r_{a_1}$ and $f(x, a_2) \geq r_{a_2}$ and $...\, f(x, a_p) \geq r_{a_p}$, then $x \in Cl_t^\geq$

(2) $D_{\leq}$ decision rules ("at most" decision rules)

$$If \ f(x,a_1) \leq r_{a_1} \ and \ f(x,a_2) \leq r_{a_2} \ and \ ... \ f(x,a_p) \leq r_{a_p}, then \ x \in Cl_t^\leq$$

The decision rule reflects a relationship between a set of conditions and a conclusion or a decision. Mark and Munakata [20] argued that the extraction of rules using rough sets is relatively simple and straightforward and that no extra computational procedures are required before rules can be extracted. Therefore, in this study, construction of the decision rules is performed based on upper and lower approximations extracted from the decision table.

4   An Empirical Example of LBS

Taiwan's Internet growth rate preserved a rate of over 70% and it also touched 77.6% in 2014. Temporarily, the development of mobile devices has stimulated the rise of the mobile Internet. We use the example of LBS because it is one of the most popular mobile services and it has already attracted over 9.87 million mobile users in Taiwan [21]. In this section, we use the JAMM software [22], which was developed by the Laboratory of Intelligent Decision Support Systems (IDSS) at Poznan University of Technology, for the rough set approach to analytical procedures, including: (1) selecting variables and data; (2) calculating the approximation; (3) finding the reductions and core attributes; and (4) generating decision rules. The results are used to understand the influence of recommendation sources on the intention to adopt LBSs.

4.1   Selection Variables and Data

To increase the accuracy of decision variables, we invited five scholars from the university information management and marketing sectors to verify whether the selected variables can be viewed as criteria for the intention to adopt LBS. Based on the literature review and on our experts' opinions, six criteria for personal and recommendation-sources attributes were identified for extracting the rules for the intention to adopt LBSs. Detailed definitions of recommendation sources for the intention to adopt LBS follow:

- Gender: Gender serves as an indicator the decision-making process [23,24] because it is gender differences might be in the intention to adopt LBSs.
- Daily internet hour usage: The daily internet usage is measured by the average number of hours a person is on the Internet in a 24-hour period.
- Past use experience: Past use experience was measured by whether login on LBSs in the past.
- WOM recommendations: WOM recommendation sources are primarily interpersonal sources, which are non-commercial personal sources used to gather information.

- Advertising recommendations: Advertising recommendations are non personal communications of information, usually paid for by the advertiser and usually persuasive in nature about products, services or ideas.
- Expert recommendations: Expert recommendations are neutral sources, and can come from experts, professionals, professors, and scholars' advices.
- Adoption intention: The adoption intention is an indication of an individual's readiness to perform a given behavior. It is assumed to be the immediate antecedent of behavior [25].

In this study, the research subjects are users and potential users of Facebook. An online survey was conducted to collect the needed responses from the subjects. A total of 382 undergraduate and graduate students from a university in northern Taiwan participated in the survey. To minimize data variation, the data collection occurred during limited periods. The participants were then asked to complete a self-reported questionnaire containing study measures for their intentions to use LBS sites such as Facebook. In addition, because daily internet hour usage, past use experience [26] and gender information [27] can also reflect the composition of the users, we also included the three variables as controls in this study.

All subjects participated in the study voluntarily. Within the sample population: 226 (59.2 %) were undergraduate students and 153 (40.8 %) were graduate students. 46% of the participants were male and 54% were female, and their ages ranged from 18 to 25 years old. More than 73% of participants spent more than an average of three hours on the Internet per day. Only a few participants (11%) had no LBS experience, while 45.2% of participants had used LBS such as Facebook more than 5 times in the last month. Finally, Most of the participants (88%) stated that they were familiar with the term "Facebook" prior to taking the survey. The statistical analyses will be conducted and their results will be reported at a later time.

4.2 Rules for the Intention to Adopt LBS

The personal attributes of the participants (gender, daily internet hour usage, and past use experience) and the attributes of the recommendation sources (WOM, advertising, and expert) were conducted. In addition, one decision attriubte, the adoption intention, is also included to pre-process the data to construct the information table, which represents knowledge in a DRSA model. The attributes of recommendation sources were measured in three dimensions: WOM (friend or classmate reviews, e.g., "Your friend/classmate talked to you about the advantage of LBS, such as Facebook"), advertising (e.g., "The platform providers presented advertisements on the web page to attract users"), and expert recommendations (professional comments, e.g., "A relevant professional introduced the benefits of LBS in the newspaper"). The respondents were asked to choose the recommendation source they would normally consult and to indicate the extent to which the source was perceived as an influence of recommendation on a 5-point Likert-type scale, with 1 = *not very important*, 3 = *neutral*, and 5 = *very important*. Furthermore, the participants were asked to evaluate their LBS usage intentions. The survey also presented statements and participants were asked to indicate their level of agreement using multi-item scales, measured on a 5-point Likert-type scale where 1 = *strongly disagree*, 3 = *neutral*, and 5 = *strongly agree*.

Based on the decision rules extraction procedures of the DRSA, a large number of rules related to the intention to use LBS can be generated. The first results obtained from DRSA analysis of the coded information table were the approximations of the decision classes and the quality of their classifications. These results revealed that the data were very well categorized and appropriate for understanding how personal and recommendation sources attributes would influence LBS usage. In addition, we classified our samples into two classes: "at least 4" (corresponds to having the intention to adopt LBSs) and "at most 3"(corresponds to having no or little intention to adopt LBSs). The accuracy of classification for the two decision classes was 99% and 98%, respectively, so most samples of the data were correctly classified. In this manner, the results represent that the 6 condition attributes play an important role in determining the intention to adopt LBSs.

To increase the classification rate and acquire the reduced attribute sets during the DRSA analysis processes, the reduction of conditional attributes through an exhaustive algorithm was employed to determine the superfluous attributes. The value of the positive region of reduction is 1.0, so there are no superfluous attributes in our analysis. Therefore, the core and the reduced set both consist of six attributes $\{a_1, a_2, a_3, a_4, a_5, a_6\}$, representing all relevant attributes in the table. This result shows that all six of these variables are importantly in understanding the intention to adopt LBSs.

In this study, the rules are divided into two classes: "at least 4" rules and "at most 3" rules. Through DRSA analysis, we generated 12 rules, of which 9 rules apply to the "at least 4" class and 3 rules apply to the "at most 3" class, as illustrated in Table 3. The coefficients of certainty, strength, and coverage associated with each rule are also illustrated. Under the different decision rules, the rule set generates relative strength and coverage.

The antecedents of the "at least 4" class of rules explain which attributes LBS companies need to attract, and the "at most 3" class of rules tells the LBS companies what attributes they should avoid. In table 3, the frequent occurrences of the variables in the "at least 4" class decision rules table include WOM recommendations (4 times), past use experience (4 times), advertising recommendations (3 times), expert recommendations (3 times), gender (2 times), and daily internet hour usage (2 times). For their part, the "at most 3" class decision rules table includes advertising recommendations (2 times), expert recommendations (2 times), daily internet hour usage (2 times), gender (1 time), and WOM recommendations (1 time). Therefore, as Table 3 shows, some of the variables had a higher degree of dependence and may impact the intention to adopt LBSs more than others. These results illustrate the different degrees of importance of variables for effecting the adoption intention, which could help managers develop marketing strategies.

5   Discussion and Managerial Implications

This investigation examined how personal variables and recommendation sources influence the users' intention to adopt LBSs. In the "at least 4" class, the analytical results showed that users who trust in recommendation source are more likely to adopt LBSs and that WOM recommendations influenced the subjects' intention to adopt LBSs more than advertising recommendations and expert recommendations did. In addition, users who have more daily internet hour usage and who are more familiar than others with LBSs rely on recommendation sources to adopt LBSs. Finally, males who trust expert recommendations are more likely to adopt LBSs. In the "at most 3" class, the analytical results showed that the intentions of females who have no confidence in expert recommendations to adopt LBSs would decrease, as would those of users who have fewer daily internet hours use and users who doubt recommendation sources.

The results of this study have implications for decision-makers. One implication is that marketers may use recommendation sources, especially WOM recommendations, to promote LBSs usage. For instance, the platform providers can design recommendation activities where users who recommend LBSs to others are rewarded. Especially in an online environment, our suggestion is consistent with Park et al. [28], who pointed out that marketers should consider providing user-generated information services and recommendations by previous users in the form of electronic word-of-mouth (eWOM).

Another implication is that different types of recommendations can attract different types of users. There are differences in how recommendation sources impact the two genders, so the platform providers can apply different recommendation strategies, such as targeting mass media (e.g., a news report) to male users and alternative media (e.g., a discussion forum) to female users.

6   Conclusions and Remarks

DRSA has not been widely used in predicting LBSs usage, especially in the context of social networks. This study uses DRSA to identify LBSs decision rules that infer the antecedents of the intent to adopt LBSs under the effects of different recommendation sources. Future research can extend this study to apply other data-mining approaches to extracting the attributes of the intention to adopt LBSs. The study is limited in that actual behavior was not assessed, so the links between intention and actual behavior in this context remain unexamined [29].

Acknowledgements. This study was supported by research funds from Ministry of Science and Technology, Taiwan (MOST 103-2410-H-468-009-).

References

1. Petrova, K., Wang, B.: Location-Based Services Deployment and Demand: A Roadmap Model. Elect. Com. Res. 11(1), 5–29 (2011)
2. Sledgianowski, D., Kulviwat, S.: Social Network Sites: Antecedents of User Adoption and Usage. In: Proceedings of the Fourteenth Americas Conference on Information Systems (AMCIS), Toronto, ON, Canada, pp. 1–10 (2008)
3. Xu, H., Oh, L.-B., Teo, H.-H.: Perceived Effectiveness of Text vs. Multimedia Location-Based Advertising Messaging. Int. J. Mob. Com. 7(2), 133–153 (2009)
4. Zhou, T.: An Empirical Examination of User Adoption of Location-Based Services. Elect. Com. Res. 13(1), 25–39 (2013)
5. Junglas, I.A., Watson, R.T.: Location-Based Services. Com. ACM. 51(3), 65–69 (2008)
6. Häubl, G., Trifts, V.: Consumer decision making in online shopping environments: the effects of interactive decision aids. Mark. Sci. 19(1), 4–21 (2000)
7. Andreasen, A.R.: Attitudes and Customer Behavior: A Decision Model. In: Kassarjian, H.H., Robertson, T.S. (eds.) Perspectives in Consumer Behavior, pp. 498–510. Scott, Foresman and Company, Glenview (1968)
8. Cox, D.F.: Synthesis: Risk Taking and Information Handling in Consumer Behavior. In: Cox, D.F. (ed.) Risk Taking and Information Handling in Consumer Behavior, pp. 604–639. Harvard University Press, Boston (1967)
9. Murray, K.B.: A Test of Services Marketing Theory: Consumer Information Acquisition Activities. J. Mark. 55(1), 10–25 (1991)
10. Greco, S., Matarazzo, B., Slowinski, R.: A New Rough Set Approach to Evaluation of Bankruptcy Risk. In: Zopounidis, C. (ed.) Operational Tools in the Management of Financial Risk, pp. 121–136. Kluwer Academic Publishers, Boston (1998)
11. Greco, S., Matarazzo, B., Slowinski, R.: Rough Set Theory for Multicriteria Decision Analysis. Euro. J. Ope. Res. 129(1), 1–47 (2001)
12. Ou Yang, Y.-P., Shieh, H.-M., Tzeng, G.-H., Yen, L., Chan, C.-C.: Business Aviation Decision-Making Using Rough Sets. In: Chan, C.-C., Grzymala-Busse, J.W., Ziarko, W.P. (eds.) RSCTC 2008. LNCS (LNAI), vol. 5306, pp. 329–338. Springer, Heidelberg (2008)
13. Rogers, E.M.: Diffusion of Innovations, 5th edn. Free Press, New York (1995)
14. Jansen, B., Zhang, M., Sobel, K., Chowdury, A.: Twitter Power: Tweets as Electronic Word of Mouth. J. Amer. Soc. Infor. Sci. Tech. 60(11), 2169–2188 (2009)
15. Richins, M.L., Root-Shaffer, T.: The Role of Involvement and Opinion Leadership in Consumer Word-of-Mouth: An Implicit Model Made Explicit. Adv. Con. Res. 15(1), 32–36 (1988)
16. Smith, D., Menon, S., Sivakumar, K.: Online Peer and Editorial Recommendation, Trust, and Choice in Virtual Markets. J. Inter. Mar. 19(3), 15–37 (2005)
17. Crisci, R., Kassinove, H.: Effects of Perceived Expertise, Strength of Advice, and Enviornmental Seting on Parental Compliance. J. Soc. Psy. 89(2), 245–250 (1973)
18. Pawlak, Z.: Rough Set. Int. J. Comp. Infor. Sci. 11(1), 341–356 (1982)
19. Liou, J.H., Yen, L., Tzeng, G.H.: Using Decision Rules to Achieve Mass Customization of Airline Services. Euro. J. Oper. Res. 205(3), 680–686 (2010)
20. Mark, B., Munakata, T.: Computing, Artificial Intelligence and Information Technology. Euro. J. Oper. Res. 136(1), 212–229 (2002)
21. Taiwan Network Information Center (TWNIC) (2014),
    `http://www.chinatechnews.com/2014/08/27/20849-taiwan-mobile-internet-users-reached-9-87-million-in-h1-2014`

22. Slowinski, R.: The International Summer School on MCDM 2006, Class note, Kainan University, Taiwan (2006),
    `http://idss.cs.put.poznan.pl/site/software.htm`
23. Crow, S.M., Fok, L.Y., Hartman, S.J., Payne, D.M.: Gender and Values: What is the Impact on Decision Making? Sex Rol. 25(3/4), 255–268 (1991)
24. Gefen, D., Straub, D.W.: Gender Differences in the Perception and Use of E-Mail: An Extension to the Technology Acceptance Model. MIS Quar. 21(4), 389–400 (1997)
25. Rogers, E.M.: The 'Critical Mass' in the Diffusion of Interactive Technologies in Organizations. In: Kraemer, K.L. (ed.) The Information Systems Research Challenge: Survey Research Methods, Harvard Business School Research Colloquium, vol. 3, pp. 245–264. Harvard Business School, Boston (1991)
26. Sung, S.K., Malhotra, N.K., Narasimhan, S.: Two Competing Perspectives on Automatics Use: A Theorectical and Empirical Comparison. Inf. Sys. Res. 17(2), 418–432 (2005)
27. Cheong, M., Lee, V.: Integrating Web-Based Intelligence Retrieval and Decision-Making from the Twitter Trends Knowledge Base. In: Proceeding of the 2nd ACM Workshop on Social Web Search and Mining, pp. 1–8 (2009)
28. Park, D.H., Lee, J., Han, I.: The Effect of On-line Consumer Reviews on Consumer Purchasing Intention: The Moderating Role of Involvement. Int. J. Ele. Com. 11(4), 125–148 (2007)
29. Bagozzi, R.P., Baumgartner, H., Yi, Y.: State Versus Action Orientation and the Theory of Reasoned Action: An Application to Coupon Usage. J. Con. Res. 18(4), 505–518 (1992)

Review of Similarities between Adjacency Model and Relational Model

Teemu Mäenpää and Merja Wanne

University of Vaasa, Faculty of Technology, Department of Computer Science, Vaasa, Finland
{teemu.maenpaa,merja.wanne}@uva.fi

Abstract. The emergence of concepts such as big data and internet of things lead into a situation where the data structures and repositories have become more complex. So, there should be a way to analyze such data, and organize it into a meaningful and usable form.

Relational model is widely used model for organizing data. Adjacency model is a data model that relies on adjacency between elements. Relational data can be represented by adjacency model. Moreover, the adjacency model can be visualized as a graph. This paper discusses the similarities between the models based on the previous studies and theories. Furthermore, this paper aims to strengthen and quantify the similarities between the models by utilizing the graph theory concepts.

This study reveals that the previous considerations between the relational model and the adjacency model can be backed up with graph theory. If a relational database is represented by adjacency model and visualized as a graph called adjacency relation system, the elements of relational database can be identified from the graph. The identification of the elements is based on the graph theory concepts such as walk, vertex degree, leaf vertex, and graph domination.

Keywords: data modeling, relational model, adjacency model, graph analysis.

1   Introduction

As concepts like big data and internet of things have emerged. There is an obvious need to analyze and organize the data into meaningful and usable patterns (such as relational databases). Prerequisite for efficient utilization of data is that the data must have clear structures and rules for manipulation. This paper provides a novel perspective for examining databases and structures representing them.

In this paper, the concepts of adjacency relation system and adjacency model are revised. Their similarities with the relational model are introduced based on previous research efforts.

The similarity perspective is strengthened by pointing out that the similarities between the models can be quantified by utilizing graph theory concepts. This paper is organized as follows. In the second and third sections adjacency model and relational model are introduced. The fourth section of this paper discusses the similarities

© Springer International Publishing Switzerland 2015   69
H.A. Le Thi et al. (eds.), *Model. Comput. & Optim. in Inf. Syst. & Manage. Sci.*,
Advances in Intelligent Systems and Computing 360, DOI: 10.1007/978-3-319-18167-7_7

between the models based on the previous research work. In the fifth section a brief example about the utilization of graph theory concepts in the analysis of the adjacency relation system based graphs is given. Finally, concluding remarks are given.

2   Adjacency Relation System and Adjacency Model

In this section, the concepts of Adjacency Relation System (ARS) and Adjacency Model (AM) are introduced. Adjacency relation systems were introduced by Wanne [1]. A generalization of ARS was given by Wanne and Linna [2] in "*A General Model for Adjacency*". Later on Töyli, Linna and Wanne [3,4] compiled the definitions under the concept of Adjacency Model. In this paper, the term Adjacency Model is used for the framework of terms and the term Adjacency Relation System is used for the formal and graph-based representation of the model.

In order to achieve understanding about the AM, the concepts of the model are introduced next. The fundamental concepts of the adjacency model are adjacency relation system, adjacency defining sets and relation combination. Adjacency relation system is based on the concepts of type, element and adjacency. Each element represents a particular type and relationship between elements is called adjacency [1].

An adjacency relation system (ARS) is represented as a pair (A, R), where $A = \{A_1, A_2, \ldots, A_n\}$ is a set of pairwise disjoint finite nonempty sets and $R = \{R_{ij} | i, j \in \{1, 2, \ldots, n\}\}$ is the set of relations. Each R_{ij} is a relation on $A_i \times 2^{A_j}$, where 2^{A_j} denotes the power set of A_j [1].

If $(x, y_1), (x, y_2), \ldots, (x, y_m) \in R_{ij}$ are all the pairs of relation R_{ij} having x as the first component, then each element $y_k (k = 1, 2, \ldots, m)$ is adjacent to the element x. Set of elements y_k is denoted by $Ad_j(x)$. ARS is symmetric if for each pair $x \in A_i, y \in A_j$ holds that $x \in Ad_i(y)$ if and only if $y \in Ad_j(x)$ for each $i, 1 \leq i \leq n$ [1].

Example 1. Consider an adjacency relation system (A, R), where $A = \{A_1, A_2, A_3\}$, $A_1 = \{x_1, x_2, x_3\}$, $A_2 = \{y_1, y_2, y_3\}$, $A_3 = \{z_1, z_2\}$ and R contains relations.

$R_{11} = \{(x_2, \{x_1\})\}$
$R_{12} = \{(x_1, \{y_1\}), (x_2, \{y_3\}), (x_3, \{y_1\})\}$
$R_{13} = \{(x_2, \{z_1\})\}$
$R_{21} = \emptyset$
$R_{22} = \{(y_2, \{y_3\})\}$
$R_{23} = \{(y_3, \{z_2\})\}$
$R_{31} = \{(z_2, \{x_2\})\}$
$R_{32} = \emptyset$
$R_{33} = \{(z_1, \{z_2\})\}.$

The adjacency relation system of example is represented by the graph shown in Fig. 1.

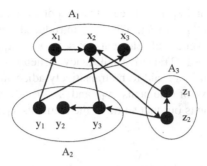

Fig. 1. Adjacency relation system for Example 1

The ARS portrayed in Fig. 1 is not symmetric because, for example, the element y_3 is adjacent to element x_2 $(Ad_2(y_2) = \{x_3\})$ but not vice versa [1].

Consider the ARS of Example 1 and redefine the relations as symmetric. Now R consists of relations.

$R_{11} = \{(x_1, \{x_2\}), (x_2, \{x_1\})\}$
$R_{12} = \{(x_1, \{y_1\}), (x_2, \{y_3\}), (x_3, \{y_1\})\}$
$R_{13} = \{(x_2, \{z_1, z_2\})\}$
$R_{21} = \{(y_1, \{x_1, x_3\}), (y_3, \{x_2\})\}$
$R_{22} = \{(y_2, \{y_3\}), (y_3, \{y_2\})\}$
$R_{23} = \{(y_3, \{z_2\})\}$
$R_{31} = \{(z_1, \{x_2\}), (z_2, \{x_2\})\}$
$R_{32} = \{(z_2, \{y_3\})\}$
$R_{33} = \{(z_1, \{z_2\}), (z_2, \{z_1\})\}.$

The ARS in Fig. 2 is symmetric. Symmetric ARS is depicted as an undirected graph.

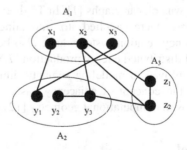

Fig. 2. Symmetric ARS

Each element in A_i represents certain entity type. The entity types are denoted with T_i, $(i = 1, \dots n)$. Each index pair $i, j \in \{1, 2, \dots, n\}$ is associated with a set $K \subseteq \{1, 2, \dots, n\} - \{i, j\}$ of indices and set of entity types $\tilde{T}_{ij} = \{T_k | k \in K\}$. The set $\tilde{T}_{ij}$ defines the adjacencies between the elements of the sets A_i and A_j. $\tilde{T}_{ij}$ is called adjacency defining set [1].

Since the elements y_3 and z_1 are adjacent to element x_2 it can be said that their adjacency is defined by set $\tilde{T}_{23} = \{T_1\}$. Töyli [5] introduced the concept of transitive adjacency which, is sequence of elements $x = x_1, x_2, \ldots x_m = y$ such that $x_i \in Ad(x_i + 1)$, $i = 1, \ldots, m - 1$. Transitive adjacency is denoted by $x \in Ad_{tr}(y)$. For example in Fig. 3 element y_1 is said to be transitively adjacent to z_2, $y_1 \in Ad_{tr}(z_2)$. The transitive adjacency of elements y_1 and z_2 is determined by sequence y_1, x_1, x_2, z_2. Töyli [5] points out that some of the elements $x_1, x_2, \ldots x_m$ can be of the same type.

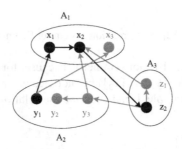

Fig. 3. Transitive adjacency

$T_i \rightarrow T_j$ denotes a relation combination which indicates that relation R_{ij} is defined on $A_i \times 2^{A_j}$. The set of relation combinations is denoted with $\{T_i \rightarrow T_j | (i, j) \in S\}$, where $S \subseteq \{1, 2, \ldots, n\} \times \{1, 2, \ldots, n\}$ [1].

The adjacency relation system has unambiguous mathematical foundation, which expands its application area. Even though, the adjacency relation system is a mathematical structure it can be visualized as a graph. Furthermore, the concept of relation combination allows the optimization of the queries on data because it can be used to restrict the search space, and thus limit the search time used by the query. [5]

The adjacency relation systems have been utilized in various application domains. The initial application area was planar graphs [1]. In Töyli et al. [3,4] and Töyli [5,6] adjacency relation systems were applied to modeling relational data and semistructured data. Adjacency relation systems have also been utilized in modeling wind power production and distributed energy production [7,8,9]. The adjacency relation systems have been employed in modeling semantic link networks [10,11]. Besides the modeling usage, the concept of adjacency has been utilized in algorithm development for applications in computational geometry [12].

3   Relational Model

In the relational model, the data is represented as a collection of relations. Each relation is represented as a table (Fig. 4). The elements of the relation (table) are tuple (a table row) and attribute (column header). The values of each column are determined by a domain of possible values [13].

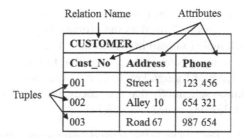

Fig. 4. Table representation of "Customer" relation

A domain represents a collection of indivisible values. Each domain has a prede-fined data type. Relation schema is used to describe the relation R. Relation schema $R(A_1, A_2, ..., A_n)$ consists of the relation name R and a list of attributes $A_1, A_2, ... A_n$. The names of the attributes are determined the by role they have in a certain domain, denoted by $dom(A_i)$. For example, a relation describing customer information can be formed as follows [13,14].

$CUSTOMER(Cust\_No, Address, Phone)$

The instance or state of the relation r of the relation schema, $r(R)$, is set of n-tuples $r = \{t_1, t_2, ..., t_n\}$. The n-tuple is defined as an ordered set of values $t = \{v_1, v_2, ..., v_n\}$. Each value v_i is an element of $dom(A_i)$. The notation $t[A_i]$ refers to the ith value of the tuple t [13].

Codd [14] states that a table representing relation R has the following properties:

1. Each row represents a tuple of R.
2. The ordering of rows is immaterial.
3. All rows are distinct.
4. The order of columns must correspond the ordering in the relation schema $R(A_1, A_2, ..., A_n)$.
5. The significance of each column is partially conveyed by labeling it with the corresponding domain.

By definition each tuple in relation must be distinct (see above item 3), which means that tuples in relation cannot have exactly same combination values of attributes. The uniqueness of the tuples is ensured by forming a superkey of the relation. Superkey is subset of values that are different for each tuple in relation of R. Typically superkey may contain redundant attributes, in these cases it is useful to select a non-redundant key for ensuring the uniqueness of the tuples in relation r of R. A key attribute must obey the following two rules [13].

1. Two distinct tuples in any state of the relation cannot have identical values for all attributes in the key.
2. It is a minimal superkey, in other words, a superkey from which it is not possible to remove any attributes and still have the uniqueness constraint of 1 hold.

Relation schema usually has more than one key. Each of these keys is called a candidate key, and one of the keys is defined as the primary key (PK) of the relation. Primary key identifies each tuple (unique) in the relation [13,14].

Sometimes it is necessary to refer from one relation to another. To maintain referential integrity, the concept of foreign key (FK) is introduced. A foreign key is a set of attributes of the relation schema R_1 that references to relation schema R_2. The foreign has to satisfy the following conditions [13,14].

1. The attributes in a foreign key have the same domain as the attributes of the primary key in R_2.
2. A value of foreign key in tuple t_1 of the relation $r_1(R_1)$ occurs as value of PK for some tuple in t_2 in the relation $r_2(R_2)$. If $t_1[FK] = t_2[FK]$, then the tuple t_1 is said to reference the tuple t_2.

4   Conceptual Similarities between AM and Relational Model

Consider a database consisting of three relations A, B and C (Fig. 5). Each relation has the id-attribute as a primary key and column a.id is defined as the foreign key for relations B and C. Databases can be modeled with Adjacency Model and represented as an ARS. ARS provides graph-based visualization for the data. Data is represented by elements, which belong to sets that represent certain entity types.

A				B		
a_id	amount	lenght		b_id	a.id	colour
a_1	1	2		b_1	a_1	red
a_2	5	56		b_2	a_2	green
a_3	6	7				

C		
c_id	a.id	weigth
c_1	a_1	30
c_2	a_1	27
c_3	a_3	2

Fig. 5. Example database

The modeling method was introduced by Töyli et al. [3], [6]. It consists of the following steps.

1. Select an attribute from a table.
2. Create a corresponding type.
3. Repeat steps one and two until all the distinct tables and attributes are handled.

4. Remove the types which have been created from the foreign keys.
5. Take a primary key and define it as an adjacency defining type.
6. Assign attribute values as elements to the corresponding type.
7. Repeat step 5 until all the distinct primary keys are defined as adjacency defining types.
8. Create an adjacency between an element of a type and an element of its corresponding adjacency defining type (i.e. between an attribute primary key of the table).
9. Repeat step 7 until all the adjacencies are created between the elements and their adjacency defining sets.
10. Create all other wanted adjacencies between the types.

The ARS representing the database of Fig. 5 is a pair (A, R), where $A = (a\_id, b\_id, c\_id, amount, length, weight, color)$, $a\_id = \{a_1, a_2, a_3\}$, $b\_id = \{b_1, b_2\}$, $c\_id = \{c_1, c_2, c_3\}$, $amount = \{1,5,6\}$, $length = \{2,56,7\}$, $weight = \{30,27,2\}$, $color = \{"red", "green"\}$ and R consists of relations.

$R_{a\_id,amount} = \{(a_1, \{1\}), (a_2, \{5\}), (a_3, \{6\})\}$
$R_{a\_id,length} = \{(a_1, \{2\}), (a_2, \{56\}), (a_3, \{7\})\}$
$R_{a\_id,b\_id} = \{(a_1, \{b_1\}), (a_2, \{b_2\})\}$
$R_{a\_id,c\_id} = \{(a_1, \{c_1, c_2\}), (a_3, \{c_3\})\}$
$R_{amount,a\_id} = \{(1, \{a_1\}), (5, \{a_2\}), (6, \{a_3\})\}$
$R_{length,a\_id} = \{(2, \{a_1\}), (56, \{a_2\}), (7, \{a_3\})\}$
$R_{b\_id,a\_id} = \{(b_1, \{a_1\}), (b_2, \{a_2\})\}$
$R_{b\_id,color} = \{(b_1, \{"red"\}), (b_2, \{"green"\})\}$
$R_{color,b\_id} = \{("red, \{b_1"\}), ("green", \{b_2\})\}$
$R_{c\_id,a\_id} = \{(c_1, \{a_1\}), (c_2, \{a_1\}), (c_3, \{a_3\})\}$
$R_{c\_id,weigth} = \{(c_1, \{30\}), (c_2, \{27\}), (c_3, \{2\})\}$
$R_{weigth,c\_id} = \{(30, \{c_1\}), (27, \{c_2\}), (2, \{c_3\})\}$.

The adjacency defining types for the ARS are defined as follows.

$\tilde{T}_{amount,length} = \tilde{T}_{amount,b\_id} = \tilde{T}_{amount,c\_id} = \tilde{T}_{length,b\_id} = \tilde{T}_{length,c\_id} = \tilde{T}_{b\_id,c\_id} = \{T_{a\_id}\}$
$\tilde{T}_{color,a\_id} = \{T_{b\_id}\}$
$\tilde{T}_{weight,a\_id} = \{T_{c\_id}\}$

The ARS-based graph of the database is shown in Fig. 6.

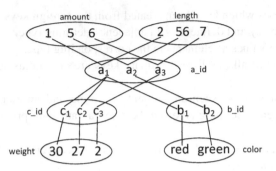

Fig. 6. An ARS representation of the database

ARS shown in Fig. 6 contains seven entity types, which corresponds attributes in the relational model. In an ARS, each entity type has elements, which represent the values of attributes in the relational model. Moreover, the concept of transitive adjacency provides an idea of a tuple in the relational model. In the relational model dependencies are carried out by primary and foreign keys, respectively in the ARS adjacency between elements depends on adjacency defining types. Thus, it can be said that adjacency defining type corresponds to the primary key in the relational model [1], [5].

5   Quantification of Conceptual Similarities

In the previous section was stated that a database can be modeled with AM. The earlier studies have shown that the models have conceptual similarities. In this section, the similarities are examined by using concepts of graph theory.

Graph is made up of set of vertices and edges connecting them. For each vertex in a graph, the degree of a vertex is the number edges incident with the vertex. In directed graphs, each vertex has an outdegree, $od(v)$, which is the number of edges directed away from the vertex and an indegree, $id(v)$, which is the number of edges directed towards the vertex. Obviously, the total degree of a vertex in directed graph is $d(v) = od(v) + id(v)$. In an undirected graph, the degree of a vertex is the number edges connected to it. If a vertex in a graph G has degree 1 it is called a leaf [15,16,17].

A vertex in a graph G dominates the vertices in G that are adjacent to it. In the graph $G = (V, E)$, where $D \subseteq V$, D is the vertex dominating set for G if every vertex of V either belong to D or is dominated by an vertex of D. Dominating number of G is the cardinality of the vertex dominating set with least number of elements. A dominating set of a graph G is minimal if none of its proper subsets are dominating. If dominating set of G contains the smallest number of elements it is said to be minimum [15].

Walk in a graph G is a sequence of edges and vertices. For example $(e_1, \dots, e_n)$ is an edge sequence in a graph. If there exist vertices $v_0, \dots, v_n$ such that $e_i = v_{i-1}v_i$ where $i = 1, \dots, n$, the sequence is called a walk [18].

The transitive adjacency can be seen as a walk that represents a tuple of a relation. For example, in Fig. 6 walk $\langle 1, a_1, 2 \rangle$ contains elements belonging to a certain tuple. In the relational model the tuples are built around the key attributes. So, it is vital to recognize from the graph vertices representing the key attributes, because walks in an ARS-based graph are formed around these vertices.

In order to determine which vertices in the graph (Fig. 6) represent the values of key attributes, the properties of the vertices are analyzed. The analysis studies total degree for each vertex. It studies if a vertex has leaf vertices and analysis also determines the vertices belonging to the minimal vertex dominating set of the graph. Table 1 lists the properties of the vertices representing the values of key attributes in Fig 6.

Table 1. Properties of the vertices representing the values of key attributes in an ARS

Vertex	Degree	Has leaves	Dominates the graph	Represents the value of primary key in database
1	1	No	No	No
5	1	No	No	No
6	1	No	No	No
2	1	No	No	No
56	1	No	No	No
7	1	No	No	No
a1	5	Yes	Yes	Yes
a2	3	Yes	Yes	Yes
a3	3	Yes	Yes	Yes
c1	2	Yes	Yes	Yes
c2	2	Yes	Yes	Yes
c3	2	Yes	Yes	Yes
30	1	No	No	No
27	1	No	No	No
2	1	No	No	No
b1	2	Yes	Yes	Yes
b2	2	Yes	Yes	Yes
red	1	No	No	No
green	1	No	No	No

In addition to the observations listed in Table 1, the properties of 25 graphs and their vertices representing databases were analyzed. The analyzed graphs contained 221 vertices that represented key attributes. The analysis results are shown in Table 2.

For analyzing purposes, each relation in a database contained exactly one tuple. This reduction provided the possibility to examine each database as single connected graphs, and thus examine the general features of vertices representing the values of key attributes. Note that typically ARS-based representations of databases are made up of multiple components.

Table 2. Analysis results

n	Average degree	Mode degree	Percentage of having leaves	Domination percentage
221	4,72	2	97 %	98 %

If the database contains relations that have only primary key or primary key and some foreign keys, then database graph would contain vertex or vertices having no leaves. Furthermore, these vertices would not belong to the minimal vertex dominating set of the graph. Thus, the percentage values listed in Table 2 are less than 100%.

Based on the observations shown in Tables 1 and 2, it can be concluded that vertices representing the values of key attributes 1) have degree of (at least) 2, 2) have leaf vertices and 3) belong to the minimal vertex dominating set of the graph.

6   Concluding Remarks

This paper focused on the conceptual similarities between adjacency model and relational model. The similarities between the model were considered first time by Wanne [1] and Töyli [5,6]. Wanne proposed that adjacency defining types of an ARS correspond to the database keys. Töyli [5,6] pointed out that relational database can be modeled with adjacency model and visualized as a graph. Töyli also provided information about the representations of relational database elements in adjacency relation system.

In this paper, the ARS-based graphs were analyzed in order to quantify the similarities proposed in earlier studies, and cement the similarities by utilizing concepts of graph theory. Transitive adjacency which is a sequence of elements in an ARS can be seen as a tuple. Respectively, in graph theory terms a tuple contains distinct elements that belong to a walk from vertex representing the value key attribute to its leaves. Furthermore, three common features for the vertices representing the values of key attributes were found. Degree of these vertices is at least two, they have leaves, and they dominate other vertices.

These findings pave the way for the following research topics. Could it be possible to reconstruct an ARS-graph back to relational database? How, for example, the dependencies could be identified from the graph? The findings of this study also bring up the question, could the quantified features be applied in graph analysis in general. Moreover, the results of this study could act as a feasibility study, which determines if a graph structure contains features, which would make it a suitable foundation for data repository.

References

1. Wanne, M.: Adjacency relation systems. Acta Wasaensia, 60: Computer Science 1. University of Vaasa, Vaasa (1998)
2. Wanne, M., Linna, M.: A General Model for Adjacency. Fundamenta Informaticae 38(1-2), 39–50 (1999)

3. Töyli, J., Linna, M., Wanne, M.: Modeling Relational Data by the Adjacency Model. In: Proceedings of the Fourth International Conference on Enterprise Information Systems, vol. 1, pp. 301–306 (2002)
4. Töyli, J., Linna, M., Wanne, M.: Modeling Semistructured Data by the Adjacency Model. In: Welzer Družovec, T., Yamamoto, S., Rozman, I. (eds.) Proceedings of the Fifth Joint Conference on Knowledge-Based Software Engineering, pp. 282–290. IOS Press, Amsterdam (2002)
5. Töyli, J.: Modeling semistructured data by the adjacency model. University of Vaasa, Vaasa (2002)
6. Töyli, J.: AdSchema – a Schema for Semistructured Data. Acta Wasaensia, 157: Computer Science 5. University of Vaasa, Vaasa (2006)
7. Heikkinen, S., Linna, M.: The Adjacency Model and Wind Power. In: Bourkas, P.D., Halaris, P. (eds.) EuroPes 2004. Acta Press, Calgary (2004)
8. Nyrhilä, V., Mäenpää, T., Linna, M., Antila, E.: Information modeling in the case of distribution energy production. WSEAS Transactions on Communications 12(4), 1325–1332 (2005)
9. Nyrhilä, V., Mäenpää, T., Linna, M., Antila, E.: A novel information model for distribution energy production. In: Proceedings of the WSEAS Conferences: 5th WSEAS Int. Conf. on Power Systems and Electromagnetic Compatibility. WSEAS, Athens (2005)
10. Mäenpää, T., Nyrhilä, V.: Framework for Representing Semantic Link Network with Adjacency Relation System. In: Giannakopoulos, G., Sakas, D.P., Vlachos, D.S., Kyriaki-Manessi, D. (eds.) Proceedings of the 2nd International Conference on Integrated Information, vol. 73, pp. 438–443. Elsevier Ltd., Oxford (2013)
11. Mäenpää, T., Nyrhilä, V.: Visualizing and Structuring Semantic Data. International Journal of Machine Learning and Computing 3(2), 209–213 (2013)
12. Zadravec, M., Brodnik, A., Mannila, M., Wanne, M., Zalik, B.: A practical approach to the 2D incremental nearest-point problem suitable for different point distributions. Pattern Recognition 41(2), 646–653 (2008)
13. Elmasri, R., Navathe, S.B.: Fundamentals of Database Systems. Addison Wesley, Boston (2007)
14. Codd, E.F.: A Relational Model of Data for Large Shared Data Banks. Communications of the ACM 13(6), 377–387 (1970)
15. Foulds, L.R.: Graph Theory Applications. Springer, New York (1992)
16. Jungnickel, D.: Graphs, Networks and Algorithms. In: Becker, E., Bronstein, M., Cohen, H., Eisenbud, D., Gilman, R. (eds.) Algorithms and Computation in Mathematics, vol. 5. Springer, Berlin (2002)
17. Newman, M.E.J.: Networks – An Introduction. Oxford University Press, Oxford (2010)
18. Jungnickel, D.: Graphs, Networks and Algorithms. In: Bronstein, M., Cohen, A.M., Cohen, H., Eisenbud, D., Sturmfels, B. (eds.) Algorithms and Computation in Mathematics, 4th edn., vol. 5. Springer, Heidelberg (2013)

Towards Fewer Seeds for Network Discovery

Shilpa Garg

Max Planck Institute for Informatics
Saarbrücken, Germany
sgarg@mpi-inf.mpg.de

Abstract. In most machine-learning problems, unlabeled data used in conjunction with a small amount of labeled data, can result in considerable improvement in learning accuracy. The goal of semi-supervised learning is to learn from these initial labeled data to predict class labels accurately. An important optimization question is to select these initial labeled instances efficiently. In this paper, we explore this problem. We propose two algorithms: one based on clustering and another based on random walk on the network. We consider four important criteria for selecting a data point as seed, with the general aim of choosing data points which are a good summary of the network. We show experimental results on different network datasets and show improvements over a recent result.

1 Introduction

Any learning problem requires an initial set of labeled data points which are then used to classify the remainder of the data points. The initial labeling requires human intervention. We call these initial labeled instances as seeds. Hence a natural question is, how do we minimize the resources spent by human experts. One popular approach is to use Active learning [1] to try and select the fewest number of initial labeled instances while still providing high classification accuracy. Also, seed selection has been attempted using various machine learning algorithms such as Conditional Random fields, Collective learning [2], etc. Recently iterative seed selection [3] approaches have been proposed that calculate the performance hit on removing nodes from a seed set. Yo Ehara at al [4] followed another approach known as bootstrapping which selects the seeds iteratively using EMR(expected model rotation) concept. The mentioned algorithms have a deficiency: there is too much information overlap in consecutive iterations; also they require a good bit of human intervention. Moreover, a natural question arises: How does the accuracy of classification vary with the number and choice of chosen seed nodes?

To this end, we formulate the problem and propose two algorithms based on hierarchical agglomerative clustering and Random walk with restart(Page Rank). We select instances which can represent the complete information while incurring minimum redundancy. This problem poses several unique challenges: first,

© Springer International Publishing Switzerland 2015
H.A. Le Thi et al. (eds.), *Model. Comput. & Optim. in Inf. Syst. & Manage. Sci.*,
Advances in Intelligent Systems and Computing 360, DOI: 10.1007/978-3-319-18167-7_8

it is an optimization problem (cost effective model) to choose the prototypical nodes(defined later) which are easy to label. Second, measuring the informativeness of the data is a NP-hard problem. Third, deciding the convergence criteria for the algorithms. Fourth, the algorithms should be efficient and scalable to be used on large datasets.

In particular, we propose the main criteria of selecting the instances to capture the notion of minimum redundancy, confusion(misclassification) and high impact. We conducted various experiments on the real-world scientific datasets. From the experiments, we observe better classification accuracy than the baseline [5] method with the same number of instances because the quality of seed nodes is better.

Seed selection plays a crucial role in several practical applications, including identifying the sources of epidemics and finding influential nodes in dynamic social networks. Each of these networks can be represented as a graph. Then we can use the algorithm proposed in this paper to find the seed nodes.

Organization of the Paper

In Section 2, we present algorithms for seed selection. In Section 3, we describe the datasets used and present the experimental results. In section 4,we discuss the performance of the algorithms. In the last section 5, we conclude the paper by showing some future directions.

2   Algorithm

In this section, we define the problem formally and provide an intuition to solve it. We represent data as an undirected weighted graph $G = (V, E)$ where V indicates the set of data points and E represents a set of connections between data points. For example, in a citation network, the vertices are the publications and there is an edge between two publications if one cites the other. Suppose there are n data points out of which l are labeled where $l \ll n$. Let p_i and q_i denote the feature vector and class label on each data point respectively. In a classification problem, we want to associate each data point with a class label. The classifier learns a mapping function f from labeled data points l to predict the label of the unlabeled data points. However, getting the initial labeled data points is time consuming and requires domain expert to get the correct labels. The goal of seed selection is to select $k \ll n$ initial unlabeled data points, whose labels are queried from the human expert while maximizing the prediction accuracy. To compare the performance of seed selection algorithm, we used a dataset with given ground truth class label on each data point. The initial seeds are labeled according to the ground truth and the prediction accuracy is measured by comparing the labels assigned by the algorithm to the true labels.

We consider following four important criteria to select the data points as seeds:

1. Low reachability - data points which don't have any label nodes in their vicinity and have very low connectivity with the remaining data points in the graph are considered as candidates to be a seed. It is hard to get a label for such data points by propagating labels from any previously labeled data points.
2. High Confusion between labels - data points which have very high uncertainty of class labels i.e. they have equal chances of being assigned either class label are good candidates for seed nodes..
3. Low redundancy - data points in the selected set must be diversely distributed over the graph to minimize information overlap.
4. High influence - we consider data points which have high influence on the unlabeled instances.

Next, we describe the two proposed algorithms formally.

2.1   Random-Walk with Restart for Seed Selection

In this section, we describe our seed-selection algorithm which is based on Random Walk with Restart (RWR) [6] and Katz centrality score [7]. We represent data as an undirected weighted graph; then we calculate the Katz centrality score [7] of each node to output the highly influential nodes. We initialize the seed set with data points with high Katz centrality score and query the human expert to label them. Next, we start random walk with restart(Page Rank) with this initial seed set and check classification accuracy. If the classification accuracy is good enough, we stop else we add more data points in the seed set on the basis of remaining three seed selection criteria and iterate over the Page Rank step.

In case, we already have some labeled instances, we initialize the seed set with such labeled instances and check classification accuracy. If the classification accuracy is not good, we check Katz centrality score and RWR and proceed ahead with the algorithm. Next, we explain the Katz centrality measure and Page Rank(RWR) algorithm briefly.

Katz centrality identifies the central nodes in the graph by taking into account the total number of walks between pairs of nodes. It computes the relative influence of a node within a network by measuring the number of the immediate neighbors (first degree nodes) and also all other nodes in the network that connect to the node under consideration through these immediate neighbors. Connections made with distant neighbors are penalized by an attenuation factor.

The idea of Random walk with restart is to do multiple walks, one for each class c, with the labeled instances from c defining the restart vector; then assign a label to an instance by comparing its scores(defined further) across different random walk distributions. This algorithm finds the probability of landing on a node by walking from labeled nodes in c, with a fixed probability of teleporting back to c at each step; this is called propagation via random walk with restart. At the steady state, each node is assigned the class which has the maximum score at that node.

Formally, we define a random walk with restart as follows: given a undirected weighted graph represented by an adjacency matrix S $(n \times n)$ where n is the number of nodes. We denote the set of initially labeled instances as T. For random walk, the transition matrix W is defined as $W_{ij} = \frac{S_{ij}}{\sum_{j=1}^{n} S_{ij}}$. The stationary distribution over the nodes is given by an n-dimensional vector v. r is the teleportation vector (for restart) with $||r|| = 1$, and α is the restart probability. r satisfies $r_i = 1$ if $i \in T$, otherwise r_i is the probability based on human expert. v is updated as $v \leftarrow \alpha \frac{r}{||r||} + (1 - \alpha)Wv$. This equation can be solved iteratively until v converges. Table 1 defines some additional notations for RWR algorithm.

At each iteration of RWR, the candidate seed set selection can be done on the basis of following criterion:

1. High influence: nodes which highly influence other nodes have high katz centrality score. The seed candidate set is initialized with nodes having high katz centrality score.
2. Confusion between labels(Conf): nodes which have very low standard deviation between probabilities of different class labels can be potential candidate seed nodes.
3. Low reachability(LR): nodes which have very low probability for all class labels can be potential candidate seed nodes.
4. Low redundancy: consecutive seed nodes in the rank list should have at least some threshold distance(path length).

Table 1. Notation

Notation	Description
W	weighted adjacency matrix, each entry gives the similarity between two vertices
r_0, r_1	initial label instances probability vectors of class label 0 and 1 respectively
v_0, v_1	stationary probability vector of class label 0 and 1 respectively after Page Rank(RWR) algorithm
$R1(LR)$, $R2(Conf)$	$R1 = v_0 + v_1$ and $R2 = v_0 - v_1$, measures for confusion and low reachability
X	set of seed candidates

Algorithm 1. Random-walk with Restart for Seed Selection

Input: G, W, r_0, r_1
Output: Seed set$|X|$
Process:

1. **Initialization**: Run Katz algorithm and take the top nodes with high score and query their labels from human expert. Initialize the seed set with these nodes.
2. **Iteration**: Run Page-Rank algorithm for each class label using $v = dr + (1-d)Wv$, where d is damping factor to get v_0 and v_1.
   Prepare a rank list as $|(R1 + R2) * (R1 - R2)|$ (increasing order)
   (a) Take top-k from ranked list mentioned above such that consecutive chosen nodes have at least some threshold distance(path length).
3. **Update Step**: Additively update the graph with the obtained seed nodes and check classification accuracy
4. **Iterate till l^2-norm of v_0 and v_1 converge**

In the next section, we describe another algorithm for seed selection.

2.2 Clustering-Based Seed Selection

In this section, we present a clustering based algorithm for seed selection. The algorithm tries to select seed candidates based on two quality parameters:

- Prototypicality: a node is said to be prototypical if it well represents a cluster of data points. Nodes which are highly connected within their clusters are good candidates for seed nodes.
- Diversity: candidate seed nodes should be well distributed over the clusters; which is to say that no two seed nodes should be close to each other.

First, we partition the graph into clusters. Since we do not know the optimal number of clusters, we use a bottom-up agglomerative clustering, where we start with the maximum number of clusters and keep merging close clusters. For example, if we have million nodes and to start with million clusters in not practical. During merging the clusters at every step, we dont know the stopping creteria of merging. Therefore, this algorithm is very inefficient in practise. Each iteration produces different clusters and subsequently, different seed candidates. Seed candidates are obtained by selecting the top scoring nodes from every partition.

3 Experiments

3.1 Dataset

We first describe the datasets in detail. We consider two scientific datasets [8].

Cora Dataset: The Cora dataset consists of 2708 scientific publications classified into one of seven classes. The citation network consists of 5429 links. Each publication in the dataset is described by a 0/1-valued word vector indicating the absence/presence of the corresponding word from the dictionary. The dictionary consists of 1433 unique words. The papers are selected in a way such that in the final corpus every paper cites or is cited by at least one other paper. After stemming and removing stopwords we were left with a vocabulary of size 1433 unique words.

CiteSeer Dataset: The CiteSeer dataset consists of 3312 scientific publications classified into one of six classes. The citation network consists of 4732 links. Each publication in the dataset is described by a 0/1-valued word vector indicating the absence/presence of the corresponding word from the dictionary. The dictionary consists of 3703 unique words. The papers are selected in a way such that in the final corpus every paper cites or is cited by at least one other paper. After stemming and removing stopwords we were left with a vocabulary of size 3703 unique words.

We represent the two scientific datasets as undirected weighted graphs, where each node represents a document and edge weights represent the similarity between documents. The edge weights are assigned by taking the normalized intersection value of the word vectors from the corresponding documents. We construct word vocabulary for each category of publications using WordNet. WordNet superficially resembles a thesaurus, in that it groups words together based on their meanings. Higher the words belonging to certain category, higher is the probability of paper to be member of that category. Also in the dataset we know the true class label/category for each document. To analyze the classification performance we compare the true and predicted labels on each document. After choosing a seed candidate node from algorithm, the human expert assigns it the label as true class label.

As already discussed above, there are several classes of documents. But we reduce the task to binary, by considering one-vs-all labeling tasks, one for each ground-truth label (i.e. each node is labeled 1 if it belongs to a particular class c and 0 otherwise, and this is repeated for all values of c). We use the standard evaluation metrics for evaluation purpose - Accuracy, Precision, Recall and f1-score.

3.2   Results

Tables 2 shows the evaluation of seed set as 700-800 data points (optimal number obtained from plot) on the two datasets using RWR algorithm. It can be seen from the tables that accuracy, precision, recall are quite good using all seed selection criteria.

Next, in the plots, we compare and contrast the labeling accuracy for different number of seeds using above mentioned algorithms and baseline paper (Page Rank Seeding) [5] on scientific datasets. It can be seen that we are able to achieve the same accuracy using less number of seed nodes compared to Page Rank Seeding. Hence choosing optimal seed nodes plays an important role to achieve good accuracy. In the baseline paper [5], the authors consider top-k high

Table 2. Left: Cora Dataset, Right: CiteSeer Dataset

	LR	Conf	LR + Conf		LR	Conf	LR + Conf
Accuracy	.612	.708	.843	Accuracy	.578	.687	.743
Precision	.602	.691	.832	Precision	.534	.601	.710
Recall	.578	.589	.778	Recall	.478	.565	.644

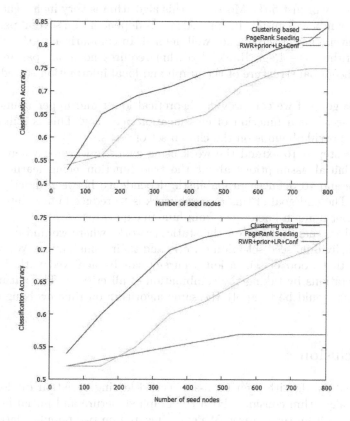

Fig. 1. Comparison of algorithms on scientific datasets Top: Cora, Bottom: CiteSeer

score nodes from rank list obtained from Page Rank. This has the disadvantage that the selected seed nodes may not be diversely distributed over the data space. Additionally, they do not handle the case of seed nodes with high confusion of labeling. In contrast, the main idea of our algorithm is to consider those nodes as seeds which are prototypical such that the information is propagated to more number of nodes and all other nodes can be labeled without any confusion between class labels. By considering both the labeled instances and graph structure/connectivity, we are able to achieve similar accuracy with less number of seeds.

4   Discussion

It can be seen from the above experiments that seed selection has an important effect on the classification accuracy. Therefore, choosing seeds wisely and with no human intervention has been an open question. We tried to reduce the human effort by getting labels from human expert only on good data points. However, we are unable to avoid human intervention completely.

From the results, we can see that the clustering algorithm did not perform well with a accuracy of just 50%. Moreover, this algorithm is very inefficient because of the high number of initial clusters. Also it is dependent on several parameters and the convergence criteria is not well defined. In comparison to the clustering based algorithm, the Page Rank algorithm requires no input parameter and considers the global structure of the graph and local information stored on each node.

It will be good if we can provide theoretical lower and upper bound on the number of seeds as a function of classification accuracy. This is still an open question to provide bounds on the chosen set of seeds.

The next step is to extend the work using agnostic learning where we dont make any initial assumptions about the true function being learned. In the dataset, we aim to define a probabilistic dependence between description and categories. The end goal of this research work is to reduce human intervention, reducing prior domain knowledge from human expert.

Our analysis mainly focus on the static networks where we find out seed set by taking individual seed selection criteria and their combination. We were able to observe that good classification accuracy can be achieved with few nodes in fewer iterations by taking the combination of all criteria. The natural road-map at hand would be to apply the same algorithm on time evolving dynamic networks.

5   Conclusion

We proposed an algorithm for semi-supervised learing based on random graph walk. This algorithm considers both the graph structure and initial labeled instances known from the dataset. We experimented on two popular datasets and showed that it significantly outperforms the previous methods. Random walk methods have an advantage over clustering based methods. With fewer number of labeled instances as training set, we were able to achieve high classification performance. It is open to see if new algorithms can be developed which combine aspects from multiple different labeling methods to achieve a labeling which is better than any individual method. In the future, we plan to perform experiments using more datasets on directed weighted graphs.

References

1. Rajan, S., Yankov, D., Gaffney, S.J., Ratnaparkhi, A.: A large-scale active learning system for topical categorization on the web. In: Proceedings of the 19th International Conference on World Wide Web, WWW 2010, pp. 791–800. ACM, New York (2010)
2. Shi, L., Zhao, Y., Tang, J.: Combining link and content for collective active learning. In: Proceedings of the 19th ACM International Conference on Information and Knowledge Management, CIKM 2010, pp. 1829–1832. ACM, New York (2010)
3. Zhu, X., Lafferty, J., Ghahramani, Z.: Combining active learning and semi-supervised learning using gaussian fields and harmonic functions. In: ICML 2003 Workshop on The Continuum from Labeled to Unlabeled Data in Machine Learning and Data Mining, pp. 58–65 (2003)
4. Ehara, Y., Sato, I., Oiwa, H., Nakagawa, H.: Understanding seed selection in bootstrapping. In: Proceedings of TextGraphs-8 Graph-based Methods for Natural Language Processing, pp. 44–52. Association for Computational Linguistics, Seattle (2013)
5. Lin, F., Cohen, W.W.: Semi-supervised classification of network data using very few labels. In: Proceedings of the 2010 International Conference on Advances in Social Networks Analysis and Mining, ASONAM 2010, pp. 192–199. IEEE Computer Society, Washington, DC (2010)
6. Tong, H., Faloutsos, C., Pan, J.Y.: Fast random walk with restart and its applications. In: Proceedings of the Sixth International Conference on Data Mining, ICDM 2006, pp. 613–622. IEEE Computer Society, Washington, DC (2006)
7. Katz, L.: A new status index derived from sociometric analysis. Psychometrika 18, 39–43 (1953)
8. Sen, P., Namata, G.M., Bilgic, M., Getoor, L., Gallagher, B., Eliassi-Rad, T.: Collective classification in network data. AI Magazine 29, 93–106 (2008)

Part II

Heuristic / Meta Heuristic Methods for Operational Research Applications

A Hybrid Optimization Algorithm for Water Production and Distribution Operations in a Large Real-World Water Network

Derek Verleye and El-Houssaine Aghezzaf

Ghent University, Department of Industrial Management, Technologiepark 903,
9052 Zwijnaarde, Belgium

Abstract. This paper is concerned with the optimization of the operational aspects related to water production and distribution in a large real-world water supply network. The main operational costs considered are those related to water production and pumping. The investigated network has some special features, such as buffer in- and outflows, which are taken into account in the optimization. The problem is naturally formulated as a non-convex mixed-integer nonlinear program (MINLP) involving a large number of variables and constraints. As a consequence, the resulting optimization model is not easily tackled by the current software packages for water systems optimization. To partially tackle this issue, a hybrid two-steps method that aims at finding a global near-optimal water systems operating plan is proposed in this article. In the first step of the method, a mixed integer linear programming model (MILP) is constructed, using the piecewise linear approximation approach proposed in [10], and then solved to optimality. The resulting solution is then used as a starting point for Newtons method, which repairs it to obtain a feasible solution for the original MINLP model. This hybrid two-steps method is tested on a realistic network and the results show that the method is very competitive in terms of computation time while providing a near-optimal implementable water production and distribution operating plan.

1 Introduction

The process of optimizing and monitoring water production and distribution in large water-supply networks involves several equally important phases, starting from the strategic design phase and ending with the operational planning phase (see among others [3], [2]). However, the network design phase and network operational planning phase are the two major and critical ones in this process. The optimal design phase of a water network considers various aspects such as optimal layout of the network, optimal dimensions of the systems components (pipes, pumps, valves, reservoirs, etc). The optimization process here involves a very large number of combinations of pipe materials, diameters, pumping stations locations and capacities which makes the underlying optimization model for this phase rather complex. The optimal operational planning phase of water

© Springer International Publishing Switzerland 2015
H.A. Le Thi et al. (eds.), *Model. Comput. & Optim. in Inf. Syst. & Manage. Sci.*,
Advances in Intelligent Systems and Computing 360, DOI: 10.1007/978-3-319-18167-7_9

production and distribution in a designed water-supply network involves operational parameters such as hydraulic pressure zone boundaries, demand patterns, control valve settings, and pump operating schedules. This paper reviews the optimization model for the real network under consideration, and discusses a two-step optimization method to determine near-optimal operational planning which minimizes the total production and distribution costs.

As already mentioned above, It is not an evident task to obtain a global near-optimal solution to the general optimization model underlying the production and distribution operational planning phase in complex water networks. Solving the entire problem in its generality is quite difficult and this complexity increases with the size and general topology of the network. In the available literature, usually some simplifications to the real network are made and then some heuristics are used to find a feasible solution to the resulting simplified optimization model. Heuristics are very popular to solve simplified versions of the water-supply operational planning problem. Moreover, these approaches usually focus on pump scheduling only ([12], [6], [1]) and do not guarantee global optimality even though the resulting solutions are in general good enough for practical purposes. The idea to use a hybrid method which couples an LP problem with an NLP is already used in [5]. However this method is limited to gravitational systems. In [4] the authors use a hybrid method in which the solution of an LP relaxation is used to trigger the subsequent greedy algorithm. This method is successfully tested on several benchmark instances. Due to the non-convexity of pressure losses in pipes, piecewise linear approximations are a good way to approximate the problem by an MILP. In [7] an MILP is proposed for multivariate piecewise functions. Unfortunately the number of breakpoints has to be quite large to assure that the approximation provides a feasible operating plan to the original problem.

In this paper we exploit the same idea of finding a good starting solution and then improving it to obtain an implementable near-optimal operating plan. The fist solution is found through the use of piecewise linear optimization and then Newton's iterative method is launched to obtain a globally near-optimal solution to the original operational optimization problem. The first piecewise-linear model was already proposed by the authors in [10]. Unfortunately, to find an an operating plan that is acceptable in practice the model has to consider a large number of variables (pieces), which make the whole solution process take a relatively large amount of computational time. In this paper, the model is slightly altered and used to provide just a good starting solution for Newton's method. This is basically done by decreasing the number of intervals. In the second step, a relatively accurate solution to the original problem is obtained. The Newton's method used is based on the method in the popular water simulation software *EPANET* [8]. Simpson and Elhay [9] revised this algorithm for the Darby-Weisbach headloss formula to gain quadratic convergence.

The remainder of the paper is organized as follows. In the next section, the mathematical model for the general operational optimization problem in water supply networks is reviewed. Next, the proposed hybrid solution approach is described in detail. Its components, the piecewise linear model and Newton's method are explained. The results showing the strength and fast convergence of the method are demonstrated on part of the network of the biggest drinking water provider of Flanders, Belgium. Finally, some concluding remarks and future research issues are discussed.

2   The Water Supply Network and Its Operational Optimization Model

This section reviews the specific features of the water supply network investigated in this paper. The model contains the basic hydraulics that define this studied network and are needed to further explain the devised algorithm. More details on this model and its physical and technical characteristics can be found in [10].

2.1   Major Parameters of the Model and Its Variables

We consider a planning horizon of T periods, where in practice the periods $t \in \{1, 2, ..., T\}$ are usually hours and the planning covers a full day (24 hours). The network topology is modeled by a directed network $G = (\mathcal{N}, \mathcal{A})$ whose node set represents junctions $\mathcal{J}$, buffers $\mathcal{B}$ and raw water sources $\mathcal{S}$, and whose arc set represents pipes $\mathcal{P}_i$, pure water pumps $\mathcal{P}_u$ and raw water pumps $\mathcal{P}_r$.

As major parameters of the model (units are in brackets), let τ_t be the length of period t (h); $c_{ij}^t(e)$ be the electricity cost for a pump in period t (€/kWh), h_i be the geographic height (in m); d_{it} be the demand at node i in period t (in m^3/h); A_i be the cross-sectional area of tank (in m^2); h_i^{fl} be the level of tank floor (in m); $c_{ij}(p)$ be the production cost in a water production center, or in short, WPC (in €/m^3); κ_{ij} be the loss coefficient (in h/m^2); h_{ij}^1, h_{ij}^2 and h_{ij}^3 be the head coefficients (varies); p_{ij}^1 and p_{ij}^2 be the power coefficients (varies); q_{ij}^{cap} be the production capacity in WPC (m^3/h); q_{ij}^{lim} be the daily extraction limit in WPC (m^3/h); f_{ij} be the maximum fluctuation in water production from one period to the next (m^3/h); and finally let g denote the gravity constant (m/s^2).

The decision variables of the model in each period t are:

- Q_{ijt} : Flow on arc i, j (m^3/h);
- $H_{it} = h_i + \dfrac{p_{it}}{\rho g}$: Piezometric head (m), with p the manometric pressure and ρ the density of water;
- I_{it}^+ : Inflow at entrance buffer (m^3/h);
- I_{it}^- : Outflow at entrance buffer (m^3/h);
- O_{it} : Outflow at exit buffer (m^3/h);
- V_{it} : Volume in tank at end of period t (m^3);
- $L_{it} = V_{it}/A_i$: Level of tank at end of period t (m);
- H_{it}^M : Mean piezometric head of tank in period t (m);
- ΔH_{ijt} : Pump head increase (m);

- P_{ijt} : Power pump (W);
- Z_{ijt} : Binary activity status pump (-);

The following restrictions are the essential constraints which are considered in this operational optimization model. These include water balance constraints, pressure losses, pumping constraints and capacity/quality constraints.

Water Balance in Nodes: At the junctions $i \in \mathcal{J}$, the flow balance equations satisfying the predicted consumption demands d_{it} are given as follows:

$$\sum_{k:(k,i)\in\mathcal{A}} Q_{kit} - \sum_{j:(i,j)\in\mathcal{A}} Q_{ijt} = d_{it}, \quad \forall i \in \mathcal{J} \tag{1}$$

Water Balance and Storage in Buffers: Buffers play a very important role in this water network. At night, when the energy tariff is low, water can be stored in the reservoir using pumps. The next day this water can flow gravitationally (or if necessary, by pumping) back into the network to meet part of the daily demand. In this way, high energy costs due to excessive pumping during the day are avoided. The buffers we consider in this study allow an inflow and two outflow possibilities, one by pumping and the second by gravitation. Constraints are placed on the in- and outflow (I^+ and I^-) such that water can only flow in the reservoir if its pressure is high enough. Similarly, water will be rerouted into the network if the pressure in the network is low. Alternatively, pumps can be used to inject an amount of water O into the network.

$$\sum_{k:(k,i)\in\mathcal{A}} Q_{kit} - \sum_{j:(i,j)\in\mathcal{A}\backslash\mathcal{P}_u} Q_{ijt} = I_{it}^+ - I_{it}^- + d_{it}, \quad \forall i \in \mathcal{B} \tag{2}$$

$$\sum_{j:(i,j)\in\mathcal{P}_u} Q_{ijt} = O_{it}^-, \quad \forall i \in \mathcal{B} \tag{3}$$

$$V_{it} = V_{i,t-1} + (I_{it}^+ - I_{it}^- - O_{it})\tau_t, \quad \forall i \in \mathcal{B} \tag{4}$$

$$V_{i0} \leq V_{iT}, \quad \forall i \in \mathcal{B} \tag{5}$$

$$I_{it}^+ \leq A_i \, l_i^{max} \, X_{it} \tag{6}$$

$$I_{it}^- \leq A_i \, l_i^{max} \, Y_{it} \tag{7}$$

$$H_{it} - h_i^{in} \geq (h_i - h_i^{in})(1 - X_{it}) \tag{8}$$

$$H_{it}^M - H_{it} \geq (h_i^{fl} - 100 - h_i)(1 - Y_{it}) \tag{9}$$

$$X_{it} + Y_{it} \leq 1 \tag{10}$$

The constraints (4) control the volume in the buffer and (5) guarantees that the total volume in the buffer at the end of the period is the same as in the beginning, which allows a cyclic production and distribution plan.

Pressure Losses in Pipes: In every pipe, hydraulic friction causes a pressure drop. This is modeled by the constraints below.

$$H_{it} - H_{jt} = \kappa_{ij} \, Q_{ijt} |Q_{ijt}|, \quad \forall (i,j) \in \mathcal{P}_i \tag{11}$$

κ is the loss coefficient, which is a function of the pipe length, diameter and the (dimensionless) friction coefficient λ. For λ we work with the (simplified) law of Prandtl-Kármán for hydraulically rough pipes.

If a control valve is present in the pipe, the restriction changes to:

$$-(1 - Z_{ijt}) \, M, \leq H_{it} - H_{jt} - \kappa_{ij} \, Q_{ijt}^2 \leq (1 - Z_{ijt}) \, M, \tag{12}$$

, where water can only flow in one direction, M is the maximum pressure difference and $Z_{ijt} = 0$ if $Q_{ijt} = 0$.

Pressure Losses in Pumps: Every pump is modeled as an arc in which the pressure can be increased by some nonnegative quantity ΔH.

$$H_{it} - H_{jt} - \kappa_{ij} \, (Q_{ijt})^2 + \Delta H_{ijt} = 0 \quad \forall (i,j) \in \mathcal{P}_u : i \in \mathcal{N} \backslash \mathcal{B} \tag{13}$$

$$H_{it}^M - H_{jt} - \kappa_{ij} \, (Q_{ijt})^2 + \Delta H_{ijt} = 0 \quad \forall (i,j) \in \mathcal{P}_u : i \in \mathcal{B} \tag{14}$$

Here, a distinction is made between pumps operating at buffers i, to account for the correct pressure H_{it}^M which is given by $h_i^{fl} + \dfrac{L_{it} + L_{i,t-1}}{2}$.

Pump Head: The extra amount of pressure ΔH that is added by the pumps can be formulated as follows:

$$\Delta H_{ijt} = h_{ij}^1 \, (Q_{ijt})^2 + h_{ij}^2 \, Q_{ijt} + h_{ij}^3 \, Z_{ijt}, \quad \forall (i,j) \in \mathcal{P}_u$$

Since we want the link to be inactive when the pump is not working ($Z = 0$), the previous constraint is potentially wrong. Therefore we change it to:

$$\Delta H_{ijt} <= h_{ij}^1 \, (Q_{ijt})^2 + h_{ij}^2 \, Q_{ijt} + h_{ij}^3 \, Z_{ijt} + (1 - Z_{ijt}) \, 100, \quad \forall (i,j) \in \mathcal{P}_u \tag{15}$$

$$\Delta H_{ijt} >= h_{ij}^1 \, (Q_{ijt})^2 + h_{ij}^2 \, Q_{ijt} + h_{ij}^3 \, Z_{ijt} - (1 - Z_{ijt}) \, 100, \quad \forall (i,j) \in \mathcal{P}_u \tag{16}$$

Power: The power is modeled by the constraints below, where the coefficients are derived from available pumping curves.

$$P_{ijt} = p_{ij}^1 \, Q_{ijt} + p_{ij}^2 \, Z_{ijt}, \quad \forall (i,j) \in \mathcal{P}_u \tag{17}$$

Capacity: The amount of raw water extracted is subject to a maximum capacity which can not be exceeded.

$$0 \leq Q_{ijt} \leq q_{ij}^{cap}(i,j), \quad \forall (i,j) \in \mathcal{P}_r \tag{18}$$

Fluctuation: The constraints below impose some transient conditions which limit the change in flow levels in between two subsequent periods, which prevent water quality problems at WPC's from occurring.

$$-f_{ij}\, q_{ij}^{cap} \le Q_{ijt} - Q_{ij,t-1} \le f_{ij}\, q_{ij}^{cap}, \quad \forall (i,j) \in \mathcal{P}_r \tag{19}$$

$$-2\, f_{ij}\, q_{ij}^{cap} \le Q_{ijT} - Q_{ij,0} \le 2\, f_{ij}\, q_{ij}^{cap}, \quad \forall (i,j) \in \mathcal{P}_r \tag{20}$$

Daily Ground Extraction Limit: Due to some contractual agreement, the restriction modeled below is imposed in some ground water extraction wells.

$$\sum_{t=1}^{T} Q_{ijt} \le q_{ij}^{lim}, \quad \forall (i,j) \in \mathcal{P}_r \tag{21}$$

Variable Bounds: Different bounds are imposed on flow in the network, minimum and maximum pressure in nodes, and the level of water in the buffers.

Objective Function : The cost function to be minimized consists of energy consumed by the pumps and the production/electricity at the water production centers:

$$\text{Minimize} \sum_{t=1}^{T} \left[\sum_{(i,j)\in\mathcal{P}_u} (p_{ij}^1\, Q_{ijt} + p_{ij}^2) \frac{c_{ij}^t(e)}{1000} + \sum_{(i,j)\in\mathcal{P}_r} Q_{ijt}\, c_{ij}(p) \right] \tau_t \tag{22}$$

3   Algorithm Outline

The proposed model is hard to solve due to the presence of nonlinear relationships and binary variables. In an earlier paper [10], the model has been approached via piecewise linear functions. However, to obtain an accurate solution to the original problem the number of segments in the piecewise linear functions has to be quite high. This leads to a higher number of binary variables and an increased computational time required to solve the problem. In the current section we provide a short summary of this PWL method. Since we want to limit the computation time, the obtained solution is not necessarily feasible to the original problem. It can however perfectly serve as a starting point for the classical Newton's method, which is covered further in this section. The final part explains which modifications have to be made in order to couple both methods and provide a final feasible near-optimal solution.

3.1   Review of the Piecewise Linear Approximation Model

In this subsection we summarize the piecewise linear approximation based model in which the nonlinear constraints (11-16) are linearized. That is, the pressure losses and pump function are over- and understimated using discrete intervals. The detailed model can be found in [10].

Q and the left hand side f of equations (11-16) are expressed as a function g of $|K + 1|$ discrete points q and the new variables λ, bounded by curves h that represent the estimation error ϵ:

$$Q_{ijt} = \sum_{k=0}^{K_{ijt}} \lambda_{ijtk} q_{ijtk} \tag{23}$$

$$f(H_i, H_j, \Delta H_{ij}) \leq \sum_{k=0}^{K_{ijt}} \lambda_{ijtk}(g(q_{ijtk}) + c_{ijtk}) + h(\alpha_{ijtk}, \epsilon_{ijtk}) \tag{24}$$

$$f(H_i, H_j, \Delta H_{ij}) \geq \sum_{k=0}^{K_{ijt}} \lambda_{ijtk}(g(q_{ijtk}) + c_{ijtk}) - h(\alpha_{ijtk}, \epsilon_{ijtk}) \tag{25}$$

$$\sum_{k=0}^{K_{ijt}} \lambda_{ijtk} = 1 \tag{26}$$

$$\lambda_{ijtk} \geq 0 \quad \forall k \in 0..K_{ijt} \tag{27}$$

Traditionally, one would then define $\alpha_{ijtk} \in \{0,1\} \quad \forall k \in 0..K_{ijt} - 1$ and add the restrictions:

- $\lambda_{ijt0} \leq \alpha_{ijt0}$
- $\lambda_{ijtk} \leq \alpha_{ijt,k-1} + \alpha_{ijtk} \quad \forall k \in 1..K_{ijt} - 1$
- $\lambda_{ijtK_{ijt}} \leq \alpha_{ijt,K_{ijt}-1}$
- $\sum_{k=0}^{K_{ijt}-1} \alpha_{ijtk} = 1$.

Implementing this approach would highly increase the amount of binary variables in our model by $|K \times (\mathcal{A}\backslash\mathcal{P}_r) \times T|$ (assuming the same number of intervals for each arc). Therefore, we instead link α and λ as follows. The function $B : \{0..K - 1\} \rightarrow \{0,1\}^{log_2(K)}$ projects every interval on a binary vector of length $log_2(K)$ [11]. Defining $\sigma(B)$ as the support vector of B we then write:

$$\sum_{k \in K^+(l,B)} \lambda_{ijtk} \leq \alpha_{ijtl} \quad \forall l \in 1..log_2(K_{ijt}) \tag{28}$$

$$\sum_{k \in K^0(l,B)} \lambda_{ijtk} \leq 1 - \alpha_{ijtl} \quad \forall l \in 1..log_2(K_{ijt}) \tag{29}$$

$$\alpha_{ijtl} \in \{0,1\} \quad \forall l \in 1..log_2(K_{ijt}) \tag{30}$$

with $K^+(l,B) = \{k \in 1..K_{ijt} - 1 : l \in \sigma(B(k)) \cap \sigma(B(k + 1))\} \cup \{0$ if $l \in \sigma(B(1))\} \cup \{K_{ijt}$ if $l \in \sigma(B(K_{ijt}))\}$ and $K^0(l,B) = \{k \in 1..K_{ijt} - 1 : l \notin \sigma(B(k))$ and $l \notin \sigma(B(k + 1))\} \cup \{0$ if $l \notin \sigma(B(1))\} \cup \{K_{ijt}$ if $l \notin \sigma(B(K_{ijt}))\}$. This effectively reduces the number of variables α.

The final MILP model is summarized in Table 1.

Table 1. PWL Model

Minimize (22)

s.t. (1-10, 17 -21, 23 - 30)

bounds on all variables

3.2   Newton's Method

This method is very popular in water simulation and is also implemented in the popular software Epanet [8]. On the basis lies the algorithm outlined here:

Algorithm 1. (Newton's method)

Initialization:
Initial configuration q^0 (vectors of Q), Nonlinear equation $f(x) = 0$.

Iteration m:
Update q^m through Newton's iterative method
$f'(x^{(m)})(x^{(m+1)} - x^{(m)}) = -f(x^{(m)})$, $x^{(0)}$ *prescribed*, $m = 0, 1, 2, ...$

Until:
$\phi(q^{(m+1)}) = \sum_{(i,j) \in \mathcal{A}, t \in 1..T} |Q_{ijt}^{(m+1)} - Q_{ijt}^{(m)}| / \sum_{(i,j) \in \mathcal{A}, t \in 1..T} |Q_{ijt}^{(m+1)}| \leq \delta_{\text{stop}}$

The nonlinear functions in the original system are the pressure losses and pump constraints (11-16). For pipes, we easily find $f'(x^{(m)}) = -2\kappa_{ij} |Q_{ijt}^{(m)}|$, such that the linearized constraint in iteration m becomes:

$$- 2\kappa_{ij} |Q_{ijt}^{(m)}| Q_{ijt}^{(m+1)} = -H_{it} + H_{jt} - \kappa_{ij} Q_{ijt}^{(m)} |Q_{ijt}^{(m)}| \tag{31}$$

For pumps, the constraints are (substituting for ΔH):

$$(-2\kappa_{ij} Q_{ijt}^{(m)} + 2h_{ij}^1 Q_{ijt}^{(m)} + h_{ij}^2) Q_{ijt}^{(m+1)} =$$
$$- H_{it} + H_{jt} - (\kappa_{ij} - 2h_{ij}^1 - h_{ij}^2)(Q_{ijt}^{(m)})^2 - h_{ij}^3 Z_{ijt},$$
$$\forall (i,j) \in \mathcal{P}_u : i \in \mathcal{N} \backslash \mathcal{B} \tag{32}$$

$$(-2\kappa_{ij} Q_{ijt}^{(m)} + 2h_{ij}^1 Q_{ijt}^{(m)} + h_{ij}^2) Q_{ijt}^{(m+1)} =$$
$$- H_{it}^M + H_{jt} - (\kappa_{ij} - 2h_{ij}^1 - h_{ij}^2)(Q_{ijt}^{(m)})^2 - h_{ij}^3 Z_{ijt},$$
$$\forall (i,j) \in \mathcal{P}_u : i \in \mathcal{B} \tag{33}$$

In order for this algorithm to work with our model formulation, it is necessary that we drop all bounds on variables and maintain only the equality constraints: flow conservation, storage volume in buffer and the newly defined linear functions.

Furthermore all binary variables are fixed. Volume in buffers is also fixed and regarded as additional demand/supply in between periods. The production rate is fixed in order to ensure that the constraints (18-21) are satisfied. The second stage model is summarized in Table 2.

Table 2. Newton model

Minimize (22)

s.t. (1-5, 31-33)

3.3   The Hybrid Algorithm

The described PWL method, when solved, provides a solution with an accuracy that depends on the number of segments K. In [10] it is shown that even for a relatively large number of intervals (16, $K = 3$) the largest error is still unacceptable.

In order to adjust the found solution such that it becomes hydraulically correct, a two-stage solution method is proposed. In the first stage, the model is solved for a certain instance of K. The second stage requires the Todini-Pilati implementation of the Newton method or a similar gradient algorithm.

The general idea of the algorithm is to solve the proposed piecewise linear model and then finetune the solution with Newton's method (Algorithm 2).

Algorithm 2. (Combined piecewise-linear approximations and Newton's method)

Step 0. (Initialization):
Start with $K = 2$

Step 1. (Piecewise-Linear Approximation):
Given K, solve the PWL model to obtain solution (Q_{LP}^n, H_{LP}^n) on the horizon $\{1, ..., T\}$.

Step 2. (Newton's Method):

We now only consider the restrictions listed in Table 2. Fix the endvolume in buffers for each time period as well as the binary variables. With this updated information, solve model 2 repeatedly until convergence is reached. Every other iteration (up until the 10th) a check is made on the pumps. If the flow would be smaller than a treshold value (1e-06) then the flow is fixed to $Q = 0$ and the binary pump variable Z becomes 0 as well. If these conditions are not present the pump becomes active again and the flow is calculated from the pump curve and the current value for H (This method is also used in [8]). If the current solution is infeasible, set $K = K \times 2$ and go to step 1. Repeat until feasibility is reached.

After testing it turned out that there can be problems with the buffers for certain values of K. More specifically, in (6-10) water can essentially flow into the buffer although there is not enough pressure. The same may be true for water flowing out of the buffer. In that case, the number of intervals is doubled and the next iteration starts.

4   Some Computational Results

To test our algorithm, we refer to the test case in paper [10]. The only difference is the allowance for pump arcs to be inactive if the pump is not working. For the solver, Gurobi 5.6.3 with standard settings is used. Presolve tolerance is *presolve_eps* $1e - 08$. Table 3 shows (in this order) the number of intervals, variables and constraints for the PWL approximation. Furthermore, the computation time and the maximum possible error (in m) on the pressure losses/pump curve for this method is displayed. The final goal value is the objective value achieved with Newton's method, whereas the error is only present after the PWL optimization (and disappears after the correction with Newton's method). All experiments were done on a i7-3770 CPU with 3.40 GHz and 32,0 GB of RAM memory. Since the model is the same as the one in [10] with exception of the corrected pump head formula, computational times are in the same order of magnitude (computation time for the Newton method is negligible). The algorithm converges for every instance of K. As mentioned previously, we note that the final solution is infeasible for small values of K. From values of $K = 16$ onwards the final solution satisfies all the restrictions from the original problem (within a tolerance of 1%). As a comparison, running the complete model with the opensource solver BONMIN gave a solution of 3740.72 in 29.80 s with the Feasibility Pump Heuristic, which was improved to 3646.82 in 778.26 s. This shows that our method is very competitive.

Table 3. Results of the hybrid method on test network

# Intervals	nvars	ncons	time (s)	error (m)	goal
n=2	1268	2432	0.45	28.75	3646.72 (INF)
4	1673	2702	3.45	7.19	3646.76 (INF)
8	2348	2972	10.91	1.80	3646.82 (INF)
16	3563	3242	53.73	0.45	3647.17
32	5858	3512	399.83	0.11	3646.82

nvars = # variables; ncons = # constraints

5   Conclusion

Finding a solution for short-term operations optimization of water supply networks can be a hard task in practice. The nonconvex MINLP model usually contains too many variables and constraints to solve to optimality in a reasonable computation time. As a consequence, most authors relax the original method or make use of heuristic methods to find local optimal solutions. Here, the nonlinear restrictions of the model are approximated by piecewise linear segments. A solution for this model can be found in short computation time, depending on the accuracy of the approximation. If the model for in- and outflow in buffers is complex, a greater accuracy (and thus larger computation time) is needed. With Newton's iterative method a feasible solution for the original network is then found within a fraction of a second. Our model provides a good alternative for other methods if a good trade-off between solution quality and computational time is needed. As prices for drinking water production and electricity are ever-increasing, such methods will keep on gaining more importance in the future.

Acknowledgement. Research funded by a Ph.D. grant of the Agency for Innovation by Science and Technology (IWT). This support is highly appreciated.

References

1. Bene, J., Selek, I., Hös, C.: Neutral search technique for short-term pump schedule optimization. Journal of Water Resources Planning and Management 136(1), 133–137 (2010)
2. D'Ambrosio, C., Lodi, A., Wiese, S., Bragall, C.: Mathematical programming techniques in water network optimization. European Journal of Operational Research (2014)
3. De Corte, A., Sörensen, K.: Optimisation of gravity-fed water distribution network design: A critical review. European Journal of Operational Research (2012)
4. Giacomello, C., Kapelan, Z., Nicolini, M.: Fast hybrid optimization method for effective pump scheduling. J. Water Resour. Plann. Manage. 139(2), 175–183 (2013)
5. Krapivka, A., Ostfeld, A.: Coupled genetic algorithm—linear programming scheme for least-cost pipe sizing of water-distribution systems. J. Water Resour. Plan. Man.-ASCE 135, 298–302 (2009)
6. López-Ibáñez, M., Prasad, T.D., Paechter, B.: Ant Colony Optimization for Optimal Control of Pumps in Water Distribution Networks. Journal of Water Resources Planning and Management 134(4), 337–346 (2008)
7. Morsi, A., Geißler, B., Martin, A.: Mixed Integer Optimization of Water Supply Networks. International Series of Numerical Mathematics, vol. 162, ch. 3, pp. 35–54. Springer (2012)
8. Rossman, L.A.: EPANET 2 Users Manual. U.S. Environmental Protection Agency, Cincinnati, OH 45268 (September 2000)
9. Simpson, A., Elhay, S.: Jacobian matrix for solving water distribution system equations with the darcy-weisbach head-loss model. Journal of Hydraulic Engineering 137(6), 696–700 (2010)

10. Verleye, D., Aghezzaf, E.-H.: Optimising production and distribution operations in large water supply networks: A piecewise linear optimisation approach. International Journal of Production Research 51(23-24), 7170–7189 (2013)
11. Vielma, J., Nemhauser, G.: Modeling disjunctive constraints with a logarithmic number of binary variables and constraints. In: Integer Programming and Combinatorial Optimization, pp. 199–213 (2008)
12. van Zyl, J.E., Savic, D.A., Walters, G.A.: Operational Optimization of Water Distribution Systems Using a Hybrid Genetic Algorithm. Journal of Water Resources Planning and Management 130(2), 160–170 (2004)

A Memetic-GRASP Algorithm for the Solution of the Orienteering Problem

Yannis Marinakis, Michael Politis, Magdalene Marinaki,
and Nikolaos Matsatsinis

School of Production Engineering and Management,
Technical University of Crete, Chania, Greece
{marinakis,nikos}@ergasya.tuc.gr, magda@dssl.tuc.gr

Abstract. The last decade a large number of applications in logistics, tourism and other fields have been studied and modeled as Orienteering Problems (OPs). In the orienteering problem, a standard amount of nodes are given, each with a specific score. The goal is to determine a path, limited in length, from the start point to the end point through a subset of locations in order to maximize the total path score. In this paper, we present a new hybrid evolutionary algorithm for the solution of the Orienteering Problem. The algorithm combines a Greedy Randomized Adaptive Search Procedure (GRASP), an Evolutionary Algorithm and two local search procedures. The algorithm was tested in a number of benchmark instances from the literature and in most of them the best known solutions were found.

Keywords: Orienteering Problem, Memetic Algorithm, Greedy Randomized Adaptive Search Procedure, Local Search.

1 Introduction

In this paper, we present an algorithm for the solution of the Orienteering Problem. The Orienteering Problem was introduced by Golden et al. [5] when they described a game played in the mountains. The idea behind this game is that we have a number of players that they start from a specified control point, they have a map that informs them about a number of checkpoints and the scores associated with each one of them and they try to pass from as many checkpoints as possible adding to their total score the score of the specific point and to return to the control point (or to go to a different control point) within a specific time period [2]. The winner is the one that maximizes its total collected score [16]. This problem is one of the first problems that belongs in a category that is called Vehicle Routing Problems with Profit, where in each node (customer) except of the demand of the point, a profit is associated and, thus, the main goal is instead of (or in addition to) the minimization of the total distance or the total travel time, the maximization of the additive profit of visiting of the most profitable customers. In all these problems, it is possible not to visit all the customers but there is a selection of the customers that we have to visit as

© Springer International Publishing Switzerland 2015
H.A. Le Thi et al. (eds.), *Model. Comput. & Optim. in Inf. Syst. & Manage. Sci.,*
Advances in Intelligent Systems and Computing 360, DOI: 10.1007/978-3-319-18167-7_10

there is usually a time limit in order to perform the service of the customers. The most known variant of the Orienteering Problem is the Team Orienteering Problem where instead of one path, a number of P paths should be determined where the total collected score should be maximized [3,12]. Other variants of the Orienteering Problem are the Orienteering and the Team Orienteering Problems with Time Windows [7,9] where, also, time windows are assigned in each arc of the graph.

The Orienteering Problem has many applications in real life problems. One of them is its use for creating a Tourist Guide for a museum or for a specific city [11,14]. When a tourist visits a museum or a city it is often impossible to visit everything, thus, he should select the most interesting exhibitions or landmarks, respectively. If the tourist has plenty of time to visit everything inside the museum, then, the problem that he has to solve is only the finding of the sequence of visiting of the paintings or the sculptures. However, if the tourist has limited time, then, he has to make a selection of the most important parts of the museum for him. Thus, an ideal way to visit as more of his interesting exhibition parts in the museum as possible is to add, initially, a score to everything that exists in the museum (based on his personal preferences) and, then, to try to maximize this score by visiting the exhibition parts that have the highest scores. Thus, it has to solve a classic Orienteering Problem. One problem that may arise from the formulation of a routing problem in a museum as an orienteering problem is that the visiting time in each exhibition part cannot be calculated exactly as either the tourist would like to stay more or there is a lot of people around it and, thus, another constraint have to be added to the Orienteering Problem. There is a number of ways to solve this problem. One way is to add a specific time to the traveling time. However, this is not the most representative way as this time is not fixed for every visitor. Another way is to add a stochastic variant in each node of the graph which will correspond to the time that a visitor will stay to this exhibition part of the museum. Finally, a third way is the way that the authors of [17] formulate their problem as an Open Shop Scheduling problem in which they divide the visitors in different groups where they all are together and spend the same time in the exhibition rooms. The visitors are the jobs while the exhibition rooms are the machines. The time each visitor group spends in an exhibition room is considered analogous to the processing time required for each job on a particular machine [17].

In this paper, we solve the Orienteering Problem (OP) as a first step for creating a Tourist Guide either for a museum or for a specific city. The city that the algorithm will be applied is the city of Chania, Crete in Greece which is a very popular tourist destination. The idea behind this paper is to develop an algorithm that it could solve satisfactory the Orienteering Problem, with very good results in classic benchmark instances and in short computational time. In this algorithm, the waiting time of the tourists was added in the traveling time of an arc and it was thought constant for every tourist. The reason that we developed the algorithm with this assumption is that as we do not have benchmark instances for the museum (or tourist) routing problem that we would like to formulate and solve, we begin with the application of the algorithm in a well known problem with well known benchmark instances and, then, we are going to proceed in the more demanding problem. Thus, we propose an algorithm that is an efficient hybridization of three algorithms, a Greedy Randomized Adaptive Search Procedure (GRASP) [4] for the creation of part of the initial solutions, an Evolutionary Algorithm as the main part of the algorithm and two local search phases (2-opt and 1-1 exchange) in order to improve each of the individuals of the population separately and, thus, to increase the exploitation abilities of the algorithm. The algorithm is denoted as Memetic-GRASP algorithm (MemGRASP). The algorithm is denoted as memetic algorithm [8] as it is an evolutionary algorithm with a local search phase. The reason that we use an evolutionary algorithm is that as the final application of the algorithm will be in the design of an tourist guide planner where the tourist will have the best option (meaning the visiting sequence of as many as possible points of interest) based on his preferences, the use of an evolutionary algorithm give us the possibility to give to the tourist alternative, very effective, options if for some reason his preferences change during his exhibition. Thus, the ability of a memetic algorithm to increase the exploration abilities of the procedure by searching in different places of the solution space give us the possibility of having good solutions in different solution space. The rest of the paper is organized as follows: In the next section a formulation of the Orienteering Problem is presented while in Section 3 the proposed algorithm is analyzed in detailed. In Section 4 the results of the proposed algorithm in the classic benchmark instances from the literature for the Orienteering Problem are given and in the last section the conclusions and the future research are presented.

2 Orienteering Problem

The Orienteering Problem (OP) can be described using a graph $G = (V, A)$, where $V, i = 1, \cdots, N$ denotes the set of nodes, each one having a score r_i and A is the set of arcs between points in V. There are two fixed points, the starting point (usually node 1) and the ending point (usually node N), where these two nodes could be the same or not and where these two nodes have zero score. There is a symmetric nonnegative cost c_{ij} associated with each arc, where c_{ij} denotes the time between point i and point j. Each node can be visited at most once and the total time taken to visit all points cannot exceed the specified limit T_{max} [16]. The main target of the solution of the OP is to determine a path, limited by T_{max}, that visits some of the nodes in order to maximise the total collected score. The scores are assumed to be entirely additive and each node can be visited at most once [16].

Making use of the notation introduced above, the OP can be formulated as an integer problem. The following decision variables are used: x_{ij} equal to 1 if a visit to node i is followed by a visit to node j, otherwise the value of x_{ij} is zero; y_i equal to 1 if node i is in the path or zero otherwise.

$$z = \max \sum_{j \in V} r_j y_j \qquad (1)$$

$$s.t.$$

$$\sum_{j \in V} x_{1j} = 1, \qquad (2)$$

$$\sum_{j \in V} x_{in} = 1, \qquad (3)$$

$$\sum_{(i,j) \in A} x_{ij} + \sum_{(j,k) \in A} x_{jk} = 2y_j \; \forall j \in V, \qquad (4)$$

$$\sum_{(i,j) \in A} c_{ij} x_{ij} \le T_{\max} \qquad (5)$$

$$\sum_{(i,j) \in S} x_{ij} \le |S| - 1 \; \forall S \subseteq V \qquad (6)$$

3 Memetic-Greedy Randomized Adaptive Search Procedure (MemGRASP) for the Orienteering Problem

In this section, the proposed algorithm is described in detail. Initially, in the MemGRASP algorithm we have to create the initial population. The first member of the population is produced by using a Greedy Algorithm. The path begins from node 1, which represents the starting point of each solution. Then, two different conditions are taken into account, the most profitable node and the nearest node to the last node inserted in the path. As there is a possibility the most profitable node not to be the nearest node to the last node inserted in the path and vice versa, we have to select which one of these two conditions is the most important. The most important condition is considered the condition of the profitable node as this is used for the calculation of the fitness function of the problem. However, as it is very important not to select a distant node, we have to add in the path a node as near as possible to the last added node in the path. In the case where more than one possible nodes have the same score (profit), which is a very possible situation in the instances the algorithm was tested, then, the nearest node is selected to be inserted in the path. Thus, the Greedy algorithm uses a hierarchical procedure where the most important condition is the profit of the node and the second most important is the distance.

At the end of the Greedy Algorithm, when no other node can be inserted in the path due to the fact that the constraint that restricts the length of the path has been violated, a complete solution is produced. We have to mention that in the formulation of the problem, described in Section 2, we mentioned that in the problem each tourist has a time limit and not a distance limit. However, in general we can consider that these two constraints could have the same role in the problem, meaning that if we have data that correspond to distance from one node to another and a limit in the length of the path, the constraint (5) could have the same role in the OP as if we had the time traveled from one node to another and a maximum time limit for the path. The solution is represented with a path representation only of the nodes that have been selected to construct the path and the 2 nodes corresponding to the starting node and the ending node (for example, the entrance and the exit of the museum (these two nodes could be the same node)). Thus, the vector corresponding to each member of the population could have different length as there is a possibility in two solutions different number of nodes to be selected. Then, for the solution the fitness function is calculated which is the summation of the profit (score) of each of the node added in the path.

Next the initial path (solution), as every other path (solution) of the population, is tried to be improved using a combination of two local search algorithms, one is a type of exchange algorithm inside the path and the other is an insertion algorithm of a node that is not in the path between two nodes that are already in the path. The first one is used in such a way that the nodes of the path do not change but only the sequence of the nodes in the selected path may change. The reason that this procedure is used is that by finding a new sequence with shorter length (or less time) it is possible to add a new node in the path without violating the constraint (5) and, thus, to increase the summation of the score (profit) and to improve the solution. The second local search procedure takes the path produced by the first local search and tries to improve the fitness function of the solution by adding new nodes in the path. This procedure is applied even if the previous local search does not improve the initial path. In this procedure, an arc is removed and between the two nodes a new node is inserted (without violation of the constraint (5)) in order to improve the solution. With this procedure we ensure that if one successful move is realized, then, the fitness function of the solution will be improved, as the only condition in order to improve a solution is to add a new node in the path without deleting some of the existing nodes. If an improvement in the solution is achieved, then, the first local search algorithm is applied again in order to see if there is a possibility of more improvement in the solution. This local search algorithm is applied in every solution during the whole iterations of the algorithm.

The next step is the calculation of the solution of the rest members of the population. From member 2 to $NP/2$, where NP is the population number, the members of the population are calculated with a random procedure in order to spread the solutions in the whole space. In the random procedure, the solution begins from the starting node and, then, a node is selected at random without violating the limit (either time or distance) constraint and is added in the path. When the constraint is violated, the path is ended in the exit node. Then, the solution is tried to be improved using the local search procedure described previously. The last members of the population are calculated using Greedy Randomized Adaptive Search Procedure (GRASP). In the GRASP algorithm, a solution is created step by step where the best node is not added in the path but a list, the Restricted Candidate List - RCL with the most possible for inclusion nodes in the path, is created and one of them is selected randomly to be included in the path. In the proposed algorithm, the RCL contains the most profitable nodes (based on their score (profit)) taking into account not to violate the limit (either time or distance) constraint and, then, in each step of the algorithm one of them is selected at random and is added in the path. Then, the RCL is updated with one node not in the path in order to keep its size (number of candidate nodes) constant. Finally, the local search described previously is applied.

All these solutions construct the initial population of the algorithm. Then, with the roulette wheel selection procedure the two parents are selected. A very interesting part of the algorithm is the crossover operator that is used in the next step. A 1-point crossover is used but it was very difficult to use it directly to the solutions as each solution contains the sequence of the nodes that a tourist will visit and, thus, two different solutions are possible to have different number of nodes. In order to solve this problem, each solution is mapped in a new vector with zeros and ones where a value equal to zero means that the node is not visited and a value equal to one means the opposite. Thus, the new vectors of two parents have the same size and the crossover operator can easily be applied and the two offspring are produced. In these vectors (either the parent vector or the offspring vector) the sequence of the nodes is not appeared. In the offspring in order to calculate the sequence of the nodes we produce a new vector using the nodes that have value equal to one and we apply the greedy algorithm described previously until the solution violates the constraint (5). Next the mutation operator is applied. The role of the mutation operator in this algorithm is either to improve a feasible solution or to transform an infeasible solution to feasible or, finally, to reject an infeasible solution that could not be transformed to a feasible one. If the limit constraint is violated, then, initially we apply the local search algorithm in order to find a solution which contains all the nodes without violating the limit constraint and if we could not find such a solution, then, we remove the less profitable nodes until the solution becomes feasible. On the other hand if all nodes of the offspring construct an feasible solution, then, all of them are selected. In order to improve the offspring, the local search described previously is applied.

Finally, the new population for the next generation is constructed. The new population contains the best solutions of the parents and of the offspring based on their fitness function taking into account not to have two solutions with the same path in order to avoid a fast convergence of the algorithm and the size of the population to remain constant in all generations. Then, the procedure continues with the selection of the new parents. All the steps of the algorithm (besides the creation of the initial population) are repeated until a maximum number of generations have been reached.

4   Computational Results

In this section, the computational results of the algorithm are presented and discussed in detail. As it was mentioned previously, we formulate the tourist (or museum) routing problem discussed in this paper as an Orienteering Problem with the waiting time in each node to be equal for all tourists and to be added in the traveling time between two nodes. Thus, as we would like to test the effectiveness of the algorithm, we have a number of sets of benchmark instances in the literature to be used for the comparisons. Thus, in the webpage http://www.mech.kuleuven.be/en/cib/op/ there is a number of benchmark instances for different variants of Orienteering Problem. We select five sets of benchmark instances, three of them proposed in [13] and two of them proposed in [1]. The three sets of Tsiligirides [13] have 18, 11 and 20 instances, respectively, with number of nodes 32, 21 and 33, respectively, each one having different value in T_{max} in the limit constraint of the problem. For example, for the first set of instances of Tsiligirides the value of T_{max} varies between 5 to 85. The increase of the value allows more nodes to be visited in the best route. For each set of benchmark instances the value of T_{max} is presented in the corresponding Table. Finally, the last two sets of benchmark instances have 26 and 14 instances with 66 and 64 nodes, respectively.

Table 1. Computational results for Tsilligirides's Set 1

	T_{max}	MemGRASP	GLS	OPT	D(R - I)	S(R - I)	Knapsack	MVP
1	5	10	10	10			10	10
2	10	15	15	15			15	15
3	15	45	45	45			45	45
4	20	65	55	65	65	65	65	65
5	25	90	90	90	90	90	90	90
6	30	110	80	110	110	110	110	110
7	35	135	135	135	135	135	125	130
8	40	155	145	155	150	150	140	155
9	46	175	175	175	175	175	165	175
10	50	190	180	190	190	190	180	185
11	55	205	200	205	200	205	200	200
12	60	225	220	225	220	220	205	225
13	65	240	240	240	240	240	220	240
14	70	260	260	260	260	245	245	260
15	73	265	265	265	265	265	255	265
16	75	270	270	270	275	275	265	270
17	80	280	280	280	280	280	275	280
18	85	285	285	285	285	285	285	285

In Tables 1-3 the results of the proposed algorithm in the three set of instances of Tsilligirides are presented. More analytically, in the first column of each one of the Tables the number of the instance is presented, in the second column the value of the T_{max} for each instance is given, in the third column the value of the objective function produced by the proposed algorithm (MemGRASP) is given, in the fourth and fifth columns the values of the objective function of the algorithm published in [15] and of the best values from the literature (OPT) published in the same paper are presented and, finally, in sixth to nine columns the results of four algorithms presented in [6] are given. In Figure 1, three different solutions having different T_{max} for each set of the Tsilligirides instances are presented. In Table 4, the results of the last two data sets are presented. The structure of the Table is as in the previous Tables, however as we have less instances from the literature to compare the results of the proposed algorithm, we separate the Table in two parts, where in the first part the results of the benchmark instances presented by [1] (denoted as Square-shaped instances) are given (columns 1 to 6) and in the second part the results of the benchmark instances presented by [1] (denoted as Diamond-shaped instances) (columns 7 to 12) are given, respectively. In both parts of the Table, besides the results of the proposed algorithm, the results of the algorithm presented in [15] (denoted as GLS), the results of the algorithm presented in [2] (denoted as Chao) and the results of the algorithm presented in [10] (denoted as DStPSO) are given. In Figure 2, three different solutions having different T_{max} for each set of the Chao instances are presented.

The proposed algorithm is tested in 87 instances in total. The best known solution from the literature is found by the proposed algorithm in 71 of them.

Table 2. Computational results for Tsilligirides's Set 2

	T_{max}	MemGRASP	GLS	OPT	D(R - I)	S(R - I)	Knapsack	MVP
1	15	120	120	120	120	120	120	120
2	20	200	200	200	200	200	200	200
3	23	210	210	210	210	210	210	210
4	25	230	230	230	230	230	230	230
5	27	230	220	230	230	230	230	230
6	30	265	260	265	265	260	260	260
7	32	300	300	300	300	300	275	300
8	35	320	305	320	320	320	305	320
9	38	360	360	360	355	355	355	360
10	40	395	380	395	385	395	380	380
11	45	450	450	450	450	450	450	450

Table 3. Computational results for Tsilligirides's Set 3

	T_{max}	MemGRASP	GLS	OPT	D(R - I)	S(R - I)	Knapsack	MVP
1	15	170	170	170	100	100	170	170
2	20	200	200	200	140	140	200	200
3	25	260	250	260	190	190	250	260
4	30	320	310	320	240	240	320	320
5	35	390	390	390	280	290	380	370
6	40	430	430	430	340	330	420	430
7	45	460	470	470	370	370	450	460
8	50	520	520	520	420	420	500	520
9	55	550	540	550	440	460	520	550
10	60	580	570	580	500	500	580	570
11	65	610	610	610	530	530	600	610
12	70	630	630	640	560	560	640	640
13	75	670	670	670	600	590	650	670
14	80	710	710	710	640	640	700	700
15	85	730	740	740	670	670	720	740
16	90	760	770	770	700	700	770	760
17	95	790	790	790	740	730	790	790
18	100	800	800	800	770	760	800	800
19	105	800	800	800	790	790	800	800
20	110	800	800	800	800	800	800	800

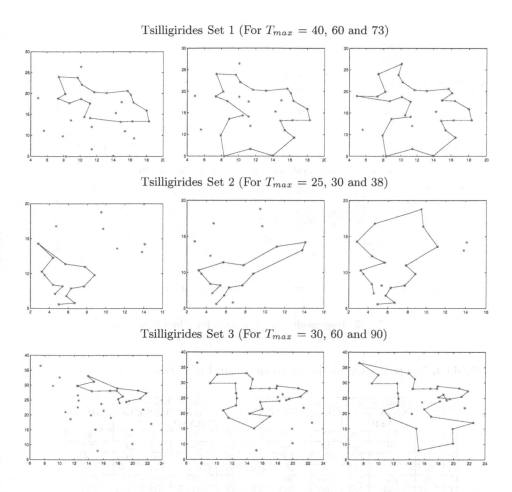

Tsilligirides Set 1 (For $T_{max} = 40$, 60 and 73)

Tsilligirides Set 2 (For $T_{max} = 25$, 30 and 38)

Tsilligirides Set 3 (For $T_{max} = 30$, 60 and 90)

Fig. 1. Representative drawings for Tsilligirides' Sets

In the other 16 the solution found by the proposed algorithm is near to the best known solution without large deviation from the best known solution. From Figures 1 and 2, we can see that as the T_{max} is increased the number of nodes that are included in the best solution is, also, increased. It should be noted that the convergence time of the algorithm was quite satisfactory as the average time (Average CPU Time) was 44 seconds with a minimum of 7 seconds and a maximum of 164 seconds for the more demanding instance. As we observed from the figures the instances are divided in two categories. In the first one the nodes are randomly scattered in the solution space (the three first sets of benchmark instances) while in the second one the nodes are placed in a diamond shape and in a square shape. These two different distributions were the reason that these instances were selected for testing the proposed algorithm. The idea behind this algorithm, as it was analyzed earlier, was the tourists that they would like to

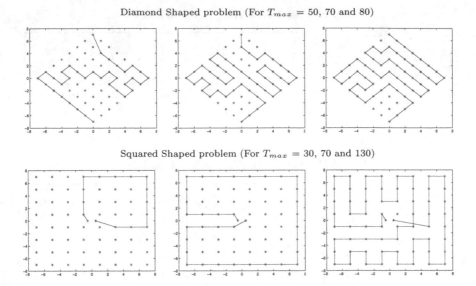

Diamond Shaped problem (For $T_{max} = 50$, 70 and 80)

Squared Shaped problem (For $T_{max} = 30$, 70 and 130)

Fig. 2. Representative drawings for Chao Sets

Table 4. Computational results for Square-shaped and Diamond-shaped test sets

		Square-shaped test set						Diamond-shaped test set			
	T_{max}	Mem-GRASP	GLS	Chao	DStPSO		T_{max}	Mem-GRASP	GLS	Chao	DStPSO
1	5	10	10	10	10	1	15	96	96	96	96
2	10	40	40	40	40	2	20	294	294	294	294
3	15	120	120	120	120	3	25	390	390	390	390
4	20	205	175	195	205	4	30	474	474	474	474
5	25	290	290	290	290	5	35	576	552	570	576
6	30	400	400	400	400	6	40	714	702	714	714
7	35	465	465	460	465	7	45	816	780	816	816
8	40	575	575	575	575	8	50	894	888	900	900
9	45	650	640	650	650	9	55	978	972	984	984
10	50	730	710	730	730	10	60	1062	1062	1044	1062
11	55	825	825	825	825	11	65	1116	1110	1116	1116
12	60	915	905	915	915	12	70	1188	1188	1176	1188
13	65	980	935	980	980	13	75	1230	1236	1224	1236
14	70	1070	1070	1070	1070	14	80	1278	1260	1272	1284
15	75	1140	1140	1140	1140						
16	80	1215	1195	1215	1215						
17	85	1260	1265	1270	1270						
18	90	1340	1300	1340	1340						
19	95	1385	1385	1380	1395						
20	100	1445	1445	1435	1465						
21	105	1515	1505	1510	1520						
22	110	1545	1560	1550	1560						
23	115	1590	1580	1595	1595						
24	120	1610	1635	1635	1635						
25	125	1655	1665	1655	1665						
26	130	1675	1680	1680	1680						

visit a town where a number of places of interest must be selected that are scattered in the whole town or a museum where all the paintings or sculptures are placed in rooms the one next to the other. Thus, we would like to present an efficient and fast algorithm that it will perform equally well in instances that describe both cases. The results of the proposed algorithm both in quality of the solutions and in computational time needed to converge to its best solution give us the possibility to proceed to the next step which is to include this algorithm to the tourist guide planner.

5   Conclusions and Future Research

In this paper, an algorithm for the solution of the Orienteering Problem is presented. The algorithm is the first step of a complete decision support system that will help a tourist to see the most important, based on his preferences, attractions of a city or a museum that he would like to visit during his vacations. The algorithm, denoted as Memetic-GRASP algorithm, is a hybridization of three well known algorithms, the Greedy Randomized Adaptive Search Procedure, an Evolutionary Algorithm and Local Search algorithms. The algorithm was tested in classic sets of benchmark instances for the Orienteering Problem and in most cases it found the best known solutions. The future steps of our research will be, initially to change the formulation of the problem and to add a stochastic variable in each node (point of interest) of the tourist where this variant will correspond to the waiting time in the specific point of interest and it will be activated only if this point of interest will be selected from the tourist, to apply this algorithm in the classic Team Orienteering Problem and in the stochastic Team Orienteering Problem and, finally, to develop a decision support system (tourist guide planner) in which the user will add, initially, his preferences and, then, the algorithm will solve either an Orienteering Problem if he would like to find one path or a Team Orienteering Problem if he would like to find multiple paths.

References

1. Chao, I.: Algorithms and solutions to multi-level vehicle routing problems. Ph.D. Dissertation, Applied Mathematics Program, University of Maryland, College Park, USA (1993)
2. Chao, I.M., Golden, B.L., Wasil, E.: A fast and effective heuristic for the Orienteering Problem. European Journal of Operational Research 88, 475–489 (1996)
3. Chao, I.M., Golden, B.L., Wasil, E.: The team orienteering problem. European Journal of Operational Research 88, 464–474 (1996)
4. Feo, T.A., Resende, M.G.C.: Greedy randomized adaptive search procedure. Journal of Global Optimization 6, 109–133 (1995)
5. Golden, B., Levy, L., Vohra, R.: The orienteering problem. Naval Research Logistics 34, 307–318 (1987)
6. Keller, C.P.: Algorithms to solve the orienteering problem: A comparison. European Journal of Operational Research 41, 224–231 (1989)

7. Montemanni, R., Gambardella, L.: Ant colony system for team orienteering problems with time windows. Foundations of Computing and Decision Sciences 34(4), 287–306 (2009)
8. Moscato, P., Cotta, C.: A gentle introduction to memetic algorithms. In: Glover, F., Kochenberger, G.A. (eds.) Handbooks of Metaheuristics, pp. 105–144. Kluwer Academic Publishers, Dordrecht (2003)
9. Righini, G., Salani, M.: Decremental state space relaxation strategies and initialization heuristics for solving the orienteering problem with time windows with dynamic programming. Computers and Operations Research 4, 1191–1203 (2009)
10. Sevkli, Z., Sevilgen, F.E.: Discrete particle swarm optimization for the orienteering problem. In: 2010 IEEE Congress on Evolutionary Computation (CEC), Barcelona, Spain (2010), doi:10.1109/CEC.2010.5586532
11. Souffriau, W., Vansteenwegen, P., Vertommen, J., Vanden Berghe, G., Van Oudheusden, D.: A personalized tourist trip design algorithm for mobile tourist guides. Applied Artificial Intelligence 22(10), 964–985 (2008)
12. Tang, H., Miller-Hooks, E.: A TABU search heuristic for the team orienteering problem. Computer and Industrial Engineering 32, 1379–1407 (2005)
13. Tsiligirides, T.: Heuristic methods applied to orienteering. Journal of Operational Research Society 35, 797–809 (1984)
14. Vansteenwegen, P., Souffriau, W., Vanden Berghe, G., Van Oudheusden, D.: Metaheuristics for tourist trip planning. In: Geiger, M., Habenicht, W., Sevaux, M., Sorensen, K. (eds.) Metaheuristics in the Service Industry. Lecture Notes in Economics and Mathematical Systems, vol. 624, pp. 15–31 (2009)
15. Vansteenwegen, P., Souffriau, W., Vanden Berghe, G., Van Oudheusden, D.: A guided local search metaheuristic for the team orienteering problem. European Journal of Operational Research 196, 118–127 (2009)
16. Vansteenwegen, P., Souffriau, W., Van Oudheusden, D.: The orienteering problem: A survey. European Journal of Operational Research 209, 1–10 (2011)
17. Yu, V.F., Lin, S.W., Chou, S.Y.: The museum visitor routing problem. Applied Mathematics and Computation 216, 719–729 (2010)

A Multi-start Tabu Search Based Algorithm for Solving the Warehousing Problem with Conflict

Sihem Ben Jouida[1], Ahlem Ouni[2], and Saoussen Krichen[1]

[1] LARODEC Laboratory, Institut Supérieur de Gestion, University of Tunis, Tunisia
benjouida_sihem@hotmail.fr, saoussen.krichen@isg.rnu.tn
[2] Faculty of Law, Economics and Management, University of Jendouba, Tunisia
ahlem.ounii@gmail.com

Abstract. In this paper, we propose a new global warehousing problem supporting advance conflicts: the Warehousing Problem with Conflict (WPC). A set of items are to be loaded into a number of warehouses. We assume a fixed capacity for all warehouses and a predefined weight for each item. The main goal of the WPC is to minimize the overall used number of warehouses under the incompatibility constraint of some pairs of items. We model this optimization problem as a Bin Packing Problem with Conflict (BPPC). As the complexity of the BPPC is NP-Hard, a Multi-start Tabu Search (MTS) is proposed as a solution approach. The empirical validation is done using a developed Decision Support System (DSS) and a comparison to a state-of-the-art approach is performed. The experimental results show that the MTS produces considerably better results than the various existing approaches to the detriment of the gap values.

Keywords: Warehousing problem, Bin packing problem, Bin packing problem with conflict, Tabu-search algorithm.

1 Introduction

The Bin Packing Problem (BPP), a fundamental combinatorial optimization problem in transportation and warehousing logistics, was introduced by David et al., 1974 [1], then extensively studied in the literature in various contexts as supply chain management.

The BPP, in its basic version, consists in selecting a minimum number of identical bins in order to load a set of items, given their weights. Several constraints, as variable bins' capacities and items' compatibilities can be added to the BPP giving rise to many variants that model real case studies. In fact, the incompatibility of items defines a relevant variant termed Bin Packing Problem with Conflicts (BPPC) [2]. The main interest is to assign a set of items to a minimum number of bins while respecting the incompatibility constraint of some pairs of items. The BPPC models a large variety of real-world applications arising in various fields of study as examination scheduling, parallel computing and database storage, product delivery and resource clustering in highly distributed parallel computing ([3], [6]).

© Springer International Publishing Switzerland 2015
H.A. Le Thi et al. (eds.), *Model. Comput. & Optim. in Inf. Syst. & Manage. Sci.*,
Advances in Intelligent Systems and Computing 360, DOI: 10.1007/978-3-319-18167-7_11

As the BPPC is the appropriate modelling of the warehousing problem in the supply chain [5], we show in this paper its ability to handle the Warehousing Problem with Conflict (WPC) where conflicts can be assumed in the storage of products. As such, the set of bins in the BPPC represent the set of warehouses in the WPC to be minimized while avoiding joint storage of products in conflict. The WPC, a potential class of optimization problems in the supply chain framework that, once optimized, it can offer cost saving solutions while avoiding joint packing of products in conflict. It handles chemical products and pharmaceuticals stocking.

Since the WPC is NP-hard, numerous approximate methods were developed and compared to achieve concurrential solution quality [6]. As it is the case, our incentive in this paper is to develop a parallel multi-start tabu search (MTS) algorithm that performs advanced features in order to well approximate the optimal solution that reports, in our case, the number of used warehouses. The MTS algorithm starts by generating an initial solution alternatively generated, randomly or using the specifically designed heuristic [7]. A Decision Support System (DSS), for the WPC, is also developed to solve the warehousing problems. The iterative process of the MTS algorithm tries to move from one neighborhood to another using threads that allow the parallel computation of alternative solutions in order to record the best configuration. In fact, extensive computational experiments across a variety of problem sizes and different loading ratios are reported. A comparison to a state-of-the-art approach is performed and the obtained results show that the MTS algorithm is computationally effective and provides high-quality solutions.

This paper is structured as follows: The WPC is described in Section 2. Section 3 states the DSS and Section 4 details the MTS algorithm for the WPC. Experimental results and computational study are reported in section 5 in order to provide the efficiency of our proposed algorithm.

2   Problem Description of the WPC

We propose in this paper an optimization problem for the WPC where a given company, disposing of a number of private warehouses and facing various customers' demands, aims to find the best warehousing strategy that minimizes the number of used warehouses and the overall holding costs. We assume that each customer demands a specific category of products. Some of the considered products' categories are incompatible and cannot be storage together in the same warehouse. We also assume that all customers' demands are available simultaneous. In order to respond to customers' demands with the adequate service level, products are stored using owned warehouses of the company. Generally, the warehousing problems are NP-hard. Hence, such problems cannot be solved exactly and thus, the development of heuristics algorithms are the viable methods to generate feasible solutions. As mentioned above, the WPC is very similar to the bin packing problem with conflict BPPC [5]. Hence, we use in this paper the mathematical formulation of the BPPC stated by [2] to model the WPC where products' demand designed items and warehouses corresponds to bins. Based on such proposal, a formal definition of the WPC could be presented as follows; given a set $N = \{1, 2, ..., n\}$ of products' demand, each demand $i = \{1, 2, ..., n\}$ is characterized by a weight w_i, and a set $K = \{1, 2, ..., m\}$ of identical warehouses with fixed capacity W that ensures the non-negative capacity consumption $w_i \leq W$. The main objective of the company is to minimize the number of used warehouses so that all customers' demand are loaded while respecting the incompatibility constraint of some products. Let $G(N, C)$ designed the conflict graph where C is a set of edges such that $(i, j) \in C$ if product of demands i and j are incompatible [8]. Figure 1 is an illustration of WPC. We assume that the products of customer 1 and n be incompatible storage items. A possible resolution of the WPC is presented in figure 1 with respecting the conflict constraint while using 2 bins to load all products.

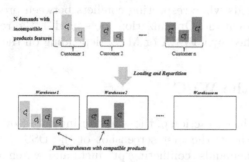

Fig. 1. An illustrative example for the WPC

Inputs	Outputs
n : the number of demands	$y_k = \begin{cases} 1 \text{ if warehouse } k \text{ is used} \\ 0 \text{ otherwise} \end{cases}$
m : the number of warehouses	
W : the capacity of warehouses	$x_{ik} = \begin{cases} 1 \text{ if demand of customer } i \text{ is assigned to warehouse } k \\ 0 \text{ otherwise} \end{cases}$
C : the set of items that are in conflict	

The mathematical formulation inspired from [2] to minimize the number of used warehouses is stated as follows:

$$Min \ \ Z(x) = \sum_{k=1}^{m} y_k \tag{1}$$

$$S.t. \ \ \sum_{k=1}^{m} x_{ik} \geq 1, i = 1, \ldots, n \tag{2}$$

$$\sum_{i=1}^{n} w_i x_{ik} \leq W y_k, k = 1, \ldots, m \tag{3}$$

$$x_{ik} + x_{jk} \leq y_k, (i, j) \in C, k = 1, \ldots, m \tag{4}$$

$$y_k \in \{0, 1\}, k = 1, \ldots, m \qquad (5)$$

$$x_{ik} \in \{0, 1\}, i = 1, \ldots, n \ \ k = 1, \ldots, m \qquad (6)$$

The objective function (1) seeks to minimize the total number of used warehouses. Constraints (2) verify that the demand of each customer is assigned to exactly one warehouse. Constraints (3) ensure that the sum the demands' weights storage in the warehouse k should respect the capacity W. Constraints (4) report the conflicts between products. Finally, constraints (5) and (6) represent the binary requirements of the decision variables. As the warehousing problem is known to be NP-hard, large scaled instances are difficult to solve using exact methods or existing optimizers. We propose to adapt a new metaheuristic based on multi-start tabu search (MTS) to test the strength of a local search method on solving WPC. A DSS that handles the warehousing problem, that tries to store arriving demands while respecting conflicts between products, is descrive in the following subsequent. Optimization tools used in the resolution step can either be accomplished by CPLEX or MTS, depending on the problem size.

3   A DSS for the WPC

The DSS starts by the extraction of the warehousing data from the supply chain database. Table 1 shows the first screenshot of the DSS that corresponds to the data entry for demands, conflicting products and warehouses' descriptions (see Figure 2). Such warehousing data and packing information constitute the inputs for the resolution step. Optimization tools used in the resolution step can either be accomplished by CPLEX or MTS, depending on the problem size. An empirical investigation of the generated solutions in terms of the gap illustrates how well the MTS operates. The DSS outputs the solution to be later visualized in a graphical format.

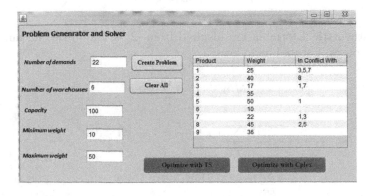

Fig. 2. The first interface of the DSS for WPC: input data

Fig. 3. The second interface of the DSS for WPC: Tabu search tuning

The first screenshot of the DSS, displayed in Figure 2, inputs the received demands followed by their weights and conflict information.

The second screenshot of the DSS, reported in Figure 3, outlines the the tuning of the computational part, designed as a metaheuristic approach named Tabu search. Herein, the stopping criteria of the resolution approach as well as all numeric parameters are defined.

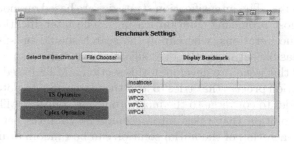

Fig. 4. The third interface of the DSS for portfolio selection : instances' running

After fixing all problem and algorithmic parameters, the DSS proceeds by running various instances to be accomplished by CPLEX or MTS, depending on the problem size, as reported in Figure 4.

4   Solution Approach: A Multi-start Tabu Search Algorithm (MTS)

The NP-hardness of the WPC makes the use of exact methods impractical to tackle large instances. Therefore, we propose to solve it using a metaheuristic method, based on MTS. The successful application of tabu search approach for solving logistics problems motivated the choice of the metaheuristic [7]. Tabu

Search(TS) algorithms are also developed by [9] and [10] which improved their ability in solving bin packing problems. Based on such results, we propose to solve the warehousing problem with conflict using a mutli-start tabu search. The algorithm starts from an initial solution, and then it explores iteratively its neighborhood. Tabu search approach, first designed by [11], is a greedy-based approach that introduces the notion of memory by recording recently visited solutions to avoid revisiting them and looping in the same search area. Its iterative design handles a single solution, one at a time, and generates its neighborhood using alternative neighborhood structures. To avoid being trapped in local minima, MTS can accept a neighbor solution that does not improve the currently observed solution.

Such movements in the search space are to be mixed with some other operators in order to lead the search to other unexplored regions. In fact, from one side, the use of the tabu list that notes already visited solutions (or moves) in order to forbid turning backs. Another stream of solution prominently including diversification and intensification operators that try improve the solution quality within the currently examined area. Based on all such proposals, the neighborhood $N(s)$ of a current solution s is closely dependent on the tabu list content and the intensification/diversification strategies. Indeed, the selected neighborhood depends on the iteration number and the selection probability computed for each iteration. As such, the set of generated neighbors are evaluated according to the number of used warehouses while conflicting products are stored in different warehouses. The intensification strategies are based on modifying choice rules to favor most promising solution features. The intensification should applied during some iterations, whereupon the search should be spread to other regions of the feasible set using the diversification strategy.

Motivated by this issue, we develop in this section a MTS based approach, an iterative exploration process that tries to lead the search area to potentially interesting area in order to converge to the most promising solution. MTS that starts from an initial solution that takes into account the conflicts between pairs of products. The generation of the initial solution can be random or using a specifically designed heuristic. Hence, the generation of the initial solution switches apply the First Fit (FF) and the First Fit Decreasing (FFD). The direct encoding is adopted to represent each solution of the TS approach. Direct encoding consists in generating a n-sized array in which the position designates the demand number and its content records the corresponding warehouse. Hence, the graphical illustration of the direct encoding is reported in figure 5. The initial solution tries to pack products in such a way to exploit the residual space of already used warehouses. The obtained solution, evaluated in terms of the number of used warehouses, is then improved using multiple neighborhoods. We adopt the probabilistic strategy to select a neighborhood for a current solution. TS operators are also applied to make some perturbations on currently generated solutions. The management of conflicts is designed in such a way to assign objects to bins while avoiding joint packing of items announced to be in conflict. The adaptation of the TS features are detailed in what follows:

$$\boxed{2}\boxed{3}\boxed{1}\boxed{1}\ldots\boxed{3}\boxed{2}\boxed{3}$$

Fig. 5. Direct encoding of a solution in the MTS approach

- **Solution encoding:** We adopt the direct encoding that assigns products to warehouses while taking into account conflicts between products. Regarding figure 5 products' demand 1 is packed in warehouse 2, products' demand 2 in warehouse 3, products' demand 3 in warehouse 1, ..., products' demand n in warehouse 3. This encoding makes the extraction of the packing positions very easy and not time consuming.
- **Initial solution:** This step is designed in such a way to start from a first feasible solution s_0 that helps converge, after applying the MTS, to high-quality results. The initial solution is generated using three different protocols:
  - Randomly: A random packing of items in warehouses is accomplished while respecting the capacity and conflict constraints.
  - First fit (FF) heuristic: It packs products in the first available bin. Note that products are inspected according to their index numbers.
  - First fit decreasing (FFD) heuristic: A more improved heuristic that tries to pack products already ranked in a decreasing order according to their sizes. The packing process starts by loading products in the first warehouse allowing such packing.
- **Tabu list:** It contains the T last generated assignments in order to prevent the reversal of the assignment as they belonging in the list. For instance, for any solution in the composite neighborhood, the tabu moves are built as follows: if we swap demand i with demand j from warehouse A to warehouse B, we forbid in later stages the initial assignment of demands i and j (i.e respectively into warehouse A and B).
- **Neighborhood:** The neighbor of an efficient solution is another efficient extreme point. The maximal number of neighbors to be explored is limited by the neighborhood size parameter. The neighborhood structure for the WPC is defined by a set S of all feasible solutions. For each feasible solution $s \in S$ is corresponding a set of neighbors $N(s)$ that can be a subset of S. The structure of neighborhood for our tabu search focuses on a permutation of the demands' assignments to the warehouses. Such permutation is realized using the swap operator and insert operators:
  Swap: Given a demand i and j stored respectively in warehouse 1 and 2, the swap operator consists in assigning demand i to warehouse 2 and demand j to warehouse 1.
  Insert: Such operator consists in removing the assignment of a demand to a given warehouse, and reflecting it to another one. The size of the Insert neighborhood is $n(n-1)$.
- **Stopping criteria:** After a predefined number of iterations Nb_{max}, experimentally depending on the problem size, the tabu search algorithm is over.

The main steps of the MTS approach are reported in the following algorithm (s^* designs the optimal solution):

```
Tabu Search Algorithm for WPC ( MTS-WPC)
 Initialize the MTS parameters:
   Set n, W and ListeT
   Set the parameter values: nb number of local iterations, nbstarts: Number of starts,
                       nbmax: number of iterations without improvement the size of TabuList}
   Generate s_0 using one of the three alternative ways.
 Iterative step:
   Generate the neighborhood N(s) of the current solution of s
     for x = 0 to nbstarts do
     Generate first solution s
     nb=0
       while  x <nb  and  nb < nbmax do
         Set s* = s
         Choose a demand  of a warehouse k to be modified (k not in ListeT)
         Apply the improvement  to i based on exchange and swap operator
         Add k to the ListeT
         If s  better than s*  then
           Set s* = s
           nb++
           else  nb = 0
         end if
       end while
     end for
   Pick up the best solution
     Update ListeT
 Termination criterion:
   If nb > nbmax then stop; otherwise go back to the iterative step
```

Table 1. Numerical results for the Uniform instances

Classes		WPC-1-Y-X					WPC-2-Y-X		
n	$d(\%)$	LB	MTS	Gap	n	d	LB	MTS	$Gap(\%)$
120	00	48.3	48.3	0.0	250	00	101.4	101.4	0.0
	10	48.3	52.3	4.64		10	101.4	108.1	6.20
	20	48.3	54.5	11.37		20	101.4	111.2	8.82
	30	48.3	55.3	12.65		30	101.4	112.5	9.86
	40	52.7	57.7	8.66		40	104.4	113.6	8.11
	50	64.3	73.2	12.65		50	125.7	136.4	7.84
	60	75.4	81.1	7.02		60	151.7	161.3	5.95
	70	87.4	99.3	11.98		70	176.2	203.2	13.28
	80	97.9	106.4	7.98		80	202.3	221.1	8.51
	90	109.5	113.1	3.18		90	226.6	239.2	5.26
Avg				**8.27**					**7.37**
Classes		WPC-3-Y-X					WPC-4 -Y-X		
n	$d(\%)$	LB	MTS	Gap	n	d	LB	MTS	$Gap(\%)$
500	00	202.6	202.6	0.0	1000	00	401.8	401.8	0.0
	10	202.6	214.2	5.41		10	401.8	430.2	6.66
	20	202.6	222.1	8.76		20	401.8	454.6	11.62
	30	202.6	225.2	10.06		30	401.8	460.2	12.69
	40	208.5	250.1	16.63		40	409.0	461.1	11.35
	50	251.5	272.2	7.62		50	504.4	590.3	14.55
	60	303.4	326.3	7.06		60	606.9	664.7	8.69
	70	349.1	396.3	11.92		70	706.4	803.2	12.05
	80	398.7	422.2	5.56		80	806.5	866.7	6.94
	90	452.2	470.2	3.82		90	905.3	950.2	4.72
Avg				**7.67**					**8.9**

5   Experimental Results

In this section, we examine the impact of the MTS algorithm on the WPC's solution quality. The first subsection details the calibration of the data used for the MTS. The second subsection concerns the comparison of results generated using MTS with other existing methods.

5.1   Data Calibration

The MTS heuristic is implemented in Java language (NetBeans IDE 6 : 9). The algorithm was run on several instances inspired from [6] which considered 8 classes of instances, each one contains 100 iterations:

- For the first four classes, we consider demands' size ranged in the interval [20, 100] and a warehouses' capacity equal to 150. The number of demands n is respectively 120, 250, 500, and 1000. Under each class, the number of demand remain constant.
- For the remaining four classes, we range the demands' size into [25, 50] and the warehouses' capacity to 100. For each warehouse, we assume a triplet assignment where an exact packing with 1000 weight is realized using three demands. The number of demands becomes respectively 60, 120, 249 and 501.

The total resulting test-bed yields 800 instances composed of 400 uniform instances and 400 triplet ones.

A conflicts graph G was generated with ten different densities for each instances of the 8 described classes. For each vertex which represents an item i in the conflicts graph, an uniform value v_i on $[0-1]$ is assigned, and an edge (i , $i\prime$) is created if $(v_i + v_{i\prime})$ / $2 \le d$ where d is the considered density of the conflicts graph which taken values on $d = 0\%, 10\%, ..., 90\%$.

Tables 1 and 2 provide the resulting solutions of the MTS applied to existing benchmark that encompasses lower bounds as reference solutions. Each table reports a set of warehousing benchmarking classes that amount to 8. Table 1 contains the results according the uniform instances (the warehouses' capacity is 150) while table 2 contains the triplet instances (the warehouses' capacity is 1000). Each $x - y - z$ designates a class of instance with the same number of items to be packed, where: x is the class number, y is the density of the conflict graph and z is the instance number in the class x ($z = 1, ..., 10$).

According to various number of demands (n), and given the density of conflict ($d\%$) as well as the best known bound, gaps values are computed using the following formula:

$$((MTS - LB)/MTS) * 100 \qquad (7)$$

Values reported in Table 1 and 2 show that fitness, generated by the MTS, are significantly near to the best known solutions and that for all uniform and triplet instances. According to these fitness, average gaps are calculated in order to be later compared to the state-of-the art methods.

Table 2. Numerical results for the Triplet instances

Classes		WPC-5-Y-X					WPC-6-Y-X		
n	$d(\%)$	LB	MTS	Gap	n	d	LB	MTS	$Gap(\%)$
60	00	20.0	20.0	0.0	120	00	40.0	40.0	0.0
	10	20.0	21.2	5.54		10	40.0	41.1	2.64
	20	20.0	22.4	10.63		20	40.0	44.8	10.68
	30	21.0	24.8	15.18		30	40.4	48.5	16.62
	40	24.7	28.3	12.90		40	48.2	57.1	15.57
	50	30.6	33.3	8.11		50	60.9	70.1	12.98
	60	36.3	41.4	12.17		60	74.3	84.6	11.98
	70	42.4	48.6	12.75		70	84.9	99.0	14.24
	80	49.0	53.6	8.52		80	96.4	109.7	12.12
	90	54.4	57.6	5.56		90	109.0	116.5	6.43
Avg				**9.11**					**10.36**
Classes		WPC-7-Y-X					WPC-8-Y-X		
n	$d(\%)$	LB	MTS	Gap	n	d	LB	MTS	$Gap(\%)$
249	00	83.0	83.0	0.0	501	00	167.0	167.0	0.0
	10	83.0	86.8	4.37		10	167.0	174.3	4.16
	20	83.0	92.8	10.56		20	167.0	188.5	11.32
	30	83.2	102.1	18.51		30	167.5	214.9	22.00
	40	98.3	120.6	18.49		40	204.8	247.8	17.3
	50	124.7	148.2	15.85		50	255.1	301.1	15.28
	60	149.7	176.9	15.37		60	305.1	365.6	16.55
	70	147.5	206.1	15.33		70	355.0	417.2	14.91
	80	200.2	228.7	12.46		80	405.2	463.9	12.65
	90	225.0	242.8	7.33		90	453.8	490.0	7.39
Avg				**11.81**					**12.16**

5.2   Comparative Study

In order to provide the performance of MTS, we propose to compare its generated results to the three following heuristics' results: The dual feasible functions (L_{DFF}) [12], the max clique based bounds (L_{MC}) [6] and the matching-based lower bound (L_{match}) [13]. Gap results for the three cited methods as well as for the MTS approach are given in table 3. Based on the experimental results, reported in table 3, one can notice that:

- gap values for MTS-WPC are always less than these for L_{DFF} and L_{MC}, which implies the efficiency of our proposed approach in studied instances(see Figure 6)
- gap results generated using MTS are greater than gaps values for L_{match} methods, for uniform instances(see Figure 6)
- MTS heuristic outputs convergent/competitive gaps regarding the L_{match} heuristic, for the triplet instances: the MTS gap is less than the L_{match} gap in 3 out of 4 instances which improve the efficiency of our proposed heuristic
- the average gap values are respectively 24.15%, 19.58%, 9.3% and 9.45% for L_{DFF}, L_{MC}, L_{match} and MTS. This fact proves that MTS gives a better approximation of the optimal solution regarding L_{DFF}, L_{MC} methods and competitive solution regarding the L_{match} method.

Table 3. MTS-WPC versus lower bounds methods

Classes	State of the art methods (Gap %)			MTS-WPC (Gap %)
	L_{DFF}	L_{MC}	L_{match}	
WPC-1-Y-X [n= 120,W= 150]	22.05	15.13	**6.29**	8.27
WPC-2-Y-X [n= 250,W= 150]	20.33	16.21	**6.99**	7.37
WPC-3-Y-X [n= 500,W= 150]	20.30	17.86	**7.32**	7.67
WPC-4-Y-X [n= 1000,W= 150]	20.54	19.25	**7.58**	8.9
WPC-5-Y-X [n= 60,W= 1000]	28.37	22.93	10.97	**9.11**
WPC-6-Y-X [n= 120,W= 1000]	27.49	22.94	11.44	**10.36**
WPC-7-Y-X [n= 249,W= 1000]	26.79	22.58	11.93	**11.81**
WPC-8-Y-X [n= 501,W= 1000]	27.35	21.93	**11.90**	12.16
Avg	24.15	19.58	9.3	9.45

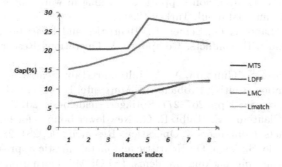

Fig. 6. Gaps behaviour regarding various approaches

6 Conclusion

Warehousing problem in the supply chain is getting more interest, as once optimized, it can offer cost saving solutions. In this paper, we studied the warehousing problem while assuming possible conflicts in the storage of some pairs of items. Furthermore, the incompatibility features, the items weights and the warehouses' capacity are an inputs. We proposed to model the warehousing problem as a bin packing problem with conflict. The solution of the problem leads to find the minimum number of warehouses while avoiding joint storage of items in conflict. We developed a decision support system , based on two optimization tools:

CPLEx and MTS, depending on the problem size. To cope with the complexity of the problem (NP-hard), we proposed to adapt the multi-start tabu search metaheuristic. The algorithm efficiency was validated by computational experiments on different instances. The algorithm demonstrated a fast convergence. A lower bound approximating the total delay was also proposed. The gap between the generated solutions and the lower bounds is reasonable. A comparison of the obtained gaps and the gaps generated by other state-of the art methods shows the performance of the MTS in solving warehousing problem with conflict. As a future work, more tests over other existing methods will be performed. Furthermore, we are considering other constraints such as warehouses' capacity. We propose also to investigate the problem in a dynamic environment.

References

1. David, S.J.: Fast algorithms for bin packing. Journal of Computer and System Sciences 3, 272–314 (1974)
2. Maiza, M.: Heuristics for Solving the Bin-Packing Problem with Conflict. Applied Mathematical Sciences 5, 739–752 (2011)
3. Christofides, N., Mingozzi, A., Toth, P.: The vehicle routing problem. In: Combinatorial Optimization. Wiley (1979)
4. Jansen, K., Öhring, S.: Approximation Algorithms for Time Constrained Scheduling. Information and Computation 132, 85–108 (1997)
5. Kemal, K., Grdal, E.: Decision support for packing in warehouses. In: 21st International Symposium, Istanbul, Turkey (2006)
6. Gendreau, M., Laporte, G., Semet, F.: Heuristics and lower bounds for the bin packing problem with conflicts. Computers & Operations Research 31, 347–358 (2004)
7. Masri, H., Krichen, S., Guitouni, A.: A Multi-start Tabu Search Approach for Solving the Information Routing Problem. In: Murgante, B., et al. (eds.) ICCSA 2014, Part II. LNCS, vol. 8580, pp. 267–277. Springer, Heidelberg (2014)
8. Khanafer, A., Clautiaux, F., Talbi, E.-G.: New lower bounds for bin packing problem with conflicts. Computers & Operations Research 206, 281–288 (2010)
9. Lodi, A., Martello, S., Vigo, D.: Heuristic and metaheuristic approaches for a class of two-dimensional bin packing problems. INFORMS Journal on Computing 11, 345–357 (1999)
10. Lodi, A., Martello, S., Vigo, D.: Heuristic algorithms for the three-dimensional bin packing problem. European Journal of Operational Research 141, 410–420 (2002)
11. Glover, F., Taillard, E., Werra, D.: A User's Guide to Tabu Search. Ann. Oper. Res. 41, 3–28 (1993)
12. Fekete, S.-P., Schepers, J.: New classes of fast lower bounds for bin packing problems. Mathematical Programming 91, 11–31 (2001)
13. Maiza, M., Radjef, M.S.: Heuristics for Solving the Bin-Packing Problem with Conflicts. Applied Mathematical Sciences 35, 1739–1752 (2011)

A Novel Dynamic Pricing Model
for the Telecommunications Industry

Kholoud Dorgham[1], Mohamed Saleh[1], and Amir F. Atiya[2]

[1] Department of Operations Research, Cairo University, Cairo, Egypt
{k.dorgham,m.saleh}@fci-cu.edu.eg
[2] Department of Computer Engineering, Cairo University, Cairo, Egypt
aatiya1@gmail.com

Abstract. Telecommunications industry is a highly competitive one where operators' strategies usually rely on significantly reducing minute rate in order to acquire more subscribers and thus have higher market share. However, in the last few years, the numbers of customers are noticeably increasing leading to more stress on the network, and higher congestion rate, i.e. worse quality of service (QoS). Because of this, pricing has emerged as a useful tool for simultaneously limiting congestion, and increasing revenue.

In this paper a dynamic pricing model is proposed for the mobile calls based on a Monte-Carlo simulation that emulates the processes of calls arrivals, calls durations and the effect of price on both. This model is then integrated with a meta-heuristic evolutionary based optimization algorithm to determine the optimal dynamic pricing schemes according to the call parameters. This integrated framework is a novel approach to dynamic pricing that aims at maximizing revenue and enhancing the QoS.

Keywords: Telecommunications, Revenue Management, Dynamic Pricing, Optimization, Monte-Carlo Simulation.

1 Introduction

In developing countries, the telecommunications market is highly competitive. All operators use price as their main competitive edge to gain market share. They rely on reducing minute rate in order to acquire more subscribers, often without regard to using scientific approaches for optimizing prices. However, pricing calls too cheaply can cause losing higher revenue from price-insensitive users, i.e. lost opportunity, while setting too high of a price could noticeably reduce demand. This calls for the application of a revenue management approach, for optimally determining this trade-off pricing.

Revenue management is the science of managing the amount of supply to maximize revenue by dynamically controlling the price (or quantity) offered. It has had a long history of research and application in several industries, for example airline companies, hotels, and retail chains, where advanced revenue optimization policies are becoming key drivers of the companies' performance. The primary aim of

© Springer International Publishing Switzerland 2015
H.A. Le Thi et al. (eds.), *Model. Comput. & Optim. in Inf. Syst. & Manage. Sci.*,
Advances in Intelligent Systems and Computing 360, DOI: 10.1007/978-3-319-18167-7_12

revenue management is based on the motto: *selling the right product to the right customer at the right time for the right price.*

There have been two general approaches for revenue optimization. The first one is performed by customer segmentation. Different prices are set for the different customer segments, and what is being controlled (for the purpose of optimizing revenue) is the amount of merchandise allocated to each segment. This approach is more prevalent in the airline industry. The other approach of revenue management is the so-called dynamic pricing approach, in which the price is continually adjusted to take advantage of demand variation. This approach is particularly prevalent in some online hotel reservations, and some online retail product sales. The online nature makes it easily possible to update the price periodically. The trend has been moving more towards applying the dynamic pricing approach or a combination between the two approaches.

There has been some previous work on price optimization in the telecommunications domain, specifically in mobile services. However, dynamic pricing is still a nascent but promising research topic in that domain, and much research is still needed to explore its benefits and prospects. Dynamic pricing implies having variable pricing of minutes that is based on traffic, network capacity, time of the day and customer segment (value based). It can be a tool for improving revenues as well as controlling congestion for better customer experience. Current trends include *"nighttime-weekend discounts"*, *"bundle-based pricing"*, *"promo-based pricing"*, *"usage-based pricing"* or *"flat pricing"*. However, true dynamic pricing is adopted in very few countries, most notably in India and in some African countries.

In the last few years, due to the aggressive competition, the numbers of customers in telecommunications industry increased noticeably, leading to more traffic on different network cells -- specifically in highly populated regions. The uneven distribution of traffic, and the congestion on some network cells can lead to an increased number of blocked calls; i.e. worse quality of service (QoS). This eventually reduces revenues as shown in figure 1.

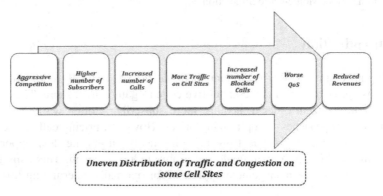

Fig. 1. Problem definition diagram

The purpose of this work is to develop a new dynamic pricing model for mobile calls that maximizes revenue and enhances QoS. This can be achieved by encouraging customers via price incentives to use network cells in their "off peak" hours.

A more uniform distribution of traffic would decrease the number of blocked calls, enhance customer experience and increase revenue. The main idea is that the pricing scheme is updated periodically, and is procedurally performed by having a fixed reference rate and a variable discount that varies within a certain pre-specified range. An improved pricing system -- even by trivial amounts per call -- can lead to huge increases in profits due to the large amount of calls at stake. In the proposed approach we assume that the pricing is based on a reference price that is set by the telecom operator multiplied by some multipliers that vary around one and produce discounts or premiums over the reference price. The multipliers are the control parameters that are optimized in order to maximize revenue, subject to a bound on the blocking rate. To obtain their effect on the revenue, a Monte Carlo simulation is developed that emulates all call processes, such as arrivals and durations, as faithfully as possible.

In Appendix I, the related work is discussed (due to paper size limitation). In summary, dynamic pricing and revenue optimization have potential prospects in the telecommunications industry, where its application by telecom operators is still in its infancy. As such, further studies and development of this topic will certainly advance its application and benefit the telecom market.

2   Proposed System

2.1   Overview

The proposed framework is based on an optimization model that sets the price of every incoming call in a dynamic setting. The idea is to maximize the revenue, and enhance the QoS i.e. reduce the number of blocked calls taking into account current demand, the demand-price sensitivity of the caller's segment (e.g. pre-paid, post-paid or enterprise), the timing slot and the overall load level. Embedded in the optimization model is a Monte Carlo simulation that simulates the incoming calls and their durations as faithfully as possible. This simulation yields an estimate of the revenue and the load for the next few time slots, and its output will therefore directly influence the pricing parameters and its function. The detailed idea will be illustrated below.

2.2   Framework and Methodology

The proposed formulation assumes that an incoming call's price is set be a reference price multiplied by two different multipliers to obtain the final price [6]. The reference price is the baseline or benchmark price set by the telecom operator. It could possibly vary according to the caller's segment such as pre-paid, post-paid or enterprise. In addition, the two multipliers represent the "control variables" that vary around the value of 1. This means that a value lower than one corresponds to a discount with respect to the reference price (such as 0.9 means the price is lower by 10 percent). On the other hand, a value that is higher than one represents a premium price regarding the reference price as well. However, the final price is strictly kept in some predefined range (within a minimum and a maximum price). The reason for price adjustments using a reference price and multipliers that vary around one is that it is sensible to have the price based on the existing price. This will keep the proposed price in check, and will make the system more readily accepted by telecom operators that probably use their long business experience in setting their prices. Each of the two multipliers corresponds to a variable that is having an influencing effect on price calculation. According to industry experts in Egypt, the most two significant variables are:

1. The remaining capacity of the incoming call cell node (cell load capacity)
2. Timing slot of the incoming call (time of the day)

This model works on any generic cell, so it could be applied within cities or villages. Therefore, in general, the relation between "cell capacity" and "time of the day" can vary. In this paper, we treat them as separate variables; nevertheless in future work, we will investigate the dependency between them.

There exists a multiplier for each of the above influencing variables. For simplicity, these multipliers are considered to be as linear or piecewise linear function, whose level and slope are determined by the optimization algorithm. The piecewise linear functions are selected based on logically expected relations identified by telecom experts. For example, if the remaining cell node capacity is high (off-peak hours), then the multiplier should be low in order to attract more calls in the network. On the other side, if the remaining capacity is low (peak hours), the call price should be higher than the norm in order to reduce the load off the network and decrease the number of blocked calls. Therefore, a linear function is used that is monotonically decreasing with the remaining cell capacity. Similar idea applies for the other multiplier, and this will be explained further. The final price of the incoming call is given by the product of the reference price and the multipliers, as follows:

$$Price = Reference\ Price * Day\ Time\ multiplier * Cell\ Capacity\ multiplier$$
(1)

The resulting price will correspond to the combined effect of the discounts and premiums of the control variables. This reflects the aggregate need for either boosting or attenuating traffic.

Figure 2 shows the overall proposed dynamic pricing framework. The components of the block diagram will be explained in detail next. Initially, the customers are segmented into three different segments according to the Telecom operators' definitions for the type of the subscriber acting as pre-paid, post-paid and enterprise. Essentially, a so-called Monte Carlo simulator is designed where its function is to emulate the calls' processes such as the arrivals, and call duration as faithfully as possible. This simulation is based on a probabilistic or stochastic modelling of each of the processes. This simulator will provide a forecast of future arrivals and durations.

Based on the simulator's forecast, the expected total future revenue in addition to the total number of blocked calls is estimated. These are passed forward to the optimization module that seeks to maximize this total revenue. The parameters of the price multipliers e.g. the line slopes and levels are the decision variables of the optimizer. Once they are obtained, they successively will determine the suggested price through the price multiplier formula (1). A price elasticity function is estimated using a linear function in order to determine how much the new suggested price will influence the demand, e.g. the arrivals rate. From this relation, a new demand factor will be obtained for the price which is suggested by the optimization algorithm. The new demand will therefore be amplified or reduced according to this demand factor. Subsequently, the new demand represented in a proportionally higher or lower arrival rate, due to respectively lower or higher pricing, will be taken into account in a new run of the Monte Carlo simulator producing a new forecast of the expected total revenue as well as the total number of blocked calls. The optimization algorithm will work on another set of price parameters, and continue in a new iteration of the whole loop as shown in figure 2. The optimizer/simulator framework keeps looping around for several iterations, until reaching the best parameter selections that lead to maximum total revenue. Once the optimal parameters are obtained, the final output is the price of every incoming call.

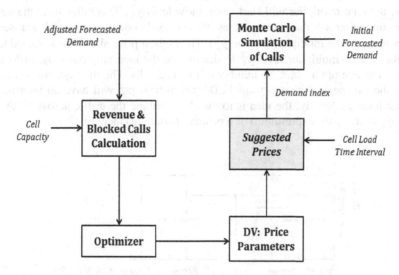

Fig. 2. Proposed framework block diagram

2.3 Price Multipliers

The following is an illustration of the idea of the reference price along with the two multipliers that determine the final suggested price.

Reference Price

According to the nature of the telecommunications market, the reference price should vary along with the subscriber type (e.g. segment) as well as the tariff model for each type as illustrated in table 1 for the telecommunications market in Egypt. Thus, for each incoming call, a mapping between the subscriber type and tariff model with its call reference price will take place.

"Time of the Day" Multiplier

The time of the day of an incoming call can be an essential controlling variable for the call's price. For example some hours coincide with peak demand, and therefore higher pricing would at the same time lead to reducing demand, and increasing revenue. Conversely, most of the cell nodes are underutilized in their off-peak hours due to the minimal flow of the call arrivals, and a boost in call volume could increase revenue. Accordingly, Figure 3 shows the proposed price multiplier (*y-axis*) against the time slots (*x-axis*). For simplicity these multipliers are considered to be a piecewise linear function. The day is divided into four time intervals whereas the peak hours are starting from 6:00 pm to 12:00 am followed by the next peak interval from 6:00 am to 12:00 pm. At the beginning of the day, the price should be low in order to encourage the customers to utilize the network on its off-peak hours e.g. from 12:00 am to 6:00 am, and thus, the time multiplier will start from a low level y_1^T. Thereafter from the second slot, the multiplier's value increases as the network congestion could get serious (corresponding to the multiplier's value y_3^T). The highest price should be assigned to the peak interval with multiplier level y_4^T to discourage the incoming calls and offload the network, in an attempt to limit the number of blocked calls. The final interval which lies between the two peak slots e.g. from 12:00 pm to 6:00 pm will have an intermediate multiplier level y_2^T. Briefly, the idea is to evenly distribute the traffic across all the day through either a negative (premium) or a positive (discount) incentive.

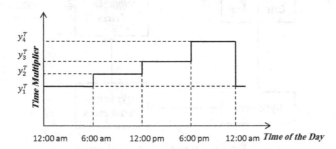

Fig. 3. Time of the day multiplier curve

"Cell-load Capacity" Multiplier

For the cell load capacity multiplier, a highly utilized cell with traffic close to its full capacity should have a higher pricing to discourage further calls. Conversely, for lightly loaded cells, pricing should be discounted. Therefore an upward sloping linear function should exist for the price, as shown in figure 4.

The *x-axis* represents the cell capacity in percentage or so-called server utilization, and the *y-axis* represents the value of the multiplier. If the network cell is under 75% from its capacity, then one has to offer some incentives, and therefore the capacity multiplier is at its lowest value y_1^C. The multiplier's value then increases sharply as the remaining capacity of the network cell node is decreased beyond 75%, until it reached the maximum value y_2^C, at which point there is no remaining capacity, and calls start to be blocked.

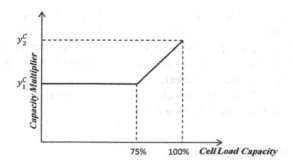

Fig. 4. Cell load capacity multiplier curve

2.4   Optimization Variables and Constraints

The average value of each multiplier function is assumed to be equals one. The reason is that these multipliers are considered correcting factors that will be multiplied by the reference price. Thus, they should vary around one, with a roughly equal average amount of discount pricing and premium pricing.

The optimization variables are as follows: *Time of the Day Multiplier:* y_1^T, y_2^T & y_3^T; and *Cell Load Capacity Multiplier:* y_1^C.

These are the variables that the optimization algorithm will determine such that the revenue is maximized. The other multiplier parameters are dependent, as they will be set such that the average value of each multiplier function equals one. The second expression of equation 2 is computed by assuming that the area under the curve is unity (as in this case, the maximum value of the x-axis is also unity)

$$y_4^T = 4 - (y_1^T + y_2^T + y_3^T)$$

$$y_2^C = 8 - 7 y_1^C$$

(2)

There are also inequality constraints that will guarantee that the multiplier functions are well behaved and produce logically accepted relation:

Time of the Day Multiplier: $0 \leq y_1^T \leq y_2^T \leq y_3^T \leq y_4^T$
$1 \leq y_4^T \leq T_0$
Cell Load Capacity Multiplier: $0 \leq y_1^C \leq y_2^C$
$1 \leq y_2^C \leq C_0$

The first and third constraints are meant to preserve the intended shape of the multiplier function. The second and the fourth constraints are bounds on the parameters that should be determined with the help of a telecom business expert. The expert should determine the maximum deviation from the reference price that is acceptable by the company from a business point of view. For the system implemented, T_0 and C_0 are set to be 1.3.

In addition to these multiplier-specific constraints, there is a global constraint for the overall price correction e.g. the product of all multipliers. Again, the maximum deviation from a reference price should not exceed a certain percentage that is determined based on business considerations. The rationale is that it is detrimental to the good will of the customer, if the price is increased too much. In our system this percentage is set to be 30 percent, which means that the final price has to be within plus or minus 30 percent of the reference price.

2.5   Proposed Model Simulator

In order to compute the future revenue that is going to be optimized, the future incoming calls as well as the calls durations need to be forecasted. For this reason, a call simulator has been developed using Matlab's SimEvent. SimEvent is a tool that provides a discrete-event simulation engine and component library for Simulink [18]. To the best of our knowledge, it is the first time that SimEvent is used for this purpose.

This Monte Carlo simulator takes the call's price parameters from the optimization algorithm and applies the simulation. The simulator is designed so that it emulates processes of call arrivals, calls durations and the effect of price on both as shown in figure 5 below:

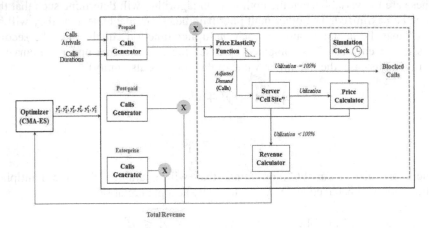

Fig. 5. Proposed System Structure

As demonstrated in the previous figure, calls are divided into three segments based on the customer type such as pre-paid, post-paid or enterprise. Besides, there are number of components in this system, as follows:

- **Calls Inter-arrival generation:** The proposed method is based on assuming that arrival rates of incoming calls follow an inhomogeneous Poisson process (IHPP) [12]. Subsequently, a Bayesian approach is applied that gives the probability density of the forecasted arrival rates. The IHPP process is defined as a Poisson process with an arrival rate $\lambda(t)$ that is a function of time. Thus, certain periods of the day have high arrival rate $\lambda(t)$ i.e. peak hours, and therefore one observes many calls initiated in this period. Conversely, some periods, such as early morning i.e. off-peak hours, have low arrival rates, and therefore a small number of calls fall in them. In this Bayesian approach a Gaussian prior is assumed for the arrival rates of the different time slots that implicitly take into account seasonal relations. Subsequently, the posterior distribution is derived, and this distribution is used to generate future arrival rate values, and then actual call arrivals (for details see [12].

- **Calls Durations:** Calls durations are assumed to be distributed as exponential. Telecom data is used to estimate the mean of the distribution. Subsequently, this distribution is used to generate calls durations.
- **Blocking Calls:** If the number of calls inside the network cell node reached the maximum cell node capacity i.e. 100 percent utility, then further calls will be precluded from entering the cell node and will be counted as blocked call. Conversely, if there is a remaining capacity in the cell node, then the call will be served. The simulator computes the number of blocked calls in a standalone block.
- In Appendix II, we explain the price elasticity formulation (due to paper size limitation).
- **Revenue Calculation:** Total revenue is then calculated after all the simulation processes. As follows:

$$Total\ Rev = \sum_{i=1}^{n} Duration_i * Price_i \tag{5}$$
$$where\ n = total\ number\ of\ calls$$

In order to have statistically accurate estimates, several Monte Carlo runs are generated, and the average total revenue is computed. Then it is passed to the optimization algorithm. The optimizer will keep exploring other sets of price parameters until it reaches the maximum revenue producing parameters.

In Appendix III, we explain the optimization solver used (due to paper size limitation).

3   Model Evaluation

In order to test the effectiveness of the proposed methodology, it is compared with the static model that is applied in a real market (in the Egyptian market). By static model we mean a model with fixed prices. We used in the benchmark the exact same price packages used in the Egyptian telecom market (see Table 1). The goal is to test whether this approach leads to an improvement in the revenue over the baseline. Hence, the test case is divided into two phases: 1- Run the overall proposed dynamic pricing model 2- Test the outcome of the dynamic model versus the static one across several days.

The final set of multipliers coming from phase one after applying thousand runs is as follows: 0.8, 0.9, 1.1, 1.2, 0.97 and 1.21 for y_1^T, y_2^T, y_3^T, y_4^T, y_1^C and y_2^C respectively. Afterwards, the dynamic price-based approach is tested in comparison with the static pricing across ten days.

As seen from table 2 and figure 8, one can observe that the proposed approach succeeds in improving the revenue by about 10 percent over the static model. Moreover, the improvement in revenue is consistent across all days. These results confirm the fact that dynamic pricing, if optimized well, should lead to better revenues.

We note that in our case the optimization is performed only once, and once the multipliers are available, the pricing is obtained in a simple and straightforward way by applying the price multiplier functions.

Table 1. Revenue improvement over the baseline tested across ten days

Days	Total Revenue (EGP Pounds)		Improvement
	Dynamic Price	Static Price	Percentage
Day 1	36,400	33,500	8.0%
Day 2	38,304	32,980	13.9%
Day 3	38,298	32,562	15.0%
Day 4	36,573	34,250	6.4%
Day 5	36,153	33,795	6.5%
Day 6	37,741	35,017	7.2%
Day 7	39,423	34,620	12.2%
Day 8	41,446	34,952	15.7%
Day 9	36,709	34,263	6.7%
Day 10	39,573	36,120	8.7%
Average	38,062	34,206	**10.1%**

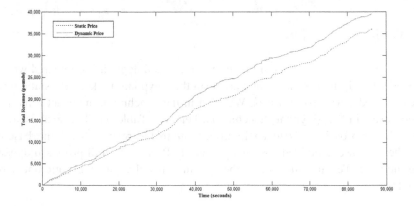

Fig. 6. Total Revenue performance in day 10

4   Conclusion and Future Work

A novel system for dynamic pricing mobile calls is developed in this study. The idea is to create a realistic model of the processes of mobile calls, and based on that determine the effect of price on demand, and hence on revenue. A pricing function is proposed, given in terms of price multipliers that frame the price as discounts or premiums over a given reference price. The proposed system gives a versatile and realistic way to assess pricing strategies, and it can also help in scenario analysis for any suggested strategy. In future work, we will investigate the relationships between various inputs; as there might be dependency between them. Moreover, the model will be tested using real data from telecom operators, and will be benchmarked with the current implemented static module.

Appendices I, II and III

Pdf file in http://is.gd/8mupme

References

1. Abdel Aziz, H., Saleh, M., El Gayar, N., El-Shishiny, H.: A randomized model for group reservations in a hotel's revenue management system. In: Proceedings of INFOS 2008 (2008)
2. Abdel Aziz, H., Saleh, M., El-Shishiny, H., Rasmy, M.: Dynamic pricing model for hotel revenue management systems. Egyptian Informatics Journal 12(3), 177–183 (2011)
3. Al-Manthari, B., Nasser, N., Ali, N., Hassanein, H.: Efficient bandwidth management in Broadband Wireless Access Systems using CAC-based dynamic pricing. In: Proceedings of the 33rd IEEE Conference on Local Computer Networks (LCN) (2008)
4. Andrawis, R., Atiya, A., El-Shishiny, H.: Forecast combination model using computational intelligence/linear models for the NN5 time series forecasting competition. International Journal of Forecasting 27(3), 672–688 (2011)
5. Andrawis, R., Atiya, A., El-Shishiny, H.: Combination of long Term and short term forecasts, with application to tourism demand forecasting. International Journal of Forecasting 27(3), 870–886 (2011)
6. Bayoumi, A.E.M., Saleh, M., Atiya, A.F., Aziz, H.A.: Dynamic pricing for hotel revenue management using price multipliers. Journal of Revenue & Pricing Management 12(3), 271–285 (2013)
7. Chen, F., Ni, R., Qin, X., Wei, G.: A novel CAC-based network pricing scheme for wireless access networks. In: Proceedings of 2010 IEEE International Conference on Wireless Information Technology and Systems (ICWIT) (2010)
8. Dhake, T., Patane, R., Jagtap, S.: Dynamic call management for congestion control in cellular networks. In: Proceedings of the International Conference & Workshop on Emerging Trends in Technology (ICWET 2011), pp. 946–947 (2011)
9. El Gayar, N., Saleh, M., Atiya, A., El-Shishiny, H., Zakhary, A., Abdel Aziz, H.: An integrated framework for advanced hotel revenue. International Journal of Contemporary Hotel Management, 2020 Vision for the Hospitality Industry Special Issue 23(1), 84–98 (2011) (Hendorson, Jauncey, Sutton (eds.))

10. Kabahuma, S., Falowo, O.: Analysis of network operators' revenue with a dynamic pricing model based on user behaviour in NGWN using JCAC. In: Proceedings of the Southern Africa Telecommunication Networks and Applications Conference (SATNAC) (2010)
11. Maille, P., Tuffin, B.: Multibid auctions for bandwidth allocation in communication networks. In: Proceedings of INFOCOM: Twenty-third Annual Joint Conference of the IEEE Computer and Communications Societies (2004)
12. Nawar, M., Atiya, A., Fayed, H.: A Bayesian approach to arrival rate forecasting for inhomogeneous Poisson processes for mobile calls: Submitted for publication (2014)
13. Pla, V., Virtamo, J., Martínez-Bauset, J.: Optimal robust policies for bandwidth allocation and admission control in wireless networks. Computer Networks 52, 3258–3272 (2008)
14. Thakurta, P., Bandyopadhyay, S.: A new dynamic pricing scheme with priority based tree generation and scheduling for mobile networks. In: Proceedings IEEE International Advance Computing Conference (IACC) (2009)
15. Yilmaz, O., Chen, I.: Utilizing call admission control for pricing optimization of multiple service classes in wireless cellular networks. Computer Communications 32, 317–323 (2009)
16. Zakhary, A., Atiya, A., El-Shishiny, H., El Gayar, N.: Forecasting hotel arrivals and occupancy using Monte Carlo simulation. Journal of Revenue & Pricing Management 10, 344–366 (2011)
17. Zakhary, A., El Gayar, N., Atiya, A.: A comparative study of the pickup method and its variations using a simulated hotel reservation data. International Journal of Artificial Intelligence and Machine Learning 8, 15–21 (2008)
18. Matlab SimEvent, http://www.mathworks.com/ (accessed December 2014)

A Practical Approach
for the FIFO Stack-Up Problem

Frank Gurski[1], Jochen Rethmann[2], and Egon Wanke[1]

[1] Heinrich-Heine-University Düsseldorf, Institute of Computer Science,
D-40225 Düsseldorf
[2] Niederrhein University of Applied Sciences, Faculty of Electrical Engineering and
Computer Science, D-47805 Krefeld

Abstract. We consider the FIFO STACK-UP problem which arises in delivery industry, where bins have to be stacked-up from conveyor belts onto pallets. Given k sequences $q_1, \ldots, q_k$ of labeled bins and a positive integer p. The goal is to stack-up the bins by iteratively removing the first bin of one of the k sequences and put it onto a pallet located at one of p stack-up places. Each of these pallets has to contain bins of only one label, bins of different labels have to be placed on different pallets. After all bins of one label have been removed from the given sequences, the corresponding stack-up place becomes available for a pallet of bins of another label. The FIFO STACK-UP problem is computational intractable [3]. In this paper we introduce a graph model for this problem, which allows us to show a breadth first search solution. Our experimental study of running times shows that our approach can be used to solve a lot of practical instances very efficiently.

Keywords: combinatorial optimization, breadth first search solution, experimental analysis.

1 Introduction

We consider the combinatorial problem of stacking up bins from a set of conveyor belts onto pallets. A detailed description of the practical background of this work is given in [6,7]. The bins that have to be stacked-up onto pallets reach the palletizer on a conveyor and enter a *cyclic storage conveyor*, see Fig. 1. From the storage conveyor the bins are pushed-out to *buffer conveyors*, where they are queued. The equal-sized bins are picked-up by stacker cranes from the end of a buffer conveyor and moved onto pallets, which are located at some *stack-up places*. Often there is one buffer conveyor for each stack-up place. Automatic guided vehicles (AGVs) take full pallets from stack-up places, put them onto trucks and bring new empty pallets to the stack-up places.

The cyclic storage conveyor enables a smooth stack-up process irrespective of the real speed the cranes and conveyors are moving. Such details are unnecessary to compute an order in which the bins can be palletized with respect to the given number of stack-up places. For the sake of simplicity, we disregard the cyclic

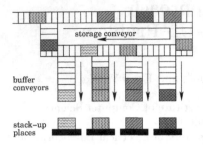

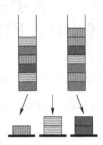

Fig. 1. A real stack-up system   **Fig. 2.** A FIFO stack-up system

storage conveyor, and for the sake of generality, we do not restrict the number of stack-up places to the number of sequences. The number of sequences can also be larger than or less than the number of stack-up places. Fig. 2 shows a sketch of a simplified stack-up system with 2 buffer conveyors and 3 stack-up places.

From a theoretical point of view, we are given k sequences $q_1, \ldots, q_k$ of bins and a positive integer p. Each bin is destined for exactly one pallet. The FIFO STACK-UP problem is to decide whether one can remove iteratively the bins of the k sequences such that in each step only the first bin of one of the sequences will be removed and after each removal of a bin at most p pallets are open. A pallet t is called open, if at least one bin for pallet t has already been removed from one of the given sequences, and if at least one bin for pallet t is still contained in one of the remaining sequences. If a bin b is removed from a sequence then all bins located behind b are moved-up one position to the front.

Our model is the second attempt to capture important parameters necessary for an efficient and provable good algorithmic controlling of stack-up systems. The only theoretical model for stack-up systems known to us uses a random access storage instead of buffer queues. Many facts are known on the stack-up system model with random access storage, see [7,8,9], where complexity results as well as approximation and online algorithms are presented.

The FIFO STACK-UP problem is NP-complete even if the number of bins per pallet is bounded, but can be solved in polynomial time if the number k of sequences or the number p of stack-up places is fixed [3]. Dynamic programming solution and parameterized algorithms for the FIFO STACK-UP problem are shown in [4] and [5]. In this paper we introduce a graph model for this problem, the so called decision graph, which allows us to give a breadth first search solution of running time $\mathcal{O}(n^2 \cdot (m+2)^k)$, where m represents the number of pallets and n denotes the total number of bins in all sequences. Since for practical instance sizes this running time is too huge for computations, we used cutting technique on the decision graph by restricting to configurations such that the number of open pallets is at most some upper bound, which is increased by 5 until a solution is found. Our experimental study of running times shows that our approach can be used to solve a lot of practical instances on several thousand bins very efficiently.

2 Preliminaries

We consider *sequences* $q_1 = (b_1, \ldots, b_{n_1}), \ldots, q_k = (b_{n_{k-1}+1}, \ldots, b_{n_k})$ of *bins*. All these bins are pairwise distinct. These sequences represent the buffer queues (handled by the buffer conveyors) in real stack-up systems. Each bin b is labeled with a *pallet symbol* $plt(b)$ which can be some positive integer. We say bin b is destined for pallet $plt(b)$. The labels of the pallets can be chosen arbitrarily, because we only need to know whether two bins are destined for the same pallet or for different pallets. The set of all pallets of the bins in some sequence q_i is denoted by

$$plts(q_i) = \{plt(b) \mid b \in q_i\}.$$

For a list of sequences $Q = (q_1, \ldots, q_k)$ we denote

$$plts(Q) = plts(q_1) \cup \cdots \cup plts(q_k).$$

For some sequence $q = (b_1, \ldots, b_n)$ we say bin b_i is *on the left of* bin b_j in sequence q if $i < j$. And we say that such a bin b_i is on the *position* i in sequence q, i.e. there are $i - 1$ bins on the left of b in sequence q. The position of the first bin in some sequence q_i destined for some pallet t is denoted by $first(q_i, t)$, similarly the position of the last bin for pallet t in sequence q_i is denoted by $last(q_i, t)$. For technical reasons, if there is no bin for pallet t contained in sequence q_i, then we define $first(q_i, t) = |q_i| + 1$, and $last(q_i, t) = 0$.

Let $Q = (q_1, \ldots, q_k)$ be a list of sequences, and let $C_Q = (i_1, \ldots, i_k)$ be some tuple in $\mathbb{N}^k$. Such a tuple C_Q is called a *configuration*, if $0 \le i_j \le |q_j|$ for each sequence $q_j \in Q$. Value i_j denotes the number of bins that have been removed from sequence q_j, see Example 1.

A pallet t is called *open* in configuration $C_Q = (i_1, \ldots, i_k)$, if there is a bin for pallet t at some position less than or equal to i_j in sequence q_j, and if there is another bin for pallet t at some position greater than i_l in sequence q_l, see Example 1. In view of the practical background we only consider sequences that contain at least two bins for each pallet.[1]

The *set of open pallets* in configuration C_Q is denoted by $open(C_Q)$, the number of open pallets is denoted by $\#open(C_Q)$.[2] A pallet $t \in plts(Q)$ is called *closed* in configuration C_Q, if $last(q_j, t) \le i_j$ for each sequence $q_j \in Q$. Initially all pallets are *unprocessed*. From the time when the first bin of a pallet t has been removed from a sequence, pallet t is either open or closed.

Let $C_Q = (i_1, \ldots, i_k)$ be a configuration. The removal of the bin on position $i_j + 1$ from sequence q_j is called a *transformation step*. A sequence of transformation steps that transforms the list Q of k sequences from the initial configuration $(0, 0, \ldots, 0)$ into the final configuration $(|q_1|, |q_2|, \ldots, |q_k|)$ is called a *processing* of Q, see Example 1.

---

[1] By the definition of open for some pallet t, using $first(q_i, t)$ and $last(q_i, t)$, we can test in time $\mathcal{O}(k)$ whether t is open within a given configuration.

[2] By performing footnote 1 for every of the m pallets, for some configuration C_Q we can compute $\#open(C_Q)$ in time $\mathcal{O}(m \cdot k)$.

It is often convenient to use pallet identifications instead of bin identifications to represent a sequence q. For n not necessarily distinct pallets $t_1, \ldots, t_n$ let $[t_1, \ldots, t_n]$ denote some sequence of n pairwise distinct bins $(b_1, \ldots, b_n)$, such that $plt(b_i) = t_i$ for $i = 1, \ldots, n$. We use this notion for lists of sequences as well. Furthermore, for some integer value n we will write $[n] := \{1, 2, \ldots, n\}$ to denote the range from 1 to n.

Example 1. Consider the list $Q = (q_1, q_2)$ of two sequences

$$q_1 = (b_1, \ldots, b_4) = [a, b, a, b] \text{ and } q_2 = (b_5, \ldots, b_{10}) = [c, d, c, d, a, b].$$

Table 1 shows a processing of Q with 2 stack-up places. The underlined bin is always the bin that will be removed in the next transformation step. The already removed bins are shown greyed out.

Table 1. A processing of $Q = (q_1, q_2)$ from Example 1 with 2 stack-up places. In this simple example it is easy to see that there is no processing of Q that needs less than 2 stack-up places, because pallets a and b as well as c and d are interlaced.

i	q_1	q_2	bin to remove	C_Q	$open(C_Q)$
0	$[a, b, a, b]$	$[\underline{c}, d, c, d, a, b]$	b_5	$(0, 0)$	$\emptyset$
1	$[a, b, a, b]$	$[c, \underline{d}, c, d, a, b]$	b_6	$(0, 1)$	$\{c\}$
2	$[a, b, a, b]$	$[c, d, \underline{c}, d, a, b]$	b_7	$(0, 2)$	$\{c, d\}$
3	$[a, b, a, b]$	$[c, d, c, \underline{d}, a, b]$	b_8	$(0, 3)$	$\{d\}$
4	$[\underline{a}, b, a, b]$	$[c, d, c, d, a, b]$	b_1	$(0, 4)$	$\emptyset$
5	$[a, b, a, b]$	$[c, d, c, d, \underline{a}, b]$	b_9	$(1, 4)$	$\{a\}$
6	$[a, \underline{b}, a, b]$	$[c, d, c, d, a, b]$	b_2	$(1, 5)$	$\{a\}$
6	$[a, b, a, b]$	$[c, d, c, d, a, \underline{b}]$	b_{10}	$(2, 5)$	$\{a, b\}$
7	$[a, b, \underline{a}, b]$	$[c, d, c, d, a, b]$	b_3	$(2, 6)$	$\{a, b\}$
8	$[a, b, a, \underline{b}]$	$[c, d, c, d, a, b]$	b_4	$(3, 6)$	$\{b\}$
9	$[a, b, a, b]$	$[c, d, c, d, a, b]$	$-$	$(4, 6)$	$\emptyset$

Definition 1. *The* FIFO STACK-UP *problem is defined as follows. Given is a list $Q = (q_1, \ldots, q_k)$ of sequences and a positive integer p. The task is to decide whether there is a processing of Q, such that in each configuration during the processing of Q at most p pallets are open.*

We use the following variables in the analysis of our algorithms: k denotes the number of sequences, and p stands for the number of stack-up places, while m represents the number of pallets in $plts(Q)$, and n denotes the total number of bins. Finally, $N = \max\{|q_1|, \ldots, |q_k|\}$ is the maximum sequence length.

In view of the practical background, it holds $p < m$, otherwise each pallet could be stacked-up onto a different stack-up place. Furthermore, $k < m$, otherwise all bins of one pallet could be channeled into one buffer queue in the original stack-up problem in practice. Finally $m \leq n/2 < n$, since there are at least two bins for each pallet.

3 The Processing Graph

Our aim in controlling FIFO stack-up systems is to compute a processing of the given sequences of bins with a minimum number of stack-up places. Such an optimal processing can always be found by computing the *processing graph*, see [3,4].

The processing graph $G = (V, A)$ contains a vertex for every possible configuration. Each vertex v representing some configuration $C_Q(v)$ is labeled by the number $\#open(C_Q(v))$. There is an arc from vertex u representing configuration $(u_1, \ldots, u_k)$ to vertex v representing configuration $(v_1, \ldots, v_k)$ if and only if $u_i = v_i - 1$ for exactly one element of the vector and for all other elements of the vector $u_j = v_j$. The arc is labeled with the bin that will be removed in the corresponding transformation step. For the sequences of Example 1 we get the processing graph of Fig. 3. Obviously, every processing graph is directed and acyclic.

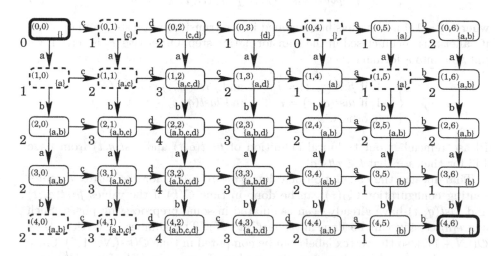

Fig. 3. The processing graph of Example 1. Instead of the bin each arc is labeled with the pallet symbol of the bin that will be removed in the corresponding transformation step. The shaded vertices will be important in the following section.

Consider a processing of a list Q of sequences. Let $B = (b_{\pi(1)}, \ldots, b_{\pi(n)})$ be the order in which the bins are removed during the processing of Q. Then B is called a *bin solution* of Q. In Example 1, we have

$$B = (b_5, b_6, b_7, b_8, b_1, b_9, b_2, b_{10}, b_3, b_4) = [c, d, c, d, a, a, b, b, a, b]$$

as a bin solution. Such a bin solution describes a path from the initial configuration $(0, 0, \ldots, 0)$ to the final configuration $(|q_1|, |q_2|, \ldots, |q_k|)$ in the processing graph. We are interested in such paths where the maximal vertex label on that path is minimal.

The processing graph can be computed in time $\mathcal{O}(k \cdot (N+1)^k)$ by some breadth first search algorithm as follows. We store the already discovered configurations, i.e. the vertices of the graph, in some list L. Initially, list L contains only the initial configuration. In each step of the algorithm we take the first configuration out of list L. Let $C_Q = (i_1, \ldots, i_k)$ be such a configuration. For each $j \in [k]$ we remove the bin on position $i_j + 1$ from sequence q_j, and get another configuration C'_Q. We append C'_Q to list L, add it to vertex set V, and add (C_Q, C'_Q) to edge set A, if it is not already contained.

For each configuration we want to store the number of open pallets of that configuration. This can be done efficiently in the following way. First, since none of the bins has been removed from any sequence in the initial configuration, we have $\#open((0, \ldots, 0)) = 0$. In each transformation step we remove exactly one bin for some pallet t from some sequence q_j, thus

$$\#open(i_1, \ldots, i_{j-1}, i_j + 1, i_{j+1}, \ldots, i_k)$$
$$= \#open(i_1, \ldots, i_{j-1}, i_j, i_{j+1}, \ldots, i_k) + c_j \tag{1}$$

where $c_j = 1$ if pallet t has been opened in the transformation step, and $c_j = -1$ if pallet t has been closed in the transformation step. Otherwise, c_j is zero. If we put this into a formula we get

$$c_j = \begin{cases} 1, & \text{if } first(q_j, t) = i_j + 1 \text{ and } first(q_\ell, t) > i_\ell \ \forall \ \ell \neq j \\ -1, & \text{if } last(q_j, t) = i_j + 1 \text{ and } last(q_\ell, t) \leq i_\ell \ \forall \ \ell \neq j \\ 0, & \text{otherwise.} \end{cases}$$

Please remember our technical definition of $first(q, t)$ and $last(q, t)$ from page 143 for the case that $t \notin plts(q)$.

That means, the calculation of value $\#open(C_Q(v))$ for the vertex v representing configuration $C_Q(v)$ can be done in time $\mathcal{O}(k)$ if the values $first(q_j, t)$ and $last(q_j, t)$ have already been calculated in some preprocessing phase. Such a preprocessing can be done in time $\mathcal{O}(k \cdot N)$. The number of vertices is in $\mathcal{O}((N+1)^k)$, so the vertex labels can be computed in time $\mathcal{O}(k \cdot (N+1)^k)$. Since at most k arcs leave each vertex, the number of arcs is in $\mathcal{O}(k \cdot (N+1)^k)$, and each arc can be computed in time $\mathcal{O}(1)$. Thus, the computing can be done in total time $\mathcal{O}(k \cdot (N+1)^k)$.

Let s be the vertex representing the initial configuration, and let $topol : V \to \mathbb{N}$ be a topological ordering of the vertices of the processing graph $G = (V, A)$ such that $topol(u) < topol(v)$ holds for each $(u, v) \in A$. For some vertex $v \in V$ and some path $P = (v_1, \ldots, v_l)$ with $v_1 = s$, $v_l = v$ and $(v_i, v_{i+1}) \in A$ we define

$$val_P(v) := \max_{u \in P}(\#open(C_Q(u)))$$

to be the maximum vertex label on that path. Let $\mathcal{P}(v)$ denote the set of all paths from vertex s to vertex v. Then we define

$$val(v) := \min_{P \in \mathcal{P}(v)}(val_P(v)).$$

The problem is to compute the value $val(t)$ where t is the vertex representing the final configuration. It holds

$$val(v) = \max\{\#open(C_Q(v)), \min_{(u,v)\in A}(val(u))\}, \tag{2}$$

because each path $P \in \mathcal{P}(v)$ must go through some vertex u with $(u,v) \in A$. So we may iterate over the preceding vertices of v instead of iterating over all paths. If $\#open(C_Q(u)) < \#open(C_Q(v))$ for all preceding configurations then a pallet must have been opened to reach configuration $C_Q(v)$.

Algorithm FIND PATH

$val[s] := 0$ ▷ Computation according to equation 2
for each vertex $v \neq s$ in order of $topol$ do
   $val[v] := \infty$
   for every $(u,v) \in A$ do   ▷ Compute $\min_{(u,v)\in A}(val(u)) =: val(v)$
      if $(val[u] < val[v])$
         $val[v] := val[u]$
         $pred[v] := u$
   if $(val[v] < \#open(v))$ ▷ Compute $\max\{\#open(C_Q(v)), val(v)\}$
      $val[v] := \#open(v)$

Fig. 4. Finding an optimal processing by dynamic programming

The value $val(t)$ can be computed by Algorithm FIND PATH given in Fig. 4. A topological ordering of the vertices of digraph G can be found by a depth first search algorithm in time $\mathcal{O}(|V| + |A|)$. The remaining work of algorithm FIND PATH can also be done in time $\mathcal{O}(|V| + |A|)$. In the processing graph we have $|V| \in \mathcal{O}((N+1)^k)$, and $|A| \in \mathcal{O}(k \cdot (N+1)^k)$.

It is not necessary to explicitly build the processing graph to compute an optimal processing, we have done it just for the sake of clarity and to enhance understanding. We combine the construction of the processing graph with the topological sorting and the path finding by some breadth first search algorithm OPTIMAL BIN SOLUTION shown in Fig. 5. Algorithm OPTIMAL BIN SOLUTION uses the following two operations.

- $head(L)$ yields the first entry of list L and removes it from L.
- $append(e, L)$ adds element e to the list L, if e is not already contained in L.

At the end of the processing of algorithm OPTIMAL BIN SOLUTION the variable $path$ contains the path of our interest described on page 145.

The computation of all at most $(N+1)^k$ values $\#open(C_Q(v))$ can be performed in time $\mathcal{O}(k \cdot (N+1)^k)$. A value is added to some list only if it is not already contained. To check this efficiently in time $\mathcal{O}(1)$ we use a boolean array. Thus, we can compute the minimal number of stack-up places necessary to process the given FIFO STACK-UP problem in time $\mathcal{O}(k \cdot (N+1)^k)$.

Algorithm OPTIMAL BIN SOLUTION

$L := ((0, \ldots, 0))$ ▷ List of uninvestigated configurations
$pred[(0, \ldots, 0)] := \emptyset$ ▷ List of predecessors of some configuration
while L is not empty do
    $C := head(L)$ ▷ let $C = (i_1, \ldots, i_k)$
    $val := \infty$ ▷ all predecessors of C are computed
    for each C_p in list $pred[C]$ do   ▷ Compute $val(C_p)$ due equation 2
        if $(val > \#open(C_p))$
        $val := \#open(C_p)$
        $path[C] := C_p$
        if $(val < \#open(C))$
        $val := \#open(C)$
    for $j := 1$ to k do
        $C_s := (i_1, \ldots, i_j + 1, \ldots, i_k)$
        if $(C_s$ is not in $L)$
        compute $\#open(C_s)$ according to equation 1
        $append(C_s, L)$
        $append(C, pred[C_s])$

Fig. 5. Construction of the processing graph and computation of an optimal bin solution at once by breadth first search

Theorem 1 ([3,4]). *Let $Q = (q_1, \ldots, q_k)$ and $N = \max\{|q_1|, \ldots, |q_k|\}$ the maximum sequence length, then the* FIFO STACK-UP *problem can be solved in time* $\mathcal{O}(k \cdot (N+1)^k)$.

4   The Decision Graph

During a processing of a list Q of sequences there are often configurations for which it is easy to find a bin b that can be removed such that a further processing with p stack-up places is still possible. This is the case, if bin b is destined for an already open pallet, see [3]. A configuration $(i_1, \ldots, i_k)$ is called a *decision configuration*, if the bin on position $i_j + 1$ of sequence q_j for each $j \in [k]$ is destined for a non-open pallet. We can restrict FIFO stack-up algorithms to deal with such decision configurations, in all other configurations the algorithms automatically remove a bin for some already open pallet.

A solution to the FIFO STACK-UP problem can always be found by computing the decision graph for the given instance of the problem. The decision graph $G = (V, A)$ has a vertex for each decision configuration into which the initial configuration can be transformed. There is an arc $(u, v) \in A$ from a vertex u representing decision configuration $(u_1, \ldots, u_k)$ to a vertex v representing decision configuration $(v_1, \ldots, v_k)$ if and only if there is a bin b on position $u_j + 1$ in sequence q_j such that the removal of b in the next transformation step and the execution of only automatic transformation steps afterwards lead to decision configuration $(v_1, \ldots, v_k)$. Arc (u, v) is labeled with the pallet symbol of the bin that will be removed in the corresponding transformation step. In Fig. 6 the decision graph for Example 1 is shown.

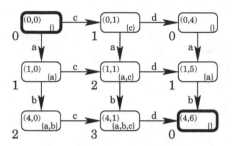

Fig. 6. The decision graph of Example 1. It consists of the shaded vertices from the processing graph in Fig. 5.

Let $T = (t_1, \ldots, t_m)$ be the order in which the pallets are opened during the processing of Q. Then T is called a *pallet solution* of Q. In Example 1 we have $T = (c, d, a, b)$ as a pallet solution. Such a pallet solution describes a path from the initial configuration to the final configuration in the decision graph.

The decision graph can be computed by some breadth first search algorithm as follows. We store the already discovered decision configurations, i.e. the vertices of the graph, in some list L. Initially, list L contains only the initial configuration. In each step of the algorithm we take the first configuration out of list L. Let $C_Q = (i_1, \ldots, i_k)$ be such a decision configuration. For each $j \in [k]$ we remove the bin on position $i_j + 1$ from sequence q_j and execute all automatic transformation steps. By this way we reach another decision configuration $C'_Q = (i'_1, \ldots, i'_k)$ or the final configuration, and we put this configuration into list L, we add C'_Q to vertex set V, and we add (C_Q, C'_Q) to edge set A, if it is not already contained.

We can combine the construction of the decision graph and the path finding by some breadth first search algorithm FIND OPTIMAL PALLET SOLUTION, as shown in Fig. 7.

The running time of algorithm FIND OPTIMAL PALLET SOLUTION can be estimated as follows. In each sequence q_j there are bins for at most m pallets. Each pallet can be opened at most once, so in a decision configuration $(i_1, \ldots, i_k)$ it must hold $i_j + 1 = first(q_j, t)$ for some pallet t. And since $i_j = 0$ and $i_j = |q_j|$ are possible as well, the decision graph has at most $(m + 2)^k$ vertices each representing a decision configuration in list L.

The automatic transformation steps can be performed for every $j \in [k]$ in time $\mathcal{O}(n + k)$ and thus for all decision configurations in L in time $\mathcal{O}((m + 2)^k \cdot k \cdot (n + k))$. The computation of $\#open(C_s)$ for one decision configuration C_s can be done in time $\mathcal{O}(m \cdot k)$ by footnote 2. Since we only compute $\#open(C_s)$ for decision configuration C_s which is not already contained in L, for all decision configurations we need time in $\mathcal{O}((m + 2)^k \cdot m \cdot k))$.

Algorithm FIND OPTIMAL PALLET SOLUTION

$L := ((0, \ldots, 0))$ ▷ List of configurations to expand
$pred[(0, \ldots, 0)] := \emptyset$ ▷ List of predecessors of some decision configuration
while L is not empty do
  $C := head(L)$ ▷ let $C = (i_1, \ldots, i_k)$
  $val := \infty$ ▷ all predecessors of C are computed
  for each C_p in $pred[C]$ do
    if $(val > \#open(C_p))$
      $val := \#open(C_p)$
      $path[C] := C_p$
    if $(val < \#open(C))$
      $val := \#open(C)$
  for $j := 1$ to k do
    $C_s := (i_1, \ldots, i_j + 1, \ldots, i_k)$
    for $i := 1$ to k do   ▷ perform automatic transformation steps
      while the first bin in q_i is destinated for an open pallet
        let C_s be the configuration obtained by removing the first bin of q_i
    if $(C_s$ is not in $L)$
      compute $\#open(C_s)$ according to footnote 2
      $append(C_s, L)$
    $append(C, pred[C_s])$

Fig. 7. Construction of the decision graph and computation of an optimal pallet solution at once by breadth first search

Thus the running time of algorithm FIND OPTIMAL PALLET SOLUTION is in $\mathcal{O}((m+2)^k \cdot k \cdot (n+k) + (m+2)^k \cdot m \cdot k)) \subseteq \mathcal{O}((m+2)^k \cdot n^2)$.

If we have a pallet solution $T = (t_1, \ldots, t_m)$ computed by any FIFO stack-up algorithm, we can convert the pallet solution into a sequence of transformation steps, i.e. a processing of Q, in time $\mathcal{O}(n \cdot k) \subseteq \mathcal{O}(n^2)$ by some simple algorithm [3]: Repeatedly, in the i-th decision configuration choose the pallet t_i, remove a bin for pallet t_i, and execute all automatic transformation steps, until the next decision configuration is reached. If no bin for pallet t_i can be removed in the i-th decision configuration, or if more than p pallets are opened, reject the input. Each transformation step of these at most n steps can be done in time $\mathcal{O}(k)$ by equation 1.

Theorem 2. *The* FIFO STACK-UP *problem can be solved in time* $\mathcal{O}(n^2 \cdot (m+2)^k)$.

Another way to describe a path from the initial configuration to the final configuration in the decision graph is as follows, see [5] for details. Let s_i denote the sequence from which a bin for pallet t_i will be removed after the i-th decision configuration. Then $S = (s_1, \ldots, s_m)$ is called a *sequence solution*. In Example 1 we have $S = (2, 2, 1, 1)$ as a sequence solution.

5   Experimental Results

Next we evaluate an implementation of algorithm FIND OPTIMAL PALLET SOLUTION.

Creating Instances Since there are no benchmark data sets for the FIFO STACK-UP problem we generated randomly instances by an algorithm, which allows to give the following parameters.

- p_{max} the maximum number of stack-up places
- m number of pallets
- k number of sequences
- r_{min} and r_{max} the minimum and maximum number of bins per pallet
- d maximum number of sequences on which the bins of each pallet can be distributed

Implementation and Evaluation For several practical instance sizes the running time of FIND OPTIMAL PALLET SOLUTION is too huge for computations, e.g. $m = 254$ and $k = 10$ lead a time of $\mathcal{O}(n^2 \cdot (m+2)^k) = \mathcal{O}(n^2 \cdot 2^{80})$. Therefor we used a cutting technique on the decision graph $G = (V, A)$ by restricting to vertices $v \in V$ representing configurations $C_Q(v)$ such that $\#open(C_Q(v)) \leq maxp$ and increasing the value of $maxp$ by 5 until a solution is found. We have implemented this algorithm in C++ on a standard Linux PC with 3.30 GHz CPU and 7.7 GiB RAM. In Tables 2-4 we list some of our chosen parameters. For each assignment we randomly generated and solved 10 instances to compute the average time for solving an instance with the given parameters. Our results show that we can solve practical instances on several thousand bins in a few minutes.

Table 2. Running times in seconds for $p_{max} = 14$, $m = 100$, $k = 8$, and $r_{min} = 10$

	r_{max}	d	time		r_{max}	d	time		r_{max}	d	time
1.	20	6	0.091	4.	30	6	0.072	7.	40	6	0.080
2.	20	10	0.061	5.	30	10	0.061	8.	40	10	0.081
3.	20	14	0.063	6.	30	14	0.054	9.	40	14	0.043

Table 3. Running times in seconds for $p_{max} = 18$, $m = 300$, $k = 10$, and $r_{min} = 15$

	r_{max}	d	time		r_{max}	d	time		r_{max}	d	time
10.	25	6	3.568	13.	35	6	2.271	16.	45	6	3.241
11.	25	10	1.652	14.	35	10	2.074	17.	45	10	1.551
12.	25	14	0.972	15.	35	14	1.729	18.	45	14	1.171

Table 4. Running times in seconds for $p_{max} = 22$, $m = 500$, $k = 12$, and $r_{min} = 20$

	r_{max}	d	time		r_{max}	d	time		r_{max}	d	time
19.	30	6	96.182	22.	40	6	83.627	25.	50	6	93.384
20.	30	10	33.340	23.	40	10	29.071	26.	50	10	39.371
21.	30	14	31.678	24.	40	14	36.943	27.	50	14	26.623

6   Conclusions and Outlook

In this paper we have considered the processing graph and the decision graph as two useful digraph models for the FIFO STACK-UP problem. A breadth first search solution and some cutting technique on the decision graph was shown to be useful to solve the problem for several practical instances.

From the point of parameterized complexity [2] our solution is a parameterized algorithm with respect to parameter m^k. In practice k is much smaller than m, since there are much fewer buffer conveyors than pallets, thus the running time of this solution is better than the sequence solution method $\mathcal{O}(n^2 \cdot k^m)$, which is shown in [5].

We are also interested in online algorithms [1] for instances where we only know the first c bins of every sequence instead of the complete sequences. Especially, we are interested in the answer to the following question: Is there a d-competitive online algorithm? Such an algorithm must compute a processing of some Q with at most $p \cdot d$ stack-up places, if Q can be processed with at most p stack-up places.

References

1. Borodin, A.: On-line Computation and Competitive Analysis. Cambridge University Press (1998)
2. Flum, J., Grohe, M.: Parameterized Complexity Theory. Springer, Berlin (2006)
3. Gurski, F., Rethmann, J., Wanke, E.: Complexity of the fifo stack-up problem. ACM Computing Research Repository (CoRR) abs/1307.1915 (2013)
4. Gurski, F., Rethmann, J., Wanke, E.: Moving bins from conveyor belts onto pallets using fifo queues. In: Proceedings of the International Conference on Operations Research (OR 2013), Selected Papers, pp. 185–191. Springer (2014)
5. Gurski, F., Rethmann, J., Wanke, E.: Algorithms for controlling palletizers. In: Proceedings of the International Conference on Operations Research (OR 2014), Selected Papers. Springer (to appear, 2015)
6. de Koster, R.: Performance approximation of pick-to-belt orderpicking systems. European Journal of Operational Research 92, 558–573 (1994)
7. Rethmann, J., Wanke, E.: Storage controlled pile-up systems, theoretical foundations. European Journal of Operational Research 103(3), 515–530 (1997)
8. Rethmann, J., Wanke, E.: On approximation algorithms for the stack-up problem. Mathematical Methods of Operations Research 51, 203–233 (2000)
9. Rethmann, J., Wanke, E.: Stack-up algorithms for palletizing at delivery industry. European Journal of Operational Research 128(1), 74–97 (2001)

Adaptive Memory Algorithm with the Covering Recombination Operator

Nicolas Zufferey

Geneva School of Economics and Management, GSEM - University of Geneva
Blvd du Pont-d'Arve 40, 1211 Geneva 4, Switzerland
n.zufferey@unige.ch

Abstract. The adaptive memory algorithm (AMA) is a population-based metaheuristics initially developed in 1995 by Rochat and Taillard. AMA relies on a central memory M and consists in three steps: generate a new solution s from M with a recombination operator, improve s with a local search operator, and use s to update M with a memory update operator. In 1999, Galinier and Hao proposed the GPX recombination operator for the graph coloring problem. In this paper, AMC, a general type of evolutionary algorithm, is formalized and called *Adaptive Memory Algorithm with the Covering Recombination Operator*. It relies on a specific formulation of the considered problem and on a generalization of the GPX recombination operator. It will be showed that AMC has obtained promising results in various domains, such as graph coloring, satellite range scheduling and project management.

Keywords: Evolutionary Algorithms, Population-Based Metaheuristics.

1 Introduction

As exposed in [12], modern methods for solving complex optimization problems are often divided into exact methods and metaheuristics. An exact method [9] guarantees that an optimal solution is obtained in a finite amount of time. However, for most real-life optimization problems, which are typically NP-hard, such methods need a prohibitive amount of time to find an optimal solution. For these difficult problems, it is preferable to quickly find a satisfying solution. If solution quality is not a dominant concern, a heuristic can be employed, but if quality is crucial, a more advanced metaheuristic procedure is warranted. There are mainly two classes of metaheuristics: local search and population based methods. The former type of algorithm works on a single solution (e.g., descent local search, tabu search), whereas the latter makes a population of (pieces of) solutions evolve (e.g., genetic algorithms, ant colonies, adaptive memory algorithms). The reader interested in a recent book on metaheuristics is referred to [4].

In Section 2, a generalized version of the GPX recombination operator (initially developed in [2] and further extended in [7,13]) is proposed within an adaptive memory algorithm (AMA) framework (initially designed in [11]). The

© Springer International Publishing Switzerland 2015
H.A. Le Thi et al. (eds.), *Model. Comput. & Optim. in Inf. Syst. & Manage. Sci.*,
Advances in Intelligent Systems and Computing 360, DOI: 10.1007/978-3-319-18167-7_14

resulting method is denoted AMC. Then, the success of some existing meta-heuristics, which can however be considered as belonging to the AMC method-ology, are discussed with a unified terminology for three NP-hard problems: the graph coloring problem GCP (Section 3 relying on [3]), a satellite range schedul-ing problem SAT (Section 4 relying on [13]), and a project management problem PM (Section 5 relying on [14]). The reader is referred to the above three refer-ences to have detailed information on the complexity issues, the literature review, and the presentation of the experiments. For each considered problem, the AMC approach is always compared with state-of-the-art methods. The main numerical results are highlighted, which allows to conclude on the excellent performance of AMC. A conclusion is provided in Section 6.

2   Adaptive Memory Algorithm

In this section, the general framework of local search techniques, the descent local search, tabu search, AMA and AMC are formally presented.

Let f be an objective function to minimize. At each step of a *local search*, a *neighbor* solution s' is generated from the *current* solution s by performing a specific modification on s, called a *move*. The set of all the neighbor solutions of s is denoted $N(s)$. First, a local search needs an initial solution s_0 as input. Then, it generates a sequence of solutions $s_1, s_2, \ldots$ in the search space such that $s_{r+1} \in N(s_r)$. The process is stopped for example when a time limit is reached. Famous local search algorithms are: the descent local search (where at each step, the best move is performed and the process stops when a local optimum is reached), variable neighborhood search and tabu search. To escape from a local optimum, in *tabu search*, when a move is performed from a current solution s_r to a neighbor solution s_{r+1}, it is forbidden to perform the reverse of that move during tab (parameter) iterations. Such forbidden moves are called *tabu* moves. Formally, at each iteration, the best non tabu move is performed.

As presented in [3], evolutionary metaheuristics encompass various algorithms such as genetic algorithms, ant systems and AMAs. They are iterative procedures that use a *central memory* where information is collected during the search process. Each iteration, called *generation*, is made of a *cooperation* phase (a *recombination operator* is used to create new offspring solutions) and a *self-adaptation* phase (the new offspring solutions are modified individually). The output solutions of the self-adaptation phase are used to update the central memory. The method stops for example when a time limit is reached. The most successful evolutionary metaheuristics are *hybrid* algorithms in the sense that a local search is used during the self-adaptation phase.

As exposed in [14], a basic version of the AMA (originally proposed in [11]) is summarized in Algorithm 1, where performing steps (1), (2) and (3) is called a *generation*. Therefore, in order to design an AMA for a problem (P), one has mainly to define: the way to initialize the population M of solutions, the recom-bination operator, the local search operator, and the memory update operator.

The considered problem (P) is supposed to have the following structure. Let $S = \{s_1, s_2, \ldots, s_n\}$ be the set of n decision variables, for which the possible

---

Algorithm 1. Adaptive Memory Algorithm (AMA)

Initialize the central memory M with solutions.

While a stopping condition is not met, **do**:

1. create an offspring solution s from M by using a *recombination operator*;
2. improve s with a *local search operator*;
3. use s to update M with a *memory update operator*.

---

values are in the set $D = \{1, 2, \ldots, k\}$. Thus, any solution s can be represented as $s = \{C_1, C_2, \ldots, C_k\}$, where C_i is the set (called the *class*) of decision variables with value i. Let $\mid C \mid$ (resp. $\mid s \mid$) be the number of decision variables belonging to class C (resp. solution $\mid s \mid$). In addition, $s \cup C$ denotes the set of decision variables belonging either to s or to C (or both), which means that a variable belonging to s and C only appears one time in $s \cup C$. Finally, $C - s$ denotes the decision variables belonging to C but not to s.

AMC is a variant of AMA which uses the covering recombination operator (CRO) to generate an offspring solution s from the memory M. In CRO, the offspring solution $s = (C_1, \ldots, C_k)$ is built class by class as follows. Assume that classes $C_1, \ldots, C_{i-1}$ are already built. In order to create C_i at step i, CRO first selects a parent solution $s^{(p)}$ in M. Such a selection can be performed at random, or with a probability which depends on the quality of the solutions of M. The class C_i of s is chosen as the class C of $s^{(p)}$ which contains the maximum number of decision variables which are not already in s. Then, solution s is augmented by setting $C_i = C - s$ (in addition, C_i might be optionally improved with a procedure depending on the problem, see procedure STABLE in Section 3). It can be observed that a *maximum covering* of the decision variables set is performed at each step of CRO. In order to avoid the offspring solution s to resemble too much to one of the solutions of M, a same solution of M cannot provide a class to s if it has already provided one during of the previous d (parameter) steps. To satisfy these restrictions, a list T is dynamically managed and contains all the solutions which cannot be selected to provide a class to s at step i, $\forall i$.

When k steps as above have been performed, let U be the set of variables without any value (i.e. $U = S - s$). It is very likely that $U \neq \emptyset$ at the end of CRO. One can even conjecture that the larger is d, the larger is U and the more different is s from any solution of M. At the end of CRO, these unassigned variables can be randomly or greedily assigned to a value in D, so that the resulting solution s is complete. Based on the above description, CRO can be formalized as proposed in Algorithm 2. In contrast with the recombination operators presented in [2,7,13], CRO does not necessarily use a fixed number of parents, and it is formulated for a class of problems (i.e. not only for the graph coloring problem).

In summary, AMC is characterized by the following features: (1) it is designed for problems having a specific structure, where each decision variable has a domain value; (2) it relies on a central memory M (containing solutions or pieces

---

Algorithm 2. Covering Recombination Operator (CRO)

Initialization: set $s = \emptyset$ and $T = \emptyset$.

For $i = 1$ **to** k, **do:**

1. select a parent solution $s^{(p)}$ in M which does not belong to T;
2. select the class C of $s^{(p)}$ such that $\mid s \cup C \mid$ is maximized;
3. remove from C the decision variables which are already in s;
4. set $C_i = C$ (and optionally improve C_i with a problem-dependent procedure);
5. update T: add $s^{(p)}$ to T, and remove its oldest element if $\mid T \mid > d$;

Sequentially assign a value in D to each decision variable of $U = S - s$.

---

of solutions) which is updated at the end of each generation (based on quality or diversity measures); (3) each offspring solution is generated with a specialized recombination operator, based on a maximum covering paradigm; (4) it requires an efficient local search operator to improve each constructed offspring solution.

3 *AMC* for the Graph Coloring Problem

3.1 Description of Problem GCP

Given a graph $G = (V, E)$ with vertex set $V = \{1, 2, \ldots, n\}$ and edge set E, GCP consists in assigning an integer (called color) in $\{1, 2, 3, \ldots, n\}$ to every vertex such that two adjacent vertices have different colors, while minimizing the number of used colors. The k-coloring problem (k-GCP) consists in assigning a color in $D = \{1, \ldots, k\}$ to every vertex such that two adjacent vertices have different colors. Thus, the GCP consists in finding a k-coloring with the smallest possible k. Note that one can tackle the GCP by solving a series of k-GCP's, starting with a large value of k (which is at most n) and decreasing k by one unit each time a k-coloring is found. Many heuristics were proposed to solve the GCP and the k-GCP. For a recent survey, the reader is referred to [8].

3.2 Structure and Initialization of the Central Memory

A *stable* is a set of vertices which are pairwise non adjacent. The central memory M contains stable sets of the graph. The number of elements in M is a multiple of k: $\mid M \mid = m \cdot k$, where $m = 10$ is a parameter of the method. In other words, M roughly contains 10 solutions.

A k-coloring s can be denoted $s = (C_1, \ldots, C_k)$, where C_i is the set of vertices with color i. To initialize M, m k-colorings are randomly built, and then improved with the local search operator (presented in Subsection 3.4). The color classes of the resulting solutions are then transformed into maximal stable sets by the use of procedure STABLE, and all these stable sets are finally introduced into M. Procedure STABLE works as follows: (1) transform the given color class

C into a stable set by sequentially removing conflicting vertices (two vertices are conflicting if they are adjacent and have the same color); (2) greedily insert vertices into C (one by one) until it becomes a maximal stable set (i.e. no other vertex can be added without creating a conflict).

3.3   Recombination Operator

The offspring solution $s = (C_1, \ldots, C_k)$ is built class by class. Assume that color classes $C_1, \ldots, C_{i-1}$ are already built. In order to create C_i, the operator first selects at random a sample $\{W_1, \ldots, W_k\}$ of k stable sets in M (which roughly corresponds to a solution). Then, the set W_r in this sample with a maximum number of uncolored vertices is chosen, and C_i is set equal to the set of uncolored vertices in W_r. Once the k color classes are built, uncolored vertices may exist: each uncolored vertex is inserted in a set C_i chosen at random. The local search operator will then take care of reducing the created conflicts (if there is no conflict, the algorithm stops).

3.4   Local Search Operator

A tabu search algorithm *Tabucol* is used for I (parameter) iterations. The search space is the set of k-colorings and the objective function f to minimize is the number of conflicts. A neighbor solution is obtained by modifying the color of a conflicting vertex. When the color of a vertex x is modified from i to j, color i is declared tabu for x for *tab* (parameter) iterations (called *tabu tenure*), and all solutions where x has color i are called *tabu solutions*. At each iteration, *Tabucol* determines the best neighbor s of the current solution s (ties are broken randomly) such that either s is a non-tabu solution, or $f(s) = 0$ (i.e., s is a legal k-coloring). The tabu tenure is set equal to $R(0, 9) + 0.6 \cdot NCV(s)$, where $R(a, b)$ returns an integer randomly chosen in $\{a, \ldots, b\}$ and $NCV(s)$ represents the number of conflicting vertices in s.

3.5   Memory Update Operator

First, each class of the solution s provided by *Tabucol* is transformed into a maximal stable set using procedure STABLE. Then, the resulting k maximal stable sets are introduced into M in replacement of k randomly chosen classes.

3.6   Results

In order to show the importance of *CRO*, *AMC* is compared to *Tabucol* with multiple restarts and different stopping conditions. More precisely, *Tabucol* is restarted as follows: 50 times with $I = 100,000$, 30 times with $I = 1.5$ millions, 10 times with $I = 10$ millions, 5 times with $I = 20$ millions. This gives different values for I and a total number of 250 millions of iterations, as it is the case for *AMC*, which ensures a fair comparison and an relevant experience to measure the benefit of *CRO*.

Table 1 reports the results on challenging benchmark graphs. The first columns indicate the name, the number n of vertices and the *density* of the considered graph G (i.e. the average number of edge between a pair of vertices). The two following columns give the *chromatic number* $\chi(G)$ (i.e. the optimal number of colors, if known) of the graph and the smallest k for which an algorithm reported in the literature has been able to find a legal k-coloring of the graph (usually obtained with an evolutionary metaheuristics relying on a kind of *CRO* recombination operator, as in [7]). The next columns, labeled *Tabucol* and *AMC*, report the best number of colors obtained with these methods. *AMC* clearly outperforms *Tabucol*, which demonstrates that *CRO* significantly contributes to the success of *AMC*. In addition, the quality of the solutions found by *AMC* is competitive with optimal or best known solutions.

Table 1. Results on the GCP instances

Graph G	n	density	$\chi(G)$	Best k	*Tabucol*	*AMC*
DSJC250.5	250	0.5	-	28	29	28
DSJC250.9	250	0.89	-	72	72	72
DSJC500.1	500	0.1	-	12	13	12
DSJC500.5	500	0.5	-	48	50	48
DSJC500.9	500	0.9	-	126	130	126
DSJR500.1c	500	0.97	-	85	86	85
DSJR500.5	500	0.47	-	122	128	125
DSJC1000.1	1000	0.1	-	20	22	20
DSJC1000.5	1000	0.5	-	83	89	84
DSJC1000.9	1000	0.9	-	223	245	224
latin_square_10	900	0.76	98	98	106	104
le450_15c	450	0.16	15	15	16	15
le450_15d	450	0.17	15	15	16	15
le450_25c	450	0.17	25	25	26	26
le450_25d	450	0.17	25	25	27	26
flat300_26_0	300	0.48	26	26	27	26
flat300_28_0	300	0.48	28	28	31	31
flat1000_50_0	1000	0.49	50	50	92	50
flat1000_60_0	1000	0.49	60	60	93	60
flat1000_76_0	1000	0.49	76	82	88	84

4   *AMC* for Satellite Range Scheduling

4.1   Description of Problem SAT

Consider a set of satellites and a set $\{S_1, S_2, \ldots, S_k\}$ of ground stations. Ground stations are communication resources (e.g. antennae). Several operations must be performed on spacecrafts, related to satellite control or payload. These operations require ground-to-space communications, called jobs. Therefore, a job is associated with some information representing the corresponding on-board operation. In SAT, a set $J = \{1, \ldots, n\}$ of jobs have to be scheduled. Each job

j is characterized by the following parameters: the (unique) satellite sat_j requested by j; the set M_j of ground stations able to process j; the duration p_j of communication j; the time r_j at which j becomes available for processing; the time d_j by which j must be completed. An important application is the one encountered by the *U.S. Air Force Satellite Control Network*. In such an context, customers request an antenna at a ground station for a specific time window along with possible alternative slots. The problem is in general oversubscribed, i.e. all the jobs cannot be performed. The goal is to schedule as many jobs as possible within its time window, such that the processing of two jobs cannot overlap on the same resource (unit capacity): in such a case, there is a conflict and one of the conflicting jobs has to be removed from the schedule. Minimizing the number of conflicting jobs (more precisely, the number of bumped jobs) is of crucial importance in a practical standpoint, because human schedulers do not consider any conflicting job worse than any other conflicting job [10]. The human schedulers themselves state that minimizing the number of conflicts reduces (1) their workload, (2) communication with outside agencies, and (3) the time required to produce a conflict-free schedule [1].

4.2   Structure and Initialization of the Central Memory

A solution is denoted $s = (C_1, \ldots, C_k; OUT)$, where C_t is the sequence of jobs scheduled on resource i (for each scheduled job, its starting time is also determined), and OUT is the set of unscheduled jobs (which has to be minimized). In order to initialize M, $m = 10$ solutions are randomly generated and improved with a tabu search TS (see Subsection 4.4) during $I = 1000$ iterations.

4.3   Recombination Operator

At each generation, an offspring solution $s = \{C_1, \ldots, C_k\}$ is built resource by resource from M. Suppose that sets $C_1, \ldots, C_{t-1}$ are already built from parent solutions, and that parent solution $s_{r'}$ (with r' in $\{1, \ldots, 10\}$) provided the set C_{t-1} to s. In addition, let A be the set of already scheduled jobs (i.e. the jobs in $C_1, \ldots, C_{t-1}$). At that moment, the next non considered time period t to deal with is chosen. Then, the set C_t is provided by the solution $s_r = \{C_1^{(r)}, \ldots, C_k^{(r)}\}$ in M (with $r \neq r'$, thus the same solution of M cannot consecutively provide two sets of the offspring solution, which means that parameter d of CRO is set to 1) such that $|C_t^{(r)} - A|$ is maximum (break ties randomly). Thus, $C_t = C_t^{(r)} - A$. At the end of such a process, each non scheduled job (considered in a random order) is scheduled in a greedy fashion.

4.4   Local Search Operator

A tabu search approach TS is used. In order to generate a neighbor solution, a job j is moved from OUT to an admissible resource C_i (within its time window and shifting the starting time of the other jobs scheduled on C_i if necessary), then, jobs in C_i which are in conflict with j (i.e. their executions overlap with the execution of j) are moved to OUT. Then, all the moves which will put j back in OUT are tabu for $tab(j)$ (parameter) iterations, where $tab(j) = 0.6 \cdot n_c + R(0,9)$, where n_c is the number of jobs in OUT in the current solution.

In order to perform a significant number of generations in AMC, TS has to be performed for a short time at each generation. This will balance the roles associated with the recombination operator and the intensification operator (tabu search). For this reason, $I = 100,000$ is used, which corresponds to a few seconds of CPU time on the used computer.

4.5   Memory Update Operator

Let s be the solution provided by TS at each generation. Let s_{worst} (resp. s_{old}) be the worst (resp. oldest) solution of M. If s is better than or equal to s_{worst}, then replace s_{worst} with s in M and update s_{worst}. Otherwise, replace s_{old} with s in M and update s_{old}. In the latter case, even if s is not able to improve the average quality of M, it is at least able to bring "fresh blood" in M.

4.6   Results

The three following metaheuristics are compared: TS (enhanced with diversification procedures), AMC (which uses TS but without diversification procedures) and GENITOR [1]. The latter method is a genetic algorithm which usually provides the best solutions for SAT up to 2007. GENITOR is based on the permutation search space, where as proposed by several researchers [1,5], a solution is encoded as a permutation π of the n jobs to schedule. Let $S^{(\pi)}$ be the search space containing all the possible permutations. From a permutation π in $S^{(\pi)}$, it is possible to generate a feasible schedule by the use of a schedule builder, which is a greedy constructive heuristic.

The most difficult benchmark instances are of size $n = 500$ with $k = 9$ resources [13]. A maximum time limit of 15 minutes on a Pentium 4 (2 GHZ, 512 MB of RAM) is considered in Table 2, which is consistent with the ones used in [1], and is appropriate in practice. In Table 2, the average numbers of bumped jobs are compared for GENITOR, TS and AMC. For each method and each instance, 30 runs were performed. It can be observed that on average, AMC slightly outperforms TS, which significantly outperforms GENITOR.

5 AMC for a Project Management Problem

5.1   Description of Problem PM

Consider a project which consists of a set V of n jobs to be performed. The project manager provides a target number k of time periods within which the project has to be performed. It is assumed that: (a) each time period has the same duration, typically a working day; (b) there is no precedence constraint

Table 2. Results on the SAT instances

Instance	GENITOR	TS	AMC	Instance	GENITOR	TS	AMC
1	115.75	55.75	55.5	11	106	51.75	52
2	101.5	54	54	12	106.25	46.5	46.75
3	125.75	65.75	63.5	13	111	56.75	56.75
4	117	49.5	50.25	14	112	58	57.5
5	113.75	52.25	51.75	15	108.75	42.5	42.75
6	121	59	58.75	16	119.75	58	58.25
7	120.5	53.25	52.5	17	107.25	50.25	50.5
8	103.5	49.5	48.75	18	112.25	53.25	52.5
9	97	38.25	38.5	19	100.25	47.5	47
10	104	50.75	51.5	20	96.75	41	41

between jobs; (c) all the jobs can have different durations; (d) each job has a duration of at most one time period. Let $c(j, j') \geq 0$ denote an *incompatibility* cost between jobs j and j', which is to be paid if both jobs are performed at the same time period. Further, for each job j, let $N(j)$ denote the set of jobs j' such that $c(j, j') > 0$. It is assumed that $c(j, j') = c(j', j)$ for all $j, j' \in V$. Hence, $j \in N(j')$ implies $j' \in N(j)$ for all $j, j' \in V$. The incompatibility costs $c(j, j')$ represent for example that the same staff has to perform jobs j and j', thus additional human resources must be hired in order to be able to perform both jobs at the same period. An *incompatibility graph* can model the incompatibility costs: a vertex represents a job and an edge indicates a positive conflicting cost. In addition, for each job j and each time period t, the cost $a(j, t)$ has to be paid if job j is performed at time period t. Such a cost is referred to as an *assignment* cost. The assignment cost $a(j, t)$ represents for example the cost of the staff and machines which have to perform job j at period t. The goal is to assign each job $j \in V$ to a time period $t \in D = \{1, \ldots, k\}$ while minimizing the total costs. This problem is new and there is no literature on it, except [14].

A solution using k periods can be generated by the use of a function *per* : $V \longrightarrow \{1, \ldots, k\}$. The value $per(j)$ of a job j is the time period assigned to job j. With each time period t can be associated a set C_t that contains the set of jobs which are performed at time period t. Thus, a solution s can be denoted $s = (C_1, \ldots, C_k)$, and the associated encountered costs are described in Equation (1). Problem PM consists in finding a solution with k time periods which minimizes these costs.

$$f(s) = \sum_{t=1}^{k} \sum_{j \in C_t} a(j, t) + \sum_{j=1}^{n-1} \sum_{j' \in \{j+1, \ldots, n\} \cap C_{per(j)}} c(j, j') \qquad (1)$$

5.2 Structure and Initialization of the Central Memory

The central memory M is of size $m = 10$ as in several AMA or genetic hybrid algorithms [2,3,13]. In order to initialize M, 10 solutions are randomly generated (i.e. assign a random value between 1 and k to each job j) and then improved with a tabu search TS (see Subsection 5.3) during $I = 1000$ iterations.

5.3   Local Search Operator

TS is applied for I (parameter tuned to 10,000) iterations. The search space is the set of k-partitions of V and the objective function to minimize is the total cost f. A move consists in changing the period assigned to a job. In order to avoid testing every possible move at each iteration, only the $q = 40\%$ most costly jobs are considered for a move at each iteration. If job j moves from C_t to $C_{t'}$ when going from the current solution s to the neighbor solution s', it is forbidden to put j back in C_t during $tab(j, C_t)$ iterations as described in Equation (2), where (u, v, α) are parameters tuned to $(10, 20, 15)$. The maximum is used to enforce $tab(j, C_t)$ to be positive. The last term of Equation (2) represents the improvement of the objective function when moving from s to s'. If s' is better than s, the improvement is positive and the reverse of the performed move will be forbidden for a larger number of iterations than if the improvement is negative.

$$tab(j, C_t) = \max \left\{ 1, R(u, v) + \alpha \cdot \frac{f(s) - f(s')}{f(s)} \right\} \qquad (2)$$

5.4   Recombination Operator and Memory Update Operator

There are exactly the same as the ones proposed in Subsections 4.3 and 4.5.

5.5   Results

The stopping condition for the compared methods is a time limit of one hour on an Intel Pentium 4 (4.00 GHz, RAM 1024 Mo DDR2). The following methods GR, DLS and TS are compared. Note that GR and DLS are restarted as long as one hour is not reached, and the best encountered solution is returned.

- GR: a greedy constructive algorithm working as follows. Let J be the set of scheduled jobs. Start with an empty solution s (i.e. $J = \emptyset$). Then, while s does not contain n jobs, do: (1) randomly select an unscheduled job j; (2) a time period $t \in \{1, \ldots, k\}$ is assigned to j such that the augmentation of the costs is as small as possible.
- TS: as described in Subsection 5.3.
- DLS: a descent local search derived from TS by setting $q = 100\%$ (i.e. considering all the possible moves at each iteration), without tabu tenures.

For a fixed value of k, Table 3 reports the average results (over 10 runs) obtained with the above methods. Let $f^{(TS)}$ be the average value of the solution returned by TS (rounded to the nearest integer) over the considered number of runs. $f^{(AMC)}$, $f^{(DLS)}$ and $f^{(GR)}$ are similarly defined. From left to right, the columns indicate: the name of the conflicting graph G, its number n of jobs, its density, the considered value of k, $f^{(TS)}$, and the percentage gap between $f^{(GR)}$ (resp. $f^{(DLS)}$ and $f^{(AMC)}$) and $f^{(TS)}$. Average gaps are indicated in the last line. The following observations can be made: (1) TS significantly outperforms GR; (2) TS is better than DLS on all but two instances; (3) AMC is the best method on each instance, which highlights the benefit to combine CRO and tabu search within an AMA framework.

Table 3. Results on the PM instances

Graph G	n	density	k	$f^{(TS)}$	Gap(GR)	Gap(DLS)	Gap(AMC)
DSJC1000.1	1000	0.1	13	241601	30.34%	6.51%	-12.61%
DSJC1000.5	1000	0.5	55	250977	25.03%	11.08%	-6.47%
DSJC1000.9	1000	0.9	149	166102	9.27%	10.31%	-5.87%
DSJC500.5	500	0.5	32	98102	42.48%	23.85%	-4.88%
DSJC500.9	500	0.9	84	64224	28.24%	27.37%	-8.35%
flat1000_50_0	1000	0.49	33	665449	21.84%	8.04%	-3.31%
flat1000_60_0	1000	0.49	40	462612	24.95%	10.90%	-4.00%
flat1000_76_0	1000	0.49	55	246157	24.39%	11.29%	-7.68%
flat300_28_0	300	0.48	19	62862	48.86%	27.19%	-0.07%
le450_15c	450	0.16	10	149041	22.40%	4.74%	-10.55%
le450_15d	450	0.17	10	146696	24.94%	7.11%	-8.02%
le450_25c	450	0.17	17	72974	4.87%	-4.78%	-9.12%
le450_25d	450	0.17	17	70852	7.98%	-2.56%	-7.17%
Average					24.28%	10.85%	-6.78%

6 Conclusion

In this paper, AMC is investigated, which is an adaptive memory algorithm using a recombination operator based on the maximum covering paradigm. It was showed that AMC was efficiently adapted to various optimization problems such as graph coloring, project management and satellite range scheduling. In addition to its effectiveness, it could be observed that AMC has a good performance according to other criteria such as speed, robustness, ease of adaptation, and ability to take advantage of the problem structure. Among the future works, one can mention new experiments with the most general version of the CRO recombination operator (e.g., by working on the optional step of Algorithm 2, or by applying AMC to another suitable problem like [6]).

References

1. Barbulescu, L., Watson, J.-P., Whitley, L.D., Howe, A.E.: Scheduling space-ground communications for the air force satellite control network. Journal of Scheduling 7(1), 7–34 (2004)
2. Galinier, P., Hao, J.K.: Hybrid evolutionary algorithms for graph coloring. Journal of Combinatorial Optimization 3(4), 379–397 (1999)
3. Galinier, P., Hertz, A., Zufferey, N.: An adaptive memory algorithm for the graph coloring problem. Discrete Applied Mathematics 156, 267–279 (2008)
4. Gendreau, M., Potvin, J.-Y.: Handbook of Metaheuristics. International Series in Operations Research & Management Science, vol. 146. Springer (2010)
5. Globus, A., Crawford, J., Lohn, J., Pryor, A.: A comparison of techniques for scheduling earth observing satellites. In: Proceedings of the Sixteenth Innovative Applications of Artificial Intelligence Conference (IAAI 2004), San Jose (2004)
6. Hertz, A., Schindl, D., Zufferey, N.: A solution method for a car fleet management problem with maintenance constraints. Journal of Heuristics 15(5), 425–450 (2009)

7. Lu, Z., Hao, J.-K.: A memetic algorithm for graph coloring. European Journal of Operational Research 203, 241–250 (2010)
8. Malaguti, E., Toth, P.: A survey on vertex coloring problems. International Transactions in Operational Research 17(1), 1–34 (2010)
9. Nemhauser, G., Wolsey, L.: Integer and Combinatorial Optimization. John Wiley & Sons (1988)
10. Parish, D.A.: A genetic algorithm approach to automating satellite range scheduling. Master's thesis, Air Force Institute of Technology, USA (1994)
11. Rochat, Y., Taillard, E.: Probabilistic diversification and intensification in local search for vehicle routing. Journal of Heuristics 1, 147–167 (1995)
12. Zufferey, N.: Metaheuristics: some Principles for an Efficient Design. Computer Technology and Applications 3(6), 446–462 (2012)
13. Zufferey, N., Amstutz, P., Giaccari, P.: Graph colouring approaches for a satellite range scheduling problem. Journal of Scheduling 11(4), 263–277 (2008)
14. Zufferey, N., Labarthe, O., Schindl, D.: Heuristics for a project management problem with incompatibility and assignment costs. Computational Optimization and Applications 51, 1231–1252 (2012)

Approximate Counting with Deterministic Guarantees for Affinity Computation

Clément Viricel[1,2], David Simoncini[1], David Allouche[1], Simon de Givry[1],
Sophie Barbe[2], and Thomas Schiex[1]

[1] Unité de Mathématiques et Informatique Appliquées UR 875, INRA, F-31320
Castanet Tolosan, France
[2] Laboratoire d'Ingénierie des Systèmes Biologiques et des Procédés, INSA, UMR
INRA 792/CNRS 5504, F-31400 Toulouse, France

Abstract. Computational Protein Design aims at rationally designing amino-acid sequences that fold into a given three-dimensional structure and that will bestow the designed protein with desirable properties/functions. Usual criteria for design include stability of the designed protein and affinity between it and a ligand of interest. However, estimating the affinity between two molecules requires to compute the partition function, a #P-complete problem.

Because of its extreme computational cost, bio-physicists have designed the K^* algorithm, which combines Best-First A^* search with dominance analysis to provide an estimate of the partition function with deterministic guarantees of quality. In this paper, we show that it is possible to speed up search and keep reasonable memory requirement using a Cost Function Network approach combining Depth First Search with arc consistency based lower bounds. We describe our algorithm and compare our first results to the CPD-dedicated software *Osprey 2.0*.

Keywords: computational protein design, protein-ligand affinity, constraint Programming, cost function network, soft arc consistency, counting problems, weighted #CSP.

Introduction

Proteins are polymer chains composed of amino-acids. Natural evolutionary processes have fashioned by means of amino-acid sequence variations (mutations, recombinations and duplications) an array of proteins endowed with functions ranging from catalysis, signaling to recognition and repair [1]. These functions are made possible by the ability of proteins to self-assemble into well-defined three-dimensional (3D) structures specified by their amino-acid sequences and hence, to interact with various types of molecular partners with high affinity and selectivity. Despite a plethora of functionalities, there is still an ever-increasing demand for proteins endowed with specific properties/functions which are not known to exist in nature. To this end, protein engineering has become a key technology to generate the proteins with the targeted properties/functions [2].

© Springer International Publishing Switzerland 2015   165
H.A. Le Thi et al. (eds.), *Model. Comput. & Optim. in Inf. Syst. & Manage. Sci.*,
Advances in Intelligent Systems and Computing 360, DOI: 10.1007/978-3-319-18167-7_15

However, despite significant advances, protein engineering still suffers from a major drawback related to the limited diversity of protein sequences that can be explored in regard to the huge potential sequence space. If one considers that each position of a small protein of 100 amino-acid can be replaced by any of the 20 possible natural amino-acids, a theoretical space size of 20^{100} sequences is reached. One can thus easily imagine that sampling such large spaces is out of reach of wet methods. Consequently, approaches aiming at rationalizing protein evolution and favoring exploration of sequence regions of particular relevance for the targeted property/function are crucially needed to increase the odds of success to tailor the desired proteins while decreasing experimental efforts. In this context, computational methods have gained a prominent place in protein engineering strategies. Structure-based Computational Protein Design (CPD) has emerged with the increasing number of protein structures available in databases and the advents in computational structural biology and chemistry to understand fundamental forces between atoms and (macro)molecules. CPD uses the tight relationships existing between the function and the structure of a protein to design novel functions by searching the amino-acid sequences compatible with a known 3D scaffold, that should be able to carry out the desired function. CPD can thus be seen as the inverse folding problem [3] which can be formulated as an optimization problem, solvable computationally. In recent years, CPD has experienced important success, especially in the design of therapeutic proteins [4], novel enzymes [5].... Nevertheless, the computational design of proteins with defined affinity for a given partner (such as a small molecule, a substrate or a peptide) which is essential for large range of applications, continues to present challenges.

Estimating the binding affinity between two molecular partners requires to compute the so-called partition function of each of the molecules and of the system formed by the two bound molecules. The partition function of a molecule that can have states $\ell \in \Lambda$ each with energy E_ℓ is defined through the Boltzmann distribution as $\sum_{\ell \in \Lambda} \exp(-\frac{E_\ell}{k_B \cdot T})$ where T is the temperature and k_B is the Boltzmann constant. Computing the partition function of a system at constant temperature therefore requires to sum a simple function of the energy over a large set of possible states Λ. This computation is easy for systems with a small number of states. It becomes extremely complex for systems that have a number of states which is exponential in the size of their description. Even with the usual simplifying assumptions adopted in the field of CPD, the problem is actually #P-complete, a computational complexity class of problems with extreme complexity.

A protein is defined by a supporting skeleton (or backbone) which is built by polymerization. Each amino-acid in the protein is composed of two parts: one part participates in the construction of the linear backbone of the protein and the remaining part is called the side-chain of the amino-acid. At each position, one may choose among 20 different possible side-chains, each defining a different amino-acid type, associated with a specific letter in the alphabet. A protein can therefore be described as a sequence of letters.

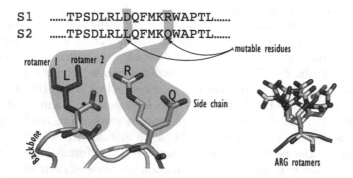

Fig. 1. A local view of a protein backbone and side-chains. Changes can be caused by amino acid identity substitutions (*for example D/L*) or by side-chain reorientations (rotamers). A typical rotamer library for one amino acid is shown on the right (ARG=Arginine).

To perform their function, proteins need to be folded into a specific 3D shape which is defined by the amino-acid composition and the effect of various atomic forces (torsion angles, electrostatic, van der Waals, solvation effects). The precise conformation of a protein is therefore defined by the conformation of the backbone and the relative conformation of all its side-chains. A usual simplification assumption in CPD is to assume that the backbone is rigid (it has only one conformation) and that the side-chains can only adopt a specific set of discrete low-energy conformations, called rotamers. Several rotamer libraries have been defined over the last two decades. For a protein with a sequence S of length n, with a known composition (side-chains), the set of reachable conformations Λ^S is then defined as the cross-product of the set of the rotamers of every amino-acid. A conformation ℓ is just a choice of rotamer for each side-chain of the protein.

The computation of the partition function Z_A of a molecule A requires the computation of the energy of every conformation. A second usual assumption of CPD states that the energy of a molecule, defined as a function of the conformation ℓ, can be decomposed as a sum of a constant term (describing the energy of the backbone), a variable term for each side-chain (describing the interaction between the side-chain and the backbone and side-chain internal interactions) and a variable term for every pair of side-chains that interact (describing the interaction between the two corresponding side-chains). Stated otherwise, the energy can be pairwise decomposed. Different such pairwise decompositions of the energy have been built over the last decades.

The affinity of 2 molecules A and B is then proportional to the ratio of the partition function of the complex formed by the two molecules denoted Z_{AB} to the product of their partition functions $Z_A \times Z_B$. One affinity estimation therefore requires to compute three partition functions. The estimated affinity being just the criteria used for molecule design, i.e. to fix the amino-acid sequence, it must be repeatedly evaluated for each possible sequence considered for design. With a choice among 20 naturally occurring amino-acids at every position, the size of the sequence space is also exponential. This is why most affinity based designs or analysis are usually done on very small or drastically simplified systems.

In this paper, we focus on the problem of computing a lower estimate of the partition function of a protein with deterministic guarantees, enabling to approximate the affinity. The algorithm we define can be seen as an evolution of a traditional approximate partition function computation algorithm developed for affinity approximation called K^* [6]. The K^* algorithm combines dominance analysis with A^* Best First Search to produce a sequence of conformations of increasing energy. This sequence can be used to compute a running sum of contributions to the partition function and prune conformations that cannot contribute sufficiently to the partition function (because of their high energy).

Our method follows the same idea but relies on the use of specific lower bounding techniques that have been developed for exact combinatorial optimization in the last decades in the field of weighted Constraint Programming [7] and more precisely Cost Function Network and local consistencies [8]. While the use of such bounds is usual for exact optimization, this is, in our knowledge, the first use of soft local consistency to prune the search space of approximate partition function computation. The second new ingredient is a shift from space-intensive best-first search to depth-first search, completed by on-the-fly variable elimination [9].

We compare our algorithm to the K^* algorithm implemented in *Osprey* [10], a CPD-dedicated platform.

1   Background and Notations

We denote by A_A the set of all 20 amino-acid types. We consider an initial protein defined by sequence of amino acids $S = s_1, \ldots, s_n$, $s_i \in A_A$. The protein S has a 3D scaffold which is selected amongst high resolution 3D structures determined either from X-ray crystallography or Nuclear Magnetic Resonance (and deposited in the Protein Data Bank). This defines an initial fixed backbone.

Among all positions of S, some positions $M \subset \{1, \ldots, n\}$ are considered as mutable and can be replaced by any amino acid in a subset A_{A_i} of A_A. For each amino-acid type t in A_A, we have a discrete library of possible conformations (or rotamers) denoted as A_t. Given the sequence S, the atomic coordinates of the associated backbone and a choice $\ell_i \in A_{S_i}$ of a given conformation for each side-chain, the energy of the global conformation can be written as a function of the vector of rotamers used, denoted as $\ell \in A^S = \prod_i A_{S_i}$:

$$E^S(\ell) = E_\varnothing^S + \sum_i E^S(\ell_i) + \sum_i \sum_{j>i} E^S(\ell_i, \ell_j)$$

where E^S is the potential energy of the protein S, $E_\varnothing^S$ is a constant energy contribution capturing interactions between fixed parts of the model, $E^S(\ell_i)$ is the energy contribution of rotamer ℓ_i at position i capturing internal interactions or interactions with the backbone, and $E^S(\ell_i, \ell_j)$ is the pairwise interaction energy between rotamer ℓ_i at position i and rotamer ℓ_j at position j [11]. Thanks to this decomposition, each energy terms, in $kcal \cdot mol^{-1}$, can be pre-computed and cached, allowing to quickly compute the energy of a conformation once the rotamer used at each position is fixed.

Assuming for the sake of simplicity that no symmetry or state degeneracy exists in the molecule considered, the partition function of a protein S for a fixed position, rotation and backbone over all side-chain conformations is defined as:

$$Z^S = \sum_{\ell \in \Lambda^S} e^{-\frac{E^S(\ell)}{k_B T}}$$

This sum contains an exponential number of terms. The exponential distribution leads to terms with sharp changes in magnitude. Most significant terms correspond to low energies. Our aim is to provide a deterministic algorithm that computes a lower approximation $\hat{Z}^S$ such that:

$$\frac{Z^S}{1+\varepsilon} \leq \hat{Z}^S \leq Z^S \tag{1}$$

The problem of exactly computing Z^S in this case is known to be #P-complete. The "simple" problem of finding a rotamer vector ℓ with energy below a value k is NP-hard [12]. Usual methods for estimating Z^S are stochastic methods using Monte-Carlo (Markov-chain) approaches with no non-asymptotic guarantees [13]. In the context of protein design, the K^* algorithm is a simple algorithm that provides an approximation to Z^S satisfying (1). We now quickly present it.

1.1 The K^* Algorithm

The K^* algorithm is an approximate *counting* algorithm that exploits optimization techniques. The first component is a dominance analysis called "Dead End Elimination" (DEE [11]). This process compares a worst case energy for a rotamer $\ell_i \in \Lambda_{S_i}$ with a best case cost for a different rotamer $\ell'_i \in \Lambda_{S_i}$ and prunes $\ell'_i \in \Lambda_{S_i}$ if its best case energy is larger than the other energy (replacing ℓ'_i by ℓ_i in any solution can only improve the energy) i.e., iff:

$$\left[E^S(\ell'_i) + \sum_{j \neq i} \min_{\ell_j \in \Lambda_{S_j}} E^S(\ell'_i, \ell_j) \right] - \left[E^S(\ell_i) + \sum_{j \neq i} \max_{\ell_j \in \Lambda_{S_j}} E^S(\ell_i, \ell_j) \right] > 0 \tag{2}$$

This is done iteratively on all positions i and all possible pairs of rotamers in Λ_i, pruning the space of conformations.

The second component is a best-first branch and bound A^* optimization algorithm exploring a tree where each node corresponds to a partially determined conformation where a subset of all positions have fixed rotamers. The sons of a node are defined by choosing a position i with a yet unfixed rotamer and creating one node per possible rotamer $\ell_i \in \Lambda_{S_i}$. Leaves correspond to completely determined conformations ℓ. At a given node at depth d with fixed rotamers

$\ell = (\ell_1, \ldots, \ell_d)$, it is possible to compute a lower bound $Lb(\ell)$ on the energy of any complete conformation below this node. Following the A^* terminology, this lower bound defines an "admissible heuristics" and is defined as $Lb(\ell) =$

$$\underbrace{\sum_{i=1}^{d} E(\ell_i) + \sum_{j=i+1}^{d} E(\ell_i, \ell_j)}_{\text{Fixed}} + \underbrace{\sum_{j=d+1}^{n} [\min_{\ell_j \in \Lambda_{S_j}} (E(\ell_j) + \sum_{i=1}^{d} E(\ell_i, \ell_j)}_{\text{Fixed-Unfixed}} + \underbrace{\sum_{k=j+1}^{n} \min_{\ell_k \in \Lambda_{S_k}} E(\ell_j, \ell_k))]}_{\text{Unfixed}}$$

Starting from the root, if the node with the best lower bound is always developed first, the sequence of *leaves* explored by A^* will produce a stream of complete conformations, starting with the optimal conformation with minimum energy (and thus maximal contribution to the partition function) and followed by sub-optimal conformations sorted by increasing energy.

For optimization alone, it is amazing to see that the simple algorithm combining DEE and A^* (DEE/A^*) is only slightly dominated by 01-Linear programming models solved using CPLEX [14] and outperforms very concise quadratic programming formulations solved using CPLEX or BiqMac [15].

For counting, the DEE preprocessing may remove sub-optimal conformations whose energy is sufficiently low to significantly contribute to the partition function. It is therefore weakened by using an energy threshold $E_w > 0$, replacing the rhs zero in equation 2. Then DEE can only prune conformations with energy at least E_w away from the optimum energy.

As A^* produces conformations in decreasing order, the space of all conformations Λ^S is split in 3 subsets: a set P of conformations that have been pruned by $DEE(E_w)$, a set V of conformations already visited by A^* and a set U of yet unexplored conformations. This splits the partition function in 3 terms: $Z^S = Z_P^S + Z_V^S + Z_U^S$, each summed on the corresponding conformation space. The value of the second term is known. It is possible to bound the value of the first term: we compute a lower bound $Lb_P = \min_{\ell_i \text{ pruned}} Lb(\ell_i)$ on the energy of every conformations pruned by $DEE(E_w)$ and conclude that $Z_P^S \leq |P| \cdot \exp(\frac{-Lb_P}{k_B T})$. Finally, if E_{A^*} is the energy of the last conformation produced by A^*, since the sequence of conformation is decreasing, we have $Z_U^S \leq |V| \cdot \exp(\frac{-E_{A^*}}{k_B T})$. Thus:

$$Z^S \leq Z_V^S + |P| \cdot e^{\frac{-Lb_P}{k_B T}} + |V| \cdot e^{\frac{-E_{A^*}}{k_B T}}.$$

Hence, if $E_{A^*} \geq -k_B T \cdot [\log(Z_V^S \cdot \frac{\varepsilon}{1+\varepsilon} - |P| \cdot e^{\frac{-Lb_P}{k_B T}}) - \log(|U|))$, we know that Z_V^S is an ε-approximation of Z_S and we can stop A^* search, possibly pruning a lot of conformations that do not contribute enough to Z^S. However, it is also possible that too many conformations have been pruned by $DEE(E_w)$ and this threshold is never reached. In this case, the A^* must exhaust all conformations and a sufficient (easy to determine) number of conformations in P must be restored and the A^* search redone. Remind that A^* is a worst-case exponential space and time algorithm. The dominance analysis of DEE is a double-edged sword: it may prune a lot of conformations but if too many conformations are pruned, a full search is required and another search needs to be done on a larger set of conformations.

1.2   Cost Function Network and Local Consistency

Following our previous work on optimization for CPD presented in [16,14], we model the distribution of energies as a Cost Function Network (CFN) or Weighted CSP. In a CFN (X, D, C, k), X is a finite set of variables, each variable $i \in X$ has a finite domain $D_i \in D$, each cost function $c_S \in C$ is a function from $\prod_{i \in S} D_i$ to $\{0, \ldots, k\}$ and k is an upper bound defining an intolerable cost. For a given assignment ℓ of X, the cost of ℓ is the sum of all cost functions (if it is less than k) or k otherwise. A complete assignment is a solution of the CFN is its cost is strictly less than k. Notice that all costs being non negative, $c_\varnothing \in C$ is a lower bound on the cost of any assignment.

The energy distribution of S can be obviously modeled in a CFN with one variable $i \in X$ for every position, the domain of i is the set of rotamers Λ_{S_i} and the cost functions include one zero-ary, unary and binary cost functions representing respectively $E_\varnothing^S$, $E_S(\ell_i)$ and $E_S(\ell_i, \ell_j)$.

If K^* relies on DEE and A^*, our new algorithm combines soft local consistencies (instead of DEE) and a Depth First Branch and Bound (DFBB) algorithm (instead of A^*). The DFBB algorithm performs counting instead of minimization and uses a dedicated dynamic pruning condition that guarantees to compute an ε-approximation of Z^S.

We quickly present the fundamental properties of soft local consistencies, without details. The reader is invited to read [8] for a precise description of existing arc consistencies for CFN. Enforcing a given soft local consistency transforms a CFN (X, D, C, k) into an equivalent CFN (X, D', C', k) that satisfies the corresponding local consistency. By equivalence, we mean that the two CFNs have the same set of solutions with the same cost. The fact that (X, D', C', k) satisfies the local consistency usually means that $c_\varnothing$ has increased (it cannot decrease in any case), providing a stronger lower bound on the optimum and that domains have been reduced (some values that cannot participate in a solution are deleted). Naturally, the lower the k is, the stronger is the pruning.

2   DFBB+Arc Consistency to Compute Z with Pruning

Traditional DFBB algorithms for CFN maintain a given level of soft arc consistency at each node (producing a non naive lower bound $c_\varnothing$ at each node) and use the cost of the current best known solution as the current upper bound k. This allows to prune efficiently as k decreases rapidly as search proceeds.

To transform this algorithm in a counting algorithm:

- each time a leaf ℓ with cost (energy) E_ℓ is encountered we contribute $\exp(\frac{-E_\ell}{k_B T})$ to a running lower estimate $\hat{Z}^S$ of Z^S (initially set to 0).
- each time we prune a node with partial assignment ℓ', we compute an upper bound $Ub(\ell')$ on the contribution to the partition function of all the leaves below the pruned node and contribute this upper bound to a running upper bound Z_+^S of the partition function over pruned conformations

- to decide when to prune, we enforce the invariant $\hat{Z}^S \geq \frac{\hat{Z}^S + Z_+^S}{1+\varepsilon}$. We prune if and only if adding $Ub(\ell')$ to Z_+^S does not break this invariant.

Theorem 1. *The algorithm is correct: it provides a final estimate $\hat{Z}^S$ such that* $Z^S \geq \hat{Z}^S \geq \frac{Z^S}{1+\varepsilon}$

Proof. Initially, $\hat{Z}^S = Z_+^S = 0$ and the invariant is satisfied. When the search finishes, all conformations have either been explored or pruned and therefore $(\hat{Z}^S + Z_+^S) \geq Z^S$. Using the invariant, we conclude that ultimately $\hat{Z}^S \geq \frac{Z^S}{1+\varepsilon}$. By construction, $Z^S \geq \hat{Z}^S$.   □

The remaining ingredient is the upper bound $Ub(\ell)$. We know that all conformations below the current node have a cost (energy) larger than $c_\emptyset$, thus a contribution to Z^S smaller than $\exp(\frac{-c_\emptyset}{k_B T})$. Denoting by $N(\ell)$ the product of the domain sizes of all unassigned variables, we can conclude that $Ub(\ell) = N \cdot \exp(\frac{-c_\emptyset}{k_B T})$ is an upper bound on the contribution of all leaves below the current node.

This general schema is improved using on-the-fly variable elimination [9]. If a variable i has only few neighbors (typically 2 or 3) at the current node, we eliminate this variable using a simple sum-product algorithm. We call the resulting algorithm Z_0^* (to leave room for future improvements).

Z_0^* can be easily improved by providing it with either a set of conformations of low energies or with any available lower bound on the partition function Z^S. These can be used to strengthen the pruning condition in the beginning of the search. Double counting of identical conformations can be easily avoided using efficient direct comparison of conformations by discrimination trees or hash tables or simply by using the maximum of this initial upper bound and the running upper bound $\hat{Z}^S$ in the pruning condition.

3   Experimental Comparison

To compare Z_0^* to K^*, we examined the binding affinity of different protein/ligand complexes. The 3D model of these molecular systems were derived from crystallographic structures of the proteins in complex with their ligands, deposited in the protein data bank (Table 1). Missing heavy atoms in crystal structures as well as hydrogen atoms were added using the *tleap* module of the *Amber 14* software package [17]. The molecular all-atom *ff14SB* was used for the proteins and the ligands while the *gaff* force field was used for the cofactor AMP (present in one studied system, PDB: 1AMU). The molecular systems were then subjected to 1000 steps of energy minimizations with the *Sander* module of *Amber 14*. Next, we have selected a portion of the proteins including residues at the interface between the protein and the ligand as well as a shell surrounding residues with at least one atom within 8 to 12 Å (according to the molecular system) of the interface.

The residues at the interface protein/ligand and the ligand were considered flexible and represented by rotamers from the Lovell's penultimate library [18].

At most 2 of the flexible residues of the protein were allowed to simultaneously mutate, while the remaining flexible residues were allowed to change their side-chain conformation. In addition to the wild-type identity for the positions allowed to mutate, a selective subset of amino-acid types (ranging from 7 to 19 amino-acids) was allowed in the redesign. All energy matrices were generated using *Osprey 2.0* [10]. The energy function is the sum of the *Amber* electrostatic, van der Waals and dihedral terms.

Table 1. For each ID, we give the ligand, the number of variable residues (flexible, mutations/mutable residues) and the number of sequences represented.

PDB ID	Ligand	Variable res.	Seq. Number
1AMU	F	(10,2/9)	1584
1TP5	KKETWV	(12,2/3)	1121
1B74	Q	(10,2/9)	1809
2Q2A	R	(13,2/12)	4716

Floating point energies were translated, multiplied by a large constant K and truncated to the nearest integer in the CFN. We implemented our algorithm in our open source solver `toulbar2`[1]. Our DFBB algorithm maintains Existential Directional Arc Consistency (EDAC) [19], on-the-fly elimination [9], a value heuristics using the existential support and a variable ordering combining weighted degree [20] and last conflict [21] heuristics.

To compare Z_0^* as implemented in our C++ solver `toulbar2` to the algorithm K^* implemented in *Osprey 2.0* in Java, we set ε to 10^{-3} and `toulbar2` and asked to estimate the affinity of each mutated proteins sequence with the substrate. For *Osprey 2.0*, we used the options `doMinimize = false, gamma = 0.0`. This last option guarantees that all mutations will be evaluated. The default values `initEw = 6.0, pruningE = 100.0, stericE = 30.0` were also used. For `toulbar2 0.9.7`, we used options `-logz -zub=0` that selects the Z_0^* algorithm and the upper bounding $Ub(\ell)$ described before and turns DEE processing off. For the 1AMU model, this represents a total of around 3,168 counting problems, each with different energy landscapes. The proxy to the affinity defined by the ratio of the partition functions is only approximated by the two methods, and these approximations may differ and rank sequences with close values differently. We compared the rank of the first 40 sequences and the two sets were identical and identically ordered.

For the 1AMU model, K^* explored 6,451,997 nodes in 76,654 seconds (CPU user time) while `toulbar2` explored 84,993 nodes in 30 seconds (CPU user time) for the same value of ε. So, Z_0^* is around 2,500 times faster and offers stronger pruning. While the difference in efficiency between C++ and Java may contribute to this difference, it can only explain a small part of this ratio. For the other models, K^* took over the time limit (1000 hours CPU user time) that we set and it is therefore impossible to compare precisely. We also studied the

---

[1] `mulcyber.toulouse.inra.fr/projects/toulbar2/`

influence of ε on the computing time and pruning. The influence is not really drastic. Notice that a value of $\varepsilon = 1$ means that our estimate is guaranteed to be only withing a factor 2 of the true value of Z^S.

Table 2. For each system, and for different values of ε we give the number of nodes and user cpu-time in seconds. The value $\varepsilon = 0$ was also tried for 1AMU. In this case, steric clashes are the only possible cause of pruning (infinite energy) and more than 340,000 nodes were explored in 72". In tight interfaces, this is an important source of pruning.

	$\varepsilon = 1$		$\varepsilon = 10^{-3}$		$\varepsilon = 10^{-6}$	
PDB ID	nodes	time	nodes	time	nodes	time
1AMU	71,779	27.123	84,993	30.209	91,046	30.741
1TP5	2,290,910	1,254.456	3,190,349	1,845.352	3,278,235	2,138.06
1B74	3,725,217	3,072.183	5,635,612	5,113.682	6,464,729	5,682.806
2Q2A	27,003,446	24,012.182	39,939,197	35,383.914	56,956,558	50,752.495

4   Conclusion

In statistical physics in general and specifically in Computational Protein Design, computing or estimating the partition function is a central but difficult problem. The computation of the simple pure conformational partition function of an amino-acid based system with fixed backbone and flexible side-chains represented as rotamer sets with a pairwise decomposed energy already defines a $\#P$-complete problem.

In general, this problem has a very dense graph which makes exact graph-structure based approaches inadequate. It also has a very sharp energy landscape that allows for efficient optimization but does not facilitate a stable probabilistic (Monte Carlo-based) estimation of the partition function. Such estimates also offer no deterministic guarantee on the quality of the estimation.

The K^* approach developed for CPD and our new Z_0^* algorithm instead exploits the fact that in some cases, a very small fraction of the conformational space may contribute enough to essentially define the partition function. In our knowledge, our algorithm is the first deterministic algorithm for weighted counting that exploits soft local consistency reformulation based lower bounds for counting with deterministic guarantees. Naturally, these results are preliminary. Several directions can be pursued to reinforce these results:

- consider larger molecular systems for comparison. The combinatorial advantage of Z_0^* compared to K^* that results from the stronger lower bound of local consistencies, the absence of DEE, together with the space effectiveness of DFBB compared to A^* should lead to even stronger speedups.
- shift to higher resolution rotamer libraries. The penultimate library has a lower resolution than alternative libraries such as [22] or [23]. In *Osprey*, another approach using continuous pre and post minimization is used instead.

We could then compare the two approaches in terms of consistency of the predicted affinity ranking.

- compare Z_0^* to exact counting algorithms such as Cachet [24], Sentential Decision Diagrams [25] both on CPD problems (with sharp changes in contributions to Z^S) and other usual probabilistic systems such as classical instances of Markov Random Fields and Bayesian Nets.
- it is known that our lower bounds have corresponding lower bounding mechanisms in Weighted Satisfiability [26]. They could therefore be used to improve Cachet or SDD solvers for approximate computations with deterministic guarantees.
- improve Z_0^* with stronger bounds and alternative search strategies that favors the early discovery of low energy conformations and possibly exploits the interaction graph structure through a tree-decomposition.

These preliminary results also show the interest of using these bounds instead of DEE dominance analysis for pruning the conformational space. Local consistency provides good lower bounds and its pruning can be easily controlled to avoid pruning sub-optimal solutions that can participate significantly to the partition function. This is a major advantage compared to DEE.

References

1. Fersht, A.: Structure and mechanism in protein science: a guide to enzyme catalysis and protein folding. W.H. Freeman and Co., New York (1999)
2. Peisajovich, S.G., Tawfik, D.S.: Protein engineers turned evolutionists. Nature Methods 4(12), 991–994 (2007)
3. Pabo, C.: Molecular technology. Designing proteins and peptides. Nature 301(5897), 200 (1983)
4. Miklos, A.E., Kluwe, C., Der, B.S., Pai, S., Sircar, A., Hughes, R.A., Berrondo, M., Xu, J., Codrea, V., Buckley, P.E., et al.: Structure-based design of supercharged, highly thermoresistant antibodies. Chemistry & Biology 19(4), 449–455 (2012)
5. Siegel, J.B., Zanghellini, A., Lovick, H.M., Kiss, G., Lambert, A.R., St Clair, J.L., Gallaher, J.L., Hilvert, D., Gelb, M.H., Stoddard, B.L., Houk, K.N., Michael, F.E., Baker, D.: Computational design of an enzyme catalyst for a stereoselective bimolecular Diels-Alder reaction. Science 329(5989), 309–313 (2010)
6. Georgiev, I., Lilien, R.H., Donald, B.R.: The minimized dead-end elimination criterion and its application to protein redesign in a hybrid scoring and search algorithm for computing partition functions over molecular ensembles. Journal of Computational Chemistry 29(10), 1527–1542 (2008)
7. Rossi, F., van Beek, P., Walsh, T. (eds.): Handbook of Constraint Programming. Elsevier (2006)
8. Cooper, M., de Givry, S., Sanchez, M., Schiex, T., Zytnicki, M., Werner, T.: Soft arc consistency revisited. Artificial Intelligence 174, 449–478 (2010)
9. Larrosa, J.: Boosting search with variable elimination. In: Dechter, R. (ed.) CP 2000. LNCS, vol. 1894, pp. 291–305. Springer, Heidelberg (2000)
10. Gainza, P., Roberts, K.E., Georgiev, I., Lilien, R.H., Keedy, D.A., Chen, C.Y., Reza, F., Anderson, A.C., Richardson, D.C., Richardson, J.S., et al.: Osprey: Protein design with ensembles, flexibility, and provable algorithms. Methods Enzymol. (2012)

11. Desmet, J., De Maeyer, M., Hazes, B., Lasters, I.: The dead-end elimination theorem and its use in protein side-chain positioning. Nature 356(6369), 539–542 (1992)
12. Pierce, N.A., Winfree, E.: Protein design is NP-hard. Protein Engineering 15(10), 779–782 (2002)
13. Rubinstein, R.Y., Ridder, A., Vaisman, R.: Fast sequential Monte Carlo methods for counting and optimization. John Wiley & Sons (2013)
14. Allouche, D., André, I., Barbe, S., Davies, J., de Givry, S., Katsirelos, G., O'Sullivan, B., Prestwich, S., Schiex, T., Traoré, S.: Computational protein design as an optimization problem. Artificial Intelligence 212, 59–79 (2014)
15. Rendl, F., Rinaldi, G., Wiegele, A.: Solving Max-Cut to optimality by intersecting semidefinite and polyhedral relaxations. Math. Programming 121(2), 307 (2010)
16. Traoré, S., Allouche, D., André, I., de Givry, S., Katsirelos, G., Schiex, T., Barbe, S.: A new framework for computational protein design through cost function network optimization. Bioinformatics 29(17), 2129–2136 (2013)
17. Case, D., Babin, V., Berryman, J., Betz, R., Cai, Q., Cerutti, D., Cheatham Iii, T., Darden, T., Duke, R., Gohlke, H., et al.: Amber 14 (2014)
18. Lovell, S.C., Word, J.M., Richardson, J.S., Richardson, D.C.: The penultimate rotamer library. Proteins 40(3), 389–408 (2000)
19. Larrosa, J., de Givry, S., Heras, F., Zytnicki, M.: Existential arc consistency: getting closer to full arc consistency in weighted CSPs. In: Proc. of the 19th IJCAI, Edinburgh, Scotland, pp. 84–89 (August 2005)
20. Boussemart, F., Hemery, F., Lecoutre, C., Sais, L.: Boosting systematic search by weighting constraints. In: ECAI, vol. 16, p. 146 (2004)
21. Lecoutre, C., Saïs, L., Tabary, S., Vidal, V.: Reasoning from last conflict(s) in constraint programming. Artificial Intelligence 173, 1592–1614 (2009)
22. Shapovalov, M.V., Dunbrack, R.L.: A smoothed backbone-dependent rotamer library for proteins derived from adaptive kernel density estimates and regressions. Structure 19(6), 844–858 (2011)
23. Subramaniam, S., Senes, A.: Backbone dependency further improves side chain prediction efficiency in the energy-based conformer library (bebl). Proteins: Structure, Function, and Bioinformatics 82(11), 3177–3187 (2014)
24. Sang, T., Bacchus, F., Beame, P., Kautz, H., Pitassi, T.: Combining component caching and clause learning for effective model counting. In: Proc. of the 7th Int. Conf. on Theory and Applications of Satisfiability Testing (SAT 2004) (2004)
25. Choi, A., Kisa, D., Darwiche, A.: Compiling probabilistic graphical models using sentential decision diagrams. In: van der Gaag, L.C. (ed.) ECSQARU 2013. LNCS, vol. 7958, pp. 121–132. Springer, Heidelberg (2013)
26. Larrosa, J., Heras, F.: Resolution in max-sat and its relation to local consistency in weighted csps. In: IJCAI, pp. 193–198 (2005)

Benefits and Drawbacks in Using
the RFID (Radio Frequency Identification) System
in Supply Chain Management

Alexandra Ioana Florea (Ionescu)

University of Economic Studies, Bucharest, Romania
sandraionescu@hotmail.com

Abstract. In 2003 Wal-Mart, the world's largest retailer announced that it would require from its top 100 suppliers to tag pallets and cases of goods with radio-frequency identification (RFID) tags. Radio-frequency identification (RFID) is the wireless non-contact use of radio-frequency electromagnetic fields to transfer data, for the purposes of automatically identifying and tracking tags attached to objects. This system was born in 1999 at MIT (Massachusetts Institute of Technology) when a consortium formed of professors, researchers and supply chain professionals created a global standard system for RFID. This paper explores the benefits and drawbacks of using this technology in supply chain management. Its ten year old history shows that, even at the beginning a lot of professionals were carried away by its implementation in supply chain, many companies found it difficult to measure the return on investment of this system as the cost of deployment is rather high. As a market survey executed by IDTechEx Research shows that the turnover of the RFID market was $7.88 billion in 2013 and no less than 5.9 billion tags were sold the same year, the RFID technology plays an important role as it is deployed in many different application areas, especially in supply chain management. This paper argues that this technology can be very successfully implemented in order to improve the supply chain system and improve the value added offered to the clients, as long as the professionals make an accurate assessment of the investment needed for the implementation.

Keywords: supply chain management, Radio Frequency Identification – RFID, logistics

1   Introduction

According to the Council of Supply Chain Management Professionals, the logistics industry contribute to the total GDP of Europe in a percentage of over 7.15% and to that of the US in a percentage of over 7.7% [1]. Logistics play an important role in economy and experience shows that logistics costs are lower in the US, due to innovation and savings made by professionals with transport, warehouse and shipping costs. Logistics is just a part of the network used in supply chain management, with the aim to plan and coordinate all the activities in the supply chain so that it achieves high

© Springer International Publishing Switzerland 2015
H.A. Le Thi et al. (eds.), *Model. Comput. & Optim. in Inf. Syst. & Manage. Sci.*,
Advances in Intelligent Systems and Computing 360, DOI: 10.1007/978-3-319-18167-7_16

levels of consumer service while keeping costs low. According to Cârstea [2], the logistic costs are important mobility barriers to strategy changing in a company.

Can innovation in logistics and supply chain keep the costs low? Or the investment necessary to implement innovation makes it to expensive so that the return on investment is not worth it? Is it the case for the RFID technology?

2   Description of the RFID Technology

Radio-frequency identification (RFID) is the wireless non-contact use of radio-frequency electromagnetic fields to transfer data, for the purposes of automatically identifying and tracking tags attached to objects [3]. The system consists of three basic components: a tag, a reader and back office data-processing equipment, the tag containing an integrated circuit chip and an antenna, having the ability to respond to radio waves transmitted from the RFID reader in order to send, process, and store information [4].

In most applications, the chip is used to store information about the object, product or shipment that the company needs to follow. The reader tracks the physical movements of the tag and thus of the object followed [5].

RFID is similar to bar-coding, as they both use labels and scanners to read the labels and a back office software to store the data, but the RFID system has the following advantages: no line of sight required, multiple parallel reads possible, individual items instead of a item class can be identified and the read/write capability [6].

There are some natural limitations to the use of RFID, due to the laws of physics, as metals and liquids effectively block the radio waves. Also, interference issues between readers may exist that prevent tags from being read. RFID can be defective and thus prevent from being read [6].

Even if the technology on which the RFID is based dates back from the 1940's and is related to radar work done by scientists and engineers, it only became of interest for the supply chain professionals in the 1990's when IBM developed and patented an ultra-high frequency RFID system. They did some pilots with Wal Mart but they never sold this technology, and in the mid 1990's they sold the patent to Intermec [7].

The true story of RFID in supply chain started when a Brand Manager of Procter & Gamble began to look into it [8]. Kevin Ashton was his name and together with three researchers cofounded MIT's Auto-ID Center. The goal of Kevin Ashton was to always have P&G's products on the shelves of supermarkets and to avoid stock outs. The challenge, at the beginning of the implementation, was to produce it cheap enough in order to use it efficiently. Procter&Gamble implemented this technology as EPC – Electronic Product Code and it had a history full of ups and downs with it. Even if in 2003 Wal-Mart, the world's largest retailer, announced that it would require from its top 100 suppliers to tag pallets and cases of goods with radio-frequency identification (RFID) tags, no later than 2009 P&G ceased placing EPC tags on promotional displays bound for Wal-Mart's RFID enabled stores. They continued to tag cases of some products being shipped to Wal-Mart.

Nowadays, this technology is used in many fields of activity: aerospace, apparel, CPG – consumer packaged goods, defense, health care, logistics, manufacturing, pharma, retail. This is due to the decreasing cost of production but also to the introduction and use of standards related to RFID use. In Europe a great impact on the use of the RFID technology had the European Commissioner, Viviane Reding, who paid a special importance to frequency management and the standards applied in the industry, starting 2006. The EU and non-EU standards were harmonized through the directive 2006/95/EC related to electrical equipment designed for use within certain voltage limits. Depending on what needs to be monitored (products, pallets), there are several frequencies used by the RFID system [9].

Also, the European Commission invited a couple of key industry players to develop their recommendations for the establishment of an European framework for RFID and they launched the initiative "Coordinating European Efforts for Promoting the European RFID Value Chain" (CE RFID) [10].

There are two standards used in the world today for RFID, the EPC Standard and the ISO standard [11]. EPCglobal is a standard developed by an international non-profit association, called GS1, governed by managers from the companies that adhered to it. The EPCGlobal is developing the industry-driven standard for the Electronic Product Code (EPC) to support the use of Radio Frequency Identification (RFID) in today's fast-moving, information rich, trading networks. The ISO standard is developed by the International Organization for Standardization and it was first created for tracking cattle.

The worldwide RFID market (Fig.1) is nevertheless very important as it reached $7.88 billion in 2013 and the forecast for 2020 is $23.4 billion, according to a survey done by ID TechEx Research, this including tags, readers, and software services for RFID cards, labels, fobs and all other form factors – for both passive and active RFID [12]. The surveyed evaluated the sales of tags in 2013 at 5.9 billion. Governments were the most important buyers as RFID improve efficiency (transit systems), safety (passport tagging) and protect industries (animal tagging).

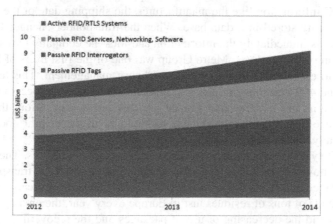

Fig. 1. Total RFID Market Projections in US$ billions, Source: IDTechEx report "RFID Forecasts, Players & Opportunities 2014-2024" www.IDTechEx.com/rfid

3   Application of RFID

Retailers in mature markets are now facing a challenge to better match the consumer needs and this is why they use technology and innovation in order to be a step ahead competition. Not only do they use the RFID technology for developing location based information services, but also for gathering data in order to acquire new data analysis capabilities [13].

Some of the world's most important retailers have implemented pilots or are already using RFID systems for different needs. Sainsbury's began with a trial in 2000 which led to the following results. They estimated the benefits achievable with a full-scale implementation without supplier participation to be at £8.5 million a year, benefits resulting from more efficient stock rotation and control [14]. But the cost at that time was very important, estimated between £18 million and £24million. Due to the decrease of the price of investment of such systems, Sainsbury's are now using the RFID system in many ways. In 2012, according to [15], Sainsbury's chose Checkpoint Systems Inc. to provide barcode labels, care labels, woven labels, and swing tickets for a clothing line, encoding RFID-based solutions. Also, in 2010, the same retail chain selected the company Harland Simon for supplying the latest temperature measuring technology to remotely monitor its food and beverage storage devices. This tag will ensure it maintains the highest levels of food safety and complies with all current food safety standards [16].

The German retailer Metro is also a keen user of the technology as they initiated in 2002 the Future Store Initiative together with more than 75 partners with the aim to modernize the retail sector [17]. International companies like Intel, SAP, Cisco Systems, Fujitsu and Siemens take part in this Initiative. They search for solutions for improving service in retail and making processes more efficient. One of the most important implementations of this Initiative is the use of RFID systems, using a transponder that carries a number, a so-called electronic product code (EPC), which allows the identification of the item and the shipping unit. Authorized users can read from the EPC information like the manufacturer, the shipping date or the weight, this information being stored in a data base. When the transponder changes location, the systems uploads immediately the information and thus the company may know exactly the location of the products. Metro Group was one of the first users of RFID, starting in 2004 to employ it in logistics and warehouse management. This technology is used in around 400 locations in Europe keeping incoming goods process running well. All the locations in Germany and France use this technology. More than 750,000 pallets are recorded each year using this technology, in the central goods depot in Unna, Germany, alone [18].

The RFID system is showing itself very useful in some industries. The grocery industry faces now several challenges: the wooden pallets used for transportation are quickly damaged and there is always a need for new pallets, the grocery industry generates thousands of tons of residues just in Europe every year, the environmental policy goals demand the decreasing of use of packages and the recovering and recycling of the waste, the temperature of perishable goods must be controlled all over the cold chain. Also, the European directive on the traceability of goods demands that any steps, used materials, manufacturing processes must be registered during the entire life-cycle of a product. The RFID system can bring an answer to all these challenges as: it has the capability to read the information at long distance, as pallets of fresh products are stored in large warehouses, it can track and monitor the temperature of the perishable goods, by incorporating a temperature sensor, it has full read/write capabilities, it ensures reader seamless integration into the customer facilities with minimum impact on their information systems [19].

The fashion industry has a lot to gain from using the RFID system, as it is characterized by short product life-cycles, high volatility and low predictability of demand. RFID helps the players in the fashion industry solve several issues: (1) ensuring that the product is delivered to the store as quickly as possible, as the life-cycle of the products are very short, (2) improving the stock management in store, helping employees identify items needed by customers more quickly and accurately, (3) preventing the sale of counterfeit and illegal products and (4) by replacing the acoustomagnetic tags or RF tags for electronic article surveillance (EAS) applications [20]. The RFID system is used successfully in logistics management, as it facilitates logistics tracking through automatic target recognition and relevant data accessing [21].

4   Benefits of RFID in Supply Chain

Many companies have already discovered that the RFID technology has a great potential for handling the complexity of globalization and that it can create a balance between cost and performance in supply chains, improving efficiencies of tracking, warehousing, loading and unloading [22]. The RFID technology improves four dimensions of business: process efficiency, accuracy, visibility and security, thus adding value to business processes [23].

Rather than analyzing the added cost of a RFID tag a company should look at the added-value that the system can bring to them, because even the cost a of tag is now only of two-cents, if it does not bring any value to the company, the investment is not worth while [5]. The price of a tag was about one dollar each in 2000 but now the prices are getting to about five cents per tag, which really makes it less expensive to use.

Many authors analyzed the advantages that the RFID system can bring to the supply chain activity of a company, as the use of new technologies represents a point of interest to all the professionals in this industry. After a thorough analysis of these papers, we have divided all the benefits shown by different authors into five big groups, according to the result brought to the activity of the company. Thus, the five main benefits are: increased profitability through operations management, visibility, security, cost savings and better client service. All the supply chain activities where the RFID can be used lead to one of these five benefits that a company can derive from implementing this technology.

4.1   Increased Profitability

The increased profitability through operations management can be achieved by using this technology in several actions of the company. The actions are not enumerated in order of importance, as each of them can have this output depending on industry, market, position on the chain of supply (Fig.2).

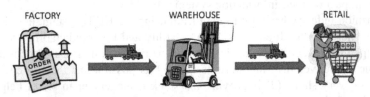

FACTORY   WAREHOUSE   RETAIL

Fig. 2. Chain of supply

Zhu [5] stated that RFID can be successfully used to remedy the problem of stock outs and excess of stock, as the discrepancies between warehouse quantities recorded in the system and stocks truly available to customers are one of the biggest problems retailers face.

In a study conducted by Chen [24], the results showed that the total operation time can be saved by 81% from current stage to future stage with the integration of RFID and lean management. Moreover, the saving in total operation time can be enhanced to 89% with cross docking, the return-on-investment (ROI) analysis showing that the proposed method is both effective and feasible.

Many benefits of RFID derive from the reduction of material handling and inspection time, as one study estimated a 40% decrease in inventory counting time by using this technology [25]. Other benefits stated by Tajima are: production tracking, quality control, supply and production continuity, space utilization, reduced stock outs. RFID could enable tracking of raw materials, work-in-process inventory, finished products and even assembly status during production and could be used to ensure quality control during production. By improving material tracking through the manufacturing process, RFID could ensure continuity in production and supply availability. In relation to better material handling, the use of RFID could improve efficiency and flexibility of space utilization and by increasing accuracy in finished goods inventories, RFID could help reduce stock outs and subsequently reduce lost sales. Lim [26] also points out the fact that knowing the location of items helps to improve space utilization and increase accuracy in product records and that the RFID system helps to reduce the amount of manual labor and material handling. Other benefits brought by the RFID systems are: better inventory management, tracking work-in progress, reducing administrative errors, reducing rework [27].

4.2   Visibility

According to Gaukler [6], the increased visibility comes from the fact that the supply chain will be more efficient due to the exact knowledge of the amount of inventory at each location in real-time. Knowing what is in the replenishment pipeline, and when it is expected to arrive, potentially allows safety stocks to be reduced, while maintaining or increasing customer service levels. Also, according to Azevedo [28], this technology was designed to allow the unique identification of every single instance of each item manufactured, as opposed to just the manufacturer and class of products, as barcodes usually do.

Two main advantages of the RFID system is increased data accuracy and increasing information visibility [5], as this technology can be used to enhance the visibility of end-to-end chain information, thus the supplier, manufacturer, logistics provider and retailer can track and trace item-level information through the entire supply chain at any time and at any location.

Roh [29] argues that RFID increases the supply chain visibility, which is the ability to track the flow of goods, inventory, and information in the supply chain in a timely manner. This can reduce bottlenecks, out-of-stock or low inventory levels in the supply chain and can significantly lower the uncertainty of goods in a supply chain. It allows companies to track and manage the flow of inventory and products, decreasing the bullwhip effect.

Two benefits stated by Tajima [25] are increased data accuracy and improved information sharing, as RFID could improve inventory records by reducing human errors in material handling, inaccuracy of inventory data being a major problem in the retail industry.

4.3   Security

Very important for the agri-food industry, RFID can bring two major benefits: traceability back and forward in food chain and securing food safety [5]. Traceability is the ability to follow the movement of a food product through the stages of production, processing, and distribution. Traceability is often needed to identify the sources of food contamination and the recipients of contaminated food in product recalls and seizures. According to Sarac [31], Marks & Spencer is also employing RFID technologies in its refrigerated food supply chain. Marks&Spencer looked for a solution that would help them move a big amount of fresh products more quickly, tracking and managing the products from the production to purchase. The RFID technology helped them capture data more quickly, improve supplier production windows, easily and accurately adjust data in transit and lower costs, providing a 85% reduction in data collection time in the distribution center, which has lead to a 15% growth in distributed volume, per year [32].

4.4   Cost Savings

According to Zhu [5], inventory shrinkages can be substantially avoided using RFID technology as items are continuously monitored. In June 2012, at the National Retail Federation's Loss Prevention Conference and Expo in New Orleans, preliminary findings of the latest National Retail Security Survey were presented. Dr. Richard Hollinger from the University of Florida quantified that the retailers lost in 2011 no less than $34.5 billion to retail theft or shrink – the loss of inventory due to employee theft, shoplifting, paperwork errors or supplier fraud [33]. Overall, that accounts for approximately 1.41 percent of retailers' sales the year before.

For Roh [29], one of the main three advantages brought by the RFID system is represented by the cost savings through reductions in theft, labor and inventory cost. According to the same paper, it takes 1 min and 40 s to scan 1000 items through RFID scanners and 33 min (20 times as long) using bar-code technology. There is no need for repeated search and scanning in the inventory tracking and storing processes.

Bottani [34] presented a list of savings on different operations that can derive from using RFID on the fast-moving consumer goods supply chain: shortening of time required to perform checking operations on pallets received, improved accuracy of process, avoiding labeling operations, reduction of shrinks, cutting down time for replenishment cycles and to perform inventory counts, shortening of time required for pallets identification, reduction of safety stocks and reductions of stock outs. Increased supply chain cost savings come from an increasing product value [35], which is influenced by factors like lost sales, theft, inventory, orders and labor costs.

According to Ginters and Martin-Gutierrez [36], this technology provides the simultaneous non-contact identification of multiple objects, thus information-processing performance is increased which in turn reduces the number of staff required.

4.5   Better Client Service

A better client service can be achieved through reduced stock outs and better inventory management on the store floor. The clothes retail chain The Gap implemented in 2001 an in-store RFID system from Texas Instruments that helped the store employees easily locate the items in stock [37]. The sales were higher in the stores where this system was implemented as almost all the merchandise was RFID –tagged and, as soon as the shipment arrived, the employees knew exactly what sizes and styles had to be on display in the store and where. They were able to better manage overall inventory. Using this system, they could avoid situations when the clients left the store without buying because a product in stock could not be found in the store.

According to Lim [26], product tracking allows for reduced shrinkages and better expiry date management. The survey lead by Bouzaabia and Boumaiza [38] showed that the logistics performance in retail has a direct impact on customer satisfaction, which depends on the availability of products and in store merchandising. Another similar study, effectuated by Silvestri [39], showed also a direct relationship between the quality perceived by customers in relation to services offered by the hypermarket, with reference to logistics, meaning the ease of finding products in the store. Mehrjerdi [27] points out that another benefit of RFID is the better management of warrantee claims, thus having a positive effect on the client service.

5   Drawbacks of RFID in Supply Chain

There still are many challenges to RFID adoption and some of them may turn into drawbacks. Wu [4] identified the following: technology challenges, standard and patent challenges, cost challenges, infrastructure challenges, ROI challenges and barcode to RFID migration challenges. This is one of the most comprehensive classifications, as other authors identified mainly the same problems. We can resume all the drawbacks into four main categories: technology challenges, infrastructure challenges, cost related issues, privacy issues.

The **technology challenges** may be related to the fact that the radio waves can be blocked by different materials and thus the information will not be transmitted correctly. The orientation of the antenna can also affect the reading of the tag information. Other technological challenges can be the collision caused by simultaneous radio transmission or the lack of consistent UHF spectrum allocation for RFID [4]. Standard challenges are also pointed out by Lim [26], as the lack of a complete and unified RFID standard has caused many companies to hesitate in adopting the RFID systems. Also, developing international standards helps managing the intellectual property and patent rights across countries [4].

The **infrastructure challenge** is related to the fact that the chains of supply can be very long nowadays, as they can involve a lot of companies and cross many countries. In order to implement a RFID tracking system on all the chain of supply, it takes a lot of effort and ready infrastructure from many companies around the globe. Also, the popularity of the bar code is not yet fading away, as it still has a lot of advantages compared to the RFID tags. The bar code is much smaller and lighter, is less expensive as it only requires the ink to be printed on paper or plastic, work with the same accuracy on various materials, it can be read anywhere in the world with a bar code reader [43]. A lot of companies use successfully bar codes and RFID together, as they can be complementary.

The **cost related issues** refer to price and return on investment. According to Carrender [40], the price of a tag was approximately one dollar in 2000 and had a spectacular drop over the years, arriving at less than 10 cents in 2009. Because the price of the tag cannot drop lower than 5 cents, the tags suppliers began to increase their functionalities and they now offer more memory compared to previous generations. This would help store essential configuration, licensing, or product information persistently, identify and serialize a device and configure devices in the warehouse or at the point of sale without taking them out of the box [41]. The financial aspects involved in any business decision, such as implementing technology in supply chain or integration of the production, are to be analyzed in comparison with the necessary investment that needs to be made [42]. The price can still be a challenge if the company does not benefit from an important return on investment.

The Return on Investment (ROI) challenges can have the most important role to play in making the managers take a decision about using RFID systems. Nowadays, reducing the cost of a particular process of the logistics and global supply chain requires a huge effort [19]. Lim [26] also points out the fact that the return on investment (ROI) is one of the crucial criteria for companies to consider when introducing a new technology and several studies have reported concerns of some industries with lack of confidence that RFID could bring a satisfactory ROI.

A challenge explained by Zhu [5] is the concern about **customer privacy.** The tags can be read by anyone with an RFID scanner. Lim [26] also points out the fact that for companies the question of privacy may be a concern in case of shared storage facility and when open loop RFID solutions are considered.

6   Conclusions

The activity of supplying material resources must be managed with the same efficiency as the activity of selling to the clients, according to Cârstea and Păun [44]. As we have seen previously, many big international companies are paying a lot of attention to improving the activity of supply chain and its efficiency, like Metro, Sainsbury's, Wal-Mart, and many players in the textile industry. The RFID system can be deployed in various fields of business, like fresh groceries or entertainment parks, due to its technical characteristics.

The benefits brought by the RFID system to the supply chain can prove to be very important to any company as it helps improve profitability through operations management, visibility, security and client service and it can bring a lot of cost savings. The cost of implementation of such a system having greatly diminished since its first deployments at the beginning of the years 2000, the companies can now focus on adding more and more functionalities and improving the return on investment.

We can conclude that as the drawbacks of this system are mainly related to the technological issues of implementation or related to the practical and administrative issues of implementation, the benefits already proven in hundred or maybe thousands of implementations around the world can trigger from the professionals in supply chain a decision to take into consideration of this technology.

In the future, RFID will be probably integrated into larger Wireless Sensor Networks (WSN) that will truly help the supply chains become more responsive. Also, the RFID technology can actively contribute to the development of green projects, by improving the energy efficiency of the production chains and the waste control, helping recycling and answering to other environmental challenges. [45]

References

1. Facts on the Global Supply Chain,
   http://cscmp.org/media-center/facts-global-supply-chain
2. Carstea, G., Deac, V., Popa, I., Podgoreanu, S.: Analiza strategică a mediului concurenţial, Ed. Economica, Bucuresti (2002)
3. Radio Frequency Identification (n.d.),
   http://en.wikipedia.org/wiki/Radio-frequency_identification
4. Wu, N.C., Nystrom, M.A., Lin, T.R., Yu, H.C.: Challenges to global RFID adoption. Technovation (26), 1317–1323 (2006)
5. Zhu, X., Mukhopadhyay, S.K., Kurata, H.: A review of RFID technology and its managerial applications in different industries. Journal of Engineering and Technology Management (29), 152–167 (2012)
6. Gaukler, G.M., Seifert, R.W.: Applications of RFID in Supply Chains. In: Trends in Supply Chain Design and Management: Technologies and Methodologies. Springer-Verlag London Ltd. (2007)
7. Roberti, M.: The History of RFID Technology (2005),
   http://www.rfidjournal.com/articles/view?1338

8. Schuster, E.W., Allen, S.J., Brock, D.L.: Global RFID – The Value of the EPCglobal Network for Supply Chain Management. Springer, Heidelberg (2007), Technology and its managerial applications in different industries
9. Todorovic, V., Neag, M., Lazarevic, M.: On the Usage of RFID Tags for Tracking and Monitoring of Shipped Perishable Goods. Procedia Engineering 69, 1345–1349 (2014)
10. Wolfram, G., Gampl, B., Gabriel, P.: The RFID Roadmap: The Next Steps for Europe. Springer (2008)
11. Hassan Ali, O.M.S.: Improved Supply Chain Performance Through TFID Technology: Comparative Case Analysis of Metro Group and Wal-Mart, Research Online (2012)
12. Global RFID market will reach $7.88 billion in 2013 (2013),
    http://www.idtechex.com/research/articles/global-rfid-market-will-reach-7-88-billion-in-2013-00005914.asp
13. Reinartz, W., Dellaert, B., Krafft, M., Kumar, V., Varadarajan, R.: Retailing Innovations in a Globalizing Retail Market Environment. Journal of Retailing 87S, S53–S66 (2011)
14. Karkkainen, M.: Increasing efficiency in the supply chain for short shelf life goods using RFID tagging. International Journal of Retail and Distribution Management 31(10) (2003)
15. UK retailer using RFID tags to reduce shrink (2012),
    http://www.fastmoving.co.za/news/retailer-news-16/uk-retailer-using-rfid-tags-to-reduce-shrink-1719
16. Keynes, M.: Sainsbury'select Harland Simon for Temperature Monitoring (2010),
    http://www.harlandsimon.com/News32.php
17. Metro Group Future Store Initiative,
    http://www.future-store.org/internet/site/ts_fsi/node/139979/Len/index.html
18. RFID in the Metro Group,
    http://www.future-store.org/internet/site/ts_fsi/node/140064/Len/index.html
19. Martinez-Sala, A.S., Egea-Lopez, E., Garcia-Sanchez, F., Garcia-Haro, J.: Tracking of Returnable Packaging and Transport Units with active RFID in the grocery supply chain. Computers in Industry 60, 161–171 (2009)
20. Bottani, E., Volpi, A., Rizzi, A., Montanari, R., Bertolini, M.: The Role of Radio Frequency Identification (RFID) Technologies in Improving Distribution and Retail Operations in the Fashion Supply Chain. Woodhead Publishing Limited (2014)
21. Mingxiu, Z., Chunchang, F., Minggen, Y.: The Application Used RFID in Third Party Logistics. Physics Procedia 25, 2045–2049 (2012)
22. Ramanathan, R., Ramanathan, U., Ko, L.W.L.: Adoption of RFID technologies in UK logistics: Moderating roles of size, barcode experience and government support. Expert Systems with Applications 41, 230–236 (2014)
23. Ustundag, A.: The Value of RFID Roadmap: Benefits vs. Costs. Springer, London (2013)
24. Chen, J.C., Cheng, C.-H., Huang, P.B.: Supply Chain Management with Lean Production and RFID Application: A Case Study. Expert Systems with Applications 40, 3389–3397 (2013)
25. Tajima, M.: Strategic Value of RFID in Supply Chain Management. Journal of Purchasing & Supply Management 13, 261–273 (2007)
26. Lim, M.K., Bahr, W., Leung, S.C.H.: RFID in the warehouse: A literature analysis (1995-2010) of its applications, benefits, challenges and future trends. International Journal of Production Economics 145(1) (2013)
27. Mehrjerdi, Y.Z.: RFID Adoption: A Systems Thinking Perspective Through Profitability Engagement. Assembly Automation 31(2) (2011)

28. Azevedo, S.G., Prata, P., Fazendeiro, P.: The Role of Radio Frequency Identification (RFID) Technologies in Improving Process Management and Product Tracking in the Textile and Fashion Supply Chain. Woodhead Publishing Limited (2014)
29. Roh, J.J., Kunnathur, A., Tarafdar, M.: Classification of RFID Adoption: An Expected benefits Approach. Information & Management 46, 357–363 (2009)
30. Zhang, M.N., Li, P.: RFID Application Strategy in Agri-Food Supply Chain Based on Safety and Benefit Analysis. Physics Procedia 25, 636–642 (2007)
31. Sarac, A., Absi, N., Dauzere-Peres, S.: A literature review on the impact of RFID technologies on supply chain management. International Journal of Production Economics 128, 77–95 (2010)
32. Marks&Spencer's RFID Supply Chain, http://www.pro-4-pro.com/en/product/4477-marks-spencer-s-rfid-supply-chain.html
33. Grannis, K.: National Retail Security Survey: Retail Shrinkage Totaled $34.5 Billion in 2011 (2012),   https://nrf.com/news/national-retail-security-survey-retail-shrinkage-totaled-345-billion-2011
34. Bottani, E., Volpi, A., Rizzi, A.: Economical Assessment of the Impact of RFID technology and EPC system on the Fast-Moving Consumer Goods Supply Chain. International Journal of Production Economics 112, 548–569 (2008)
35. Ustundag, A., Tanyas, M.: The Impact of Radio Frequency Identification (RFID) technology on Supply Chain Costs. Transportation Research Part E45, 29–38 (2009)
36. Ginters, E., Martin-Guttierez, J.: Low Cost Augmented Reality and RFID Application for Logistics Items Visualization. Procedia Computer Science 26, 3–13 (2013)
37. The Gap Tests Texas Instruments RFID Smart Label Technology for Tracking Denim Clothing form the Factory to Store Shelves (2013), http://www.thefreelibrary.com/The+Gap+Tests+Texas+Instruments+RFID+Smart+Label+Technology+for...-a080008646
38. Bouzaabia, O., Boumaiza, S.: Le role de la performance logistique dans la satisfaction des consommateurs: investigation dans la grande distribution. La Revue Gestion et Organisation 5, 121–129 (2013)
39. Silvestri, C.: Quality and Customer Satisfaction: Relationships and Dynamics. A case study. Business Excellence and Management 4(1) (2014)
40. Carrender, C.: Focus on RFID's Value, Not Tag Cost (2009), http://www.rfidjournal.com/articles/view?5339
41. Schoner, B.: Five-Cent Wireless Networking – The Most Important Invention in RFID Yet (2012), http://rfid.thingmagic.com/rfid-blog/bid/91343/Five-Cent-Wireless-Networking-The-Most-Important-Invention-in-RFID-Yet
42. Corbos, R.A.: Integration and Competition – Appropriate Approaches for Achieving Excellence in Management. Business Excellence and Management 1(1) (2011)
43. RFID Vs Barcodes: Advantages and Disadvantages Comparison, http://www.aalhysterforklifts.com.au/index.php/about/blog-post/rfid_vs_barcodes_advantages_and_disadvantages_comparison
44. Carstea, G., Paun, O.: Increasing the Role of the Function of Ensuring Material Resources as an Active Participant in the Development of Strategies under Current Conditions. In: Proceedings of the 7th International Management Conference "New Management for the New Economy", Bucharest (2013)
45. Duroc, Y., Kaddour, D.: RFID potential impacts and future evolution for green projects. Energy Procedia 18, 91–98 (2012)

Exploring the Repurchase Intention of Smart Phones

Chiao-Chen Chang

National Dong Hwa University, No. 1, Sec. 2, Da Hsueh Rd., Hualien, Taiwan
aka@mail.ndhu.edu.tw

Abstract. This study targets on smart phone users to investigate the brand image from consumer's point of view. The purpose of this research study on brand image, satisfaction, trust, repurchase intention, and it will enhance the marketing strategy of future firms to function as the criterion. Hence, we conducted a questionnaire survey asked respondents for their smart phone use behavior by an online questionnaire survey. The sample of 313 respondents in Taiwan. The results show that brand image about smart phones and brand image about operating systems will lead to satisfaction and trust, then influence the repurchase intention of smart phones.

Keywords: brand image, satisfaction, trust, repurchase intention.

1   Introduction

According to Gartner, Inc., global mobile voice and data revenue will exceed one trillion dollars a year by 2014. Mobile will generate revenue from a wide range of additional services such as context, advertising, application and service sales, and so on. The congregation of mobile telephony, Internet services, and personal computing is emerging of the smartphone and the "mobile Internet" [1]. Therefore, smart phone is the product of combination of regular mobile phone and PDA (personal digital assistant), and positioned as a notebook computer and PDA replacement. This computer communication has become an emerging phenomenon for personal and business voice, data, e-mail, and Internet access because mobile networks and communication is making task easier and available at any place and any time. Power-efficient processors, the modern operating system, broadband Internet access, and productivity-enhancing application will propel the popularity of smart phones. Each of these will be a significant business worth several tens of billions of dollars per year. The main drivers for smart phones that enable repurchase intention are customer satisfaction and trust. To achieve research goals, it is necessary to build brand image on established brand about smart phones, infrastructure platforms, and then add specific mobility value. In addition, when users perceive that there are no significant influences between satisfaction and trust associated with different providers, what encourages subscribers to stay with their original providers is brand image strength.

Therefore, the key drivers of repurchase intention of smart phones are brand image about smart phones and brand image about operating systems, as well as satisfaction and trust. However, few studies have discussed a link between satisfaction, trust, and

H.A. Le Thi et al. (eds.), *Model. Comput. & Optim. in Inf. Syst. & Manage. Sci.*,
Advances in Intelligent Systems and Computing 360, DOI: 10.1007/978-3-319-18167-7_17

the repurchase intention in the smart phone industry. Moreover, in order to fill this research gap, the aim of this paper was to develop and test a research model to investigate how brand image about smart phones or platforms influences revisiting intentions via satisfaction and trust in a smart phone setting.

2   Literature Review and Research Hypotheses

2.1   Smart Phones and Computer Communications

The smart phone can be perceived to be an everywhere communication device. Following the definition of Smart Phone from IDC, it is the Converged Mobile Device combining mobile phone and PDA. It not only has essential function of voice communication but also can access wireless internet, e-mail and Personal Information Management (PIM), and can expand the function with the application software of games and multi-media.

Smartphones normally include the geographies of a phone with those of other general mobile devices, such as personal digital assistant, media player and GPS navigation part. Most have a touchscreen border and can perform 3rd-party apps, and are also camera phones. Advanced smartphones add broadband internet web browsing, Wi-Fi, motion sensors and mobile payment mechanisms.

2.2   Repurchase Intention

A study of post-purchase processes as primary to consumer decisions to repurchase products or services has attracted more and more attention from consumer researchers [2]. Jen and Hu [3] indicated that passengers repurchase intentions are determined by their perception of service value and the attractiveness of alternative modes.

2.3   Satisfaction

Customer satisfaction, which means that "the summary psychological state resulting when the emotion surrounding disconfirmed expectations is coupled with the consumer's prior feelings about the consumption experience" [4], is often considered as an important determinant of repurchase intention [5]. Lee [6] revealed that satisfaction is positively correlated with continuance intentions. Satisfaction is one of the antecedents of continuance intentions. Besides, in information systems, including electronic commerce sites, the intention to continuously use is related to satisfaction. As a result, the following hypothesis was:

H1. Satisfaction has a positive impact on repurchase intentions.

2.4   Trust

Trust reflects a willingness to be in vulnerability based on the positive prospect towards another party's future behavior [7]. Trust consists of three beliefs: ability, integrity and benevolence. Ability means that mobile service providers have the essential ability and knowledge to fulfill their tasks. Integrity refers to mobile service providers keep their promises and do not deceive users. Benevolence is mobile service providers are considered with users' interests, not just their own [8].

Also, satisfaction leads to a progress in trust [9]. In other words, trust is an outcome of customer satisfaction [10]. Satisfaction is also the preceding factor of trust level, whereas the trust level acts as a consequential factor. Indeed, satisfaction is the antecedent factor of trust level. Therefore, the hypotheses were as follows:

H2. Trust has a positive impact on repurchase intentions.

H3. Satisfaction has a positive impact on trust.

2.5   Brand Image

Brand image is defined as the reasoned or emotional perception consumers attach to specific brands [11]. It also refers to a series of brand associations stored in a consumer's memory [12]. Besides, brand attributes consist of 'bits' of information that are related to a brand name in consumer memory and that, when joined with the brand name, make up a brand's image [13]. Keller suggested that brand image is the sum the total of brand associations held in the memory of the consumers that led to perceptions about the brand. It consists of functional and symbolic brand beliefs. Then, Keller classified the associations of brand image into quality dimension and affective dimension. That is, brand image meant that the reasoned or emotional perceptions consumers attach to a specific product or service's brand [14]. Extended the concept of the above research, brand image should be discussed in the two dimensions of hardware and software in the smart phone setting. Brand image about smart phones refers to the brand perceptions from smart phone brands, whereas brand image about operating systems refers to the brand association from operating systems. Moreover, mobile operating systems are installed on smart phones and communicators to ensure their quality work. Thus, users choose brands of smart phones may consider the brand of mobile operating systems. For the analysis of Google Android and Windows Mobile Smart Phone which are mainly used for Smart Phone, Spec. and O.S. analysis as well as Data analysis are conducted, and evidence data are created by extracting Forensic data of Google Android and Windows Mobile Smart Phone. Among above descriptions, brand image can be discussed by two dimensions brand image about smart phones vs. brand image about operating systems. Therefore, we hypothesized that:

H4a. Brand image about smart phones has a positive impact on satisfaction.

H4b. Brand image about smart phones has a positive impact on trust.

H5a. Brand image about operating systems has a positive impact on satisfaction.

H5b. Brand image about operating systems has a positive impact on trust.

3　　Methodology

3.1　　Research Framework

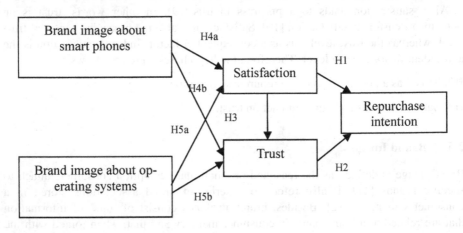

Fig. 1. Research model

3.2　　Questionnaire Design

Using the questionnaire for smart phone users. The users background which fit in the smart phone consumption included gender, age, education, occupation, and the monthly income level were collected. A 5-point Likert scale was used for all of the measurement scale items, where options ranged from "strongly disagree" (1) to "strongly agree" (5).

3.3　　Data Collection

An online questionnaire survey was administered via Internet to collect users' opinions. In total, 400 surveys were distributed, 313 data were replied, showing a response rate of 78.3 %. The questionnaire was also delivered to customers during the period of September 10th, 2014 through November 25th, 2014. About the user characteristics, 73.2% of respondents were single; while 26.8% were married. The number of male respondents was more than the female. Overall, the proportion of males was 61.3%. Most of the respondents were 21-30 years old (38.7%). Over 88.2% of the respondents had at least a college level education. The primary occupation of the respondents were: students (43.1%), business (25.6%), manufacturing (12.6%), and others (18.7%). 21.2% of the respondents made an income of $10,001~$30,000; 40.6% made $30,001~$50,000.

4   Results

The research model was performed through structural equation modeling (SEM) with AMOS 19.0. Following the two-step approach suggested by Anderson and Gerbing [15], our data analysis included two steps. First, examined the measurement model to test reliability and validity. Then we performed structural equation modeling analysis to test the research hypotheses empirically.

4.1   Measurement Model Analysis

Convergent validity was examined through composite reliability. The composite reliability of 0.81-0.87 for both constructs is above the recommended guideline of 0.70 [16]. Fornell and Larcker [17] suggested that discriminant validity is established if the square root of the average variance extracted (AVE) for an individual construct is greater than the correlation of that construct with other constructs. Findings show that this condition is met in all cases. The measure indications of this model fit with the experts' suggestion. The statistics of fit indications are acceptable compared to the ideal level.

4.2   Structural Model Analysis

In a structural model, it is essential to determine the significance and association of each hypothesized path and the variance explained (R^2 value). The result shows the standardized path coefficient and the significance of each path reported by AMOS 19.0 software. The first step in model testing is to estimate the goodness-of-fit, which is typically done using a χ^2 test; however, such tests are sensitive to sample sizes, and the probability of rejecting any model increases as sample size increases, even when the model is minimally false. Bentler and Bonnett [18] suggested the $\chi^2 / d.f.$ ratio ($d.f.$: degrees of freedom) as a more appropriate measure of model fit. This ratio, which should not exceed 5.0 for models with good fit [19], was estimated at a value of 3.13 in our hypothesized model (χ^2). Also, the recommended and actual values of fit indices of the goodness-of-fit (GFI), the adjusted goodness-of-fit (AGFI), the normalized fit index (NFI), the comparative fit index (CFI), and the root mean square error of approximation (RMSEA) are fitted in our findings. For all indices, the actual values are better than the recommended values. Therefore, the model has a good fitness [20].

The path significance of each hypothesized was also demonstrated. The direct influence of satisfaction has a significant impact on repurchase intentions ($\beta = 0.67; p < 0.001$), and trust also produced positive effects on the repurchase intentions ($\beta = 0.56; p < 0.001$). Thus, H1 and H2 are supported. To validate the key mediating role of satisfaction, a mediation test was performed following Baron and Kenny's guidelines [21]. Results show that, in terms of variance explained, trust adds 17 percent of variances in repurchase intentions ($\beta = 0.23$; $R^2 = 0.47 \rightarrow R^2 = 0.64$), thus providing support for H3. The path coefficient from brand image about smart phones to satisfaction and trust are positively significant ($\beta = 0.73$; $\beta = 0.62$), supporting H4a and H4b. The positive links from brand image about operating phones to satisfaction and satisfaction are also significant ($\beta = 0.55$; $\beta = 0.43$), lending support to H5a and H5b. As a result, all of the hypothesized associations are supported and strongly significant at $p < 0.001$.

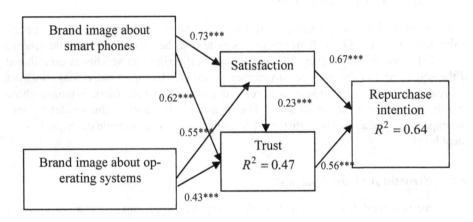

Fig. 2. Research model results

Notes. Numbers presented in the figure are standardized path coefficients, and R^2 showed the proportion of the variations of the variable that can be explained by its causing components; $* p < 0.05$; $** p < 0.01$; $*** p < 0.001$.

5   Discussion

Although repurchase intentions or behavior was investigated in the service context (e.g., [22], little studies have discussed the repurchase intention of the setting of smart phones, especially from a brand image perspective. Overall, our finding filled up the gap and showed that both satisfaction and trust have significant effects on repurchase intentions of smart phones. This research finding was consistent with past IT/IS research' suggestions [23], continuance intention was affected by satisfaction and trust.

Also, the analysis for the mediating role of trust shows support for H3. This finding is interesting because, according to one of our other preliminary studies conducted in the context of IT/IS, trust positively mediates the relationship between satisfaction and repurchase intentions. In the smart phone context, whether user trust in the smart phone using background is believed to strengthen the satisfaction-repurchase intention path. This is also consistent with Fang et al's suggestion [10], the level of trust helps predict repurchase intentions, and implies that whether user use the smart phone in addition to their usage intention, trust play a critical role to keep users remaining in the smart phone brands.

Our study indicates that brand image will lead to satisfaction and trust, this was consistent with Nam, Ekinci and Whyatt's suggestions [24]. Further, our results indicated that people use a smart phone to consider brand image about smart phones rather than brand image about operating systems, suggesting that the hardware and software in brand image of smart phones are slightly different than those of brand image in general.

6   Implications

The present study investigated that brand image about smart phones and brand image about operating systems together help to build up satisfaction and trust in the smart phone market. An implication is that the effects of brand image about smart phones are more than those of brand image about operating systems. That is, brand image about smart phones have significant influence on satisfaction and trust, in line with Lee et al.'s propositions [9], physical brand image enhances satisfaction and trust. In addition, the SEM analysis found that the relationship among satisfaction, trust, and repurchase intentions and indicated that trust play a critical mediator in the link of satisfaction and repurchase intentions. It implies that satisfaction as a consumer's overall emotional response to the entire brand experience following the last purchase. Thus, service providers should consider that how to increase user satisfaction which created repurchase intentions.

Acknowledgements. This study was supported by research funds from College of Management, National Dong-Hwa University, Taiwan (MOST 103-2410-H-259-023-).

References

1. Ishii, K.: Internet use via Mobile Phone in Japan. Telecommunications Policy 28(1), 3–58 (2004)
2. Oliver, R.L.: Cognitive, Affective, and Attribute Bases of the Satisfaction Response. Journal of Consumer Research 20(3), 418–430 (1993)
3. Jen, W., Hu, K.C.: Application of Perceived Value Model to Identify Factors Affecting Passengers' Repurchase Intentions on City Bus: A Case of the Taipei Metropolitan Area. Transportation 30(3), 307–327 (2003)
4. Oliver, R.L.: Measurement and evaluation of satisfaction processes in retail settings. Journal of Retailing 57(3), 25–48 (1981)
5. Deng, Z., Lu, Y., Wei, K.K., Zhang, J.: Understanding Customer Satisfaction and Loyalty: An Empirical Study of Mobile Instant Messages in China. International Journal of Information Management 30(4), 289–300 (2010)

6. Lee, M.C.: Explaining and Predicting Users' Continuance Intention toward E-Learning: An Extension of the Expectation–Confirmation Model. Computers & Education 54(2), 506–516 (2010)
7. Mayer, R.C., Davis, J.H., Schoorman, F.D.: An Integrative Model of Organizational Trust. The Academy of Management Review 20(3), 709–734 (1995)
8. Zhou, T.: An Empirical Examination of User Adoption of Location-Based Services. Electronic Commerce Research 13(1), 25–39 (2013)
9. Lee, D., Moon, J., Kim, Y.J., Mun, Y.Y.: Antecedents and Consequences of Mobile Phone Usability: Linking Simplicity and Interactivity to Satisfaction, Trust, and Brand Loyalty. Information & Management (2014) (in press)
10. Fang, Y., Qureshi, I., Sun, H., McCole, P., Ramsey, E.: Trust, Satisfaction, and Online Repurchase Intention: The Moderating Role of Perceived Effectiveness of E-Commerce Institutional Mechanisms. MIS Quarterly (in press, 2015)
11. Dobni, D., Zinkhan, G.M.: In Search of Brand Image: A Foundation Analysis. In: Goldberg, M.E., Gorn, G., Pollay, R.W. (eds.) Advances in Consumer Research, pp. 110–119. Association for Consumer Research, Provo (1990)
12. Aaker, D.A.: Managing Brand Equity. The Free Press, New York (1991)
13. Keller, K.L.: Conceptualizing, Measuring, and Managing Customer-Based Brand Equity. Journal of Marketing 57(1), 1–22 (1993)
14. Koubaa, Y.: Country of Origin, Brand Image Perception, and Brand Image Structure. Asia Pacific Journal of Marketing and Logistics 20(2), 139–155 (2007)
15. Anderson, J.C., Gerbing, D.W.: Structural Equation Modeling in Practice: A Review and Recommended Two-Step Approach. Psychological Bulletin 103(3), 411–423 (1988)
16. Sharma, S.: Applied Multivariate Techniques. John Wiley & Sons, New York (1996)
17. Fornell, C., Larcker, D.F.: Evaluating Structural Equation Models with Unobservable Variables and Measurement Error. Journal of Marketing Research 18(1), 39–50 (1981)
18. Bentler, P.M., Bonett, D.G.: Significance Tests and Goodness of Fit in the Analysis of Covariance Structures. Psychological Bulletin 88(3), 588–606 (1980)
19. Bentler, P.M.: EQS Structural Equations Program Manual. BMDP Statistical Software, Los Angeles (1989)
20. Gefen, D., Straub, D.W., Boudreau, M.C.: Structural Equation Modeling and Regression: Guidelines for Research Practice. Communications of the Association for Information Systems 4(7), 1–70 (2000)
21. Baron, R.M., Kenny, D.A.: The Moderator-Mediator Variable Distinction in Social Psychological Research: Conceptual, Strategic, and Statistical Considerations. Journal of Personality and Social Psychology 51(6), 1173–1182 (1986)
22. Haverila, M.: Mobile Phone Feature Preferences, Customer Satisfaction and Repurchase Intent among Male Users. Australasian Marketing Journal 19(4), 238–246 (2011)
23. Lee, M.C.: Understanding the Behavioral Intention to Play Online Games: An Extension of the Theory of Planned Behavior. Online Information Review 33(5), 849–872 (2009)
24. Nam, J., Ekinci, Y., Whyatt, G.: Brand Equity, Brand Loyalty and Consumer Satisfaction. Annals of Tourism Research 38(3), 1009–1030 (2011)

Particle Swarm Optimization with Improved Bio-inspired Bees

Mohammed Tayebi[1] and Ahmed Riadh Baba-Ali[2]

[1] UHBC University Hassiba Ben Bouali Chlef Algeria
[2] LRPE Laboratory USTHB University Alger Algeria

Abstract. To solve difficult problems, bio-inspired techniques have been developed to find solutions. We will propose in this paper, a new hybrid algorithm called BEPSO (Bees Elitist Particle Swarm Optimization) which is an algorithm based on the PSO, and inspired by the behavior of bees when searching for food. BEPSO uses an iterative process where each iteration integrates a random search followed by an exploration phase and an intensification phase. Three parameters allow the user to control the duration of each phase which gives the algorithm the flexibility to adapt the different kind of problems. After each iteration, the best obtained result is stored, which gives BEPSO an elitist behavior. Besides, random fluctuation [1] avoids the premature convergence problem. Then we will present the tests on 25 well known Benchmark functions in the literature. We will compare BEPSO with six other variants of the PSO algorithm. The results have shown the superiority of our approach compared to the other approaches both in terms of quality of solutions and convergence speed.

Keywords: *Swarm* intelligence, Bio-inspired, optimization, PSO.

1   Introduction

Biology is full of resources and models that can be used in technology. One may cite in the field of computer science, algorithms such as Particle Swarm Optimization algorithms (PSO), Ant Colony Optimization (ACO) and Artificial Bee Colony (ABC).

Our main work consists of two major changes to the PSO algorithm inspired by the behavior of bees observed in the experience of Bonabeau [1]:

1   Alternate between the global and local displacement of particles during the execution of the algorithm, to take advantage of the characteristics of the two methods namely the exploration and exploitation,

2   An application of one of the fundamental principles of a self-organized system which is the amplification of random fluctuations [1].

In this paper we will first introduce the theoretical foundations of our approach, then will describe the basic PSO method, and introduce the proposed BEPSO (Bees

© Springer International Publishing Switzerland 2015   197
H.A. Le Thi et al. (eds.), *Model. Comput. & Optim. in Inf. Syst. & Manage. Sci.*,
Advances in Intelligent Systems and Computing 360, DOI: 10.1007/978-3-319-18167-7_18

Elitist Particle Swarm Optimization) algorithm. Finally, we will compare our results with those obtained by six other PSO based algorithms.

2   Particle Swarm Optimization

2.1   Introduction

The particle swarm optimization (PSO) is an evolutionary computation technique inspired by the social behavior of flocks of birds. Kennedy, J., and Eberhart, R., introduced [2] PSO method in 1995. It consists of a set of particles moving in a search space, where each particle represents a possible solution of the optimization problem.

2.2   The Equation of Movement

The PSO algorithm is generally based on the following equation [4]:

$$v_{ij}(t+1) = \chi *[v_{ij}(t) + c_1R_1 (p_{ij}(t)-x_{ij}(t)) + c_2R_2 (p_{gj}(t)-x_{ij}(t))] \tag{1}$$

$$x_{ij}(t+1) = x_{ij}(t) + v_{ij}(t+1) \tag{2}$$

As c1 and c2 are positive constants;
R1 and R2 are random numbers generated in [0,1];
$X_i = (x_{i1}, x_{i2}, ..., x_{iD})$ represents the ith particle in the swarm;
$P_i = (p_{i1}, p_{i2}, ..., p_{iD})$ is the best position found by the particle i;
g is the index of the best particle of the swarm;
t the iteration counter;
$V_i = (v_{i1}, v_{i2}, ..., v_{iD})$ is the velocity of the particle i;
D is the dimension of the search space;
N the size of the swarm (number of particles in the swarm)
i = 1, 2, ..., N. j = 1, 2, ..., D.
χ is a parameter called the coefficient of constriction or constriction factor in order to increase the algorithm convergence, and to avoid the explosion of the swarm[4].

With these two equations (1), (2) we calculate the new position of the particles. The first equation describes the way in which it modifies the speed. The second equation calculates the next position from the current position, and the calculated speed (1). The basic PSO algorithm is described in [9].

2.3   Approaches L-Best and G-Best

There are two approaches generally used in PSO: the L-best and the G-best approaches. According to the considered neighborhood, each particle chooses a limited number of particles which provides information on their own experience (the places they have visited), these particles are called informants [9] of a particle.

For G-best, it is assumed that the set of informants of a particle is formed by all the swarm, while for L-best the set of informants is formed by a set of particles constituting a random neighborhood topology. The difference between these two approaches is in the third term of the motion equation (1). Where for G-best, g indicates the index of the best particle of the whole swarm; while for L-best, g means the best particle among the considered neighborhood.

2.4   Neighborhood Topology

Several topologies are described in the literature [10]. In our paper we adopt a randomly chosen particle to constituted neighborhood of particles informants. This approach is easier to implement and is less computer resources consuming.

3   BEPSO the Proposed Algorithm

3.1   Introduction

Clerc said [9] that the behavior of particles in the PSO algorithm is similar to the behavior of bees rather than that of the birds, the original inspiration model. Referring to biology and the foundation of the theory of self-organizing biological systems, we have proposed improvements to the original PSO algorithm, that we will detail in what follows.

3.2   Bonabeau Experience (Biology as Inspiration)

Bonabeau et al. [1] in his book "Swarm intelligence from natural to artificial systems" refers to an experiment on a bee hive to see the behavior of individuals of that hive.

"The experience is to put two food sources at equal distances from a bee hive at 8: 00 am. The two sources are not identical in their nutrient values. Site A is characterized by a sugar concentration of 1.00mol/l, and the B site has characterized a sugar concentration of 2.50 mol/l. 8: 00 until midday Source A is visited 12 times, and source B 91 times. At noon the two sources have been modified and the source A is now characterized by a sugar concentration of 2.50 mol / l and the source B is characterized by a sugar concentration of 0.75 mol/l. From noon to 16: 00, Source A is visited 121 times and source B 10 times".

He concluded: "It has been shown experimentally that a bee has a relatively high probability to dance for a good food source and abandoning a poor source of food. These simple behavioral rules allow the colony to choose the better quality source. "

Bonabeau [1] noticed that the bees are able to select the best source of food in a short time. In the other hand, the least concentrated sites in sugar are also visited by some bees.

3.3   Exploration versus Exploitation

Bees are able to find a balance between the operation of a food source, and the exploration of the search space. So when a superior food foraging site appears in the search space, bees detect and exploit it.

BEEKMAN et al. [12] said: "As soon as conditions are not stable, however, which is mostly the case in nature, it becomes important to have a mechanism that allows the change-over to another food source or food sources when they have become more profitable or when the initial food source has been depleted. This means that in order to do well in a dynamically changing environment, insect colonies should allow the storage of information about food patches which are currently exploited but at the same time allow the exploration for new sites".

"The key to keeping track of changing conditions is the trade-off between exploitation and exploration: the use of existing information (exploitation) versus the collection of new information (exploration)".

3.4   Discussion

1.1. Bees Communicate with each other by the mean of dances inside the hive. This contributes to spread site information. This allows a transmission of information to all the swarm, thereby the best site will be identified and used by all the bees.

1.2. For the least concentrated sources, we notice that they are not abandoned. They are visited perhaps by bees who have lost [1] their way, or who did not attend the dance. But these bees can bring to the hive useful information on another much better site. In fact as soon as the concentration of the two sources has changed in favor of site B (see the experience of Bonabeau mentioned earlier), some bees bring this information to the hive on the new concentration of this source. So the bees begin to exploit this new site.

This is one of the principles of biological systems random fluctuations [1] (random walks, errors, random switching-task, and so on). Not only are the structures they emerge despite the randomness but randomness is often crucial and vital to a colony, since it allows the discovery of new solutions.

These two points are the basis of our BEPSO as described below.

3.5   Proposed Algorithm

We have introduced three new parameters: *random_max, glob_max* and *local_max*. They are used in combination with another PSO parameter, that is *eval_max* (the maximum number of evaluations of the objective function at each attempt). We modified the PSO algorithm to aggregate two PSO approaches G-best, and L-best in the same algorithm as follows:

- The algorithm consists of three nested loops, one inside the other.
- The loop at the top level is a random reset location of the particles except the best particle already found, in order not to lose the optimum during the research phase. This method allows to keep track of the best position found and to explore other parts of the search space (random exploration). This loop is performed a number of times equal to *random_max* (a parameter introduced in the BEPSO method). This loop incorporates two nested loops, the first performs a global search, while the second (innermost) performs a local search.

- The algorithm perform a *glob_max* number of global searches. For each particle displacement carried by G-best, a *local_max* number of searches using the L-best algorithm are performed. In the meantime, each displacement to a new position is evaluated.

The two inner loops are modeling the behavior of foraging bees. The innermost loop that runs a *local_max* number of times (L-best displacement), modelises site exploitation. When a bee returns to the hive to deposit the nectar and pollen, it can be informed by the means of her sister's dance that a better site exists. In this case it changes direction to the new site, which is modeled by a movement of the G-best exploration algorithm.

The loop of random motion (the uppermost) provides more diversification to the algorithm without losing the best site. The algorithm always keeps track of the best site found so far by the swarm. This gives our algorithm an elitist behavior as well as keeping random fluctuation properties.

The algorithm is guided with the following arguments, linked by the equation (3):

$$eval\_max = random\_max*(1 + glob\_max * (local\_max + 1)) \tag{3}$$

The maximum number of allowed evaluations is set by the user. Then the *random_max*, *global_max* and, *local_max* are accordingly set depending we want to intensify the randomness of particle motion, or local or global research. For the same number of evaluations, the algorithm allows several choices of *random_max*, *glob_max* and *local_max* parameters.

3.6   Program: The BEPSO Algorithm

```
Step1  t ← 0
Step2  as (target is not reached) do /* the loop executes random_max
            once, or until the threshold is reached*/
Step3  initialize S, P, and g, in the way to not lose the g of the
            previous execution (t-1) only in favor of a much better
            position (S: the position of all particles, P: the best
            personal location found for each particle, g: the best
            particle in the whole swarm.)
Step4  as (target is not reached) do /*the loop executes glob_max once,
            or until the threshold is reached */
Step5  S updated using the equations of motion with a global neighborhood
            (G-best)
Step6  evaluate S
Step7  update P and redefine the index g
Step8  as (the goal is not reached) do /*the loop executes local_max
            once, or until the threshold is reached */
Step9  S updated using the equations of motion with a local neighborhood
            (L-best)
Step10 evaluate S
Step11 update P and redefine the index g
Step12 end do
Step13 end do
Step14 t ← t + 1
Step15 end do
Step16 print the best position found.
```

4   Quality of the Results

4.1   Experimental Parameters

We have compared our results with those of six PSO variants that are BPSO [3] CLPSO [5] GPSO-J [6], MPSO including VAT [13] and ATM-PSO [7] ATM-PSO* [7]. This choice was made due to the fact that:

i.   The published results are of good quality compared to other previous PSO variants.

ii.   The used functions contain hard functions of different types such as: convex, non-convex, uni-modal, multimodal, non-separable, ill-conditioned, discontinuous and noisy.

iii. Among the 25 functions, the tests include 10 Benchmark functions from the CEC2005 session on optimization [11]. These functions have undergone various transformations such as the shift and rotation. They were also used to evaluate the performance of BEPSO on transformed functions.

To facilitate the reading of the results we have noted the best published results [7] in bold characters. Besides we draw in gray, the cells which contain the best result compared to BEPSO in order to highlight the BEPSO contribution.

To be objective in our comparison, we have used the same parameters used in [7], for all our tests. Our swarm consists of 20 particles. k, the number of informants particles, is equal to 2. All experiments were performed 30 times. The maximum number of function evaluations was set to $eval\_max = 5000 * D = 150000$ (D the size of the search space), with D equal to 30 for all functions. For both L-best, and G-best methods, we adopted a constriction factor as proposed by Clerc et al. [4], with the parameters $\chi = 0.7289$; $c1 = c2 = 2.05$. This choice is motivated by the fact that Clerc et al. [4] have mathematically demonstrated the stability and the convergence of the PSO algorithm for these values of the three parameters χ, $c1$, and $c2$. Besides, for L-best we chose a random topology neighborhood.

As mentioned previously the parameter $eval\_max$, and the parameters that we have introduced in our approach ($random\_max$, $glob\_max$ and $local\_max$) are related by the equation (3). To objectively evaluate our algorithm, we set the total number of iterations of the algorithm to be equal to those of other compared algorithms. Given their linear combination according to the equation (3), BEPSO has a wide range of possible parameters values for $random\_max$, $glob\_max$ and $local\_max$. As a consequence, we performed the tests with 35 different values of these parameters. Table 1 contains the best result and the worst results among those obtained by BEPSO for the 35 tests.

The parameters used for the 35 tests are:

$random\_max$ takes the following values: {1, 10, 30, 50, 100}. All PSO algorithms initialize random particle locations at the beginning of the program. When *random_max* = 1, BEPSO carry out a single random initialization of particle locations at the beginning of the algorithm. Other values indicate that during the execution phase of the algorithm, other random initializations of particles are performed with *random_max* equal to {10, 30, 50, 100}. For each value of *random_max*, *glob_max* takes the following values {10, 30, 50, 100, 300, 500, 1000}, which makes 35 tests to perform. The *local_max* parameter is deduced from equation (3). For example when *random_max*=10 and *glob_max*=100 then *local_max* is approximately equal to 150, in order to keep the total number of evaluations equal to 150000.

4.2   Results and Comparison

The results in Table 1 show that BEPSO outperforms the other algorithms in the majority of tests. We will discuss the results for each type of functions.

4.2.1   Uni-modal Functions

- The "Sphere" is a simple unimodal function. All algorithms are able to optimize it with a more or less acceptable error. Unlike the other algorithms, BEPSO is able to reach the global optimum without any error.
- "Quadric" is a unimodal function. Some algorithms such as Basic PSO (BPSO) CLPSO, GPSO-J, are not able to find good solutions. The best published result is the one of ATM-PSO. BEPSO finds a better solution even in the worst case.
- None of the six tested algorithms could optimize the "Rosenbrock" function which is an unimodal function. Note that BEPSO was the only one to do, for all the runs with different parameters. Even in the case of the worst result, BEPSO was able to optimize this difficult function.
- Again BEPSO for all the 35 tests, BEPSO outperforms the results published for the optimization of the uni-modal "Schewfel.21" function.
- BEPSO comes close to the global optimum (which is equal to zero for all functions) with an error of about 10^{-324} for the "Schewfel.22" function, while the best result obtained by the concurrent algorithms which is the ATM-PSO is with an error of order of 10^{-54}.
- For the "Elliptic" function, our algorithm reaches the global optimum, while basic PSO (BPSO) cannot optimize this function. The best result obtained by the other algorithms is the one of ATM_PSO with an error of the order of 10^{-95}. The six functions just mentioned are all uni-modal. They are optimized successfully by BEPSO, and in all cases we obtained the best results.

Table 1. Mean Results and Standard Deviations (Values in Parentheses) Over 30 Independent Trials, for Six PSO Algorithms on 25 Benchmark Functions

| Functions | BPSO | CLPSO | GPSO-J | MPSO-TVAC | ATM-PSO | ATM-PSO* | BEPSO | |
							best case	worst case
Sphere	3.57E-103	8.99E-014	3.85E-007	2.87E-041	**8.90E-104**	6.87E-034	0	8.71E-53
	(-1.00E-102)	(-4.66E-14)	(-7.97E-07)	(-5.94E-41)	**(-3.18E-103)**	(-2.01E-33)	(0)	(1.19E-51)
Quadric	388.89	1657.34	6.15	4.05E-005	**3.46E-10**	3.43E-009	2.63E-80	1.85E-15
	(-1470.91)	(-356.09)	(-2.07)	(-1.81E-04)	**(-5.07E-10)**	(-4.47E-09)	(1.70E-73)	(1.74E-14)
Rosenbrock	17.62	20.88	27.11	19.36	**15.13**	15.99	1.38E-25	2.71E-14
	(-2.2)	(-2.58)	(-16.91)	(-19.18)	**(-8.71E-01)**	(-0.77)	(1.49)	(0.99)
Schewfel.21	7.70E-003	6.74	6.04E-001	4.28E-002	**4.52E-04**	2.50E-003	5.36E-53	1.76E-13
	(-5.43E-03)	(-9.37E-01)	(-7.56E-01)	(-4.33E-02)	**(-4.05E-04)**	(-8.98E-04)	(6.11E-46)	(9.78E-13)
Schewfel.22	1.48E-046	3.17E-009	5.16E-003	5.77E-014	**2.02E-54**	1.05E-017	4.9406564584e-324	1.43E-31
	(-7.94E-46)	(-8.34E-10)	(-9.80E-03)	(-3.09E-13)	**(-1.09E-53)**	(-3.13E-17)	(0)	(1.10E-30)
Step	27.20	**0**	**0**	**0**	**0**	**0**	0	0
	(-75.24)	**(0)**	**(0)**	**(0)**	**(0)**	**(0**	(0)	(1.62)
Elliptic	620726.36	3.68E-011	546.50	5.39E-027	**1.33E-95**	3.83E-029	0	2.15E-51
	(-2413133.56)	(-1.89E-11)	(-2888.8)	(-2.90E-26)	**(-7.14E-95)**	(-1.14E-28)	(0)	(4.25E-50)
Rastrigin	55.12	1.34E-006	15.86	11.54	**0**	15.02	0	34.82
	(-14.21)	(-1.66E-06)	(-4.52)	(-6.83)	**(0)**	(-6.91)	(0)	(17.24)
Schewfel.26	5041.88	**1.76E-12**	1336.29	2017.19	2.36E-12	3289.29	593.71	2019.52
	(-931.93)	**(-3.27E-13)**	(-290.81)	(-434.29)	(-9.57E-13)	(-534.01)	(500.23)	(410.03)
Ackley	1.64	8.45E-008	1.43E-003	4.21E-014	**2.59E-14**	3.61E-014	4.00E-15	4.00E-15
	(-8.69E-01)	(-1.96E-08)	(-1.36E-03)	(-6.53E-15)	**(-6.12E-15)**	(-4.88E-15)	(6.38E-16)	(5.11E-01)
Greiwank	2.04E-02	**1.95E-009**	4.02E-02	1.21E-02	2.22E-02	1.06E-02	0	0
	(-2.56E-02)	**(-4.35E-09)**	(-4.02E-02)	(-1.39E-02)	(-2.03E-02)	(-1.02E-02)	(5.95E-03)	(1.79E-02)
Penalized1	1.00E-001	4.83E-015	4.47E-009	9.94E-031	**1.57E-32**	2.07E-002	1.57E-32	1.57E-32
	(-1.98E-01)	(-2.31E-15)	(-1.16E-08)	(-5.20E-30)	**(-5.47E-48)**	(-4.15E-02)	(5.47E-48)	(5.47E-48)
Penalized2	1.10E-001	6.52E-014	1.26E-008	3.75E-032	**1.35E-32**	2.87E-32	1.35E-32	1.35E-32
	(-3.27E-01)	(-2.95E-14)	(-3.39E-08)	(-5.55E-32)	**(-2.74E-48)**	(-8.18E-32)	(5.47E-48)	(1.79E-02)
Weierstrass	4.36	1.54E-007	9.65E-001	5.00E-002	**0**	6.08E-001	0	0
	(-2.33)	(-7.92E-08)	(-9.42E-01)	(-2.69E-01)	**(0)**	(-7.44E-01)	(0)	(9.60E-01)
Quartic	5.12E-003	8.18E-03	**4.45E-003**	4.99E-003	9.77E-03	1.07E-02	1.95E-04	1.28E-03
	(-1.97E-03)	(-2.39E-03)	**(-9.95E-04)**	(-2.13E-03)	(-3.16E-03)	(-3.18E-03)	(5.14E-04)	(5.59E-04)
f16	454.02	1.93E-014	5.38E-006	8.87E-007	**0**	**0**	0	0
	(-467.68)	(-8.17E-15)	(-1.01E-05)	(-4.57E-06)	**(0)**	**(0)**	(0)	(0)
f17	455.98	3293.06	6.62	2431.03	**1.22E-09**	3.29E-008	0	2.27E-13
	(-1276.99)	(-636.98)	(-1.75)	(-2359.08)	**(-1.87E-09)**	(-3.50E-08)	(10.02)	(1.21E-13)
f18	3496895.52	21679625.67	3307270.28	10901812.29	**783621.61**	866916.18	19529.70	60953.07
	(-4155615.8)	(-6280161.88)	(-1255078.87)	(-8058205.86)	**(-389786.34)**	(-258915.76)	(66454.19)	(90299.12)
f19	7383.76	**4587.19**	7679.08	11647.23	7500.35	8069	56.27	1553.43
	(-2651.99)	**(-570.43)**	(-2197.39)	(-2528.08)	(-1399.01)	(-2149.02)	(946.92)	(994.73)
f20	31900030.75	**19.76**	258.15	126085341.41	49.79	81.45	3.41E-13	6.37E-12
	(-73445513.2)	**(-16.21)**	(-328.26)	(-296818886.2)	(-60.84)	(-25.59)	(1.83)	(0.99)
f21	32.47	1.06	1.69E-001	86.54	3.84E-02	**2.04E-002**	2.84E-14	5.68E-14
	(-49.08)	(-6.62E-02)	(-6.49E-02)	(-126.21)	(-3.38E-02)	**(-1.26E-02)**	(1.13E-02)	(1.38E-01)
f22	20.94	20.96	20.99	20.78	20.32	**20.05**	20.26	20.72
	(-6.53E-02)	(-6.71E-02)	(-3.12E-02)	(-7.71E-02)	(-9.11E-02)	**(-2.11E-02)**	(1.14E-01)	(4.91E-02)
f23	130.65	1.16E-008	19.11	80.88	**0**	17.82	0	32.20
	(-29.15)	(-1.06E-08)	(-7.82)	(-13.61)	**(0)**	(-7.01)	(5.06)	(13.12)
f24	**26.96**	27.41	29.67	29.89	28.41	27.45	10.40	24.65
	(-3.28)	(-1.29)	(-3.6)	(-2.51)	(-4.37)	(-2.52)	(3.45)	(2.00)
f25	27018.97	26820.05	32171.70	131217.49	**5340.49**	10356.02	0.36	3041.81
	(-28554.63)	(-7255.38)	(-23616.06)	(-50911.32)	**(-5729.76)**	(-8716.89)	(23976.16)	(28648.97)

4.2.2 Multimodal Functions

- "Rastrigin" is one of the multimodal functions that are difficult to optimize. Thus, only BEPSO and ATM-PSO were able to reach the global optimum. These results were obtained with a *random_max* > 1, which means that the dispersion of particles is enhanced by resetting particles locations.
- The only function that BEPSO couldn't optimize is "Schewfel.26" which is a multimodal function very difficult to optimize. However, BEPSO reached an optimum with an error e=593.71, while in the literature of PSO [8], a permissible value of e = 2000 is acceptable.
- BEPSO also outperformed the other algorithms in the optimization of the multimodal functions, "Ackley," "Greiwank", "Penalized1", "Penalized2" and "Weierstrass". The results of 35 runs with BEPSO for these functions are better than the results found by concurrent algorithms, with the exception of ATM-PSO which shows a performance equal to that of the last three BEPSO functions. However BEPSO is the only one who can optimize "Greiwank" and the results found by BEPSO for "Ackley" function are the best.
- "Quartic" is a noisy function, which contains a random term. The best results are those of BEPSO. Note that even in the worst case of BEPSO, it outperforms the results found by concurrent algorithms.

4.2.3 The Transformed Functions

The ten remaining functions are the functions that have undergone different transformations such as shifts and rotations.

- BEPSO in all cases again outperforms the other competing algorithms for seven functions out of ten that are: f17, f18, f19, f20, f21, f24, and f25.
- Only BEPSO and ATM-PSO with its two variants reached the global minimum of the function f16. It is reported that 35 BEPSO runs optimize it.
- f23 is the "Rastrigin Rotated" function. This function is notoriously difficult even without rotation. Only BEPSO, and ATM_PSO could optimize it. We noticed that the best results were found with a value of *random_max* parameter greater than 1. The qualities of the results are improved when increasing *random_max*. For *random_max* values above 30, the majority of the tests give errors below 3. This result is important because the admissible error used in the literature for this function without rotation is 50.
- The function f22 is optimized by all the algorithms to approximate values between [20, 21], the best result is the one found by ATM-PSO* with an error equal to 20.03. BEPSO optimizes this function with a small difference since the error is equal to 20.26.

The best results obtained with BEPSO are better for 23 of 25 test functions, and even the worst outcomes of BEPSO outperformed those of the other algorithms in the majority of the test. This indicates that the BEPSO algorithm is efficient even without parameters adjustment.

5   Comparison of Algorithms Speed Convergence with Q-Measure Parameter

Table 2. Q-Measure of Convergence: Mean Function Evaluations Needed to Reach Thresholds Divided by the Corresponding Success Ratio. Percentages in Parentheses are Success Ratios

Functions	BPSO	CLPSO	GPSO-J	MPSO-TVAC	ATM-PSO	ATM-PSO*.	BEPSO best case	Worst case
Sphere	**7837**	67977	19343	45736	8101	10254	903.4	1252.9
	(100.00%)	(100.00%)	(100.00%)	(100.00%)	(100.00%)	(100.00%)	(100.00%)	(100.00%)
Quadric	86095	N/A	N/A	114732	**61433**	69029	10047.8333	19312.8667
	(93.00%)	(0.00%)	(0.00%)	(100.00%)	**(100.00%)**	(100.00%)	(100.00%)	(100.00%)
Rosenbrock	1049	34654	6537	33447	**980**	1875	300.7	935.833333
	(100.00%)	(100.00%)	(100.00%)	(97.00%)	**(100.00%)**	(100.00%)	(100.00%)	(100.00%)
Schewfel.21	94368	N/A	N/A	121625	69280	**64932**	10423.6667	33305.1333
	(100.00%)	(0.00%)	(0.00%)	(97.00%)	(100.00%)	**(100.00%)**	(100.00%)	(100.00%)
Schewfel.22	8531	68696	40598	46968	**8349**	9538	1003.8	1396
	(100.00%)	(100.00%)	(83.00%)	(100.00%)	**(100.00%)**	(100.00%)	(100.00%)	(100.00%)
Step	N/A	51251	17563	47345	11131	**9225**	943.066667	2640.8
	(0.00%)	(100.00%)	(100.00%)	(100.00%)	(100.00%)	**(100.00%)**	(100.00%)	(100.00%)
Elliptic	29563	87792	132040	66225	**12930**	16478	1443.7	2337.73333
	(43.00%)	(100.00%)	(50.00%)	(100.00%)	**(100.00%)**	(100.00%)	(100.00%)	(100.00%)
Rastrigin	N/A	87201	1795500	176747	**15187**	125388	16840.0333	280783.491
	(0.00%)	(100.00%)	(7.00%)	(50.00%)	**(100.00%)**	(30.00%)	(100.00%)	(43.33%)
Schewfel.26	N/A	24439	47987	134117	**9726**	N/A	13912.5	183429.6
	(0.00%)	(100.00%)	(100.00%)	(40.00%)	**(100.00%)**	(0.00%)	(6.76%)	(16.67%)
Ackley	58238	62659	42857	45808	**9664**	13021	1042.33056	4943
	(13.00%)	(100.00%)	(100.00%)	(100.00%)	**(100.00%)**	(100.00%)	(96.67%)	(100.00%)
Greiwank	**6811**	65437	18309	44381	7140	8835	812.433333	1159.03333
	(97.00%)	(100.00%)	(90.00%)	(100.00%)	(100.00%)	(100.00%)	(100.00%)	(100.00%)
Penalized1	16772	58792	31043	54780	**14030**	18104	308.133333	718.533333
	(67.00%)	(100.00%)	(100.00%)	(100.00%)	**(100.00%)**	(80.00%)	(100.00%)	(100.00%)
Penalized2	22987	66865	32460	51712	**15443**	17291	1752.1875	13908.5333
	(57.00%)	(100.00%)	(100.00%)	(100.00%)	**(100.00%)**	(100.00%)	(80.00%)	(100.00%)
Weierstrass	N/A	72818	N/A	58958	**14530**	47027	2753.42593	22072.6333
	(0.00%)	(100.00%)	(0.00%)	(97.00%)	**(100.00%)**	(100.00%)	(60.00%)	(100.00%)
Quartic	**5070**	38094	11137	39563	5988	10919	1277.93333	2173.16667
	(100.00%)	(100.00%)	(100.00%)	(100.00%)	(100.00%)	(100.00%)	(100.00%)	(100.00%)
f16	19575	30115	6577	125568	**3446**	3810	329.133333	823.9
	(13%)	(100%)	(100%)	(100%)	**(100%)**	(100%)	(100.00%)	(100.00%)
f17	63576	N/A	65467	2202300	**23174**	36901	4981.36667	11402.2
	(47%)	(0%)	(100%)	(7%)	**(100%)**	(100%)	(100.00%)	(100.00%)
f18	N/A	N/A	N/A	N/A	N/A	N/A	N/A	N/A
	(0%)	(0%)	(0%)	(0%)	(0%)	(0%)	(0.00%)	(0.00%)
f19	N/A	N/A	N/A	N/A	N/A	N/A	784020	1816890
	(0%)	(0%)	(0%)	(0%)	(0%)	(0%)	(3.33%)	(3.33%)
f20	80083	101362	94454	4339200	**38284**	38998	2090.5	9174.23333
	(20%)	(100%)	(47%)	(3%)	**(73%)**	(100%)	(100.00%)	(100.00%)
f21	1455	21163	4617	182278	1425	**1379**	145.1	463.7
	(90%)	(100%)	(100%)	(77%)	(100%)	**(100%)**	(100.00%)	(100.00%)
f22	**20**	43	100	40	**20**	20	3	300
	(100%)	(100%)	(100%)	(100%)	**(100%)**	(100%)	(100.00%)	(100.00%)
f23	30675	24859	6107	54823	**4275**	17544	635.7	6459.43333
	(13%)	(100%)	(100%)	(90%)	**(100%)**	(100%)	(100.00%)	(100.00%)
f24	**20**	43	100	40	**20**	20	3	300
	(100%)	(100%)	(100%)	(100%)	**(100%)**	(100%)	(100.00%)	(100.00%)
f25	N/A	N/A	N/A	N/A	N/A	N/A	107880	4365390
	(0%)	(0%)	(0%)	(0%)	(0%)	(0%)	(3.33%)	(3.33%)

To examine the BEPSO convergence rate, we used the admissible error ε which is set for each reference function [8]. For each algorithm, the Q-measure [7] (defined below) was used to compare the rate of convergence.

To measure the speed of convergence and the robustness of an algorithm, we assigned to each function a threshold to be achieved by the algorithms. If one run among the 30 attempt is completed by reaching the threshold, this attempt is considered as successful and the number of function evaluations necessary (FE_i) to reach this threshold is used to calculate the Q-measure parameter.

Let n_t and n_s denoting the total number of attempts, and the number of attempts made successfully, respectively. Success rate SR is defined: $SR = n_s / n_t$.

The average function used to evaluate successful attempts to reach the threshold is:

$$MFE = \frac{\sum_{i=1}^{n_s} FE_i}{n_s}.$$

Then the Q-measure is defined as follows:

$$Q_m = \frac{MFE}{SR} = \frac{n_t \sum_{i=1}^{n_s} FE_i}{n_s^2}.$$

The performance of BEPSO is shown in Table 2. BEPSO is the fastest to converge to the thresholds, it is ahead of the other algorithms in 22 of the 25 functions Benchmark tests. BEPSO reaches even the thresholds of the functions f19, f25 that have not been optimized from other algorithms. The f18 function has not been optimized by any algorithm. For Rastrigin function BEPSO ranked second, and the results found by BEPSO for this function and the best published result are very close.

We define the improvement of convergence of BEPSO as the number of function evaluations necessary to reach the function thresholds of the best algorithm divided by the number of function evaluations needed to BEPSO to reach this threshold. We found the average of improvement order of 896%.

6   Conclusion and Perspective

BEPSO is a PSO algorithm inspired by the behavior of bees. BEPSO combines the two PSO standard variants that are L-best and G-best, in order to take advantage of the main features of these two variants, namely the exploitation and exploration. The aggregation of these two variants in a single algorithm is done in a manner inspired by the behavior of bees in nature. Another main feature of our approach is the integration into the algorithm of the random fluctuation as well as elitism also observed in the behavior of bees (as shown by the experience of Bonabeau).

While remaining simple as PSO, BEPSO outperforms the algorithms tested in the majority of the tested functions and converges faster towards the optimum. The results of BEPSO are excellent for different types of functions (uni-modal, multimodal, noisy, and even the functions that have undergone transformations, such as shifts and rotations).

BEPSO is able to reach the global minimum in nine test functions, while the algorithm that gives the best results among the comparison algorithms could not reach the global optimum for five functions. The results obtained with BEPSO are better for 23 of 25 functions. For the remaining two functions BEPSO ranks second in optimizing the function f22 with a slight difference (BEPSO reached the optimum with an error of 20.26 while the best result found is an error of 20.05) . For the other function (schewfel2.26) BEPSO ranks third among all compared algorithms.

The contribution of BEPSO is remarkable from the viewpoint of quality result, for example Griewank is optimized by BEPSO, for the 35 runs carried out with different parameters. BEPSO even outperformed all algorithms in terms of speed in 22 of 25 function tests. Besides, for both f19 and f25 functions, only the 35 BEPSO runs could meet the thresholds. For the Rastrigin function BEPSO ranked second, and the results found by BEPSO and the best published result are similar (difference 1%). Moreover, on the average, BEPSO provides also a significant improvement in terms of convergence speed is in the average in the order of 896%.

References

1. Bonabeau, E., Théraulaz, G., Dorigo, M.: Swarm Intelligence From Natural to Artificial Systems (1999)
2. Kennedy, J., Eberhart, R.: Particle swarm optimization. In: Proc. IEEE Int. Conf. on Neural Networks, vol. IV, pp. 1942–1948. IEEE Service Center, Piscataway (1995)
3. Yuhui, S., Eberhart, R.: A modified particle swarm optimizer. In: Proceedings of the IEEE International Conference on Evolutionary Computation, pp. 69–73 (1998)
4. Clerc, M., Kennedy, J.: The particle swarm: explosion stability and convergence in a multi-dimensional complex space. IEEE Transactions on Evolutionary Computation 6, 58–73 (2002)
5. Liang, J.J., Qin, A.K., Suganthan, P.N., Baskar, S.: Comprehensive learning particle swarm optimizer for global optimization of multimodal functions. IEEE Trans. Evol. Comput. 10(3), 281–295 (2006)
6. Krohling, R.A.: Gaussian particle swarm with jumps. In: Proceedings of the IEEE Congress on Evolutionary Computat., pp. 1226–1231 (2005)
7. Wang, Y.-X., Xiang, Q.-L., Zhao, Z.-D.: Particle Swarm Optimizer with Adaptive Tabu and Mutation: A Unified Framework for Efficient Mutation Operators. ACM Transactions on Autonomous and Adaptive Systems 5(1), Article 1 (2010)
8. Wang, Y.-X., Xiang, Q.-L., Zhao, Z.-D.: Online Appendix to: Particle Swarm Optimizer with Adaptive Tabu and Mutation: A Unified Framework for Efficient Mutation Operators. ACM Transactions on Autonomous and Adaptive Systems 5(1), Article1 (2010)
9. Clerc, M.: Particle Swarm Optimization. ©ISTE Ltd. (2006) ISBN-13:978-1-905209-04-0
10. Kennedy, J.: Small Worlds and Mega-Minds: Effects of Neighborhood Topology on Particle Swarm Performance. IEEE 0-7803-5536-9/99/ (1999)
11. Suganthan, P., Hansen, N., Liang, J., Deb, K., Chen, Y., Auger, A., Tiwari, S.: Problem definitions and evaluation criteria for the CEC 2005 special session on real-parameter optimization. Tech. rep. 2005-05, Nanyang Technological University and KanGAL (2005)
12. Beekman, M., Sword, G.A., Simpson, S.J.: Biological Foundations of Swarm Intelligence published in book swarm intelligence Introduction and application. Natural Computing Series. Springer (2008), Series Editors: Rozenberg, G.
13. Ratnaweera, A., Halgamuge, S., Andwatson, H.: Self-organizing hierarchical particle swarm optimizer with time-varying acceleration coefficients. IEEE Trans. Evol. Comput. 8(3), 240–255 (2004)

The Impact of a New Formulation When Solving the Set Covering Problem Using the ACO Metaheuristic

Broderick Crawford[1,2,3], Ricardo Soto[1,4,5], Wenceslao Palma[1],
Fernando Paredes[6], Franklin Johnson[1,7], and Enrique Norero[8]

[1] Pontificia Universidad Católica de Valparaíso, Valparaíso, Chile
{broderick.crawford,ricardo.soto,wenceslao.palma}@ucv.cl
[2] Universidad Finis Terrae, Providencia, Chile
[3] Universidad San Sebastián, Santiago, Chile
[4] Universidad Autónoma de Chile, Santiago, Chile
[5] Universidad Central de Chile, Santiago, Chile
[6] Escuela de Ingeniería Industrial, Universidad Diego Portales, Santiago, Chile
fernando.paredes@udp.cl
[7] Departamento de Computación e Informática, Universidad de Playa Ancha,
Valparaíso, Chile
franklin.johnson@upla.cl
[8] Escuela de Ingeniería, Facultad de Ingeniería,
Universidad Santo Tomás, Viña del Mar, Chile
enorero@santotomas.cl

Abstract. The Set Covering Problem (SCP) is a well-known NP hard discrete optimization problem that has been applied to a wide range of industrial applications, including those involving scheduling, production planning and location problems. The main difficulties when solving the SCP with a metaheuristic approach are the solution infeasibility and set redundancy. In this paper we evaluate a state of the art new formulation of the SCP which eliminates the need to address the infeasibility and set redundancy issues. The experimental results, conducted on a portfolio of SCPs from the Beasley's OR-Library, show the gains obtained when using a new formulation to solve the SCP using the ACO metaheuristic.

Keywords: Set Covering Problem, Ant Colony Optimization, Meta-heuristics.

1 Introduction

The Set Covering Problem (SCP) has a wide popularity between researchers and practicioners because it arises in many real world problems such as crew scheduling [12, 22], production planning in industry [27], facility location problems [28] and others important real life situations [23]. The SCP is NP hard in the strong sense and it consists in finding a subset of columns in a $0 - 1$ matrix such that they cover all the rows of the matrix at minimum cost.

© Springer International Publishing Switzerland 2015
H.A. Le Thi et al. (eds.), *Model. Comput. & Optim. in Inf. Syst. & Manage. Sci.*,
Advances in Intelligent Systems and Computing 360, DOI: 10.1007/978-3-319-18167-7_19

Many algorithms have been developed to solve the SCP. Exact algorithms are mostly based on branch-and-bound and branch-and-cut [2, 18]. However, these algorithms suffer from two major drawbacks; they are rather time consuming and can only solve instances of very limited size. For this reason, heuristics and metaheuristics have been developed to find good or near-optimal solutions faster when applying to large problems. Classical greedy algorithms are very simple, fast, and easy to code in practice, but they are designed and applicable to a particular problem, and rarely produce high quality solutions [9]. Heuristics based on Lagrangian relaxation with subgradient optimization [8, 6] are much more effective compared with classical greedy algorithms. As top-level general search strategies, metaheuristics are applicable to a large variety of optimization problems, they are simple to design and implement. Genetic algorithms [4], simulated annealing [5], tabu search [7], evolutionary algorithms [13], ant colony optimization (ACO) [26, 10], electromagnetism (unicost SCP) [24], gravitational emulation search [1] and cultural algorithms [14] are metaheuristic that have been also successfully applied to solve the SCP.

ACO is a swarm intelligence metaheuristic inspired from the cooperative behaviour of real ants which perform complex tasks in order to find shortest paths to the food sources. An ant deposits pheromone on the ground in order to guide the other ants towards a target point (the food source). Thus, a larger amount of pheromone on a particular path is more desirable to ants selecting a path [16]. ACO has attracted the attention of researchers and many successful applications of ACO have been developed to solve optimization problems [23]. In particular, some ACO based approaches have been proposed to tackle the SCP. In [19] an approach based on the ant system combined with a local search procedure is proposed. Several ACO variants, combined with a fine-tuned local search procedure, are studied in [21] focusing on the influence of differents ways of defining the heuristic information on the performance of the ACO algorithms. In [11], a lookahead mechanism is added to ACO algorithms in order to generate only feasible solutions. A novel single-row-oriented solution construction method is proposed in [26] generating solutions in a faster way and exploring broader solutions areas. Recently, an hybrid approach [10] including a constraint programming mechanism based on arc consistecy is proposed in order to generate feasible partial solutions.

As has been noted in state of the art approaches, when solving the SCP with metaheuristics two issues must be tackled: solution infeasibility and set redundancy. A solution is unfeasible if one or more rows remains uncovered and a set is redundant if all the elements covered by a colum subset are also covered by other column subset in the solution. Trying to solve larger instances of SCP with the original Ant System (AS) [17] or Ant Colony System (ACS) [15] implementation of ACO generates a lot of unfeasible labeling of variables, and the ants can not obtain complete solutions using the classic transition rule when they move in their neighborhood. The root of the problem is that the transition rule does not check for constraint consistency when constructing a solution, thus it generates unfeasible solutions. Moreover, when a solution is founded the ACO metaheuristic does not verify the inclusion of redundant columns in the solution.

In order to improve this aspect of ACO, a new formulation [25] is proposed to overcome these difficulties. The objective function naturally penalizes redundant sets and infeasible solutions. The SCP is solved as a unconstrained optimization problem using a maximization objective that replaces both the cost minimization objective and the full coverage constraint of the classical formulation. Thus, the need for addressing infeasibility and redundant issues is avoided. In this paper, we evaluate the impact of a new formulation when solving the SCP using the ACO metaheuristic in terms of accuracy of results and convergence speed. We perform extensive experiments on a portfolio of SCPs from the Beasley's OR-Library that show the gains when using this alternative SCP formulation.

This paper is organized as follows: In Section 2, we explain the problem. In Section 3, we describe the ACO algorithm used in this work. In Section 4, we present the experimental results obtained. Finally, in Section 5 we outline some concluding remarks.

2   Problem Description

In this Section, we present both the classical and the new formulation of the SCP.

2.1   The Classical Formulation

The Set Covering Problem (SCP) can be formally defined as follows. Let $A = (a_{ij})$ be an m-row, n-column, $0-1$ matrix. We say that a column j covers a row i if $a_{ij} = 1$. Each column j is associated with a nonnegative real cost c_j. Let $I = \{1, ..., m\}$ and $J = \{1, ..., n\}$ be the row set and column set, respectively. The SCP calls for a minimum cost subset $S \subseteq J$, such that each row $i \in I$ is covered by at least one column $j \in S$. A mathematical model for the SCP is

$$Minimize \ \sum_{j=1}^{n} c_j x_j \tag{1}$$

subject to

$$\sum_{j=1}^{n} a_{ij} x_j \geq 1, \quad \forall i \in I \tag{2}$$

$$x_j \in \{0, 1\}, \quad \forall j \in J \tag{3}$$

The goal is to minimize the sum of the costs of the selected columns, where $x_j = 1$ if the column j is in the solution, $x_j = 0$ otherwise. The constraints imposes the requirement that all the rows i must be covered by at least one column.

2.2   The New Formulation

The new formulation [25] views the covering of an element as a gain at a given cost. Thus, the objective function has a maximization objective. Let $A = (a_{ij})$ be an m-row, n-column, $0 - 1$ matrix. We say that a column j covers a row i if $a_{ij} = 1$.

$$Maximize \quad \sum_{i=i}^{m} g_i y_i - \sum_{j=1}^{n} c_j x_j \tag{4}$$

subject to

$$y_i \leq \sum_{j=1}^{n} a_{ij} x_j, \quad \forall i \in I \tag{5}$$

$$x_i, y_i \in \{0, 1\} \tag{6}$$

where $x_j = 1$ if the column j (with cost $c_j > 0$) is in the solution, 0 otherwise, $y_i = 1$ if the row (with gain $g_i > 0$) is covered in the solution, $y_i = 0$ otherwise. The gain $g_i = c_{min}(y_i) + \epsilon$ is attributed to each row y_i, where ϵ is a small positive constant and $c_{min}(y_i)$ is the cost of the cheapest column among the columns that cover the row y_i.

3   Ant Colony Optimization for the Set Covering Problem

In this section, we briefly present ACO algorithms [16] and give a description of their use to solve the SCP.

ACO algorithms mimics the capability of real ants to find shortest paths between the nest and food source. From a optimization point of view, the ants are looking for *good solutions*. Real ants deposit pheromone on the ground while they searching for food. Thus, an artificial ant colony implements artificial ants whose role is to build solutions using a randomized constructive search driven by pheromone trails and heuristic information of the problem.

During the execution of the algorithm the pheromone trails are updated in order to take into account the cumulated search experience. This allows to reinforce the pheromone associated with good solutions and to *evaporate* the pheromone trails over time to avoid premature convergence. ACO can be applied in a very straightforward way to SCP. A solution is composed by columns which have associated a cost and a pheromone trail [16]. Each column can be visited by an ant only once and then a final solution has to cover all rows. Each ant starts with an empty solution and adds columns until a cover is completed. Each eligible column j has a pheromone trail τ_j and a heuristic information η_j. A column is added to a partial solution calculating a probability that depends of pheromone trail and the heuristic information. The most common form which the ACO decides to include a column is using a (*Transition Rule Probability*) as follows:

$$p_j^k(t) = \frac{\tau_j^{\alpha} \eta_j^{\beta}}{\sum_{l \notin S^k} \tau_l [\eta_l]^{\beta}} \quad \text{if } j \notin S^k \tag{7}$$

Where a partial solution of the ant k is represented by S^k. α and β are two parameters which determine the relative influence of the pheromone trail and the heuristic information in the probabilistic decision [16, 21].

3.1   Pheromone Trail τ_j

An important design decision in ACO algorithms is the setting of the pheromone trail. In the original ACO implementation for TSP the choice is put a pheromone quantity on every link between a pair of cities, but for other combinatorial problems the pheromone is a value assigned to the decision variables (first order pheromone values) [16]. In this work the pheromone trail is put on the problem's component (each eligible column j). The quantity of pheromone trail assigned to columns is based on the idea: *the more pheromone trail on a particular item, the more profitable that item is* [20]. Experimentally, we decide that the pheromone deposited in each component will be a constant value.

3.2   Heuristic Information η_j

In this work we use a heuristic information that is simply calculated as $\eta_j = 1/c_j$, where c_j is the cost of column j. In other words, the heuristic information gives preference to columns having a low cost. An ant ends the construction of a solution when all rows are covered.

3.3   ACS

In this work, we implement the Ant Colony System (ACS) [15] algorithm, the more efficient algorithm of the ACO metaheuristic.

ACS exploits the search experience accumulated by the ants more strongly than others ACO algorithms does through the use of a more aggressive action choice rule. Pheromone evaporation and pheromone deposit take place only on the columns belonging to the best so far solution. Each time an ant chooses a column j, it removes some pheromone from the component increasing the exploration.

ACS exploits a pseudo-random transition rule in the solution construction; ant k chooses the next column j with criteria:

$$Argmax_{l \notin S^k} \quad \{\tau_l [\eta_l]^\beta\} \quad \text{if } q \leq q_o \tag{8}$$

and following the Transition Rule Probability (equation 7) otherwise. Where q is a random number uniformly distributed in $[0, 1]$, and q_0 is a parameter that controls how strongly the ants exploit deterministically the pheromone trail and the heuristic information.

4   Experimental Evaluation

We have implemented ACS using the classical formulation (ACS-C) and the alternative formulation (ACS-A). The effectiveness of the proposed algorithms was

Table 1. Parameter Settings for the ACS Algorithm

	α	β	ρ	q_0	ϵ
ACS	1	0,5	0,99	0,5	3×10^{-6}

evaluated experimentally using SCP test instances from Beasley's OR-library [3]. Each instance was solved 30 times, the algorithms have been run with 100 ants and a maximum number of 10000 iterations. Table 1 shows the value considered for each standard ACO parameter: α is the relative influence of the pheromone trail, β is the relative influence of the the heuristic information, ρ is the pheromone evaporation rate, q_0 is used in ACS pseudo-random proportional action choice rule, ϵ is used in ACS local pheromone trail update. For each instance the initial pheromone τ_0 was set to 10^{-6}.

Algorithms were implemented using Java and the experiments were conducted on a 2.0GHz Intel Core2 Duo T5870 with 1Gb RAM running Debian 3.2.46-1.

Tables 2 describes de problem instances, they show the problem code, the number of rows m (constraints), the number of columns n (decision variables), the *Density* (i.e. the percentage of non-zero entries in the problem matrix), and the best known cost value for each instance *Opt* (IP optimal) of the SCP instances used in the experimental evaluation.

Table 2. Details of the SCP test instances

Problem	m	n	Density	Opt
scp410	200	1000	2	514
scpa1	300	3000	2	253
scpa2	300	3000	2	252
scpa3	300	3000	2	232
scpa4	300	3000	2	234
scpa5	300	3000	2	236
scpb1	300	3000	5	69
scpb2	300	3000	5	76
scpb3	300	3000	5	80
scpc1	400	4000	2	227
scpc2	400	4000	2	219
scpc3	400	4000	2	243
scpc4	400	4000	2	219
scpcyc07	672	448	2	144
scpcyc08	1792	1024	2	344
scpcyc09	4608	2304	2	780
scpcyc10	11520	5120	2	1792
scpd1	400	4000	5	60
scpd2	400	4000	5	66
scpd3	400	4000	5	72

Table 3. Experimental results of SCP benchmarks using ACS-C and ACS-A.

(a) ACS-C experimental results.

Problem	ACS-C	RPD	RPI	MIC	Secs	Itrs.
scp410	568	10.51	99.73	0.06	1659.02	5862
scpa1	289	14.23	99.64	0.04	2802.08	8762
scpa2	346	37.3	99.04	0.04	2593.78	8798
scpa3	356	53.45	98.63	0.11	926.51	2568
scpa4	257	9.83	99.75	0.02	4929.82	8997
scpa5	256	8.47	99.78	0.03	3210.15	9350
scpb1	87	26.09	99.33	0.03	3113.3	9179
scpb2	102	34.21	99.12	0.05	2084.27	7860
scpb3	84	5	99.87	0.05	2028.83	7202
scpc1	259	14.1	99.64	0.06	1802.12	5237
scpc2	258	17.81	99.54	0.03	2995.24	9795
scpc3	268	10.29	99.74	0.04	2232.93	6728
scpc4	236	7.76	99.8	0.12	933.83	2289
scpcyc07	167	15.97	99.59	0.06	1791.53	6193
scpcyc08	379	10.17	99.74	0.03	3195.54	9959
scpcyc09	815	4.49	99.88	0.04	2486.71	7974
scpcyc10	1875	4.63	99.88	0.03	3388.81	9945
scpd1	68	13.33	99.66	0.03	3130.9	9273
scpd2	86	30.3	99.22	0.04	2470.74	9183
scpd3	89	23.61	99.39	0.04	2580.69	8671
avg		17.58	99.55	0.05		

(b) ACS-A experimental results.

Problem	ACS-A	RPD	RPI	MIC	Secs	Itrs.
scp410	539	4.86	99.88	0.11	936.47	3598
scpa1	275	8.7	99.78	0.13	762.2	2563
scpa2	345	36.9	99.05	0.21	462.14	1354
scpa3	356	53.45	98.63	0.22	452.39	1675
scpa4	269	14.96	99.62	0.05	2132.98	4683
scpa5	278	17.8	99.54	0.07	1481.5	4073
scpb1	102	47.83	98.77	0.09	1129.27	3688
scpb2	94	23.68	99.39	0.13	780.94	2659
scpb3	84	5	99.87	0.07	1350.54	4020
scpc1	359	58.15	98.51	0.13	772.08	2295
scpc2	305	39.27	98.99	0.11	868.9	2746
scpc3	298	22.63	99.42	0.09	1058.92	3824
scpc4	268	22.37	99.43	0.2	499.3	1406
scpcyc07	167	15.97	99.59	0.08	1279.41	3944
scpcyc08	379	10.17	99.74	0.05	1979.97	5958
scpcyc09	879	12.69	99.67	0.07	1379.88	5132
scpcyc10	1987	10.88	99.72	0.04	2237.57	7989
scpd1	154	156.67	99.98	0.06	1537.1	4374
scpd2	71	7.58	99.81	0.06	1694.44	6606
scpd3	84	16.67	99.57	0.05	1826.91	5813
avg		29.31	99.25	0.09		

Computational results (best cost obtained) are shown in Tables 3a and 3b. The quality of a solution is evaluated using the relative percentage deviation (RPD) and the relative percentage improvement (RPI) measures [26]. The RPD value quantifies the deviation of the objective value Z from Z_{opt} which in our case is the best known cost value for each instance (see the third column), and the RPI value quantifies the improvement of Z from an initial solution Z_I (see the fourth column).

These measures are computed as follows:

$$RPD = (Z - Z_{opt})/Z_{opt} \times 100 \tag{9}$$

$$RPI = (Z_I - Z)/(Z_I - Z_{opt}) \times 100 \tag{10}$$

For all the implemented algorithms the solution quality and computational effort ($Secs$) are related using the marginal relative improvement (MIC) measure (see the fifth column). This measure quantifies the improvement achieved per unit of CPU time ($RPI/Secs$). The solution time is measured in CPU seconds and it is the time that each algorithm takes to first reach the final best solution.

The results expressed in terms of the numbers of iterations needed to obtain the best solution and the solving time show the effectiveness of ACS-A over ACS-C to solve SCP (see Tables 3a and 3b). Thus, ACS-C provides high quality near optimal solutions and it has the ability to generate them for a variety of instances.

From Tables 3a and 3b, it can be observed that ACS-A offers an an excellent tradeoff between the quality of the solutions (see average RPD and average RPI) obtained and the computational effort required (see MIC, Secs and Itrs columns). The results show that ACS-A is a bit worse in the average MIC but it can obtain optimal solutions in some instances w.r.t. ACS-C. Moreover, in all the cases the ACS-A converges faster than the ACS-C.

5   Conclusion

Highly constrained combinatorial optimisation problems such as the SCP have proved to be a challenge for constructive metaheuristic. In this paper, we evaluate the impact of a new formulation when solving the SCP using the ACO metaheuristic. This new formulation eliminates the need for addressing infeasibility and redundant issues.

The effectiveness of the proposed approach was tested on benchmark problems, we solved SCP with ACO using the classical and the new formulation, results were compared w.r.t. accuracy of results and convergence speed. The experimental results shows the gains obtained when using a new formulation to solve the SCP using the ACO metaheuristic.

Acknowledgements. The author Broderick Crawford is supported by grant CONICYT/FONDECYT/ REGULAR/1140897, Ricardo Soto is supported by grant CONICYT /FONDECYT/INICIACION/11130459 and Fernando Paredes is supported by grant CONICYT/FONDECYT/REGULAR/1130455.

References

[1] Balachandar, S.R., Kannan, K.: A meta-heuristic algorithm for set covering problem based on gravity 4(7), 944–950 (2010)

[2] Balas, E., Carrera, M.C.: A dynamic subgradient-based branch-and-bound procedure for set covering. Operations Research 44(6), 875–890 (1996)

[3] Beasley, J.E.: Or-library: distributing test problems by electronic mail. Journal of the Operational Research Society 41(11), 1069–1072 (1990)

[4] Beasley, J., Chu, P.: A genetic algorithm for the set covering problem. European Journal of Operational Research 94(2), 392–404 (1996)

[5] Brusco, M., Jacobs, L., Thompson, G.: A morphing procedure to supplement a simulated annealing heuristic for cost- and coverage-correlated set-covering problems. Annals of Operations Research 86, 611–627 (1999)

[6] Caprara, A., Fischetti, M., Toth, P.: A heuristic method for the set covering problem. Operations Research 47(5), 730–743 (1999)

[7] Caserta, M.: Tabu search-based metaheuristic algorithm for large-scale set covering problems. In: Doerner, K., Gendreau, M., Greistorfer, P., Gutjahr, W., Hartl, R., Reimann, M. (eds.) Metaheuristics, Operations Research/Computer Science Interfaces Series, vol. 39, pp. 43–63. Springer US (2007)

[8] Ceria, S., Nobili, P., Sassano, A.: A lagrangian-based heuristic for large-scale set covering problems. Mathematical Programming 81(2), 215–228 (1998)

[9] Chvatal, V.: A greedy heuristic for the set-covering problem. Mathematics of Operations Research 4(3), 233–235 (1979)

[10] Crawford, B., Soto, R., Monfroy, E., Castro, C., Palma, W., Paredes, F.: A hybrid soft computing approach for subset problems. Mathematical Problems in Engineering, Article ID 716069, 1–12 (2013)

[11] Crawford, B., Castro, C.: Integrating lookahead and post processing procedures with ACO for solving set partitioning and covering problems. In: Rutkowski, L., Tadeusiewicz, R., Zadeh, L.A., Żurada, J.M. (eds.) ICAISC 2006. LNCS (LNAI), vol. 4029, pp. 1082–1090. Springer, Heidelberg (2006)

[12] Crawford, B., Castro, C., Monfroy, E.: A hybrid ant algorithm for the airline crew pairing problem. In: Gelbukh, A., Reyes-Garcia, C.A. (eds.) MICAI 2006. LNCS (LNAI), vol. 4293, pp. 381–391. Springer, Heidelberg (2006)

[13] Crawford, B., Lagos, C., Castro, C., Paredes, F.: A evolutionary approach to solve set covering. In: ICEIS (2), pp. 356–363 (2007)

[14] Crawford, B., Soto, R., Monfroy, E.: Cultural algorithms for the set covering problem. In: ICSI (2), pp. 27–34 (2013)

[15] Dorigo, M., Gambardella, L.M.: Ant colony system: A cooperative learning approach to the traveling salesman problem. IEEE Transactions on Evolutionary Computation 1(1), 53–66 (1997)

[16] Dorigo, M., Stutzle, T.: Ant Colony Optimization. MIT Press, USA (2004)

[17] Dorigo, M., Maniezzo, V., Colorni, A.: Ant system: optimization by a colony of cooperating agents. IEEE Transactions on Systems, Man, and Cybernetics, Part B 26(1), 29–41 (1996)

[18] Fisher, M.L., Kedia, P.: Optimal solution of set covering/partitioning problems using dual heuristics. Management Science 36(6), 674–688 (1990)

[19] Hadji, R., Rahoual, M., Talbi, E., Bachelet, V.: Ant colonies for the set covering problem. In: Dorigo, M., et al. (eds.) ANTS 2000, pp. 63–66 (2000)

[20] Leguizamón, G., Michalewicz, Z.: A new version of ant system for subset problems. In: Angeline, P., Michalewicz, Z., Schoenauer, M., Yao, X., Zalzala, A. (eds.) Proceedings of Congress on Evolutionary Computation (CEC 1999), July 6-9. IEEE Press, Washington, DC (1999)

[21] Lessing, L., Dumitrescu, I., Stützle, T.: A comparison between ACO algorithms for the set covering problem. In: Dorigo, M., Birattari, M., Blum, C., Gambardella, L.M., Mondada, F., Stützle, T. (eds.) ANTS 2004. LNCS, vol. 3172, pp. 1–12. Springer, Heidelberg (2004)

[22] Mesquita, M., Paias, A.: Set partitioning/covering-based approaches for the integrated vehicle and crew scheduling problem. Computers and Operations Research 35(5), 1562–1575 (2008), part Special Issue: Algorithms and Computational Methods in Feasibility and Infeasibility

[23] Mohan, B.C., Baskaran, R.: A survey: Ant colony optimization based recent research and implementation on several engineering domain. Expert Systems with Applications 39(4), 4618–4627 (2012)

[24] Naji-Azimi, Z., Toth, P., Galli, L.: An electromagnetism metaheuristic for the unicost set covering problem. European Journal of Operational Research 205(2), 290–300 (2010)

[25] Nehme, B., Galinier, P., Guibault, F.: A new formulation of the set covering problem for metaheuristic approaches. ISRN Operations Research, Article ID 203032, 1–10 (2013)

[26] Ren, Z.G., Feng, Z.R., Ke, L.J., Zhang, Z.J.: New ideas for applying ant colony optimization to the set covering problem. Computers and Industrial Engineering 58(4), 774–784 (2010)

[27] Vasko, F.J., Wolf, F.E., Stott, K.L.: Optimal selection of ingot sizes via set covering. Operations Research 35(3), 346–353 (1987)

[28] Vasko, F.J., Wilson, G.R.: Using a facility location algorithm to solve large set covering problems. Operations Research Letters 3(2), 85–90 (1984)

Part III

Optimization Applied to Surveillance and Threat Detection

A Multi-level Optimization Approach for the Planning of Heterogeneous Sensor Networks

Mathieu Balesdent and Hélène Piet-Lahanier

Onera - The French Aerospace Lab, F-91123 Palaiseau, France

Abstract. Optimal sensor network problems prove to lead to complex optimization problems under constraints for which regular solutions are difficult to find. This applies even more when numerous types of heterogeneous sensors and relays have to be handled in order to fulfill the network mission requirements, in the presence of various obstacles in the zone to cover. In this article, a new formulation of the heterogeneous wireless sensor network planning problem is described. The proposed optimization strategy decomposes the global optimization problem in two steps in order to reduce the combinatorial complexity and to handle heterogeneous sensors. First, virtual sensor nodes are defined and optimized in order to handle the relay node planning. Then, a multi-level optimization strategy allows to allocate the available sensors to the different relays and to optimize their position in order to satisfy the coverage requirements while minimizing the global network cost. The proposed approach is illustrated on a large scale sensor network planning problem and the results are discussed.

1 Introduction

A sensor is a device used to transform physical stimuli into signals and sensing data. A sensor network consists of various sensors connected via sensor nodes that encapsulate one or more sensors and their communication protocol. Uses of sensor networks have known a sharp increase since the last decades and cover wide range of applications such as battlefield surveillance [4], environmental monitoring and so on [28]. The main issues to be addressed in the definition of a sensor network is how can its performances be evaluated. For answering this crucial question, an adequate metrics is the amount of monitored zone relative to the area where the network is established [27]. This value is classically defined as the coverage and the design of at least good or even optimal sensor networks can be searched for using the coverage value as a performance criterion. Coverage can be classified into three categories [10]: the coverage of targets, *e.g.* sensors collecting data relative to a set of specific points, the coverage of entire area (or blanket coverage) or the coverage of barriers, *e.g.* the surveillance of movement between segregated areas [25].

The search for the optimal network definition has been intensively studied in the literature as the *art gallery* problem [19]. In this problem, the objective is to define the minimal number of guards to cover a given zone. This problem and

---

its variants are known to be NP-hard [26]. The majority of the articles registered in the literature considers the coverage issue as one of the main function of the designed sensor network [27]. Numerous solutions and algorithms have been proposed in order to enhance the sensor network coverage while minimizing the overlapping between the different sensors [14], [24]. Optimal network problems prove to lead to complex optimization problems under constraints for which regular solutions are difficult to find. Additional complexity relies on the fact that the set of sensors to be located and connected may be composed of various types of sensors leading to a great variability in the amount of coverage provided by each component [23]. This heterogeneous sensor planning optimization problem thus involves mixed integer / continuous variables and non linear objective function and constraints. The integer variables may be related to the allocation of the different sensor nodes (*i.e.* choice of location of the node in the zone) and the continuous variables may include the different positions (and orientations if required) of the nodes (sensors or relays). In this case, the integer variables may have a direct influence on the typology of the optimization problem. Indeed, if a node is activated (*i.e.* is positioned on the map), its position has to be taken into account and optimized, and consequently 3 additional variables (and corresponding constraints) have to be included in the sensor positioning optimization problem. In this context, specific optimization approaches have to be defined in order to allow the modification of the optimization problem typology during the network optimization process. Moreover, this category of optimization problems does not comply with the classical optimization algorithms by relaxation (*e.g.* Branch and bound, *etc.*) and specific heuristics have to be employed in order to handle the continuous and categorical optimization variables.

Due to non-linearities, multimodality and discontinuous objective function, suitable optimization methods become scarce. Genetic algorithms seem efficient tools and have been widely used to optimize the sensor network planning for detection [8], [9], [16]. These algorithms allow to take into account multiple objective for the optimization [22] and can be adapted to the discrete variable handling. However, the amount of computation required for obtaining a stable solution may prove prohibitive when the network must be tailored for short-term deployment. In this paper, a multilevel optimization approach is described in order to handle the heterogeneity of involved sensors efficiently and to provide a network configuration in reduced time. The remainder of the paper is organized as follows. The network model, its objective and constraints are first defined. The multi-level optimization approach is described afterwards and examples of simulation results are presented and discussed.

2   Problem Description

A set of n sensors, consisting of l different types of sensor technology, is considered. Each sensor provides a measurement of some state components of a target when it is located within its detection zone with a detection probability. This detection zone can be described as a range R_i and field of view $[\theta_{min_i}, \theta_{max_i}]$ for the ith sensor. A target located in this zone is detected with probability p_i when it is measured by the associated sensor. Sensor relays are defined as connection links between two or more sensors. The relay is assumed to be efficient if the sensors are located at a distance to the relay location lower than a threshold $d_{connect}$. The coverage function is the area where a target is assumed to be detected by at least one sensor with a probability p greater than a threshold p_{min}. Example of heterogeneous sensor network is depicted in Figure 1, with several sensors of 3 different types, relays, obstacles and connectivity links between the sensors and relays.

The goal here is to find the minimum cost network (*e.g.* minimum number of sensors) that is able to comply with coverage targets (*i.e.* probability of detection), and subject to various constraints concerning the visibility, connectivity and presence of obstacles in the zone to cover.

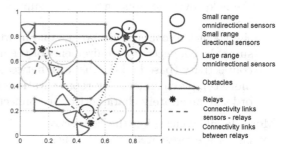

Fig. 1. Example of wireless heterogeneous sensor network

3   Multi-level Optimization Approach

In order to solve the heterogeneous wireless sensor network planning problem, a multi-level optimization approach is proposed. It aims at limiting the combinatorial complexity induced by the planning (*i.e.* allocation and position) of various network nodes (*e.g.* heterogeneous sensors, relays). This problem is proved to be a NP-hard problem [2], [12]. The goal here is to find the minimum cost network (*e.g.* minimum number of sensors) that is able to comply with coverage targets (*i.e.* probability of detection), and subject to various constraints concerning the visibility, connectivity and presence of obstacles in the zone to cover. In the following sections, we describe the global approach and the different levels of optimization.

3.1   Description of the Proposed Approach

We propose here a two-step approach for solving the network planning problem. In the first step, we define the concept of virtual sensor nodes (see Section 3.2), in order make a partition of the initial zone to cover Z into different subzones $Z_i, i = 1, \ldots, N$ in which the network node locations will be optimized. The second step consists in finding the optimal sensor node allocation and positions in order to minimize the global cost of the network (see Section 3.4). In order to deal with this global MINLP optimization problem, a bi-level optimization scheme is used in order to handle the induced categorical variables (allocation of the sensors to the different zones) and continuous variables (position and orientation of the different sensors).

3.2   Definition of the Virtual Sensor Nodes

The first step of the proposed approach consists in defining and optimizing virtual sensor nodes. A virtual sensor node (VSN) is a virtual object composed of one relay node (R) and heterogeneous sensors (S). We define here N virtual nodes composed of different sensors and we suppose that a relay node will be placed at the center of a VSN. The number of sensors might vary from one VSN to another. The concept of virtual nodes allows to handle the heterogeneity of sensors through the use of generic objects of analog properties, which are:

- the field coverage, zone in which we suppose that a target is detected by the VSN,
- the geometric connectivity area, zone in which we suppose that a VSN ensures the connectivity requirements of the network. Note that the geometric connectivity (based on the distance between the VSN) can be used to derive more complex connectivity representations such as the adjacency matrix in order to accurately characterize the connectivity graph formed by the VSN.

More details concerning the connectivity issues are presented in [18]. These two zones are represented through convex polytopes with variable geometry. To this end, the Voronoï centroidal tesselation is used to define the zone of influence of each VSN [15]. Figure 2 illustrates a VSN planning, with 6 VSN (in blue), the corresponding Voronoï zones (in green) and four obstacles of different geometry (in red).

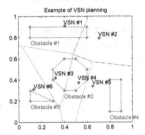

Fig. 2. VSN (blue), VSN Voronoï zones (green) and obstacles (red)

3.3   Optimization of the Virtual Sensor Nodes

The VSN are supposed to be equivalent, and one aims at uniformly covering the considered zone Z, with respect to a set of obstacles. Let:
- Z, a zone to cover,
- $w(Z)$, a ranking of subzones of Z in terms of coverage target,
- $\mathbf{O}$, a set of obstacles (or subzones which are not relevant to the coverage requirements) in Z,
- N, the number of available VSN.

The optimization problem to solve in order to handle the VSN planning is:

$$\text{Minimize} \quad \sigma(A(\mathbf{x}, \mathbf{O}, w(Z))) \tag{1}$$
$$\text{With respect to} \quad \mathbf{x} = \{\mathbf{x}_1, \ldots, \mathbf{x}_N\} \tag{2}$$
$$\text{Subject to} \quad \mathcal{O}(\mathbf{x}, \mathbf{O}, Z) = 0 \tag{3}$$
$$\mathcal{C}(\mathbf{x}, \mathbf{O}, Z) = 0 \tag{4}$$
$$\mathcal{K}_0(\mathbf{x}) \leq 0 \tag{5}$$
$$\mathcal{K}_i(\mathbf{x}_i, Z_i(\mathbf{x}_i, Z)) \leq 0, \forall i = 1, \ldots, N \tag{6}$$

$\mathbf{x}_i = (x_i, y_i)$ is the location of the ith VSN, $A(\mathbf{x}, \mathbf{O}, w(Z))$ is a function that computes the area of the Voronoï zone corresponding to the ith VSN, and restricted to the zone without obstacles. $\sigma(\cdot)$ is the measure of dispersion (*e.g.* standard deviation) applied to the set formed by the area of the Voronoï zones.

$\mathcal{O}(\mathbf{x}, \mathbf{O}, Z)$ is the constraint function which is equal to 0 if the VSN located at $\mathbf{x}_i$ is outside an obstacle. This constraint function is generic and has to be adapted to each obstacle, *e.g.* if the obstacle is described by a polytope, the constraint is defined by a series of linear constraints relative to the sides of the polytope; if the obstacle is a circle, the constraint can be modeled using the obstacle radius and distance between the sensor and the center of the obstacle.

$\mathcal{C}(\mathbf{x}, \mathbf{O}, Z)$ is the constraint function which is equal to 0 if the restricted Voronoï zone of a VSN (*i.e.* the VSN Voronoï zone taken away from the intersection of this zone to the obstacles) is connected. For example, in Figure 2, the Voronoï zone associated to VSN #4 is not connected because this zone is broken by obstacle #2. The the other VSN Voronoï zones are connected.

$\mathcal{K}_0(\mathbf{x})$ is the constraint function which checks the connectivity of the graph composed of the VSN. This scalar constraint is based on the difference between the maximal distance between the different unconnected VSN connected subgraphs and a distance of connectivity $d_{connect}$.

$$\mathcal{K}_0(\mathbf{x}) = \max_{\substack{k = 1, \ldots, T \\ l = 1 \ldots, T \\ l \neq k}} \left[\min_{\substack{n_k = 1, \ldots, N_k \\ n_l = 1, \ldots, N_l}} d(n_k, n_l) \right] - d_{connect} \tag{7}$$

with T the number of connected subgraphs, N_k (respectively N_l) the number of VSN of the subgraph k (respectively l), and $d(n_k, n_l)$ the euclidean distance between VSN n_k of the subgraph k and VSN n_l of the subgraph l. One can note that other techniques can be used to evaluate the connectivity of the graph (*e.g.* the algebraic connectivity of the VSN graph *i.e.* the 2nd smallest eigenvalue of the Laplacian matrix associated to the VSN graph, see [13]) but have not been chosen because of their binary nature.

The chosen modeling of connectivity constraint has been selected because it allows to guide more easily the optimizer towards reduction of the constraint vector Eq.(6). In Figure 3, 11 VSN are divided into 3 connected subgraphs *i.e.* subgraphs for which the distance between each node is lower than $d_{connect}$ (here 0.2). The distances between the three groups (*i.e.* d_{12}, d_{13} and d_{23} in Figure 3) can be computed and their maximum is then compared to $d_{connect}$ in order to calculate $\mathcal{C}(\mathbf{x}, \mathbf{O}, Z)$.

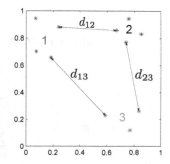

Fig. 3. Distances of connection between the three different connected subgraphs (in red, blue and green)

Finally, $\mathcal{K}_i(\mathbf{x}_i, Z_i(\mathbf{x}_i, Z))$ are the constraint functions which ensure that for each VSN i, $i = 1, \ldots, N$, the entire zone of the associated Voronoï area $Z_i(\mathbf{x}_i, Z)$ is connected (*i.e.* its distance with respect to the VSN node is smaller than the connectivity range of the relay node associated to the VSN i). This optimization problem can be then solved using classical non linear programming optimization algorithms. In this article, we use NOMAD (Nonlinear Optimization by Mesh Adaptive Direct Search) [6].

3.4   Strategy for the Sensor Allocation and Placement

The second step of the proposed approach consists in finding the sensor placement and allocation for each of the zones Z_i, $i = 1, \ldots, N$ defined previously by the VSN optimization. Since the sensors can be heterogeneous (*e.g.* radar, optic, acoustic, seismic, *etc.*), each of them involving a specific modeling in terms of performance estimation and number of variables required in their planning, the proposed approach itself involves two nested optimization levels. The first level, called the system-level, performs the allocation of the different sensors for each of the zones, subject to several constraints concerning the global stock of sensors and the capacity of the different relay nodes. The second level, called the subsystem-level, determines the positions (and orientations if required) of the sensors in order to maximize the coverage of the local sensor network in the considered VSN subzones.

Hierarchical System-Level Optimization. At the system-level, the optimization problem corresponding to the cost minimization of the sensor network is:

$$\text{Minimize} \qquad \sum_{i=1}^{N} \sum_{j=1}^{l} A_{i,j} c(T_j) \tag{8}$$

$$\text{With respect to} \qquad \mathbf{A} = \bigcup_{i=1}^{N} \{A_{i,1}, A_{i,2}, \ldots, A_{i,l}\} \tag{9}$$

$$\text{Subject to } P_d(\mathbf{A}, \mathbf{X}^*) = \frac{1}{N} \sum_{i=1}^{N} P_d(\mathbf{A}_i, \mathbf{X}_i^*) \geq P_T \tag{10}$$

$$\sum_{j=1}^{l} A_{i,j} \leq K_i, \forall i = 1, \ldots, N \tag{11}$$

$$\sum_{i=1}^{N} A_{i,j} \leq C_j, \forall j = 1, \ldots, l \tag{12}$$

with:
- $c(T_i)$: the cost of a sensor of type T_i
- $A_{i,j} \in \mathbb{N}$: the number of sensors of type j allocated to the zone Z_i,

- $\mathbf{A} = \bigcup_{i=1}^{N} A_{i,.}$: the vector describing the allocation of sensors for all the zones defined at the VSN optimization step (vector of integers),
- $\mathbf{X} = \bigcup_{i=1}^{N} \mathbf{X}_i$: the position (and orientation if needed) of the sensors in each zone (are only involved at the subsystem-level),
- $P_d(\mathbf{A}, \mathbf{X}^*)$: the probability of detection of the sensor network. $\mathbf{X}^*$ stands for the optimized positions and orientations of the sensors, determined by the subsystem-level optimization process. For conservative design, the mean constraint Eq.(10) can be replaced by a worst-case coverage, defined by:

$$P_d(\mathbf{A}, \mathbf{X}^*) = \min_{i=1,\ldots,N} [P_d(\mathbf{A}_i, \mathbf{X}_i)] \geq P_T$$

- P_T: the target probability of detection,
- K_i: the maximal capacity of the relay node of the ith VSN,
- C_j: the global stock of sensors of type j.

We consider in this problem N zones and l different types of sensors. This problem involves discrete optimization variables ($\mathbf{A}$) which cannot be relaxed during the optimization, and linear and non linear constraints. Thus, specific optimization solvers have to be used in order to efficiently solve this problem. In this article, the evolutionary CMA-ES (Covariance Matrix Adaptation - Evolution Strategy) algorithm [21] is used because of its ability to find global optimum for MINLP problems. Note that other algorithms such as NOMAD [1] can also be used to address this problem.

Subsystem-Level Optimization Dedicated to Sensor Placement Using Evolutionary Algorithm. For each zone $Z_i, i = 1, \ldots, N$, defined by the Voronoï diagrams associated to the optimized VSN, the sensor planning optimization problem (*i.e.* maximization of the probability of detection) is expressed as follows, from the sensor allocation determined at the system-level:

$$\text{Maximize} \qquad P_d(\mathbf{A}_i, \mathbf{X}_i) \qquad (13)$$

$$\text{With respect to } \mathbf{X}_i = \{x_{ji}, y_{ji}, \theta_{ji}\}, \forall j = 1, \ldots, \sum_{j=1}^{l} A_{i,j} \qquad (14)$$

$$\text{Subject to} \qquad \mathcal{I}(\mathbf{X}_i, Z_i) \leq 0 \qquad (15)$$

$$\mathcal{O}(\mathbf{X}_i, \mathbf{O}, Z_i) = 0 \qquad (16)$$

with:

- Z_i: the subzone to cover,
- $x_{ji}, y_{ji}, \theta_{ji}$: the abscissa, ordinate and orientation of sensor #j in zone #i,
- $\sum_{j=1}^{l} A_{i,j}$: the number of sensors assigned to the zone Z_i,
- $\mathcal{I}(\mathbf{X}_i, Z_i)$: a linear constraint vector function that defines the geometry of the subzone Z_i. This constraint is lower than 0 if the sensors are located in Z_i, defined at the previous step (VSN optimization).
- $\mathcal{O}(\mathbf{X}_i, \mathbf{O}, Z_i)$: a constraint vector function that checks the compatibility of the sensor positions with respect to the obstacles $\mathbf{O}$. This constraint is equivalent to Eq. (3) for the VSN optimization.

Since the global probability of detection is obtained by integration over the zone Z_i, the computation of the probability of detection P_d is performed through its Monte Carlo estimation. For that purpose, one places K targets t_k in the zone using a uniform distribution. Then, for each target t_k, its probability of detection is defined using the probability of non detection [17] as follows:

$$p(t_k, \mathbf{A}_i, \mathbf{X}_i) = 1 - \prod_{m=1}^{\sum_{j=1}^{l} A_{i,j}} (1 - p(t_k, s_m)) \tag{17}$$

with s_m the mth sensor in the zone. Then, the global probability of detection in the zone is defined by

$$P_d(\mathbf{A}_i, \mathbf{X}_i) = \frac{1}{K} \sum_{k=1}^{K} p(t_k, \mathbf{A}_i, \mathbf{X}_i^*) \tag{18}$$

This representation of the probability of detection is very generic and can be used with heterogeneous modelings of probability of detection (*e.g.* binary [11], exponential [14], probabilistic [3]) for the different types of sensors involved in the subsystem-level optimization. Since a Monte Carlo estimation of P_d is used, the subsystem optimization problem has to be solved by optimization solvers able to handle noisy functions. For practical issues, CMA-ES [20] is used in this article. This algorithm turns out to be efficient for continuous optimization in terms of convergence speed and robustness to initialization as shown in [7]. We use a penalized objective function strategy in order to handle the constraints. For more details concerning noisy optimization using evolutionary algorithms, see [5].

4 Numerical Experiments

4.1 Presentation of the Test Scenario

In order to evaluate the efficiency of the proposed approach, the following planning problem of heterogeneous sensor network has been proposed. The sensor network is composed of 6 relay nodes and 100 sensors. The sensors are generic and can be of three types (Table 1): high-range omnidirectional (*e.g.* radar), small-range omnidirectional (*e.g.* seismic) and small-range directional (*e.g.* infra-red camera). The relay nodes are similar, their characteristics are summarized in Table 1, we suppose that the maximal number of sensors which can be connected to a relay node is 15. All the data are adimensioned. The zone to cover Z is a 1×1 square in which 4 obstacles of different geometries are present (Figure 4). The probability of detection model used in this scenario is the binary model [29]. The target probability of detection considered in the system-level constraint (mean constraint) Eq.(10) is equal to 0.9. The global optimization problem involves 342 variables used to characterize 70 omnidirectional sensors (described by their 2D position), 30 directional sensors (described by their 2D position and pan angle), and 6 relays (described by their 2D position).

Table 1. Characteristics of sensor and relay nodes

	Type	Range (sensing angle)	Cost	Stock
Sensors	Small-range omnidirectional	0.1	1	40
	High-range omnidirectional	0.2	2	30
	Small-range directional	0.1 (60°)	1.5	30
Relays	Generic	0.4	1	6

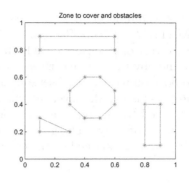

Zone to cover and obstacles

Fig. 4. Zone to cover and obstacles

4.2 Results

At the first step, 6 VSN are involved, each of them corresponding to a relay node. The result of the VSN optimization step is described in Figure 5, and the connectivity network formed by the relays is depicted in Figure 7. The evolutions of the objective function and maximum of constraint vector over the iterations are given in Figure 6. As depicted in this Figure, the dispersion of Voronoï area (objective function) can be decreased by a factor 400 during the optimization. The first feasible design (*i.e.* satisfying the constraints) is obtained at the 214th iteration. The global optimization of the different VSN converges in this example in 2500 iterations, that represents a total computation time of 5 minutes (MATLAB, 2.7Ghz Pentium with 8 cores / Windows 7).

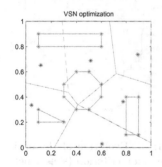

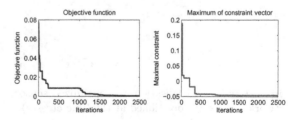

Fig. 5. Optimal configuration of VSN network

Fig. 6. Evolutions of the objective function and maximal of constraint vector over the iterations during the VSN optimization

Concerning the second step, the connected graph formed by the relay nodes is depicted in Figure 7. The initial (random) and optimized allocations of sensors are depicted in Figure 8. The total cost of the sensor network has been decreased from 135 to 62 (Figure 9), and the probability of detection under the zone to cover is above the threshold (0.9). The convergence has been achieved in

2000 system-level iterations, that represents a total computation time of 4 hours (MATLAB, 2.7Ghz Pentium with 8 cores / Windows 7). In this scenario, all the directional sensors have been preferably removed (Table 2). This can be explained by the fact that these sensors present a smaller coverage area comparing to both omnidirectional sensors and are more costly than small-range omnidirectional sensors. Moreover, all the high-range omnidirectional sensors have been allocated to the different relay nodes. The constraints relative to the maximal capacity of relays (11), global stocks of sensors (12) and the obstacle avoidance (16) are also satisfied at the convergence of the algorithm.

Table 2. Optimal allocation of the different sensors

VSN	Small omnidir.	Large omnidir.	Small dir.
1	0	5	0
2	0	5	0
3	0	5	0
4	0	5	0
5	2	5	0
6	0	5	0
Total	2	30	0
Total cost	62		

Fig. 7. Connectivity links between the relays and associated sensors (solid lines) and between the different relays (dash lines)

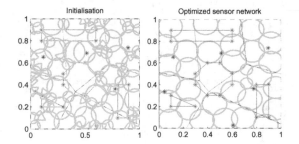

Fig. 8. Random initialisation and optimized sensor network (relays are positioned at the blue crosses)

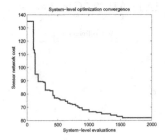

Fig. 9. Evolution of the optimization criterion (sensor network cost)

4.3   Robustness Analysis

In order to evaluate the robustness to initialization of the proposed approach, 5 optimizations of the sensor network have been performed from different random initializations at both system and subsystem levels. The results of this study are illustrated in Figure 11.

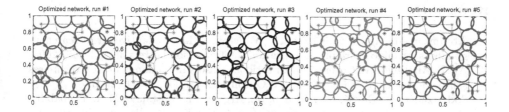

Fig. 10. Optimized networks for different random initializations

As depicted in this Figure, the proposed method converges to the same optimum in terms of objective function (62) for all the different initializations, even if the optimal sensor allocations can be quite different (Figure 10). The low dispersion of the optima in terms of objective function can be explained by the use of efficient evolutionary global optimization algorithm (CMA-ES) at both system and subsystem levels.

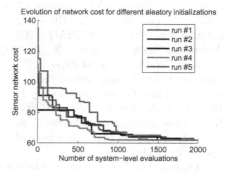

Fig. 11. Evolution of the optimization criterion for different random initializations

5   Conclusion

Determination of optimal sensor network is well-known to be a complex optimization problem especially in the case when the coverage is sought for using heterogeneous types of sensors. The proposed two-layer solution described in this paper provides a flexible framework reducing combinatorial complexity. It enables to take into account heterogeneous sensor characteristics and fulfilling the mission requirements in terms of limited number of relays, distance of communications, obstacle avoidance and restricted detection area. Its application to a large scale sensor network planning problem illustrates its efficiency in terms of robustness to initialization and reduced computing time. The perspectives of this work are now to introduce robustness of the design regarding to potential sensor faults or breakdowns and potential reallocation procedure when the occurrence of a fault is identified.

Acknowledgments. The authors thank M. Brevault for fruitful discussions.

References

1. Abramson, M., Audet, C., Chrissis, J., Walston, J.: Mesh adaptive direct search algorithms for mixed variable optimization. Optimization Letters 3(1), 35–47 (2009)
2. Ai, J., Abouzeid, A.: Coverage by directional sensors in randomly deployed wireless sensor networks. J. of Combinatorial Optimization 11(1), 21–41 (2006)
3. Akbarzadeh, V., Gagné, C., Parizeau, M., Argany, M., Mostafavi, M.A.: Probabilistic sensing model for sensor placement optimization based on line-of-sight coverage. IEEE Transactions on Instrumentation and Measurement 62(2), 293–303 (2013)
4. Akyildiz, I., Su, W., Sankarasubramaniam, Y., Cayirci, E.: Wireless sensor networks: a survey. Computer Networks 38(4), 393–422 (2002)
5. Arnold, D.: Noisy Optimization with Evolution Strategies. Springer (2002)
6. Audet, C., Le Digabel, S., Tribes, C.: NOMAD user guide. Technical report G-2009-37, Les cahiers du GERAD (2009)
7. Auger, A., Finck, S., Hansen, N., Ros, R.: Comparison Tables: BBOB 2009 Noisy Testbed. INRIA Technical Report N°384 (2010)
8. Barrett, S.: Optimizing sensor placement for intruder detection with genetic algorithms. In: Intelligence and Security Informatics, pp. 185–188. IEEE (2007)
9. Bhondekar, A., Vig, R., Singla, M., Ghanshyam, C., Kapur, P.: Genetic algorithm based node placement methodology for wireless sensor networks. In: Proceedings of the International Multiconference of Engineers and Computer Scientists, vol. 1, pp. 18–20 (2009)
10. Cardei, M., Wu, J.: Coverage in wireless sensor networks. In: Handbook of Sensor Networks, pp. 422–433 (2004)
11. Chakrabarty, K., Iyengar, S., Qi, H., Cho, E.: Grid coverage for surveillance and target location in distributed sensor networks. IEEE Transactions on Computers 51(12), 1448–1453 (2002)
12. Cheng, X., Du, D., Wang, L., Xu, B.: Relay sensor placement in wireless sensor networks. Wireless Networks 14(3), 347–355 (2008)
13. de Abreu, N.M.M.: Old and new results on algebraic connectivity of graphs. Linear Algebra and its Applications 423(1), 53–73 (2007), Special Issue devoted to papers presented at the Aveiro Workshop on Graph Spectra A
14. Dhillon, S.S., Chakrabarty, K.: Sensor placement for effective coverage and surveillance in distributed sensor networks. IEEE Wireless Communications and Networking 3, 1609–1614 (2003)
15. Du, Q., Faber, V., Gunzburger, M.: Centroidal voronoi tessellations: applications and algorithms. SIAM Review 41(4), 637–676 (1999)
16. Feng, G., Liu, M., Wang, G.: Genetic algorithm based optimal placement of pir sensors for human motion localization. Optimization and Engineering 15(3), 1–14 (2013)
17. Ghosh, A., Das, S.: Coverage and connectivity issues in wireless sensor networks: A survey. Pervasive and Mobile Computing 4(3), 303–334 (2008)
18. Godsil, C., Royle, G., Godsil, C.D.: Algebraic graph theory. Springer (2001)
19. González-Baños, H.: A randomized art-gallery algorithm for sensor placement. In: 17th Annual Symposium on Computational Geometry, pp. 232–240 (2001)
20. Hansen, N.: The CMA evolution strategy: A comparing review. In: Lozano, J.A., Larrañaga, P., Inza, I., Bengoetxea, E. (eds.) Towards a New Evolutionary Computation. STUDFUZZ, vol. 192, pp. 75–102. Springer, Heidelberg (2006)
21. Hansen, N.: A CMA-ES for Mixed-Integer Nonlinear Optimization. Research Report RR-7751, INRIA (October 2011)

22. Jourdan, D., de Weck, O.: Layout optimization for a wireless sensor network using a multi-objective genetic algorithm. In: IEEE 59th Vehicular Technology Conference, vol. 5, pp. 2466–2470. IEEE (2004)

23. Lazos, L., Poovendran, R.: Stochastic coverage in heterogeneous sensor networks. ACM Transactions on Sensor Networks 2(3), 325–358 (2006)

24. Li, J., Wang, R.-C., Huang, H.-P., Sun, L.-J.: Voronoi based area coverage optimization for directional sensor networks. In: 2nd International Symposium on Electronic Commerce and Security, vol. 1, pp. 488–493. IEEE (2009)

25. Mulligan, R., Ammari, H.: Coverage in wireless sensor networks: a survey. Network Protocols and Algorithms 2(2), 27–53 (2010)

26. O'Rourke, J.: Art gallery theorems and algorithms, vol. 57. Oxford University Press (1987)

27. Wang, B.: Coverage problems in sensor networks: A survey. ACM Computing Surveys (CSUR) 43(4), 32 (2011)

28. Yick, J.R., Mukherjee, B., Ghosal, D.: Wireless sensor network survey. Computer Networks 52(12), 2292–2330 (2008)

29. Zhu, C., Zheng, C., Shu, L., Han, G.: A survey on coverage and connectivity issues in wireless sensor networks. Journal of Network and Computer Applications 35(2), 619–632 (2012)

Contribution to the Optimization of Data Aggregation Scheduling in Wireless Sensor Networks

Mohammed Yagouni[1,2], Zakaria Mobasti[3],
Miloud Bagaa[3], and Hichem Djaoui[2]

[1] Laromad, USTHB, BP 32 El-Alia, Bab-Ezzouar, Algiers, Algeria
mohammed.yagouni@univ-lorraine.fr
[2] RECITS Laboratory, USTHB
{zakariausthb,hichemdjaoui}@gmail.com
[3] Research Center on Scientific and Technical Information (CERIST) Algiers, Algeria
bagaa@mail.cerist.dz

Abstract. Wireless Sensor Network (WSN) is a very important research area. It has wide range of applications in different domains, such as industrial production processes and military fields. The sensing of events and forwarding of sensed information, through multi-hops wireless communications, toward the base-station is the main functionality of a WSN. In this paper, we study the problem of how gather and aggregate data from the whole sensor nodes to the base-station, with efficient way, such that the energy consumption is reduced, messages collision are prevented and the time latency is minimized. The problem of data aggregation scheduling is mathematically modeled, and then a heuristic to resolve the problem are proposed. The simulation results show the efficiency of our solution compared to the best approaches existed in literature.

Keywords: Sensors, Wireless Sensor Network, data aggregation scheduling, Heuristic, mathematical modeling.

1 Introduction

In recent years more and more applications are provided on multimedia devices possibly nomads (PC, Laptop, Phone, PDA). However, these applications are unable to account for their surroundings. Further technological development in fields of micro-electronics, micro-mechanical, and wireless communication technologies wire have created, with a reasonable cost, small communicating devices equipped with sensors. These new objects, called "sensor nodes" or more commonly "sensor" feature capture units, calculation, storage and communication. To feed their need current, these nodes are equipped with a battery or a system of energy recovery from Environmental. Thus, the deployment of these entities, to collect and transmit environmental data to one or more collection points, independently, form a Wireless sensor network (WSN). This network has many application perspectives in various fields, such as industrial monitoring, military operations, monitoring and management natural phenomena. The Wireless

Sensor Networks (WSN) have unsealed a search scientific enough, not only information security but jointly issue optimization of energy consumption sensors, extending the life of WSNs, remains a fundamental limitation for such sensors. Energy limitation is the main concern of any application of a WSN. indeed communication between nodes is "energy intensive" transmission data of a sensor represents the largest share of energy consumption [9]. "An important mechanism to reduce energy consumption is the aggregation of data. The aggregation of data in a network removes redundancy as well as shipment of unnecessary data and hence reduces the energy used in communications. Several aggregation protocols that minimize consumption energy have been proposed. However, they suffer from the increase in "data latency "because prior to the aggregation process, each node should wait for a time WT called preset ("Waiting Time") for receiving data from other nodes. Reducing data latency helps to increase the network output and the early detection of events. latency data (resp, accuracy) is reduced (respectively, greater), if the network nodes are well planned by optimum distribution of slots [10]. Several aggregation protocols which minimize the energy consumption suffer from increased latency, because of the waiting time a reader can expect from his "son" for collect data and send it to the well.

2   Related Works

Since the appearances of wireless sensor networks new research theme are open they have been involved in the resolution of problems that prevent the best operation of wireless sensor networks. After a lot of consisting work in this area and especially the so-called problem of data aggregation scheduling in wireless sensor network, researchers have started serious and as a result in 2005 Xujin Chen & al [1] have proved that this problem is NP-hard and proposed the SDA method based on the construction of the shortest path tree then to progressively applying some special test which gives us an aggregation tree planned, then Huang & al [2] in 2007 , have developed the NCA method based on dominating set using the first fit algorithm for planning, Bo Yu et al [3] in 2009 was detected the NCA vulnerabilities by correcting and proposing a new algorithm that is also based on the dominating set but other types of planning. The works are not stopped at this point so Malhotra & al [4] in 2010, proposed ACSWS method simply takes it the tree of shortest paths and plans smartly each node in order to minimizing latency. In 2011 the first mathematical model for the problem was appeared by Tian & al [5] respecting of the problem constraints and which gave us the optimal solution to 10 nodes in a long lasting but it 'is already a very advanced step then authors have proposed a new solving approach with changing all the foregoing reasoning and introducing new criteria for the selection of the tree and planning. Miloud Bagaa & al [6] have made a study of all existing methods by taking advantage of each strength methods and avoiding limitations of the others,which guided them to found a very well method of data aggregation that s' called DASUT in 2012. Finally and after analyzing the work wholes have arrived us build a conclusion exists and we have first proposed a model that gave us the solution over 30 node and an heuristic with a performance proved by compared tests.

3   Network Model and Problem Description

Let a wireless sensor network application be defined as an undirected connected graph G = (V;E), where V represents the set of nodes and E represents the set of edges. Let $|V| = n$ be the number of nodes on the network and $|E| = m$ be the number of links wireless connecting nodes in the network. Let i be the unique identifier of each node. All nodes have precise knowledge about time, such that each node u transmits exactly on its allowed time slot r. There is no partial transmission or segmentation, such that every transmission from node u to node v begins and ends during the time slot r. Each time slot r has a duration of one time unit. The time used for the execution of the in-network data processing operation is negligible. The aggregation is executed during the time slot when a node is receiving a transmission. Every node must transmit in one and only one time slot during a snapshot collection.

Precedence Constraints

with t'≤ t

A node j must wait for data from all its predecessors so that i can transmit its data to its successor k therefore it can not receive and transmit at the same time $(t = t')$, or transmit pre-arrival data its predecessors $(t' < t)$.

Constraints of Overlap

A node j can not receive at the same slot **t** more than one transmission.

Collision Constraints

If the node k transmits to the neighbor l at the time slot t all nodes j which have a communication link with the node k can not receive a transmission at time slot t.

4   Lower Bound

We consider a network of N sensor nodes, where the node 0 represents the sink node (base station).

To develop the appropriate mathematical model to our problem, the definition of some notations that we will use throughout this modeling is required.

Notation	Definition
N	the number of nodes
E	the set of edges
ν (i)	the set of neighbors of node i

Definition of variables:

T is the variable which express the latency.

$$X_{<i,j>,t}= \begin{cases} 1 \text{ if the edge (i, j) takes the slot t,} \\ 0 \text{ otherwise} \end{cases} \quad (i,j)\in E,\ t=\overline{1,N}$$

Note: the number of nodes **N** is an upper bound for latency.

The optimization problem of planning data aggregation in WSN can be modeled as the following mixed integer problem :

Minimize T

U.c

$$\left\{ \begin{array}{l} t * X_{<i,j>,t} \qquad \leq T, \\ \forall\, (i,j) \in E \; ; \; \forall\, t = \overline{1,N} \qquad\qquad\qquad\qquad\qquad (1) \\[2em] \displaystyle\sum_{j\in\ \nu(i)} \sum_{t=1}^{N} X_{<i,j>,t} = 1, \\ \forall\, i = \overline{1,N} \qquad\qquad\qquad\qquad\qquad\qquad\qquad (2) \\[2em] X_{<i,j>,t} + X_{<j,k>,t'} \leq 1, \\ \forall\, (i,j) \in E \; ; \; \forall(j,k) \in E \; ; \; \forall\, t,t' = \overline{1,N}/t' \leq t \quad (3) \\[2em] \displaystyle\sum_{i\in\nu(j)} X_{<i,j>,t} \qquad \leq 1, \\ \forall\, j = \overline{0,N} \; ; \; \forall\, t = \overline{1,N} \qquad\qquad\qquad\qquad (4) \\[2em] X_{<i,j>,t} + X_{<k,l>,t} \leq 1, \\ \forall\, t = \overline{1,N} \; ; \; \forall(j,k) \in E \; ; \; i \in \nu(j) \; ; \; i \neq k \; ; \; l \in \nu(k) \; ; \; j \neq l \quad (5) \\[2em] X_{<i,j>,t} \in \{0,1\}, \qquad \forall\, (i,j) \in E \; ; \; \forall\, t = \overline{1,N} \end{array} \right.$$

The first type of constraints states: whatever the slot chosen for the edge (i, j) must be less than or equal to the latency **T**. The second category of constraints: each node must transmit exactly once . The third family of constraints indicate: a node **j** must await the arrival of the data from all its predecessors in that it can transmit its data to its successor **k** so it can not receive and transmit at the same time (t = t'), or transmit data before the arrival of its predecessors (t'<t). The fourth set of constraints: a node **j** can not receive at the same slot **t** over one transmission. The fifth type of constraints: if the node **k** transmits to the neighbor l at time slot **t**, all nodes which have a communication link with the node **k** can not receive a transmission at time **t**.

4.1   Size of the Model

Total number of constraints

$|E|*N + N + |E|^2 * N^2 + (N+1) * N + |E| * N$

The number of variables

$|E| * N + 1$

5   Proposed Heuristic

After the studying of [1] [2] [3] [4] [5] [6] [11] and the implementation of different algorithms and protocols, we got to offer resolution method that takes the positives of each method and avoids the limitations. In this method, the construction of the aggregation tree and the scheduling run simultaneously. An aggregation tree is first created , then the resulting tree is used as an input to the scheduling algorithm. Notice that the tree created will not be considered as a data aggregation tree, it will only be realized to sort the network nodes in the scheduling process.

5.1   Tree Construction

This method proceeds level by level it starts from the top level to the base station and a node is ready to be executed after the completion of the planning of all nodes of level preceding, knowing that nodes the same level runs in parallel. At each iteration we try to characterize each node in the manner shown in the following schema in order to planned it.

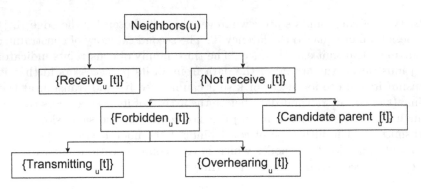

- **Receive**$_u$**[t]** = the neighbors of **u** who receive a transmission at slot **t**.
- **Not receive**$_u$**[t]** : set of **u**'s neighbors who does not receive a transmission at slot **t**.
- **Candidate parent**$_u$**[t]** : set of **u**'s neighboring candidate to become a parent of node **u** at time slot **t** and among these candidates is chosen the one with the minimum of neighbors in order to minimize the number of nodes forbidden at this time.
- **Forbidden**$_u$**[t]**: set of forbidden **u**'s neighbors to be parents at slot **t**.
- **Overhearing**$_u$ **[t]**: set of **u**'s neighbors that are on a field of transmission of a node which transmitted at the slot **t**.
- **Transmitting**$_u$**[t]**: **u**'s neighbors that transmits at slot **t** or before.

The choice of the planning when a collision is established or for example **u** transmit to **v** and **v** transmits **u** in the same slot is done using a comparison of the triplets for each node **u** constructed as follows: (**Slot of the node u, u's parent neighbors number without u, u's neighbors number without u).**

The valid schedule is that of the smallest triplet, and for the others same as the previous pattern changing the parent or slot.

The figure 1 shows how we built an aggregation tree , we start with the lowest level, then level by level we schedule each nodes.

In (a), node 7 has chosen the parent who has the minimum degree i.e node 4, then (b) the choice of node 7 will be validated and the next level that contains the nodes 4,5 and 6 will be activated, here, node 4 selected node 1 to slot 2, node 5 has selected node 4 slot 2, node 6 and node 3 has chosen to slot 1.

Three triplet will be built, for node 4 (2,3,3) 5 for the node (2,3,5) and to the node 6 (1,3,3). What qualifies the node 4 and 6 to be validated for this iteration and node 5 will change his relative and slot for the next iteration and so on until the end of the construction of the shaft as shown (f).

5.2   Nodes Scheduling

In this method, the construction of the aggregation tree and the scheduling run simultaneously. The previously resulting tree is used as an input to the Nodes

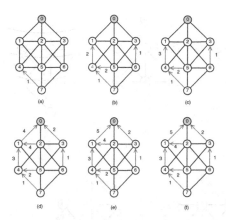

Fig. 1. The tree Construction

scheduling algorithm. It will operate as a sort of network nodes in the planning process. It is run in several iterations, designed to determine the set of nodes that can be scheduled in the same slot. The leaf nodes of the resulting tree are classified according to the number of non-leaf neighbors, where the node that has the smallest number of non-leaf neighbors is first considered. Let S and $\bar{S}$ respectively denote the sorted set of leaf nodes and not leaves nodes in the resulting tree. For all nodes u in S, the following test is done:

If $\exists\, v \in \bar{S}$ where v is the private neighbor of u (ie,$\nexists$ w $\in$ S - $\{u\}$, wherein w is a neighbor of (v), then u is planned and v is selected as a parent in this slot. Here u is removed from the resulting tree. Otherwise u will be integrated throughout $\bar{S}$.

When all the nodes in S are tested, the nodes leaves that are not planned as well as new leaf nodes (in the tree of resulting) are considered for the next iteration, and they will be sorted and tested using the same approach.

In Figure 2, we apply the «nodes scheduling» methods for the construction and the schedule of the aggregation's tree. As shown in (g), we start by building and sorting the set S that contains the leaves of the resulting tree, S = $\{6,7,5,2\}$ then $\bar{S}$ that contains the non-leaf nodes, $\bar{S}$ = $\{0,1,3,4\}$. node 6 has not a private neighbor in $\bar{S}$ so it will be removed from S and added in $\bar{S}$ the same for 7. The node 5 has a private neighbor "the node 7" which has been integrated in this iteration in $\bar{S}$ node 2 also has a private neighbor in $\bar{S}$ "the node 0". so nodes 5 and 2 take as a parent nodes 7 and 0 respectively in time slot 1. in (b), updates the sets S and $\bar{S}$, after that the slot becomes 2 and similarly the remaining nodes are planned until the aggregation's tree planned in (k).

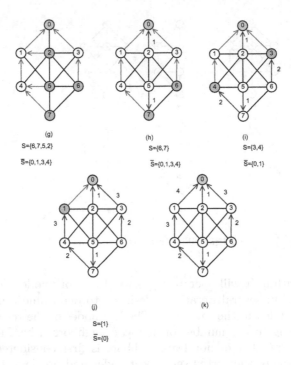

Fig. 2. Nodes scheduling

6   Simulation Results

In this section, we discuss the performance of our solution and compare them to that of DAS-UT , GGT, and Exact solution given by the optimal solution of the mathematical modeling of the problem:

Time latency (last time slot): is defined as the time required for the base station B to receive the aggregated data from all the sensor nodes; In the simulation experiments, N nodes are randomly deployed according to a uniform random distribution. The types of experiments:

- Vary Number of nodes and fix D ;
- Vary D and fix N;

The algorithms has been implemented in Python using the cplex12.5 . It was tested on a PC with 2.60GHz, 4,00 Go Ram , running under Windows.

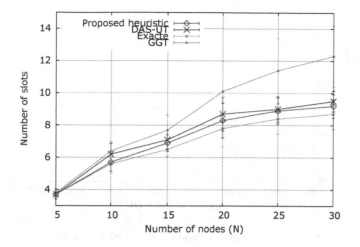

Fig. 3. Var number of nodes 5 − 30 with degree 5

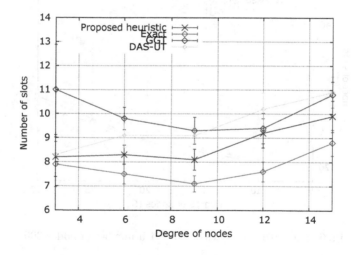

Fig. 4. Var degree of nodes 3 − 15 with number of nodes 20

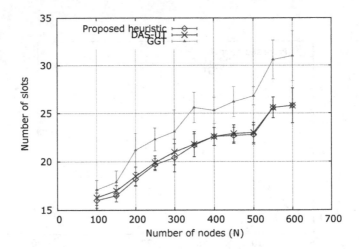

Fig. 5. Var number of nodes 100 − 600 degree 10

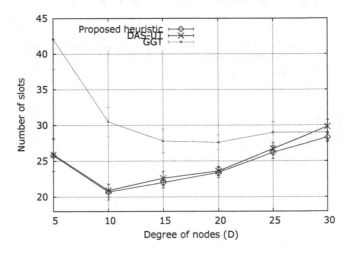

Fig. 6. Var degree of nodes 5 − 30 with number of nodes 300

7 Conclusion

In this paper we have studied the problem of planning data in wireless sensor network and we have proposed an adequate mathematical modeling to obtain an optimal solution and it gives the solution for more than 30 nodes knowing that the existing gives solution for only 10 nodes and then we proposed an heuristic. The idea of our method is to build a planned tree that guide us to build the solution of the problem as and to the measurement by applying the test. The effectiveness of our method and the standard differential of its result shows very clearly the contribution of the work realized in this paper. These conclusions remain valid in the whole of the tested instances.

References

1. Chen, X., Hu, X., Zhu, J.: Minimum data aggregation time problem in wireless sensor networks. In: Jia, X., Wu, J., He, Y. (eds.) MSN 2005. LNCS, vol. 3794, pp. 133–142. Springer, Heidelberg (2005)
2. Huang, S., Wan, P., Vu, C., Li, Y., Yao, F.: Nearly constant approximation for data aggregation scheduling in wireless sensor networks. In: Proc. IEEE Infocom 2007 (2007)
3. Bo, Y., Li, J., Li, Y.: Distributed data aggregation scheduling in wireless sensor networks. In: Proc. IEEE Infocom 2009 (2009)
4. Malhotra, B., Nikolaidis, I., Nascimento, M.: Aggregation convergecast scheduling in wireless sensor networks. Wireless Networks (2010)
5. Tian, C., Jiang, H., Wang, C., Wu, Z., Chen, J., Liu, W.: Neither Shortest Path Nor Dominating Set: Aggregation Scheduling by Greedy Growing Tree in Multi-hop Wireless Sensor Networks. IEEE Transactions on Vehicular Technology 60(7) (2011)
6. Bagaa, M., Derhab, A., Lasla, N., Ouadjaout, A., Badache, N.: Semistructured and unstructured data aggregation scheduling in wireless sensor networks. In: Proc. IEEE Infocom 2012 (2012)
7. Nikolaidis, Ioanis., de Souza, Evandro: An exploration of aggregation convergecast scheduling, Department of Computing Science, University of Alberta, Edmonton, Canada (2013)
8. Bagaa, M., Younis, M., Badache, N.: Efficient Data Aggregation Scheduling in Wireless Sensor Networks with Multi-Channel Links. In: MSWiM 2013, Barcelona, Spain (2013)
9. Somov, A., Minakov, I., Simalatsar, A., Fontana, G., Passerone, R.: A methodology for power consumption evaluation of wireless sensor networks. In: Proc. IEEE ETFA 2009 (2009)
10. Bagaa, M., Challal, Y., Derhab, A., Badache, N.: Data Aggregation Scheduling Algorithms in Wireless Sensor Networks: solutions and challenges, survey (2014)
11. Mohammed, Y.: Using metaheuristics for Optimizing Satellite Image Registration. International Journal of Combinatorial Optimization and Informatics 3(3), 69–80 (2012) ISSN:2007-1558

References

Optimizing a Sensor Deployment with Network Constraints Computable by Costly Requests

Frederic Dambreville

DGA MI, Rennes, France & ENSTA Bretagne, Brest, France
submit@fredericdambreville.com
http://www.fredericdambreville.com

Abstract. This paper deals with the planning of networked sensors. In such sensors deployment, two main aspects have to be considered: cover most strategic areas, and ensure the connectivity between sensors and network nodes. The connectivity is computed by expensive simulations, and these computation requests should be optimized as an experiment design in order to enhance the constraint models, while the sensor deployment itself is solved by usual optimizer. We propose a simulated approach, inspired from the known algorithm, EGO, in order to address this two-level problem.

Keywords: Sensor management, Networks, Optimal design.

Notations

- $\{1 : N\}$ means the set $\{1, 2, \cdots, N\}$ and $n = 1 : N$ means $n \in \{1 : N\}$,
- $\mathbb{I}[X]$ is defined by $\mathbb{I}[\text{true}] = 1 - \mathbb{I}[\text{false}] = 1$,
- $\mathcal{P}(X)$ is the probabilistic density function on measurable set X.

1 Introduction

The main background of this paper is the optimal planning of networked sensors. In the context of our application, a main concern is the ability of the network to mobility, in the sense that it could be quickly removed and deployed. These constraints have resulted in some practical choices for the design of our system. Especially, the resulting implementation consists in insulated groups of sensors with their processing devices, and these groups are exchanging data by radio with connected network nodes.

In order to speed up the sensor deployment, planning tools are of main interest. In an idealistic world, such planning may be described as a precomputed graph-based problem, and some classical, but efficient [1], algorithms may be applied. However, we have to face a difficult issue here: the connectivity between the sensors and the nodes is difficult to model and has to be computed by means of a specific and heavy simulation process. Requests to this process are quite expensive and are restricted. On the other hand, these requests are necessary in order to ensure a better modelling of the planning optimization problem.

H.A. Le Thi et al. (eds.), *Model. Comput. & Optim. in Inf. Syst. & Manage. Sci.*,
Advances in Intelligent Systems and Computing 360, DOI: 10.1007/978-3-319-18167-7_22

This problem is formalized very generally as the maximization of an unknown criterion, defined a priori by sampling a known random function (law of model error). In such case, each actual evaluation of the function increases the knowledge about the criterion, and subsequently the efficiency of the maximization. The issue is to optimize the sequence of value to be evaluated, in regards to the evaluation costs. There is here a fundamental link with the domain of experiment design. Jones, Schonlau and Welch proposed a general method, the Efficient Global Optimization (EGO) [2,3], for solving this problem in the case of additive functional Gaussian law (Krigging). This approach is based on iterative maximization of the expected improvement of the best actual evaluations. Since the functional noise is modeled by Krigging, it may be subject to the curse of dimentionality, in the case of some complex models.

In this work, an implementation of the EGO is proposed, based on a rare event simulation approach. This simulated approach does not request the linear Gaussian hypothesis on the functional noise, thus implementing non Gaussian functional noise, and even simulated functional noise. This work itself extends a previous work [4], by implementing now multimodal laws for the simulation. It is applied to the aforementioned networked sensor planning.

In the first section 2 of this paper, we propose an abtract formalisation of our networked sensor planning. In section 3, a rare-event simulation approach is proposed for solving this bi-level sensor planning. Section 4 presents a scenario and numerical results. Section 5 concludes.

2   Networked Sensor Planning with Experiment Sub-processes

There is a consequent litterature, which addresses the problem of connectivity [5,6,7] in a sensors network. Many works, especially considering large networks, focus on optimizing the structural connectivity of the networks (its topology, cost, robustness [8]), together with more or less elaborated communication models. The inclusion of simplified terrain models are addressed by some papers; dynamic planning approaches may also be considered in order to adapt to resulting connectivity issues [9].

In our application, the network size in not the main issue: the network is build of a limited number of groups of sensors. However, the connectivity among these groups, owing to the complex environmental context, is computed for each network configuration by means of complex simulations. This results in the specific problem simplified and illustrated subsequenlty.

2.1   A Model-Noised Networked Sensor Planning

We present first, as a preliminary of the theoretical subsequent developments, the simplified and abstracted networked sensor planning, which will be solved in section 4. This simplified problem is easily generalizable, typically in terms of sensor ranges, number of nodes and connectivity geometry.

In this problem, we plan both the sensors deployment modeled by their position and radius:

$$(\sigma_s, R_s) \in (\mathbb{R}^2) \times \mathbb{R}_+ , \text{ for } s = 1 : S,$$

and the position of the node:

$$\eta \in \mathbb{R}^2 ,$$

to which the sensors are connected (on real networks, their may be many nodes).

We face a difficult issue here: the connectivity between the sensors and the node, a consequence of the intervisibility between the tools, has to be computed by means of a specific simulation process. Actually, the connectivity has to be recomputed, each time the nodes positions are changed. But this computation is expensive and restricted. In this paper, we will consider a quite simplified model for the connectivity. First at all, we consider a map of obstacle in the form of a set of points and radius segments:

$$(o_d, [A_d, A_d + \Delta]) , \quad \text{with } o_d \in \mathbb{R}^2 \text{ and } [A_d, A_d + \Delta] \subset \mathbb{R}_+ , \text{ for } d = 1 : D.$$

Associated with an obstacle d is defined a connectivity attenuation between sensor s and node η which is dependant of the intersection of the sensor-to-node segment with the obstacle:

$$a_{d,s} = 1, \text{ if } \frac{\overrightarrow{\eta\sigma_s} \cdot \overrightarrow{\eta o_d}}{||\overrightarrow{\eta\sigma_s}||^2} \notin [0,1],$$

$$a_{d,s} = \min\left\{1, \max\left\{0, \frac{1}{\Delta}\left(\frac{||\overrightarrow{\eta\sigma_s} \wedge \overrightarrow{\eta o_d}||}{||\overrightarrow{\eta\sigma_s}||} - A_d\right)\right\}\right\}, \text{ if } \frac{\overrightarrow{\eta\sigma_s} \cdot \overrightarrow{\eta o_d}}{||\overrightarrow{\eta\sigma_s}||^2} \in [0,1].$$

The attenuations are multiplicative and the connectivity for sensor s is:

$$a_s = \prod_{d=1:D} a_{d,s} ,$$

and this connecivity will affect the performance of a sensor.

The networked sensors are deployed for monitoring some areas of interest, which are represented by points in this simplified model:

$$z_i \in \mathbb{R}^2 , \text{ for } i = 1 : I.$$

The whole networks is then evaluated from the monitored areas and against the connectivity attenuation:

$$f(\sigma, \eta; o, A) = \sum_{i=1:I} \max_{s=1:S} a_s \mathbb{I}[||\overrightarrow{\sigma_s z_i}|| \leq R_s] . \tag{1}$$

The purpose of the deployment planning is to optimize (σ, η) in order to maximize f. However, the parameters of the problem are not entirely known:

- R, Δ, are known,
- $\nu = (o, A)$ are unknown,

but the prior probability $p(o, A)$ is known.

In order to enhance this prior knowledge, the planner may request simulation processes in order to compute the true value of a plan $x = (\sigma, \eta)$. The issue is to choose the shortest sequence of requests in order to save expensive computation resources.

Next section presents a general formalism to this problem.

2.2   The Theoretical Problem

A system is characterized by means of a noisy criterion function:

$$f : (x, \nu) \in X \times N \mapsto f(x, \nu) \,,$$

where:

$x \in X$ is a parameter to be optimized ,

$\nu \in N$ is a model noise ,

$p_\nu \in \mathcal{P}(N)$ is a known probabilistic noise prior .

and by an unknown *actual* model noise:

$\widehat{\nu} \in N$ is the actual value of the model noise .

The purpose is to optimize x so as to maximize the criterion. The noise on f is a model noise and *it is always possible to evaluate the actual value of the criterion for any specific actual parameter $\widehat{x}$*. In other word, the purpose here is not exactly to optimize a mean criterion, but rather to choose a *good* sequence of actual parameters so as to optimize the actual function.

The application described in section 2.1, and characterized by the criterion (1) where $x = (\sigma, \eta)$ and $\nu = (o, A)$, is typically a particular case of this problem.

It comes that each actual evaluation of the criterion is costly, while, in comparison, the evaluation of the criterion model is considered free. Since each actual evaluation of the criterion provides also some knowledge about the actual model noise $\widehat{\nu}$, the issue is to balance optimally between actual evaluation and model evaluation, so as to find a near optimal solution to the actual criterion.

So as to deal with this problem, Welch proposed [2] the famous Efficient Global Optimization method, which is based on an iterative optimization maximizing the Expected Improvement. More precisely, Welch considered the case of a (spatial) Gaussian noise combined with a linear model, and derived exact computation of the sequence. Our main contribution is to extend Welch algorithm to any cases by means of simulation approaches. Rare event simulation methods are quite instrumental here.

From a general point of view, Welch approach takes the form of the following recursive computation:

[Expected Improvement Maximization (EIM)]
[Purpose: Optimize actual criterion $f(\cdot, \widehat{\nu})$ by evaluating most promising parameter values]
1. Set $n = 0$,
   // n is the number of actually evaluated parameters
2. Repeat:
   (a) Compute $\widehat{x}_{n+1}$, the next candidate for an actual evaluation:

$$\widehat{x}_{n+1} \in \arg\max_{x \in X} \int_{\nu \in N} p_\nu[n](\nu) f[n](x, \nu) \, d\nu , \qquad (2)$$

   where:

$$p_\nu[n](\nu) = p_\nu\left(\nu \mid \forall k = 1:n, \ f(\widehat{x}_k, \nu) = \widehat{y}_k\right) , \qquad (3)$$

$$f[n](x, \nu) = \max\left\{f(x, \nu), \max_{k=1:n} \widehat{y}_k\right\} . \qquad (4)$$

   // The quantity $\widehat{y}_k$ is the actual evaluation of $\widehat{x}_k$
   (b) Request the actual evaluation of $\widehat{x}_{n+1}$:

$$\widehat{y}_{n+1} = f(\widehat{x}_{n+1}, \widehat{\nu}) ,$$

   // This computation is costly
   (c) Set $n \leftarrow n + 1$,
   until the convergence of $(\widehat{x}_{1:n}, \widehat{y}_{1:n})$ is sufficient.
   [Output:]
   – The sequence $(\widehat{x}_{1:n}, \widehat{y}_{1:n})$,
   – The model noise estimation $p_\nu[n]$.

The function $f[n] - \max_{k=1:n} \widehat{y}_k$ evaluates the improvement of f at step n. The conditional probability $p_\nu[n]$ is the posterior knowledge of ν obtained after the n first measurements.

From these considerations, it appears that we need to:

– Evaluate the conditional probability $p_\nu[n]$,
– Compute the optimal parameter $\widehat{x}_{n+1}$.

We will see that first task is performed by rare event simulations.

3   Cross-Entropic Approaches

3.1   A Cross-Entropy Method

The cross-entropy method (CE) has been pioneered by Rubinstein [10], and was initially settled for the simulation of rare event. It is based on a recursive importance sampling driven by a family of sampling densities. We present subsequently a variant of the CE, based on mixtures of laws. Contrary to the hierarchical models extensively used in [11], the mixture weights are not explicitly sampled, and the CE has to simulate them in a similar way than the EM algorithm. In this section, the following notations are considered:

- Ω is a probabilistic space,
- $p_\omega \in \mathcal{P}(\Omega)$ is a probabilistic distribution on Ω,
- $\phi : \omega \in \Omega \mapsto \mathbb{R}$ is a measurable function,
- $\phi^{-1}([\gamma, +\infty[)$ is a p_ω-rare event, that is such that:

$$\int_{\omega \in \Omega} \mathbb{I}[\phi(\omega) \geq \gamma] p_\omega(\omega) \, d\omega \ll 1 \,,$$

where:

$$\mathbb{I}[\text{true}] = 1 - \mathbb{I}[\text{false}] = 1 \,,$$

- $\pi(\cdot|\Theta) = \big(\pi(\cdot|\theta)\big)_{\theta \in \Theta}$ is a family of probabilistic distributions on Ω. Mixtures of the pdfs $\pi(\cdot|\theta)$ will be used as importance samplers:

$$\sum_{k=1:K} \tau_t^k \pi(\cdot|\theta_t^k) \,,$$

is the pdf mixture used as sampler at step t,
- N is the number of samples generated at step t of the algorithm, and ω_t^n such that $n \in \{1 : N\}$ are the samples generated at step t,
- $R : \varphi \in \mathbb{R} \mapsto R(\varphi) \in [0,1]$ is a selective function which controls how the importance sampling is samples-driven. The function $R : \varphi \in \mathbb{R} \to [0,1]$ select the best samples in regards to their evaluation. The definition of R is adaptive and depends on the generated samples. Typically, R may be the quantile selection as defined in [10],
- $\alpha \in]0,1]$ is a smoothing parameter of the sampler update, which defines the balance between the current importance sampler and the selected samples,

CE for Simulation. The CE simulation is an adaptive importance sampling driven by the samples, their evaluations and a family of sampling distribution. Here, we propose an upgrade scheme implementing a mixture $\sum_{k=1:K} \tau^k \pi(\cdot|\theta^k)$ as sampling distribution. This upgrade scheme, based on a prediction of the mixture index for each sample, shares some similarities with the EM algorithm:

[CE simulation]
[Purpose: Build an Importance sampler for the rare event $\phi^{-1}([\gamma, +\infty[)]$
1. Initialize $t = 0$, θ_0 and τ_0,

2. Repeat:
   (a) Generate the samples $\omega_t^n \in \Omega$, for $n \in \{1 : N\}$, according to the probabilistic density function (pdf) $\sum_{k=1:K} \tau_t^k \pi(\cdot|\theta_t^k)$,
   (b) Compute the evaluations $\phi(\omega_t^n)$ of the samples for $n \in \{1 : N\}$,
   (c) Compute the selective parameters:

$$\rho_t^n = \frac{p_\omega R(\phi(\omega_t^n))}{\sum_{k=1:K} \tau_t^k \pi(\omega_t^n|\theta_t^k)} \,, \quad \text{for all } n \in \{1 : N\} \,,$$

(d) Compute $\delta_t^{n,k}$, the probability of the mixture index conditionally to the sample ω_t^n :

$$\delta_t^{n,k} = \frac{\tau_t^k \pi(\omega_t^n | \theta_t^k)}{\displaystyle\sum_{k=1:K} \tau_t^k \pi(\omega_t^n | \theta_t^k)} \ , \quad \text{for all } n \in \{1:N\} \text{ and } k \in \{1:K\} \ ,$$

(e) Update the mixing weights:

$$\tau_{t+1|t} = \frac{\displaystyle\sum_{n=1:N} \rho_t^n \delta_t^{n,k}}{\displaystyle\sum_{k=1:K} \sum_{n=1:N} \rho_t^n \delta_t^{n,k}} \ ,$$

(f) Update each sampler by maximizing the cross-entropy with the selected samples weighted by δ :

$$\theta_{t|t+1}^k \in \arg\max_{\theta^k \in \Theta} \sum_{n=1:N} \rho_t^n \delta_t^{n,k} \log\left(\pi(\omega_t^n | \theta)\right) \ , \quad \text{for all } k = 1:K \ , \quad (5)$$

(g) Mix $\pi(\cdot|\theta_t^k)$ and $\pi(\cdot|\theta_{t|t+1}^k)$ into $\pi(\cdot|\theta_{t+1}^k)$ according to their weights:

$$\theta_{t+1}^k \in \arg\max_{\theta^k \in \Theta} \int_{\omega \in \Omega} \left(\alpha\pi(\omega|\theta_t^k) + (1-\alpha)\pi(\omega|\theta_{t|t+1}^k)\right) \log\left(\pi(\omega|\theta^k)\right) \ d\omega \ , \tag{6}$$

(h) Mix τ_t and $\tau_{t|t+1}$ into τ_{t+1} according to their weights:

$$\tau_{t+1} = \alpha\tau_t + (1-\alpha)\tau_{t|t+1} \ ,$$

(i) Set $t \leftarrow t+1$,

until the convergence is sufficient.

[Output:]

– The final importance sampler $\displaystyle\sum_{k=1:K} \tau_{\overline{t}}^k \pi(\cdot|\theta_{\overline{t}}^k)$,

– The likelihood ratio $\omega \in \Omega \mapsto W(\omega|\theta_{\overline{t}}) = \dfrac{p_\omega(\omega)}{\displaystyle\sum_{k=1:K} \tau_{\overline{t}}^k \pi(\omega|\theta_{\overline{t}}^k)}$.

It is noticed that optimizations (5) and (6) are both solved on many classical law familly; among them are the the the discrete and the Gaussian laws. The criterion for convergence is, as in the classical CE [10], achieved when a sufficient ratio of samples is within the rare event $\phi^{-1}([\gamma, +\infty[)$.

The final importance sampler combined with the likelihood ratio allows an efficient simulation of p_ω around the rare event $\phi^{-1}([\gamma, +\infty[)$. From such simulations could be estimated both the probability $p_\omega(\phi(\omega) \geq \gamma)$ of the rare event and the conditional probability $p_\omega(\cdot|\phi(\omega) \geq \gamma)$ on the rare event.

3.2   Implementation of the EIM Process by Means of the CE

Estimating the Conditional Law. A main ingredient of the [EIM] process is to estimate the conditional probability:

$$\nu \in N \mapsto p_\nu[n](\nu) = p_\nu(\nu \mid \forall k = 1 : n, \ f(\widehat{x}_k, \nu) = \widehat{y}_k) \ . \tag{7}$$

General approach: Let be defined a multivariate function Ψ (typically a paraboloid) such that:

- $\Psi : \varphi_{1:n} \in \mathbb{R}^n \mapsto \mathbb{R}_-$, is continuous ,
- $\max\limits_{\varphi_{1:n} \in \mathbb{R}^n} \Psi(\varphi_{1:n}) = 0$ and $\arg\max\limits_{\varphi_{1:n} \in \mathbb{R}^n} \Psi(\varphi_{1:n}) = \{0\}$.

Then, the event $\{\nu \in N \ / \ \forall k = 1 : n, \ f(\widehat{x}_k, \nu) = \widehat{y}_k\}$ may be approximated as follows:

$$\{\nu \in N \ / \ \forall k = 1 : n, \ f(\widehat{x}_k, \nu) = \widehat{y}_k\} = \arg\max\limits_{\nu \in N} \Psi\left(\left(f(\widehat{x}_k, \nu) - \widehat{y}_k\right)_{k=1:n}\right) \ . \tag{8}$$

Then, the conditional probability $p_\nu[n]$ is reformulated as follows:

$$p_\nu[n] = p_\nu\left(\cdot \left| \nu \in \arg\max\limits_{\nu \in N} \Psi\left(\left(f(\widehat{x}_k, \nu) - \widehat{y}_k\right)_{k=1:n}\right)\right.\right) \ . \tag{9}$$

Considering back the notation of section 3.1 about the cross-entropy method, the computation of $p_\nu[n]$ is equivalent to:

Estimate:

$$p_\omega\left(\cdot \left| \omega \in \arg\max \phi(\Omega)\right.\right) \ , \tag{10}$$

Where:

$$\omega = \nu, \ \Omega = N, \ p_\omega = p_\nu, \tag{11}$$

$$\phi(\omega) = \Psi\left(\left(f(\widehat{x}_k, \nu) - \widehat{y}_k\right)_{k=1:n}\right) \ . \tag{12}$$

Considering the fact:

$$\arg\max \phi(\Omega) = \bigcap\limits_{\gamma < \max \phi(\Omega)} \phi^{-1}([\gamma, \infty[) \ , \tag{13}$$

the algorithm of section 3.1 is not far from a suitable solution. The principle is to sample rare events, $\phi^{-1}([\gamma_t, +\infty[)$, which are increasing approximations of the set of maximizers $\arg\max \phi(\Omega)$ (this set may be multimodal). When the [CE simulation] process ends at step $\overline{t}$, typically when stationnarity occurs, the value:

$$\mu_{\overline{t}} = \frac{1}{N} \sum\limits_{n=1:N} \phi(\omega_{\overline{t}}^n) \ ,$$

is computed and the distance $\mu_{\overline{t}} - \max \phi(\Omega)$ is an evaluation of the quality of the importance sampler. Moreover, the event $[\mu_{\overline{t}}, +\infty[$ is also used as an approximation of $[\max \phi(\Omega), +\infty[$. Then, the sampling of law $p_\omega(\cdot \mid \omega \in \arg\max \phi(\Omega))$ is approximated by sampling $p_\omega(\cdot \mid \phi(\omega) \geq \mu_t)$ by means of the following process:

[Conditional simulation]
[Purpose: Generate samples by means of the computed sampler

$$\sum_{k=1:K} \tau_{\bar{t}}^k \pi(\cdot|\theta_{\bar{t}}^k)]$$

1. For $n = 1 : N$, do:
   (a) Generate the samples ω_n according to the pdf mixture $\sum_{k=1:K} \tau_{\bar{t}}^k \pi(\cdot|\theta_{\bar{t}}^k)$,
   (b) Repeat from step (1a) until $\phi(\omega_n) \geq \mu_{\bar{t}}$,
2. Compute the weights $V_n = W(\omega_n|\theta_{\bar{t}})$ for each $n = 1 : N$,
3. Compute the weights:

$$W_n = \frac{V_n}{\displaystyle\sum_{n=1:N} V_n},$$

for each $n = 1 : N$,
[Output:]
   – The weighted particle cloud $((\omega_n, W_n))_{n=1:N}$.

The Whole EIM Process. The **[EIM]** algorithm is implemented by means of the CE:

[CE EIM]
[Purpose: Optimize actual criterion $f(\cdot, \widehat{\nu})$ by evaluating most promising parameter values]

1. Set $n = 0$,
2. Repeat:
   (a) Compute an approximating sampler $\widehat{p}_\nu[n]$ of the conditional law $p_\nu[n]$ by means of the algorithms **[CE simulation]** and **[Conditional simulation]**,
   (b) Define $f[n]$ by (4),
   (c) Compute $\widehat{x}_{n+1}$, the maximizer of $\int_{\nu \in N} p_\nu[n](\nu)f[n](x,\nu)\,d\nu$, by means of dedicated algorithm (this algorithm choice is a parameter of the method). The integration is computed by a Monte Carlo method based on the sampler $\widehat{p}_\nu[n]$,
   (d) Compute $\widehat{y}_{n+1} = f(\widehat{x}_{n+1}, \widehat{\nu})$,
   (e) Set $n \leftarrow n + 1$,
   until the convergence is sufficient.
[Output:]
   – The sequence $(\widehat{x}_{1:n}, \widehat{y}_{1:n})$,
   – The model noise approximation $\widehat{p}_\nu[n]$.

4   Practical Implementation and Numerical Results

In this section, the algorithm **[CE EIM]** defined in previous section is applied to the networked deployment formalized by criterion 2.1. In order to run this algorithm, two sampling families are considered for approaching the conditionnal law, one being purely Gaussian, and the second being a mixture of two Gaussian. In order to solve the optimization (2), a Gaussian-based CE approach is also used [10] but this well known method is not detailed here.

4.1   Practical Implementation of the Problem

In our implementation, the scenarii are generated randomly. We present the solve of a sampled scenario, subfigures 1 to 3 of figure 4.1, which is charaterized by:

- Areas of interest, and map object are generated uniformly within the space $[0, 1] \times [0, 1]$,
- 10 areas of interest are generated randomly:

i	0	1	2	3	4	5	6	7	8	9
Xz_i	−0.49	0.42	0.50	0.79	0.24	−0.98	−0.38	0.90	−0.91	−0.23
Yz_i	−0.85	0.19	0.20	0.83	0.31	−1.00	0.07	0.74	0.67	−0.74

- 8 map objects are generated randomly with $\Delta = 0.25$ and radius A_d generated uniformly within $[0.25, 0.75]$:

d	0	1	2	3	4	5	6	7
Xo_d	0.87	−0.95	0.27	−0.83	−0.94	0.86	0.78	−0.86
Yo_d	0.51	−0.98	0.21	−0.99	−0.27	−0.08	−0.19	0.86
A_d	0.63	0.67	0.35	0.39	0.30	0.68	0.28	0.43

- 5 sensors with radius R_s generated uniformly within $[0.25, 0.75]$:

s	0	1	2	3	4
R_s	0.71	0.69	0.39	0.27	0.73

The dimension of the optimized varaible is 12. Different algorithm runs are presented subsequently, while considering two cases for the simulation of the conditional laws:

- Importance samplers are multivariate Gaussian laws on vector $\nu = (o, A)$,
- Importance samplers are a mixture of two multivariate Gaussian laws on vector $\nu = (o, A)$.

The following parameters are used for the **[CE EIM]**:

- 100 samples are used for estimating the expectation in (2),
- The CE process, which builds the sampler for the conditional simulation, is implemented as follows:
  - 1000 samples are used for each step,
  - The selecting function R is linear, mapping onto $[0, 1]$,
  - The smoothing parameter is $\alpha = 0.9$,
  - The CE is processed during 100 steps.

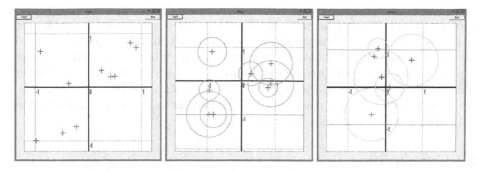

Fig. 1. Network: Areas of interest / Map objects / Sensors & node

4.2   Numerical Results

First at all, the best optimal solution obtained with *god view*, *ie.* full knowledge on o, A, is 8.6 and is described in subfigures 3 and 4 of figure 4.1. This result has been obtained by means of several runs of a meta-heuristic, a CE Gaussian based optimization algorithm [10], which is not presented here. It is noticed that at least 1000 samples are used by this method in order to reach the suboptimal value 8. The following table summarize 10 step of the EGO algorithm, for two testing run for multivariate Gaussain generators:

step n	0	1	2	3	4	5	6	7	8	9
$\max_x E_\nu f[n]$	6.56	7.72	8.36	8.47	8.31	8.68	8.41	8.41	8.24	8.64
$\widehat{y}_n$	5	7.45	4.33	7.69	6.23	3.93	6	5.06	6	7.31
$\mu_{\overline{t}}$	0	1.68	2.25	2.42	2.67	2.71	2.69	2.65	2.66	2.55
$\max_x E_\nu f[n]$	6.54	7.22	7.77	7.69	7.90	8.60	8.63	8.11	8.61	8.86
$\widehat{y}_n$	4	2	4.29	4	4	5.72	6.71	4	8	2.30
$\mu_{\overline{t}}$	0	2.83	3.51	3.16	3.04	3.01	2.96	2.85	2.77	2.79

The following table summarize 10 step of the EGO algorithm, for two testing run for mixtures of 2 multivariate Gaussain generators:

step n	0	1	2	3	4	5	6	7	8	9
$\max_x E_\nu f[n]$	6.23	8.74	7.48	9.00	9.90	8.87	10	9.91	9.99	9.98
$\widehat{y}_n$	5	7.12	0	6.16	5	7	1	0.13	8	2.32
$\mu_{\overline{t}}$	0	1.62	2.11	3.53	3.34	3.19	3.00	3.43	3.64	3.48
$\max_x E_\nu f[n]$	6.78	9.48	9.25	8.22	9.53	9.79	8.22	9.43	9.87	9.37
$\widehat{y}_n$	6	8	5.65	2	6.12	2	3.01	4.33	3.22	5
$\mu_{\overline{t}}$	0	1.25	1.63	2.31	2.57	2.70	2.70	2.79	2.80	2.87

It appears in both cases that the CE-EGO reaches the near-optimal value 8 (or at least the suboptimum 7) by using quite few sampling test requests. The EGO is thus confirmed on this example. However, there are clearly some points to be improved. The main improvement is needed on the conditional estimator.

The quality of the conditionnal estimator, evaluated by means of $\mu_{\bar{t}}$, is rather poor (this value should be near 0). Since the law is conditionned by the set of noises mapping to some values, the conditionner may be typically a curve (or a hyper-curve) and should be multimodal. For this reason, mixtures were expected to produce better simulations. But the enhancement is not convincing however. At last, owing to the expectation error (estimated from 100 samples), the convergence becomes noisy, especially when the best request nears to the optimum. But this is a secondary drawback.

5   Conclusion

In this paper, we addressed a networked sensor planning, with two concurrent objectives: optimize the covering of interest areas and maximize the saving of the requests to the costly simulation processes needed for evaluating the nertwork connectivity. In order to solve the formulated planning problem, we proposed an original implementation of the EGO, based on a rare event simulation approach. This simulated approach does not require the functional noise to be additive Gaussian. The work itself extends a previous work [4], by implementing multimodal laws for the simulation. The simulation results have confirm the efficiency of the EGO approach for optimizing a criterion with limited requests to an evaluator, and implemented it on a new application. We identified some points to be improved in future works, especially related to the conditional simulation. Sampling conditionally to a tight constraint is by the way a topic of interest in other domains, such as Bayesian filtering for example.

References

1. Nguyen, D.M., Dambreville, F., Toumi, A., Cexus, J.-C., Khenchaf, A.: A column generation based label correcting approach for the sensor management in an information collection process. In: Nguyen, N.T., van Do, T., Le Thi, H.A. (eds.) ICCSAMA 2013. SCI, vol. 479, pp. 77–89. Springer, Heidelberg (2013)
2. Jones, D.R., Schonlau, M.J., Welch, W.J.: Efficient global optimization of expensive black-box function. J. Glob. Optim. 13(4), 455–492 (1998)
3. Marzat, J., Walter, E., Piet-Lahanier, H.: Worst-case global optimization of black-box functions through Kriging and relaxation. J. Glob. Optim. (2012)
4. Dambreville, F.: Bi-level Sensor Planning Optimization Process with Calls to Costly Sub-processes. In: Nguyen, N.T., Attachoo, B., Trawiński, B., Somboonviwat, K. (eds.) ACIIDS 2014, Part II. LNCS, vol. 8398, pp. 382–391. Springer, Heidelberg (2014)
5. Younis, M., Akkaya, K.: Strategies and techniques for node placement in wireless sensor networks: A survey. Ad Hoc Networks 6, 621–655 (2008)
6. Ghosh, A., Das, S.K.: Coverage and connectivity issues in wireless sensor networks: A survey. Pervasive and Mobile Computing 4, 303–334 (2008)
7. Zhu, C., Zheng, C., Shu, L., Han, G.: A survey on coverage and connectivity issues in wireless sensor networks. Computer Communications 29(4), 490–501 (2006)
8. Tang, J., Hao, B., Sen, A.: Relay node placement in large scale wireless sensor networks. Computer Communications 29(4), 490–501 (2006)

9. Tan, G., Jarvis, S., Kermarrec, A.-M.: Connectivity-Guaranteed and Obstacle-Adaptive Deployment Schemes for Mobile Sensor Networks. IEEE Trans. on Mobile Computing 8(6), 836–848 (2009)
10. De Boer, P.T., Kroese, D.P., Mannor, S., Rubinstein, R.Y.: A tutorial on the cross-entropy method. Annals of Operations Research 134 (2002)
11. Dambreville, F.: Cross-entropic learning of a machine for the decision in a partially observable universe. J. Glob. Optim. 37, 541–555 (2007)

Simulation–Based Algorithms
for the Optimization of Sensor Deployment

Yannick Kenné[1], François Le Gland[2], Christian Musso[3], Sébastien Paris[1],
Yannick Glemarec[4], and Émile Vasta[4]

[1] Université de Toulon/LSIS, Toulon, France
yan_kenne@yahoo.fr, sebastien.paris@lsis.org
[2] INRIA, Rennes, France
francois.le_gland@inria.fr
[3] ONERA, Palaiseau, France
christian.musso@onera.fr
[4] DGA Techniques navales, Toulon, France
{yannick.glemarec,emile.vasta}@intradef.gouv.fr

Abstract. Two simulation–based algorithms are presented, that have
been successfully applied to an industrial optimization problem. These
two algorithms have different and complementary features. One is fast,
and sequential: it proceeds by running a population of targets and by
dropping and activating a new sensor (or re–activating a sensor already
available) where and when this action seems appropriate. The other is
slow, iterative, and non–sequential: it proceeds by updating a population
of deployment plans with guaranteed and increasing criterion value at
each iteration, and for each given deployment plan, there is a population
of targets running to evaluate the criterion. Finally, the two algorithms
can cooperate in many different ways, to try and get the best of both
approaches. A simple and efficient way is to use the deployment plans
provided by the sequential algorithm as the initial population for the
iterative algorithm.

1 Introduction and Context

The problem considered here can be described as follows: a limited number of
sensors should be deployed by a carrier in a given search area, and should be
activated at a limited number of time instants within a given time period, so as to
maximize the probability of detecting a target (present in the given search area
during the given time period). The criterion to be maximized is a probabilistic
criterion, because the target behavior and its initial position are not known
exactly, however a probabilistic model is available, as a Markov process [2]. There
is also an information dissymmetry in the problem: if the target is sufficiently
close to a sensor position when this sensor is activated, then the target can learn
about the presence and exact position of the sensor, and can temporarily modify
its trajectory so as to escape away before it is detected. This is referred to as
the target intelligence. Of course, if the target is too close to the sensor position

© Springer International Publishing Switzerland 2015
H.A. Le Thi et al. (eds.), *Model. Comput. & Optim. in Inf. Syst. & Manage. Sci.*,
Advances in Intelligent Systems and Computing 360, DOI: 10.1007/978-3-319-18167-7_23

when this sensor is activated, then it is detected and any attempt to escape away is pointless. Conversely, if the target is sufficiently far from the sensor position when this sensor is activated, then nothing happens (the target is not detected, and it cannot learn anything about the sensor position). This is summarized in Figure 1.

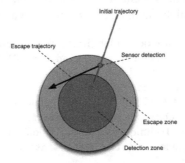

Fig. 1. Detection and target intelligence

Notice that because of this information dissymmetry, activating a sensor permanently is a costly and also very poor strategy, since the target would necessarily learn this sensor position and would manage to escape away before it is detected. Activating a sensor at a few suitably chosen time instants is a more sensible strategy. This is why the optimization problem is not only about where (in which positions) to deploy the sensors but also when (at which time instants) to activate them: in other words, one should solve a space–time optimization problem, and the deployment plan is a sequence of positions and activation times. Finally, notice that not all deployment plans are admissible: for instance, if one sensor position is significantly remote from another sensor position, then the second sensor could not be deployed shortly after the first sensor has been activated, i.e. some constraints should be satisfied. Indeed, if a sensor at position r has been activated at time t, and if the next action is a sensor at position r' to be activated at time t', then the average speed $|r' - r|/(t' - t)$ should be compatible with (less than) the carrier speed, or equivalently the distance $|r' - r|$ should be compatible with (less than) the maximum possible distance travelled by the carrier in the time length $(t' - t)$.

To summarize, a deployment plan is a sequence of positions and activation times, that should respect some physical constraints. The criterion to be maximized over all admissible deployment plans is the probability of a target to be detected. The criterion is evaluated by running a Monte Carlo simulation with a large number of target trajectories under a given probabilistic model. Notice however that the probabilistic model for the target behavior depends on the deployment plan, because of the target intelligence.

Two different simulation–based algorithms have been designed to solve jointly this optimization problem, and are described in the next two Sections 2 and 3. Two different scenarii are presented in Section 4 to illustrate the behaviour and performance of the two algorithms.

2 Sequential Algorithm

The first algorithm is sequential with a discrete time model and a tree structure: at each time instant, each branch in the tree corresponds to a proposed deployment plan running from the initial time instant. For each branch of the tree separately, a population of surviving (not already detected) targets is propagated according to the probabilistic model for the target behavior, a model that depends on the information (stored along the branch) about the position of sensors that have already been deployed so far. Whenever the empirical probability distribution associated with this population exhibits some significant modes, then the tree branches as follows. For each newly detected mode separately, a new edge is created which consists of (i) a sensor deployed at this time instant and at the position of this detected mode, or (ii) an already deployed sensor reactivated, provided its position is close enough to the position of this detected mode. This mechanism extends the branch, hence the associated deployment plan: the probability of detection along this extended branch, i.e. the probability of detection for this extended deployment plan, is re–evaluated, and the population of surviving targets is updated. This is summarized in Figure 2.

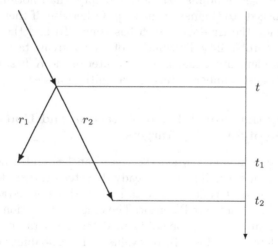

Fig. 2. In the deployment plan running from the initial time instant and represented by the incoming branch, two modes are detected at time t at positions r_1 and r_2 respectively, that can both be reached by the carrier, and two new edges are created accordingly, that correspond to two different concurrent extensions of the same deployment plan. Notice that an additional extension of the same deployment plan could also be considered, in which no new sensor would be deployed at time t.

To summarize, associated with any branch in the tree is a deployment plan, seen as a sequence of sensor positions (attached to edges), sensor activation times (attached to vertices), and probabilities of detection (attached to edges).

Internally, associated with any branch in the tree there is also a population of surviving (not already detected) targets and its empirical probability distribution, propagated along edges and updated at vertices. To specify the algorithm completely, it remains to explain how to

- decide that the time to branch has come,
- find the modes positions,
- re–evaluate the probability of detection,
- update the population of surviving targets.

2.1   Deciding to Branch and Finding the Modes

At each time instant after the last activation time, a population of N surviving (not already detected) targets is available, and a two–dimensional histogram with bins of size $\Delta x \times \Delta y$ is constructed [6] in the search area D. A bin is said admissible if it contains more than $\lambda \geq 1$ times as many targets as the average number per bin of uniformly distributed targets over the search area D. Notice that the total number of bins in the search area D is simply area$(D)/(\Delta x \times \Delta y)$, hence a bin is admissible if it contains a number of targets greater than $\lambda\, N\, \Delta x \times \Delta y/$area$(D)$. If the set of admissible bins is empty, than nothing happens and the algorithm proceeds to the next time step. Otherwise, if there is at least one admissible bin, then the time to branch has come. To find the position of the local modes of the probability distribution of the surviving target, a mean–shift algorithm [4] is implemented, that uses the center of the K best admissible bins (containing the largest number of targets) as initial guesses.

2.2   Evaluating the Probability of Detection and Updating the Population of Surviving Targets

Just before the time instant t when the tree branches and a new edge is created, a population of surviving (not already detected) targets is available. For each new edge created at time t, some targets in this population are detected by the activated (or re–activated) sensor. The empirical fraction of the detected targets is an estimate of the probability of detecting a target at time t that was not already detected before. To re–evaluate the probability of detection, or equivalently the complement probability of no–detection, notice that the probability of no–detection at time t can be simply expressed as the product of two probabilities: the probability of no–detection just before time t, i.e. just before the new edge is created, and the probability of not detecting a target at time t that was not already detected before, easily estimated as the complement of the empirical fraction considered above. To be more specific, if x denotes the deployment plan running from the initial time instant until time t, i.e. a branch in the tree, if $P_{\mathrm{d}}(x, t-)$ denotes the probability of detection just before time t, using the deployment plan x, i.e. the probability of detection at the last activation time $t-$ before time t in this deployment plan, if $P_{\mathrm{d}}(x, t \mid t-)$ denotes the probability of detecting a target at time t that was not already detected before, easily estimated as the empirical fraction of the detected targets, and if $P_{\mathrm{d}}(x,t)$ denotes the probability of detection at time t, using the deployment plan x, then

$$1 - P_{\mathrm{d}}(x,t) = (1 - P_{\mathrm{d}}(x,t-))\,(1 - P_{\mathrm{d}}(x,t\mid t-))\,,$$

and iterating this relation yields

$$1 - P_{\mathrm{d}}(x,t) = \prod_{s\in T(x,t)} (1 - P_{\mathrm{d}}(x,s\mid s-))\,,$$

where $T(x,t)$ denotes the ordered set of activation times in the deployment plan x up to (and including) t. This is summarized in Figure 3.

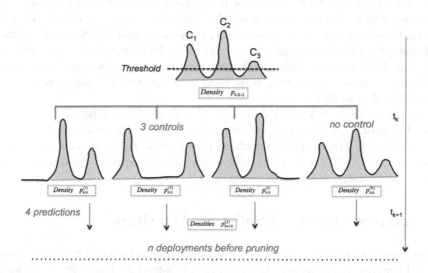

Fig. 3. In the deployment plan running from the initial time instant and represented by the probability density of surviving targets, three modes are detected at time t_k, that can all be reached by the carrier, hence four different concurrent extensions of the same deployment plan are proposed, and the probability density of surviving target is updated accordingly

The population of surviving target is updated by discarding the detected targets. Notice that some targets in the updated surviving population learn about the presence and exact position of the activated (or re–activated) sensor, and by construction are not detected. The intelligence obtained by these targets will impact their future behavior, since they will make sure to keep away from the sensor.

2.3   Pruning the Tree

Repeating this procedure as time goes on, an increasingly broad tree grows where each branch is seen as an unique deployment plan, and to each terminal vertex is associated the probability distribution of the surviving target exposed to this deployment plan running from the initial time instant. There is clearly a need to pruning the tree, in order to reduce the computationnal burden and the memory load. Two criteria have been used to select more promising deployment plans and discard others. The first criterion is the probability of detection, which seems a natural idea. The second criterion is the entropy (seen as a measure of concentration) of the probability distribution of the surviving target, with the idea that a target whose probability distribution is concentrated in some small region is more easy to detect than a target whose probability distribution is widespread all over space, and should be preferred. An empirical estimate of the entropy can be obtained from the available population of surviving targets [5]. Of course, other criteria could be used, e.g. the number of sensors deployed or the number of activation times, with the idea that a conservative deployment plan that has used only a few sensors is less expensive and still has many sensors available, and should be preferred. In practice, deployment plans are selected along the Pareto front. This simple selection strategy allows one to continue deeply in the tree. At the final time instant, a variety of deployment plans is produced and is available.

This algorithm is fast, and it usually provides various efficient deployment plans. However, there is no guarantee that the maximum probability of detection is achieved. The iterative algorithm presented in the next section enhances the detection probability in using these approximate, sequentially obtained, suboptimal deployment plans as initial guesses.

3   Iterative Multilevel Splitting Algorithm

The second algorithm is global (non–sequential) and iterative. A reference, somehow arbitrary, probability distribution $\mu(dx)$ is introduced initially in the space of solutions (deployment plans).

At a given iteration of the algorithm, say at iteration $(r - 1)$, a population $(x_{r-1}^1, \cdots, x_{r-1}^N)$ of N deployment plans is available, that all have a probability of detection $P_d(x_{r-1}^i)$ above some threshold q_{r-1} for $i = 1, \cdots, N$. Actually, this population is distributed according to $\mu_{r-1}(dx) = \mu(dx \mid P_d(x) \geq q_{r-1})$, the reference probability distribution in the space of deployment plans, conditionnal on the criterion (probability of detection) being above the threshold q_{r-1}. Let the deployment plans be sorted according to non–increasing values of the probability of detection, i.e.

$$P_d(x_{r-1}^{(1)}) \geq P_d(x_{r-1}^{(2)}) \geq \cdots \geq P_d(x_{r-1}^{(N)}) \geq q_{r-1} .$$

A new threshold is defined as the empirical $(1 - \alpha)$–quantile $q_r = P_d(x_{r-1}^{(m)})$ of the evaluated probabilities of detection, where $m = \lfloor \alpha N \rfloor$. By construction, notice that $q_r \geq q_{r-1}$. The $(N - m)$ deployment plans that have a probability of detection below the new threshold q_r are eliminated, and the m best deployment plans $(x_{r-1}^{(1)}, \cdots, x_{r-1}^{(m)})$ that all have a probability of detection above the new threshold

q_r are selected and replicated. Not only do the N selected and replicated deployment plans $(\widehat{x}_r^1, \cdots, \widehat{x}_r^N)$ have a probability of detection above the new threshold q_r, but they are distributed according to $\mu_r(dx) = \mu(dx \mid P_d(x) \geq q_r)$, the reference probability distribution in the space of deployment plans, conditionnal on the criterion (probability of detection) being above the new threshold q_r. Unfortunately, the population $(\widehat{x}_r^1, \cdots, \widehat{x}_r^N)$ of selected and replicated deployment plans contains only a small number m of really different deployment plans, and replicas of these. To bring in more diversity in the population, i.e. to recover a population of N genuinely all different deployment plans with a probability of detection above the new threshold q_r, another mechanism is introduced. Actually, the population of selected and replicated deployment plans is repeatedly shaked under the action of (a few steps of) a Gibbs sampler that preserves the probability distribution $\mu_r(dx)$ of the population. In other words, at each step of the Gibbs sampling the shaked population of deployment plans is distributed according to $\mu_r(dx)$, and in particular, at each step of the Gibbs sampling all the deployment plans in the shaked population have a probability of detection above the new threshold q_r. When the action of the Gibbs sampler has been completed, a sufficiently diverse population $(x_r^1, \cdots, x_r^N)$ of N deployment plans is obtained, that all have a probability of detection $P_d(x_r^i)$ above the new threshold q_r for $i = 1, \cdots, N$, and a new iteration of the algorithm starts again following the same cycle of operations

- definition of the new threshold as an empirical quantile,
- selection and replication of the deployment plans with a probability of detection above the new threshold,
- Gibbs sampling.

The procedure is iterated, until it becomes impossible to increase the threshold. When the algorithm terminates, the current value of the threshold is the proposed approximation of maximum probability of detection, and the current population of deployment plans provides a set of maximizers. To specify the algorithm completely, it remains to explain how to design the Gibbs sampler.

3.1 Gibbs Sampling

A deployment plan is given by

- a number $s = 1, \cdots, s_{max}$ of sensors actually deployed, where s_{max} denotes the maximum number of sensors that could possibly be deployed, i.e. the total number of available sensors,
- for each $i = 1, \cdots, s$, the position r_i where the i-th sensor is actually deployed in the search area $D \subset \mathbb{R}^2$,
- for each $i = 1, \cdots, s$, a number $n_i = 1, \cdots, n_{max}$ of time instants when the i-th sensor can be activated (for simplicity, the maximum number n_{max} of activation times is the same for each sensor),
- for each $i = 1, \cdots, s$ and for each $j = 1, \cdots, n_i$, the time instant t_{ij} when the i-th sensor is activated for the j-th time.

The set of all possible deployments plans is defined as

$$\mathcal{X}_0 = \bigcup_{s=1}^{s_{\max}} \left(\{s\} \times \prod_{i=1}^{s} \left(D \times \bigcup_{n_i=1}^{n_{\max}} \left(\{n_i\} \times [0,T]^{n_i}\right)\right)\right),$$

and an arbitrary element in this set is of the form

$$x = (s, (r_i, n_i, (t_{i,j}, j = 1, \cdots, n_i), i = 1, \cdots, s)) .$$

An initial probability distribution is introduced in the space $\mathcal{X}_0$. Without further information, this reference probability distribution should be as less informative as possible. A possible solution is

$$\mu_0(dx) = \frac{1}{s_{\max}} \prod_{i=1}^{s} (1_{(r_i \in D)} \frac{dr_i}{\text{area}(D)} \frac{1}{n_{\max}} \frac{1}{T^{n_i}} dt_{i1} \cdots dt_{in_i}) , \qquad (1)$$

which is easily interpreted. The proposed Gibbs sampler consists of several elementary moves

- deploy a new sensor, with its position and activation time instant,
- remove a sensor,
- change the position of a given sensor,
- introduce a new activation time for a given sensor,
- delete an existing activation time for a given sensor.

The transition kernel associated with any of these elementary moves, is reversible for the reference probability distribution $\mu_0(dx)$.

Recall that a deployment plan is admissible if it satisfies the constraints induced by the carrier speed. Indeed, let the activation times $((t_{i,j}, j = 1, \cdots, n_i), i = 1, \cdots, s)$ be sorted in ascending order. For any ordered pair $t \le t'$ of successive activation times corresponding to sensor positions r and r' respectively, the average speed $|r' - r|/(t' - t)$ should be less than the carrier speed. The projection of the probability distribution defined in (1) on the set $\mathcal{X} \subset \mathcal{X}_0$ of all admissible deployment plans, is defined as

$$\mu(dx) \propto 1_{(x \in \mathcal{X})} \mu_0(dx) , \qquad (2)$$

up to a multiplicative normalizing constant. To produce a transition kernel that is reversible for the probability distribution $\mu(dx) \propto 1_{(x \in \mathcal{X})} \mu_0(dx)$, it is sufficient to consider any of the same elementary moves, and check that the deployment plan obtained after the move does also satisfy the constraints. To produce a transition kernel that leaves invariant the probability distribution $\mu'(dx) = \mu(dx \mid P_d(x) \ge q)$, i.e. the reference probability distribution in the space of deployment plans, conditionnal on the probability of detection being above some threshold q, the following simple strategy can be used and is easily implemented: starting from a deployment plan with a probability of detection above the threshold q

- apply the transition kernel associated with any of the elementary moves,
- check that the deployment plan obtained after the move does also satisfy the constraints,
- evaluate the probability of detection associated with the proposed deployment plan,
- if this value is still above the threshold q, then accept the proposed deployment,
- otherwise, if this value falls below the threshold q, then reject the proposed deployment, i.e. keep the original deployment plan.

As with many iterative optimization algorithms with a high dimensional state and with a possibly strongly multimodal criterion, an important issue is the initial condition. Without further information, this reference probability distribution should be as less informative as possible, which yields (1) and (2). However, the number of iteration in this case turns out to be very large and the performance of the algorithm can be enhanced if an educated initial condition is used. This is why it has been proposed to use the various efficient deployment plans produced by the sequential algorithm presented in Section 2 as an initial population of deployment plans for the iterative multilevel splitting algorithm.

4 Numerical Results

Two different scenarios are presented to illustrate the behaviour and performance of two algorithms: the sequential algorithm presented in Section 2 and the fused algorithm, i.e. the multi–level splitting algorithm presented in Section 3 with the various efficient deployment plans produced by the sequential algorithm as an initial population of deployment plans.

4.1 Barrier Scenario

At the initial time instant, the target is located somewhere in the Z1 area, and its goal is to reach the Z3 area. Therefore it needs to cross the Z2 aera where the sensors are deployed.

The multi–level splitting (Split) is intialized by using the various efficient deployment plans produced by the sequential algorithm (Seq), see Figure 4. The number of iterations is drastically reduced when using this initial population of deployment plans. Typically, 10 iterations are sufficient for the algorithm to converge. Otherwise, if a random initialization is used, the multi–level splitting algorithm needs much more iterations.

For each algorithm (sequential and multilevel–splitting), the activation times (first line) and the corresponding activated sensor labels (second line) are shown in Table 1. For example, the sensor #1 is activated at 1400 s and reactivated at 3400 s and 3731 s. Notice also that the fused algorithm has eliminated a sensor. This illustrates the shaking effect of elementary moves of the Gibbs sampler used in the multilevel–splitting algorithm.

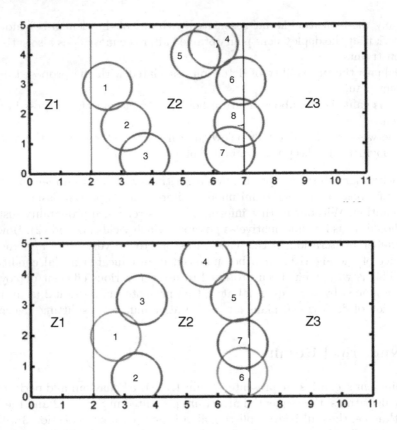

Fig. 4. Top: an efficient deployment plan proposed by the sequential algorithm, with sensors positions, and their respective order of deployment. — Bottom: the (almost) optimal deployment plan obtained by the fused algorithm with sensors positions, and their respective order of deployment.

Table 1. Activation times (and the corresponding activated sensor labels) for the sequential algorithm and for the fused algorithm

seq	1400	3000	3400	5200	5600	7800	9200	9400	11000	12800	14400		
	1	2	1	3	2	4	5	4	6	7	8		
split	1400	3000	3400	3731	5223	6987	8529	8933	11000	11600	13400	13800	14314
	1	2	1	1	3	4	5	4	6	5	7	6	6

The final probability of detection obtained with the sequential algorithm is 0.9509, and the fused algorithm provided an enhanced deployment plan with an improved probability of detection of 0.9830.

4.2   Flaming Datum Scenario

A target as been reported seen, then it has dived in order to escape radially with an unknown heading. Some time later, a carrier arrives on the spot and the search begins. In this case the optimal solution is a spiral. The solution obtained by the fused algorithm is shown in Figure 5.

The final detection probability is close to one as one target only has managed to survive among a large population of 10000 targets.

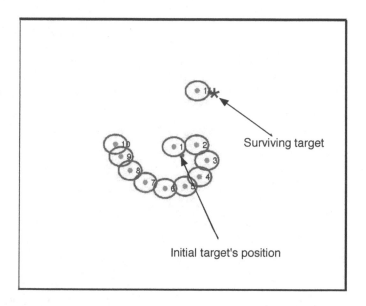

Fig. 5. The (almost) optimal deployment plan obtained by the fused algorithm with sensors positions, and their respective order of deployment

References

1. Botev, Z.I., Kroese, D.P.: An efficient algorithm for rare-event probability estimation, combinatorial optimization, and counting. Methodology and Computing in Applied Probability 10(4), 471–505 (2008)
2. Chouchane, M., Paris, S., Le Gland, F., Musso, C., Pham, D.T.: On the probability distribution of a moving target. Asymptotic and non-asymptotic results. In: Proceedings of the 14th International Conference on Information Fusion, Chicago, pp. 99–101. ISIF (July 2011)
3. Chouchane, M., Paris, S., Le Gland, F., Ouladsine, M.: Splitting method for spatio-temporal sensors deployment in underwater systems. In: Hao, J.-K., Middendorf, M. (eds.) EvoCOP 2012. LNCS, vol. 7245, pp. 243–254. Springer, Heidelberg (2012)
4. Comaniciu, D., Meer, P.: Mean shift: a robust approach toward feature space analysis. IEEE Transactions on Pattern Analysis and Machine Intelligence PAMI-24(5), 603–619 (2002)

5. Pham, D.T.: Fast algorithm for estimating mutual information, entropies and score functions. In: Proceedings of the 4th International Symposium on Independent Component Analysis and Blind Signal Separation, Nara, pp. 17–22 (April 2003)
6. Scott, D.W.: Multivariate density estimation. Theory, practice, and visualization. Wiley Series in Probability and Mathematical Statistics. John Wiley & Sons, New York (1992)

Part IV

Maintenance and Production Control Problems

Impact of the Corrective Maintenance Cost on Manufacturing Remanufacturing System Performance

Sadok Turki, Zied Hajej, and Nidhal Rezg

LGIPM / Université de Lorraine, Ile du Saulcy, 57045 Metz Cedex, France
{Sadok.Turki,zied.hajej,nidhal.rezg}@univ-lorraine.fr

Abstract. This paper studies a manufacturing remanufacturing system composed by two parallel machines, a serviceable inventory, a remanufacturing inventory and a customer which demands a constant quantity of part. Taking into account the return of the used part and the maintenance action, discrete flow model is used to model and to simulate the system. The goal of this work is to evaluate the optimal serviceable inventory level which allows minimizing the sum of inventory, lost sales costs and the corrective maintenance cost. Numerical results are presented to show the impact of the corrective maintenance cost on the optimal serviceable inventory level.

Keywords: Manufacturing-remanufacturing system, discrete flow model, corrective maintenance, returned products.

1 Introduction

The return of the used products brings an interesting economic burden to the producers, while many of the used products have economic value and could be remanufactured to satisfy the market demand. Hence, research on supply chain management has been paying attention on the recovery processes of end of life products for remanufacturing. To study such system, discrete flow model is usually used. Indeed, discrete flow model (Hazelton [1], Vasic and Ruskin [2]) is widely used for design, optimal control and optimization of manufacturing systems. This model has recently been considered as an alternative paradigm to queuing network for analysis and synthesis of discrete event systems. Indeed, discrete flow model is more realistic for discrete manufacturing systems than stochastic fluid model (Yao and Cassandras [3], Xie, . Turki et al. [4])), and it allows tracking individual parts part by part either in performance evaluation or real-time flow control and is generally easier to simulate. Also, this model is used for the network system design or for modelling manufacturing-remanufacturing systems. In this paper we use a discrete flow model for describing a manufacturing remanufacturing system and to take into account returned products and remanufacturing products. In the literature we find many published works on reverse logistics and which focus on aspects of inventory management and production planning. Shi, Zhang, and Sha [5] presented a stochastic model for determining the optimal production and remanufacturing quantities for a product portfolio. The authors assumed that the product demands are independent, for each product new and

© Springer International Publishing Switzerland 2015
H.A. Le Thi et al. (eds.), *Model. Comput. & Optim. in Inf. Syst. & Manage. Sci.*,
Advances in Intelligent Systems and Computing 360, DOI: 10.1007/978-3-319-18167-7_24

remanufactured units are perfect substitutes, the returns are of unknown quality and the amount of returned cores is a function of their acquisition price, which is also a decision variable. Zhou, Naim, Tang, and Towill [6] considered a manufacturing and remanufacturing system and then adopted an inventory control strategy in the manufacturing loop which is an automatic pipeline, inventory and order based production control system and in the remanufacturing loop they employ a Kanban policy to represent a typical pull system and to control the remanufacturing process. Indeed, the authors analyzed the dynamic performance of the system which has implications on total costs in terms of inventory holding, capacity utilization and customer service failures. Turki et al. [7] studied a manufacturing- remanufacturing system with constant demand and then they evaluated the optimal manufacturing buffer which allows minimizing the total cost. The authors adopted stochastic fluid model to formulate the problem. Indeed, the advantage of this model comparing to discrete flow model, that it allows more easily to apply an interesting gradient method for optimizing the proposed system; this method called perturbation analysis (PA). Besides, as we mentioned earlier, stochastic fluid model is not realistic as the discrete flow model, thus in this paper we are interested to adopt the latter for modeling our system.

2   Manufacturing-Remanufacturing System with Maintenance Actions

We study a manufacturing-remanufacturing system (*Fig.1*) composed by two parallel machines which are subject to random failures and repairs denoted M_1 and M_2 for manufacturing and remanufacturing, respectively. We assume that both machines are producing the same type of product. Production activity in forward direction and reverse logistics is considered in this system (i.e. activity of the remanufacturing of the used products). The customers demand denoted D is supposed known and constant and which is satisfied from a serviceable inventory S which can be filled up by the machines M_1 and M_2. The Inventory R is operated for the stock keeping of the returned products ahead of the remanufacturing process. Returned products will be then remanufactured by the machine M_2 and then stoked in the serviceable inventory S with the manufactured products. We denote B the number of returned products (i.e. the number of the remanufacturable products) and which is constant and proportional to the customers demand D.

We assume that the machine M_1 is never starved. The machine M_1 is either down or up. The state of the machine at time t, denoted $\alpha(t)$, is given by:

$$\alpha(t) = \begin{cases} 1 & \text{machine } M_1 \text{ is up} \\ 0 & \text{machine } M_1 \text{ is down} \end{cases} \tag{1}$$

The state of the machine M_2 at time t, denoted $\beta(t)$, is given by:

$$\beta(t) = \begin{cases} 1 & \text{machine } M_2 \text{ is up} \\ 0 & \text{machine } M_2 \text{ is down} \end{cases} \tag{2}$$

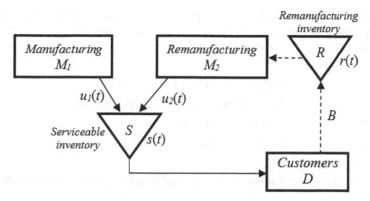

Fig. 1. Studied manufacturing-remanufacturing system*

When the machine is up, the production rate of M_1 denoted by $u_1(t)$, could take a value between 0 and its maximum U_1 (the machine capacity)., i.e., $0 \leq u_1(t) \leq U_1$. When the machine is down $u_1(t)=0$. The times to repair and times to failure are exponentially distributed with rate μ_1 and λ_1 respectively. We have the same state for the machine M_2, with $u_2(t)$ is the production rate, U_2 is the maximal production rate, μ_2 and λ_2 are times to repair and times to failure of the machine M_2. The failure/repair process is an independent random process. It does not depend on the system parameters.

The following assumptions are considered:

- The maximal production rate of the machine M_1 permits to satisfy the demand, i.e. $U_1 > D$. This assumption allows avoiding having always the serviceable inventory empty.
- The maximal production rate of the machine M_2 is upper to the number of returned products, i.e. $U_2 > B$. This assumption allows avoiding having always the remanufacturing inventory very full.
- If the demand is unsatisfied, the demand is lost with a corresponding cost (lost sales cost).
- The remanufactured products can be considered meeting the same quality level as the new products so that both type of products can be distributed like new.

We assume the case of infinite capacity for S and R. The serviceable inventory level and the remanufacturing inventory level are denoted, respectively, by $s(t)$ and $r(t)$ and are given by the following equations:

$$s(t) = s(t - dt) + u_1(t - dt) + u_2(t - dt) - D \qquad (3)$$

$$r(t) = r(t - dt) + B - u_2(t - dt) \quad \text{(With } dt \text{ is the simulation time step)} \qquad (4)$$

Remark 1: In the first step of the approach which determines the optimal inventory level, the capacity of serviceable inventory S is supposed infinite, and then the optimal inventory level will be determined according to the hedging point.

The control policy is a hedging point policy (Akella and Kummar [12]) and which ensures that the part does not exceed a given number of products, denoted by h. The control policy is defined as follows:

For the machine M_1:

$$u_1(t) = \begin{cases} U_1 & \text{if } \alpha(t) =1 \text{ and } s(t) \leq h+D-(U_1+u_2(t)) \\ h-s(t)-u_2(t) & \text{if } \alpha(t) =1 \text{ and } h+D-(U_1+u_2(t)) < s(t) < h \\ D-u_2(t) & \text{if } \alpha(t) =1 \text{ and } s(t) = h \\ 0 & \text{if } \alpha(t) =0 \text{ or } s(t) > h \end{cases} \qquad (5)$$

For the machine M_2:

When R is empty, the machine M_2 is supplied directly by the returned products.

$$u_2(t) = \begin{cases} U_2 & \text{if } \beta(t) =1, r(t) >0 \text{ and } s(t) \leq h+D-(U_2+u_1(t)) \\ h-s(t)-u_1(t) & \text{if } \beta(t) =1, r(t) >0 \text{ and } h+D-(U_2+u_1(t)) < s(t) < h \\ D-u_1(t) & \text{if } \beta(t) =1, r(t) >0 \text{ and } s(t) = h \\ B & \text{if } \beta(t) =1, r(t) =0 \text{ and } s(t) \leq h+D-(B+u_1(t)) \\ 0 & \text{if } \beta(t) =0 \text{ or } s(t) > h \end{cases} \qquad (6)$$

According to this policy, when S become full (i.e. $s(t) =h$) the sum of $u_2(t)$ and $u_1(t)$ is equal to the demand rate (i.e. $u_1(t) + u_2(t) =D$). We assume in this case that the machines M_1 and M_2 share the production. In other word, each machine produces half of the production (i.e. $u_1(t) =D/2= u_2(t)$). Therefore, as our system is studied in a discrete setting, we assume that the demand is pair number in order to have always the both production rates are integer numbers. Besides, in the case when $u_1(t) = h - s(t) - u_2(t)$ we have two cases:

Case 1: if the difference $h-s(t)$ is a pair number then we have $u_1(t) = u_2(t) = h - s(t)/2$

Case 2: if the difference $h-s(t)$ is an impair number, in this case we assume that $u_1(t) = \lfloor (h-s(t)/2) \rfloor +1$ and $u_2(t) = \lfloor (h-s(t)/2) \rfloor$.

We assume that the return rate B is proportional to the demand D. We denote by p ($0<p<1$) the percentage of sales which are returned for remanufacturing. Indeed, the customer returns rate may be as high as 15% of sales in the coming years, and in sectors such as catalogue sales and e-commerce it could reach as much as 35% (Rubio and Corominas [13]). Then we have:

$$B = p. D \qquad (7)$$

We denote $w(t)$ the number of unsatisfied demands (lost) at time t and which is defined as follows:

$$w(t) = \begin{cases} 0 & \text{if } s(t) \geq D \\ D-s(t) & \text{if } s(t) < D \end{cases} \qquad (8)$$

We denote by $F(t)$ the cost function at time t and which is composed by the inventory cost and the lost sale cost. $F(t)$ is given by:

$$F(t) = cs.s(t) + cr.r(t) + cs^-.w(t)$$ (9)

Where:

- cs and cr : are the units inventory cost respectively for S and R;

- cs^- : unit lost sale cost.

The total cost function, denoted by $FT(h)$ depending on h and the number of corrective maintenance is given by :

$$FT(h) = \sum_{t=0}^{t=T} F(t) + N.cm$$ (10)

Where T is the total simulation time, N is the number of corrective maintenance actions and cm the cost of the corrective maintenance.

In what follows, numerical results are presented to show the interest of our proposal method and to study the impact of the percentage of the returned products p on the value of the optimal serviceable inventory level h.

3   Impact of the Corrective Maintenance Cost on the Optimal Serviceable Inventory Level

In this part, we are interested to study the impact of the corrective maintenance cost on the value of the optimal serviceable inventory level h. While the cost function depends of the number of the corrective maintenance, the value of h which minimizes this cost function will certainly also depend on the percentage of the returned products. Therefore, we vary the values of the corrective maintenance cost and we determine the value of h by using the simulation of the discrete flow model.

The following parameters are used for the simulation:

- U_1 = 12 products /time unit;
- U_2 = 6 products /time unit;
- D= 10 products /time unit;
- The total simulation time is equal to T=10E+07 time units;
- The times to failure or repair are given by exponential distribution, the mean time between failures MTBF is equal to 3 and the mean time to repair MTTR is equal to 1;
- The unit inventory cost cs is equal to 1 monetary unit;
- The unit lost sales cost cs- is equal to 50 monetary units;
- The unit inventory cost cr is equal to 1 monetary unit.

Table 1. Impact of the corrective maintenance cost on the optimal serviceable inventory level

cm	h
10	25
20	18
30	14
40	11
50	9

As we see, the optimal serviceable inventory level decreases when the cost of the corrective maintenance increases.

4   Conclusion

In this paper, we studied a manufacturing-remanufacturing system composed by two parallel machines, a serviceable inventory, a remanufacturing inventory and customers who demand a constant quantity of product. Discrete flow model is adopted to describe the system and to take into account machine failure, returned products, remanufacturing products and maintenance actions. The times to failure and times to repair are random variables with exponential distribution. Using simulation the impact of the corrective maintenance cost on the optimal serviceable inventory level is studied.

References

[1] Hazelton, M.L.: Bayesian inference for network-based modes with a linear inverse structure. Transportation Research Part B 44, 674–685 (2010)
[2] Vasic, J., Ruskin, H.J.: A discrete flow simulation model for urban road networks, with application to combined car and single-file bicycle traffic. In: Murgante, B., Gervasi, O., Iglesias, A., Taniar, D., Apduhan, B.O. (eds.) ICCSA 2011, Part I. LNCS, vol. 6782, pp. 602–614. Springer, Heidelberg (2011)
[3] Yao, C., Cassandras, C.G.: Using infinitesimal perturbation analysis of stochastic flow models to recover performance sensitivity estimates of discrete event systems. Discrete Event Dynamic Systems 22, 197–219 (2012)

[4] Turki, S., Bistorin, O., Rezg, N.: Optimization of stochastic fluid model using perturbation analysis: a manufacturing-remanufacturing system with stochastic demand and stochastic returned products. In: 11th IEEE International Conference on Networking, Sensing and Control (ICNSC 2014), Miami, FL, USA, April 7-9 (2014)

[5] Shi, J., Zhang, G., Sha, J.: Optimal production planning for a multi-product closed loop system with uncertain demand and return. Computers and Operations Research 38(3), 641–650 (2011)

[6] Zhou, L., Naim, M.M., Tang, O., Towill, D.R.: Dynamic Performance of a Hybrid Inventory System with a Kanban Policy in Remanufacturing Process. OMEGA 34(6), 585–598 (2006)

[7] Turki, S., Bistorin, O., Rezg, N.: Infinitesimal Perturbation Analysis Based Optimization for a Manufacturing-Remanufacturing System. In: 18th IEEE Conference on Emerging Technologies and Factory Automation (ETFA 2013), Cagliari, Italy, September 10-13 (2013)

[8] Wardi, Y., Giua, A., Seatzu, C.: IPA for continuous stochastic marked graphs. Automatica 49(5), 1204–1215 (2013)

[9] Yao, C., Cassandras, C.G.: Using infinitesimal perturbation analysis of stochastic flow models to recover performance sensitivity estimates of discrete event systems. Discrete Event Dynamic Systems 22, 197–219 (2012)

[10] Yu, H., Cassandras, C.G.: Perturbation analysis for production control and optimization of manufacturing systems. Automatica 40(6), 945–956 (2004)

[11] Turki, S., Hennequin, S., Sauer, N.: Perturbation analysis-based optimisation for a failureprone manufacturing system with constant delivery time and stochastic demand. Int. J. Advanced Operations Management 4(1/2), 124–153 (2012)

[12] Akella, R., Kummar, P.R.: Optimal control of production rate in failure prone manufacturing system. IEEE Transaction on Automatic Control 31(2), 116–126 (1986)

[13] Rubio, S., Corominas, A.: Optimal Manufacturing - Remanufacturing Policies in a Lean Production Environment. Computers and Industrial Engineering 44, 345–361 (2008)

[14] Turki, S., Hennequin, S., Sauer, N.: Perturbation analysis for continuous and discrete flow models: a study of the delivery time impact on the optimal buffer level. International Journal of Production Research 51(13), 4011–4044 (2013)

Integration of Process Planning and Scheduling with Sequence Dependent Setup Time: A Case Study from Electrical Wires and Power Cable Industry

Safwan Altarazi[1,2] and Omar Yasin[2]

[1] Applied Science Private University, Mechanical and Industrial Engineering Department, Amman, Jordan
s_altarazi@asu.edu.jo
[2] German-Jordanian University, Industrial Engineering Department, Amman, Jordan
{safwan.altarazi,o.yasin}@gju.edu.jo

Abstract. This paper addresses the integration of process planning and scheduling (IPPS) with sequence dependent setup times for a case from the electrical wires and power cables industry. While the IPPS problem has been a subject of researchers' attention in recent years, majority of research in this field neglect setup time or assume it as a part of processing time. The objective is to simultaneously select the most feasible process plan and schedule with minimum makespan. The problem is modeled as a mixed integer linear programming problem and an example is presented to demonstrate the applicability and effectiveness of the proposed modeling approach.

Keywords: Integrated process planning and scheduling, Sequence dependent setup time, Electrical Wires and Power Cable Industry, Mixed Integer Programming.

1   Introduction and Related Research

Process planning is a manufacturing system function which prepares detailed operation instructions to transform an engineering design into a final part [10]. The outcome of process planning is the information required for manufacturing processes, including the identification of the machines, tools and fixtures. The information provided by process planning routes the workpiece through the individual manufacturing stages until its final product stage. Computer-aided process planning (CAPP) systems is typically used to generate candidate process plans from which one is selected based on some criteria [2]. In fact, CAPP acts as a bridge between computer aided design (CAD), which is primarily concerned with the effective use of computers to support the design engineering function by creating, modifying or documenting an engineering design; and computer aided manufacturing (CAM), defined as the effective use of computer technology in planning, management and control of the manufacturing activities. Figure 1 illustrates this functional relationship.

© Springer International Publishing Switzerland 2015
H.A. Le Thi et al. (eds.), *Model. Comput. & Optim. in Inf. Syst. & Manage. Sci.*,
Advances in Intelligent Systems and Computing 360, DOI: 10.1007/978-3-319-18167-7_25

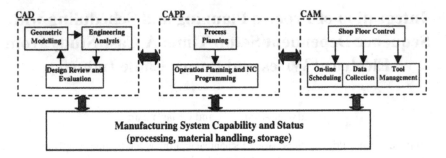

Fig. 1. CAPP as a bridge between CAD and CAM [19]

Scheduling is a decision-making tool that deals with the allocation of resources to tasks over given time periods in order to optimize one or more objectives such as the minimization of the completion time of the last task, named usually as the makespan, and the minimization of the number of tasks completed after their respective due dates [16]. Characteristically, a scheduling problem is specified by a set of decision variables representing the time and machine assignments of a set of jobs (the term "job" is usually referred to "product" in scheduling problems), together with a set of constraints or restrictions on the decision variables. The problem is to find the best schedule from some set of available alternatives with regards to some optimality criteria such that the constraints are satisfied. Among the classic scheduling problems are flow shop, job shop, and open shop which can be considered as the three basic scheduling problems [18]. Among the challenging attributes of scheduling problems which has been considered negligible for long time, but has significant impact on many industrial applications, is the consideration of tasks' setup time. In sequence dependent setup time (SDST) scheduling problems setup time depends on the sequence of jobs processed on each machine, i.e., the switch between jobs. For example, if job j is processed before job k, 2 time units have to be considered between the processing of the jobs. However, if job k is processed before job j, 3 time units should be considered. The importance and applications SDST problems have been discussed in several researches such as in [1] and [16].

It was just until the 1980s when researchers have started recognizing the advantages of integrating process planning and scheduling (IPPS) functions. Basically, IPPS unifies the solution space of process planning and scheduling by postponing the process plan selection and feature sequencing decisions until the scheduling stage [24]. IPPS takes into consideration that each job may have more than one process plan, i.e., alternative sequence of operations, and each operation can be performed on multiple machines. Then, IPPS simultaneously generates a production schedule through the selection of the best process plan from a set of available process plans for each job, and selects the suitable machines for the operations in order to satisfy one or more than one optimality criterion. The integration of these two functions has several advantages, including:

- Performance measures of a manufacturing system such as throughput time, resource utilization, work in process inventory, average flow time, and average tardiness can be improved significantly [24].

- The dynamic nature of real manufacturing systems which is caused by several events such as machine breakdown, new job arrival, job cancellation, and adding new machine can make a plan infeasible. Consequently, frequent alteration of the plans due to the changed status of the shop floor is required. IPPS reduces the time delay between production planning and plan execution, and thus, a more realistic plan without frequent need for modification is generated [9].
- Process planning emphasizes the technological requirements of a job with no regard for the competition that exists between jobs for the resources, while scheduling involves the timing aspects and resource sharing of all jobs [15]. Therefore, these functions may have different and sometimes conflicting objectives. IPPS offers the possibility of considering these conflicting criteria simultaneously, which can result in production cost reduction, bottleneck elimination, balanced level of machine utilization, and improved facility productivity [15].

Different versions and characteristics of the IPPS problem associated with various perspectives and solution techniques have been reported in literature. Kumar and Rajotia [12] proposed a framework for the IPPS problem in a job shop environment for axisymmetric components. Design specifications, availability of machine tools, and shop floor status were considered in generating feasible process plans for incoming products. Tan and Khoshnevis [23] proposed a linearized polynomial mixed integer model which assures that no overlapping in scheduling machines of operations from different process. Garcia-Sabater et al. [6] presented a two-stage sequential planning scheme for integrated operations planning and scheduling system using a mixed-integer programming model. Altarazi [2] presented proposed a novel modeling for the IPPS problem in job shop environment that simultaneously allocates operational tolerances while minimizing its manufacturing cost, minimizing work in process inventory, and figuring operation-machine assignments. The proposed mixed-integer nonlinear model minimized two objectives: the total operational tolerance-manufacturing cost and the WIP inventory. A preemptive method was used to optimize the two associated objectives. Baykasoğlu and Özbakır [4] also addressed the IPPS problem with two objectives: total flow time and total cost of process plans. In the proposed model, the generic process plan was represented as a grammar, dispatching rules were applied for sequencing operations, and a multi-objective Tabu search was employed to generate alternative solutions. Li and McMahon [14] utilized simulated annealing (SA) to facilitate the integration and optimization of process planning and scheduling. The balanced level of machine utilization, makespan, job tardiness and manufacturing cost were used to evaluate the approach. Brandimarte and Calderini [5] implemented a two-phase hierarchical Tabu search for the integration and optimization of the process planning and scheduling. Shao et al. [20] presented a modified Genetic Algorithm (GA) approach for the IPPS problem where efficient genetic representations and operator schemes have been developed. The experimental results indicated the superiority and adaptability of the method. In addition, Amin-Naseri and Afshari [3] presented a GA-based algorithm for the IPPS problem with precedence constraints. Lee and Kim [13] introduced a simulation based GA approach for solving the IPPS problem. In the presented approach, a simulation module computes performance measures based on process plan combinations, then, these measures are fed into a GA in order to improve the solution quality represented by scheduling objectives such as makespan or lateness. Haddadzade, Razfar, and Zarandi [8] considered stochastic processing times in their modeling of the IPPS problem and solved it with a hybrid SA-Tabu algorithm. Guo et al. [7] used the particle swarm optimization algorithm to solve the IPPS problem. Wong et al. [26] considered the IPPS problem with rescheduling and used a hybrid multi-agent approach to solve this problem in job shop environments. Kim et al. [11] utilized symbiotic evolutionary algorithm, an artificial intelligence search technique, to handle the process planning and scheduling functions simultaneously.

The majority of IPPS research neglect setup time consideration and or assume it as a part of processing time. Recently, Wan et al. [25] implemented an Ant Colony algorithm to solve the IPPS-SDST problem. Also, Nourali et al. [17] presented a mathematical model to solve the IPPS-SDST problem. This research presents a new mathematical modeling for the IPPS-SDST considering a case from the electrical wires and power cable industry.

The next section describes the production process of electrical wires and power cables. Section three is the statement of the problem while section four presents the mathematical modeling for the problem. Implementation is presented and discussed in section five. Finally, concluding remarks are presented in section six.

2   The Production Process of Electrical Wires and Power Cables

The electrical wires and cables industry is a crucial industry with hundreds of billions investment around the world. Its' products resemble the raw material for many other industries business activities such as construction, telecommunication, power transmission, electronics, etc. The products of the industry are usually classified into electrical wires and cables. An electrical wire is a single flexible strand or solid rod of metal, usually cylindrical, and surrounded by an insulator. Electrical cables on the other hand, frequently called power cables, consist of two or more wires running next to each other and bonded, braided, or twisted together, and commonly insulated to form a single assembly.

An extruded cable production line is a sophisticated manufacturing process. It consists of many sub processes that must work in concert with each other. The conductor rod, usually from copper or aluminum, is first drawn to the specified diameter. After drawing, the wire is softened, or annealed in a water bath. To raise the temperature for the annealing process, a large electrical current passes through the wire for a fraction of a second, raising its temperature briefly to about 1000F. Next, the wire, now soft and flexible, is passed through an extruder, where either a single or double coating of plastic insulation is applied. High-density polyethylene or polyvinyl chloride (PVC) polymers, colored in one of ten industry-standard colors, are two typical polymers used for this purpose. Exiting the extruder, the coated wire travels and passes through another cooling trough and is coiled on take-up reels. The manufacturing steps for single-rod wire ends by this stage. For two-rod wire and cables manufacturing two or more conductors of the same gauge are twisted together forming what is called the core. Next, a metal or non-metal mesh is braided around the cable. If the unit is to form part of a larger cable, it next goes to the cabling operation. At cabling, multi cores are twisted together on a rotating rod to form a multi-unit cable core. Depending on the cable design and application, a protective metal sheathing of either aluminum or aluminum and steel combined may be added in a manufacturing step called jacketing or armor. Then, the outer cable jacket is insulated by black-color low-density polyethylene, or cross-linked polyethylene (XLPE), through an extrusion process. Finally, the jacketed cable then passes through a temperature- controlled water trough, which cools the jacket. The cable is dried, and the top layer of the jacket is heated slightly so that printer markings can be imprinted on it. Figure 2 shows the production process flow of different wire and cable types.

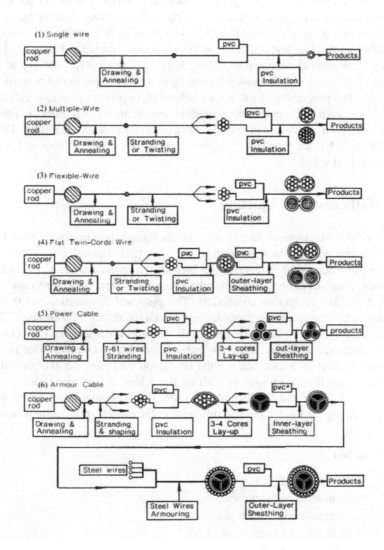

Fig. 2. The manufacturing process flowcharts for examples of electrical wires and power cable [22]

3   Statement of the Problem

Typically, the manufacturer of electrical wires and power cables faces a scenario where multiple products, some with more than one feasible process plan, are to be allocated on limited number of available machines and be produced within some time limit. Furthermore, different allocation sequences of products (wires and cables) on machines require a lot of setup efforts in terms of die changes in the drawing operation, extruders' cleaning and changes of insulation material or color for the insulation operation, parameters changes of the twisting and annealing operations, and other setup time-consuming activities. The above scenario in fact perfectly fits within the IPPS-SDST scope. The IPPS-SDST considers a set of n number of jobs $J = \{J_j\}_{1 \leq j \leq n}$ and a set of m number of machines $M = \{M_i\}_{1 \leq i \leq m}$. Each job J_j is defined as a set of operations that have to be processed according to a set of r number of process plans $L = \{L_l\}_{1 \leq l \leq r}$. The processing of job J_j on machine M_i in process plan L_l is called operation O_{jil}. Operation O_{jil} requires the exclusive use of M_i for deterministic processing time p_{ji} , i.e., each machine can process, at most, one operation at a time. This problem is strongly NP-hard since its simpler version with 0 setup times and 1 process plan per job is also NP-hard.

4   Mathematical Modeling

The proposed mathematical model is a mixed integer programming model (MIP). MIP is a natural way to attack scheduling problems [21]. The objective is to minimize the makespan whereas the constraints set include the operations sequence constraints, the process plan selection constraints, the operations' start time constraint, and the non-negativity and integrality constraints. The proposed formulation has the following assumptions: there are no precedence constraints among operations of different jobs, all operations of the same job require distinct machines and are subject to precedence constraints, operations cannot be interrupted (non-preemptive), that is, once an operation is started on a machine it must be completed before another operation can begin on that machine; each job can visit each machine at most once, that is, recirculation is not allowed; and finally only one machine is available for each operation. Before presenting the model, the notation used in the model formulation is defined.

4.1   Notation

N	total number of jobs
M	total number of machines
R	total number of process plans
n_j	total number of operations of job j
O_{ji}	ith operation of job j
O_{jil}	ith operation of job j in process plan l
p_{ji}	processing time of O_{ji}

S_i the set of operations assigned to machine M_i

x_{ji} the starting time of the processing of operation O_{ji}

x_{jil} the starting time of the processing of operation O_{jil} on process plan l

C_{max} makespan of a schedule

K An arbitrary large positive number

S_{jki} setup time between job j and job k if job j precedes job k on machine i

$$y_{jki} = \begin{cases} 1, \text{if operation } O_{ki} \text{ precedes operation } O_{ji} \text{ on machine } i \\ 0, \text{otherwise} \end{cases}$$

$$y_{jkilf} = \begin{cases} 1, \text{if operation } O_{kif} \text{ on process plan } f \text{ precedes operation } O_{jil} \text{ on process plan } l \\ 0, \text{otherwise} \end{cases}$$

$$d_{jl} = \begin{cases} 1, \text{if process plan } l \text{ is selected for job } j \\ 0, \text{otherwise} \end{cases}$$

Note that the x_{ji} and x_{jil} are the continuous decision variables and y_{jki}, y_{jkilf} and d_{jl} are the binary decision variable.

4.2   The Proposed MIP Model

The proposed MIP model for the IPPS-SDST is given as follows:

$$Minimize \quad C_{max} \tag{1}$$

Subject to

$$x_{j,i+1,l} - x_{jil} + K(1 - d_{jl}) \geq p_{jil} \qquad \forall j = 1, \ldots, n; i = 1, \ldots, n_j - 1; l = 1, \ldots, r \tag{2}$$

$$\sum_{l=1}^{r} d_{jl} = 1 \qquad \forall j = 1, \ldots, n \tag{3}$$

$$x_{kif} - x_{jil} + K y_{jkilf} + K(1 - d_{jl}) \geq p_{jil} + S_{jki} \qquad \forall (k,i), (j,i) \in S_i, (k,i) \neq (j,i), i = 1, \ldots, m; l, f = 1, \ldots, r \tag{4}$$

$$x_{jil} - x_{kif} + K(1 - y_{jkilf}) + K(1 - d_{jl}) \geq p_{kif} + S_{kji} \qquad \forall (k,i), (j,i) \in S_i, (k,i) \neq (j,i), i = 1, \ldots, m; l, f = 1, \ldots, r \tag{5}$$

$$x_{jil} + p_{ji} \leq C_{max} \qquad \forall j = 1, \ldots, n; i = 1, \ldots, m; l = 1, \ldots, r \tag{6}$$

$$x_{jil} \geq S_{0ji} \qquad \forall j = 1, \ldots, n; i = 1, \ldots, n_j; l = 1, \ldots, r \tag{7}$$

$$x_{ji} \geq 0 \qquad\qquad \forall j = 1, \dots, n; i = 1, \dots, n_j \tag{8}$$

$$y_{jkilf} \in \{0,1\} \qquad\qquad \forall j, k = 1, \dots, n; j \neq k; i = 1, \dots, m; l, f = 1, \dots, r \tag{9}$$

$$d_{jl} \in \{0,1\} \qquad\qquad \forall j = 1, \dots, n; l = 1, \dots, r \tag{10}$$

The objective function (1) is the minimization of makespan C_{max}. Constraints set (2) ensures that jobs are processed according to precedence relationship between operations of the same job, stating that an operation $O_{j,i+1,l}$ cannot start before the end of its preceding operation O_{jil} in job J_j. Constraints set (3) ensures that only one process plan is selected for job j. Constraints sets (4) and (5) are called disjunctive constraints because one or the other alone must hold for the selected process plan only, i.e., the two constraints hold only for one process plan per job and, each two distinct operations O_{jil} and O_{kif} sharing the same machine i on in the selected process plan cannot be scheduled simultaneously. Also, these two constraints are used to incorporate the sequence dependent setup times. Constraints set (6) defines the maximum completion time of all jobs, i.e., the makespan of the schedule. Constraints set (7) enforces the start time of any operation to occur after the initial setup time on its first assigned machine ,i.e., setup time for each job j on when it comes first on the sequence of jobs on machine i. Constraints sets (8-10) imply non-negativity and integrality of the corresponding variables.

5   Example

The applicability of the proposed modeling methodology is tested by implementing it to a case from a Jordanian manufacturer of electrical wires and power cables. The case has four products (jobs) and four machines (4×4). Each product can be processed according to two different process plans. Table 1 and 2 shows the two process plans for each product and the processing times for the products on the machines, respectively. Also, Table 3 shows the sequence dependent setup times where the rows with 'job 0' represent the initial setup times for the products. For example, if job 2 is processed first on machine 2, it requires 2 time units as an initial setup time.

Table 1. The two process plans for the four products

1st Process plan (operations sequence)				2nd Process plan (operations sequence)			
Job				Job			
1	4	3	2	1	1	2	3	4
2	4	1	2	3	2	3	2	1	4
3	3	2	4	1	3	3	4	2	1
4	1	2	3	4	4	4	3	2	1

Table 2. Processing times for the four products on the four machines

Job/machine	Processing Time			
	1	2	3	4
1	2	3	2	3
2	3	2	7	2
3	4	3	6	4
4	10	3	4	5

Table 3. Setup times for the four products on the four machines

Machine 1					Machine 2				
Job ▶	1	2	3	4	Job ▶	1	2	3	4
▼					▼				
0	1	1	1	1	0	2	2	2	2
1	0	1	2	0	1	0	0	1	1
2	1	0	1	0	2	0	0	1	0
3	1	0	0	1	3	0	2	0	0
4	1	0	2	0	4	0	2	1	0

Machine 3					Machine 4				
Job ▶	1	2	3	4	Job ▶	1	2	3	4
▼					▼				
0	3	3	3	3	0	4	4	4	4
1	0	0	2	1	1	0	3	1	3
2	1	0	1	1	2	2	0	2	2
3	0	2	0	1	3	1	4	0	3
4	0	2	0	0	4	1	1	2	0

Using the inputs shown in Tables 1-3, the MIP model shown in equations (1-10) was developed under AMPL IDE and solved using CPLEX 12.6.0.0 solver running on a PC with a core i5 2.27 GHz and 4GB RAM. The AMPL code is given by the appendix. The resulted optimal solution, represented by the starting time of each job on each machine along with the selected process plan for each job, is given by Table 4. As can be seen, the first process plan was selected for jobs 1, 2 and 4, and the second process plan was selected for job 3. A Gantt chart for the optimal schedule, including processing and setup times for each job according to the selected process plans, is also shown in figure 3. As can be seen, the makespan is 27 time units.

For comparisons purposes, the above case was solved as a scheduling problem only in which only single process plan (operations sequence) for each job was assumed available. After running the model, the makespan was found to be 28 and 34 time units when considering separately the first and the second process plans, respectively. This indeed approves the benefits of integrating the IPPS and the SDST problems and shows the superiority which the IPPS-SDST modeling can provide over separate implementation of the SDST problem.

Table 4. The optimal schedule for the studied case

Starting Time					Selected process plan
Operation					
Job	1	2	3	4	
1	19	14	11	8	First
2	14	17	20	4	First
3	23	20	3	12	Second
4	1	11	14	22	First

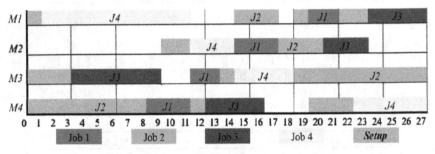

Fig. 3. The Gantt chart for the optimal schedule

6   Conclusions

This research presented the formulation and implementation of a MIP mathematical model that integrates the IPPS and SDST problems. The implementation results of a case from the electrical wires and cable industry indicated that the model can perform well in its related activities. However, the model complexity would limit its applicability as the number of operations increases. Improvements in the mixed integer programming algorithms may help to overcome this limitation. For possible future research extension, issues such as including alternative machines for each operation would result in a more integrated model.

References

1. Allahverdi, A., Ng, C.T., Cheng, T.E., Kovalyov, M.Y.: A survey of scheduling problems with setup times or costs. Europ. J. Oper. Res. 187, 985–1032 (2008)
2. Altarazi, S.: An optimization tool for operational tolerances allocation, work in process inventory minimization, and machines assignment in a discrete part manufacturing environment. Int. J. Adv. Manuf. Technol. 55, 1069–1078 (2011)
3. Amin-Naseri, M.R., Afshari, A.J.: A hybrid genetic algorithm for integrated process planning and scheduling problem with precedence constraints. Int. J. Adv. Manuf. Technol. 59, 273–287 (2012)

4. Baykasoğlu, A., Özbakır, L.: A grammatical optimization approach for integrated process planning and scheduling. J. Intell. Manuf. 20, 211–221 (2009)
5. Brandimarte, P., Calderini, M.: A hierarchical bicriterion approach to integrate process plan selection and job shop scheduling. Int. J. Prod. Res. 33(1), 161–181 (1995)
6. Garcia-Sabater, J.P., Maheut, J., Garcia-Sabater, J.J.: A two-stage sequential planning scheme for integrated operations planning and scheduling system using MILP: the case of an engine assembler. Flex. Serv. Manuf. J. 24, 171–209 (2012)
7. Guo, Y.W., Li, W.D., Mileham, A.R., Owen, G.W.: Applications of particle swarm optimization in integrated process planning and scheduling. Robot. Comput.-Integr. Manuf. 25, 280–288 (2009)
8. Haddadzade, M., Razfar, M.R., Zarandi, M.F.: Integration of process planning and job shop scheduling with stochastic processing time. Int. J. Adv. Manuf. Technol. 71, 241–252 (2014)
9. Jain, A., Jain, P.K., Singh, I.P.: An integrated scheme for process planning and scheduling in FMS. Int. J. Adv. Manuf. Technol. 30, 1111–1118 (2006)
10. Kang, M., Han, J., Moon, J.G.: An approach for interlinking design and process planning. J. Mater. Process. Technol. 139, 589–595 (2003)
11. Kim, Y.K., Park, K., Ko, J.: A symbiotic evolutionary algorithm for the integration of process planning and job shop scheduling. Comput. Oper. Res. 30, 1151–1171 (2003)
12. Kumar, M., Rajotia, S.: Integration of process planning and scheduling in a job shop environment. Int. J. Adv. Manuf. Technol. 28, 109–116 (2006)
13. Lee, H., Kim, S.S.: Integration of process planning and scheduling using simulation based genetic algorithms. Int. J. Adv. Manuf. Technol. 18, 586–590 (2001)
14. Li, W.D., McMahon, C.A.: A simulated annealing-based optimization approach for integrated process planning and scheduling. Int. J. Comput. Inetll. Manuf. 20, 80–95 (2007)
15. Li, X., Gao, L., Shao, X., Zhang, C., Wang, C.: Mathematical modeling and evolutionary algorithm-based approach for integrated process planning and scheduling. Comput. Oper. Res. 37, 656–667 (2010)
16. Mandahawi, N., Al-Shihabi, S., Altarazi, S.: A max-min ant system to minimize total tardiness on a single machine with sequence dependent setup times implementing a limited budget local search. Int. J. Res. Rev. App. Sci. 6, 30–40 (2011)
17. Nourali, S., Imanipour, N., Shahriari, M.R.: A Mathematical Model for Integrated Process Planning and Scheduling in Flexible Assembly Job Shop Environment with Sequence Dependent Setup Times. Int. J. Math. Analy. 6, 2117–2132 (2012)
18. Pinedo, M.L.: Scheduling: theory, algorithms, and systems. Springer Science & Business Media, NewYork (2012)
19. Saygin, C., Kilic, S.E.: Integrating flexible process plans with scheduling in flexible manufacturing systems. Int. J. Adv. Manuf. Technol. 15, 268–280 (1999)
20. Shao, X., Li, X., Gao, L., Zhang, C.: Integration of process planning and scheduling-A modified genetic algorithm-based approach. Comput. Oper. Res. 36, 2082–2096 (2009)
21. Stafford, E.F., Tseng, F.T., Gupta, J.N.D.: Comparative evaluation of MILP flowshop models. J. Oper. Res. Soc. 56, 88–101 (2005)
22. Taiwan Turnkey Project Association, http://www.tpcc.org.tw/wholeplant2/mainchoose/1-4.htm
23. Tan, W., Khoshnevis, B.: A linearized polynomial mixed integer programming model for the integration of process planning and scheduling. J. Intell. Manuf. 15, 593–605 (2004)
24. Tan, W., Khoshnevis, B.: Integration of process planning and scheduling—a review. J. Intell. Manuf. 11, 51–63 (2000)

25. Wan, S., Wong, T., Zhang, S., Zhang, L.: Integrated process planning and scheduling with setup time consideration by ant colony optimization: Doctoral dissertation. The University of Hong Kong, Hong Kong (2012)
26. Wong, T.N., Leung, C.W., Mak, K.L., Fung, R.Y.K.: Integrated process planning and scheduling/rescheduling—an agent-based approach. Int. J. Prod. Res. 44, 3627–3655 (2006)

Appendix: AMPL Code for the IPPS-SDST Proposed Model

```
option solver cplex;
option cplex_options 'mipdisplay=2 mipinterval=100' 'integrality = 1e-06'
'absmipgap=0 mipgap=1e-9' 'mipemphasis=2'  ;
param n, integer, > 0;
param m, integer, > 0;
param r, integer, > 0;

set J := 1..n;
set J0 :=0..n;
set M := 1..m;
set L := 1..r;

param route{j in J, t in 1..m,l in L}, in M;
param K := 1e+6;
param p{j in J, i in M}, >= 0;
param s{j in J0, k in J,i in M}, >= 0;

var x{j in J, i in M, l in L}, >= 0;
var d{j in J, l in L}, binary;
var Y{j in J, k in J, i in M,l in L,f in L}, binary;
var z;
subject to const_1{j in J, t in 1..m-1,l in L}:
x[j, route[j,t+1,l],l] - x[j, route[j,t,l],l]+K*(1-d[j,l]) >= p[j, route[j,t,l]];
subject to const_2{j in J}:sum{l in L} d[j,l]=1;
subject to const_3{k in J, j in J, i in M, l in L , f in L: k <> j }:
        x[k,i,f] - x[j,i,l]+ K * Y[j,k,i,l,f]+K*(1-d[j,l]) >= p[j,i]+s[j,k,i] ;
subject to const_4{k in J, j in J, i in M, l in L , f in L: k <> j }:
        x[j,i,l] - x[k,i,f]+ K *(1- Y[j,k,i,l,f])+K*(1-d[j,l]) >= p[k,i]+s[k,j,i] ;
subject to const_5{j in J,i in M, l in L}: z >= x[j,i,l] + p[j,i];
subject to const_6{j in J,i in M, l in L}:  x[j,i,l] >= s[0,j,i];

minimize obj: z;
```

Modeling of a Management and Maintenance Plan for Hospital Beds

Wajih Ezzeddine, Jérémie Schutz, and Nidhal Rezg

Université de Lorraine, LGIPM, EA 3096, Metz, F-57045, France
{wajih.ezzeddine,jeremie.schutz,nidhal.rezg}@univ-lorraine.fr

Abstract. Beds management is a complex problem in hospital centers. Many proposed models are based on reducing hospitalization cost and optimizing the use of resources. In this paper, a new model for bed planning is presented. This model takes into account the capacity of beds shared between two types of patients: scheduled and non-scheduled patients. Otherwise, the model manages also preventive maintenance actions during the planning horizon.

Keywords: Optimization, Bed management, Scheduling, Replanning.

1 Introduction

Hospital are complex systems that connecting a large number of technical and human resources. Nowadays, health care systems are looking for more efficiency. To satisfy this need, new strategies are developed, especially in bed planning in context of increasing requests of care. In fact, bed resource is one of the most expensive resources in hospitals. In addition, hospital capacity is usually defined as bed capacity.

This paper is structured in three main parts: the first presents a brief state of the art on research work related to beds planning and sizing. In the second part, we exhibit a model for beds planning and discuss obtained results. The third part focuses on the integration of maintenance planning as a constraint in the model. The obtained results in this section will be compared to those of the second part.

2 State of the Art

Several approaches were proposed to increase beds allocation and management. Several research work [1],[6],[4] focused on mathematic modeling to express the problem in terms of variables and constraints. From Lapierre et al. work, Murray [5] proposes to develop a predictive model to optimize resources allocation based on simulation.

Neural network theory has been used to estimate the length of stay for every patient coming from emergency. In other work [7], simulation helps to understand and study the hospital operating system and evaluate the different decisions of

© Springer International Publishing Switzerland 2015
H.A. Le Thi et al. (eds.), *Model. Comput. & Optim. in Inf. Syst. & Manage. Sci.*,
Advances in Intelligent Systems and Computing 360, DOI: 10.1007/978-3-319-18167-7_26

beds management. Bechar [2] proposed a method based on linear programming to integrate patients coming from emergency (called nonscheduled patients) in bed planning. The non-scheduled patient method uses two algorithms: one to plan patients coming in the order and a second program for re-planning patients whose hospitalization day is not confirmed.

3   Model Description

In this part, a planning of beds allocation is presented. In this case, the maintenance actions plan is not considered. Bechars model [2] consists in modeling the problem in two programs: the first program consists to plan scheduled and nonscheduled patients in available beds. Earliest and latest hospitalization must be respected for each patient and the model objective is to optimize costs of refusing patients. After each assignment, beds availability matrix is updated.

The second program aims to find availability for refused scheduled patients from the first assignment, to make it possible, earliest and latest hospitalization dates have to be discussed and modified.

Ben Bachouch [3] used this model and tried to introduce new constraints: non-mixed patients (a man and woman cannot be in the same room during their care duration) and considering double rooms (with two beds).

Our proposition is to continue the development of these two models by adding new and real important constraints to make the model more consisting. We propose in our model the following new constraints: room can be single or double. Indeed, we add a new constraint for patient in need to stay isolated. This type of patient has to be in a single room during this hospitalization.

The proposed model is based on following assumptions:

- The length of stay is considered as known in advance.
- Patients need to be isolated or not are also known in advance.
- Patients cannot occupy more than one bed during his stay care duration (bed change is not permitted).
- Each patient is associated to two dates: earliest hospitalization date and latest hospitalization date. Real hospitalization date must be included between these two dates. For scheduled patients, these dates could be modified (usually for replanning).
- Two type of rooms: double rooms with two beds and single rooms with just one bed.
- Mixed rooms is forbidden.

Thus, the main objective consists to assign the maximum of present patients. Assignment has to respect constraints and not to disturb current planning. Availability (empty and occupied beds) are determined and updated after each assignment. If a patient is refused, he can be redirected to other establishment. This decision depends on many parameters (age of patient, emergency degree, availability in other establishment). If it is a scheduled patient, the decision before redirecting patient is to try to negotiate and modify his hospitalization dates in order to accept him.

Mathematical modeling of beds planning is based on a linear program with mixed variables. Before presenting the model, used notations are defined.

3.1 Notations

- N: number of patients
- T: planning horizon
- L: number of beds
- D_i: earliest hospitalization date for patient i
- Ti: latest hospitalization date for patient i
- Si: gender of patient i
- H: hospitalization daily cost
- LOS_i: length of stay for patient i
- M: positive constant
- NC: number of rooms
- t: time elementary period (1 day)
- $B_{l,t} = 1$ if bed l is free during the period t, 0 otherwise.
- $M_{j,k,l} = 1$ if beds j and k are in the same room, 0 otherwise
- $S_i = -1$ if patient i is a man, 1 if it is a woman
- $SP_{l,t} = -1$ if bed l is occupied by a man during t, 1 if bed is occupied by a woman and 0 if it is free.
- $I_i = 0$ if patient i need to be isolated, 1 otherwise
- $R_i = 0$ if the room i is single, 1 if it is double

3.2 Decision Variables

- $X_{i,l,t} = 1$ if patient i is assigned in bed l during the period t, 0 otherwise.
- $F_i = 0$ if patient i is accepted, 0 otherwise
- $A_{i,l} = 1$ if patient i is assigned in bed l, 0 otherwise
- J_i indicates the real hospitalization date
- End_i indicates the end hospitalization date: $(End_i = J_i + LOS_i - 1)$

Variables i, l and t have to be included respectively in $[1, N]$, $[1, L]$ and $[1, T]$. Based on these notations, modeling algorithm is developed in the next part.

3.3 Model Algorithm

Objective Function The objective function is given by the following relation.

$$\text{Min}\left(H \cdot \sum_{i=1}^{N} [(J_i - D_i) + (F_i \cdot LOS_i)] \right) \tag{1}$$

The first part of this function focuses on optimizing costs resulting to assignment delay. The expression reflects the delay costs between earliest hospitalization date and real hospitalization date. The second part focuses on minimizing costs resulting to refuse a patient which expressed by the profit if he had been accepted.

Constraints

$$\sum_{l=1}^{L} X_{i,l,t} \leq 1 \tag{2}$$

The constraint (2) guarantees, that a patient cannot occupy more than one bed during the period t.

$$\sum_{i=1}^{N} X_{i,l,t} \leq 1 \tag{3}$$

The constraint (3) guarantees that a bed is assigned at most for one patient during the period t.

$$\sum_{l=1}^{L} \sum_{t=D_i}^{T} X_{i,l,t} + F_i \cdot LOS_i = LOS_i \tag{4}$$

The constraint (4) guarantees that for each assigned patient, his length of stay in the hospital is equal to his occupation duration.

$$\sum_{l=1}^{L} A_{i,l} \leq 1 \tag{5}$$

$$LOS_i \cdot A_{i,l} = \sum_{t=D_i}^{T_i} X_{i,l,t} \tag{6}$$

Constraints (5) and (6) guarantee that a patient can occupy at most one bed during his length of stay.

$$J_i \leq (t \cdot X_{i,l,t} + (-X_{i,l,t} + 1) \cdot M \tag{7}$$

The constraint (7) guarantees that the hospitalization date of the patient i cannot exceed the first period for which a bed was assigned to him.

$$J_i \geq D_i \tag{8}$$

$$J_i \leq T_i \tag{9}$$

Constraints (8) and (9) guarantee that patient hospitalization date is between his earliest and latest hospitalization date.

$$J_i + LOS_i - 1 \geq \sum_{l=1}^{L} X_{i,l,t} \tag{10}$$

Constraint (10) allows to determine the end hospitalization date for each patient.

$$\sum_{t=D_i}^{T} X_{i,l,t} \cdot B_{l,t} = A_{i,l} \cdot LOS_i \tag{11}$$

The constraint (11) guarantees that the bed assigned to a patient is free during his length of stay.

$$R_i \cdot M_{j,k,l} \cdot \left(\sum_{i=1}^{N} S_i \cdot (X_{i,j,t} - X_{i,k,t}) \right) \leq 1 \qquad (12)$$

$$R_i \cdot M_{j,k,l} \cdot \left(\sum_{i=1}^{N} S_i \cdot (X_{i,j,t} - X_{i,k,t}) \right) \geq -1 \qquad (13)$$

$$R_i \cdot M_{j,k,l} \cdot \left(SP_{j,t} - \sum_{i=1}^{N} (X_{i,j,t} \cdot S_i) \right) \leq 1 \qquad (14)$$

$$R_i \cdot M_{j,k,l} \cdot \left(SP_{j,t} - \sum_{i=1}^{N} (X_{i,j,t} \cdot S_i) \right) \geq -1 \qquad (15)$$

Constraints (12), (13), (14) and (15) guarantee the non-mixed rooms

$$X_{i,j,t} \cdot \sum_{k=1}^{NC} (R_k \cdot M_{l,l,k}) \leq I_i \qquad (16)$$

Constrain (16) guarantees that patient in need to be isolated has to be assigned to a single room.

For this mathematical model, the ingretity constraints are :

- $X_{i,j,t}$, F_i, $A_{i,l}$ are boolean variables.
- J_i, END_i are in the interval $[1, T]$.

3.4   Model Resolution

For the numerical example, we consider a care unit with 10 beds in 6 bedrooms (4 double and 2 singles). In addition, we will consider the following data:

- A planning horizon of 28 days (4 weeks).
- The service is continuous and operates 7 days.
- The cost of a day's stay in the unit is on average 365 euro.

The resolution of this model is presented in three tables. The first expresses the availability of resources and the empty beds occupied during the horizon stud- ied. The second table expresses the characteristics of each patient: gender, length of stay, earliest and latest hospitalization date, need to be isolated and if he is scheduled or nonscheduled.

The following table shows the initial availability of beds for the planning horizon. Letter H expresses that the bed is occupied by a man and F expresses that the bed is occupied by a woman. An empty box is a free bed. For example, the bed one is occupied by a woman during the day 6 and day 7, a man by day 7 and 8.

Table 1. Initial availability of beds during the studied horizon

	Beds	Planning horizon													
		1	2	3	4	5	6	7	8	9	10	11	12	13	14
Room 1	1						W	W	M	M					
	2										W	W			
Room 2	3														
	4						W	W	M	M					
Room 3	5										W	W			
	6		M	M	M	M	M	W	W	W	W	W	W		
Room 4	7						W	W	M	M					
	8										W	W			
Room 5	9														
Room 6	10						W	W	M	M					

	Beds	Planning horizon													
		15	16	17	18	19	20	21	22	23	24	25	26	27	28
Room 1	1				W			M	W	W				M	M
	2														
Room 2	3														
	4	14			W			M	W	W				M	M
Room 3	5														
	6	W	W						M	M	W	W	W	W	W
Room 4	7				W			M	W	W				M	M
	8	W	W	W	W						M	M			
Room 5	9														
Room 6	10				W			M	W	W				M	M

We will try now to include patients in the table 1, while considering the updated table after each assignment. We will consider a number of 35 patients shared between scheduled and nonscheduled patients.

The following table summarized the data of these 35 patients:

The implementation of the proposed model seeks to insert the patients according to their data. In Table 2, we summarized the most characteristic data of the simulation results on the date of admission, date of discharge from hospital, the bed reserved for the length of stay and a variable that indicates acceptance or refusal of the patient. Simulation results of this model give the schedule shown in the table 3.

With the solver Xpress, this optimal solution is obtained after 1.9 seconds. The objec-tive function is estimated to 23360 euros. From a total of 35 patients, the program declined 11 patients.The value of this unit efficiency is only 0.685 despite the existence of free beds. 9 patients are mostly rejected due to the constraint of earliest and latest hospitalization dates.

In the next section, we will try to increase the unit efficiency by fitting into the planning refused patients. This decision does not concern unscheduled patients (i.e. coming from the emergency).

Table 2. Patient characteristics: data and simulation results

Patient(i)	Characteristic (i)	I (i)	D(i)	T(i)	S(i)	LOS(i)	J(i)	End(i)	bed(i)	F(i)
1	Not scheduled	Yes	1	2	W	4	1	4	10	accepted
2	Not scheduled	No	3	6	W	5	3	7	3	accepted
3	Not scheduled	No	2	3	M	3	2	4	7	accepted
4	Not scheduled	No	3	4	M	2	3	6	5	accepted
5	Not scheduled	No	1	4	W	7	1	7	2	accepted
6	Scheduled	No	4	4	M	6	4	9	0	refused
7	Scheduled	Yes	4	4	W	4	4	7	0	refused
8	Not scheduled	No	6	7	M	4	6	9	0	refused
9	Not scheduled	No	1	5	M	3	1	3	8	accepted
10	Not scheduled	No	3	6	W	6	3	8	9	accepted
11	Not scheduled	Yes	9	10	M	3	10	12	10	accepted
12	Not scheduled	No	7	11	M	4	8	11	3	accepted
13	Not scheduled	No	8	11	W	5	8	12	0	refused
14	Scheduled	No	11	11	M	5	11	15	4	accepted
15	Scheduled	No	11	11	W	6	11	16	1	accepted
16	Not scheduled	Yes	12	14	M	4	12	15	0	refused
17	Not scheduled	No	8	12	M	2	8	9	8	accepted
18	Not scheduled	No	10	13	W	6	12	17	5	accepted
19	Scheduled	No	12	12	M	3	12	14	8	accepted
20	Not scheduled	No	12	13	W	2	12	13	2	accepted
21	Scheduled	No	9	9	W	4	9	12	9	accepted
22	Not scheduled	Yes	15	16	W	5	15	19	0	refused
23	Not scheduled	No	14	20	W	2	14	15	2	accepted
24	Not scheduled	No	12	17	M	4	12	15	3	accepted
25	Not scheduled	No	12	18	M	3	12	14	7	accepted
26	Not scheduled	Yes	13	18	W	7	13	19	9	accepted
27	Scheduled	No	18	18	M	6	18	23	5	accepted
28	Not scheduled	No	14	18	W	8	14	21	0	refused
29	Scheduled	No	21	21	M	4	21	24	0	refused
30	Not scheduled	No	15	19	M	4	18	21	6	accepted
31	Not scheduled	Yes	17	20	W	5	17	21	0	refused
32	Not scheduled	No	14	19	M	4	14	17	10	accepted
33	Not scheduled	No	19	28	M	5	20	24	1	accepted
34	Not scheduled	No	24	28	W	7	24	30	0	refused
35	Scheduled	No	25	25	M	5	25	29	0	refused

Table 3. Planning for assigned patients in the studied horizon

Room	Beds	| Planning horizon 1	2	3	4	5	6	7	8	9	10	11	12	13	14
Room 1	1						W	W	M	M			15	15	15	15
	2	5	5	5	5	5	5	5			W	W	18	18	18
Room 2	3			2	2	2	2	2	12	12	12		24	24	24
	4						W	W	M	M		14	14	14	14
Room 3	5		9	9	9						W	W	20	20	23
	6			M	M	M	M	M	W	W	W	W	W	W	
Room 4	7		3	3	3	W	W	M	M				19	19	19
	8			4	4				17	17	W	W	25	25	25
Room 5	9			10	10	10	10	10	10	21	21	21	21	26	26
Room 6	10	1	1	1	1	1	W	W	M	M	11	11	11		32

Room	Beds	| Planning horizon 15	16	17	18	19	20	21	22	23	24	25	26	27	28
Room 1	1	15	15		W			M	W	W				M	M
	2	18	18	18											
Room 2	3	24													
	4	14			W			M	W	W				M	M
Room 3	5	23				27	27	27	27	27	27				
	6	W	W	30	30	30	30			M	M	W	W	W	W
Room 4	7				W			M	W	W				M	M
	8	W	W	W	W							M	M		
Room 5	9	26	26	26	26	26	33	33	33	33	33				
Room 6	10	32	32	32	W			M	W	W				M	M

3.5 Replanning of Refused Patients

The re-planning step consists to assign scheduled patients already rejected by the first plan. To do so, earliest and latest hospitalization dates must be negotiated. Table 4 shows the results of re-planning refused patients:

AD_i, ND_i, AT_i, NT_i define respectively the first earliest hospitalization date for patient i, the new earliest date, the latest hospitalization date and finally the new latest date.

Modifying dates allowed accepting patients 28, 31 and 34 in the new planning. The solution gives the optimal value of the objective function fixed at 20075 euros which is less than the previous solution. The new value of the unit efficiency is 0.771. However, the results of the allocation of other patients do not remain the same in this step. If the hospital wants to keep the old solution for one reason or another, rescheduling will be manual.

To accept other scheduled or non-scheduled patients, there are two options: the first is to add extra beds and vary capacity of beds and the second is to transfer patients to other units of the same hospital if unit is specialized care is saturated.

Table 4. Results of re-planning scheduled patients

Patient(i)	Emergency(i)	I(i)	D(i)		T(i)		S(i)	LOS(i)	J(i)	End(i)	Bed(i)	F(i)
			AD_i	ND_i	AT_i	NT_i						
8	Scheduled	No	6	5	7	8	M	4	6	9	0	R
13	Scheduled	No	8	7	11	12	W	5	8	12	0	R
16	Scheduled	Yes	12	11	14	15	M	4	12	15	0	R
22	Scheduled	Yes	15	14	16	17	W	5	15	19	0	R
28	Scheduled	No	14	13	18	19	W	8	13	20	2	A
31	Scheduled	Yes	17	16	20	21	W	5	17	21	9	A
34	Scheduled	No	24	23	28	29	W	7	22	28	9	A

In the next section, a new constraint will be added in or study the effect of associating the maintenance plan to our planning.

4 Effects of Maintenance on Beds Planning

In this part, a new model including maintenance actions on beds is presented. Hospital system is considered as an industrial system and beds are considered as machines. Machines are always in need to preventive and corrective maintenance actions to maintain their productivity. In our study, we propose the sequential strategy which consists in planning one activity between production (in our case, patient planning) and maintenance activities. Then, we integrate this planning as a strong constraint of resources unavailability in the second planning. In our case, First maintenance planning is realized and then integrated in beds planning.

4.1 Model Description

Our model is based on the following assumptions:

- Beds must be maintained after 180 working days. This duration is not only based on the bed life duration but also the set of equipment around (oscilloscope, sensors...).
- The deterioration of the equipment follows an exponential distribution.
- If failure is detected then the equipment is replaced by a new one.
- The time replacement of the deteriorated equipment is fixed at one day.

4.2 Model Algorithm

The same notations mentioned in the previous part are used. Comparing to the previous model, a new constraint will be added in order to include the maintenance actions plan.

$$\sum_{t=D_i}^{T} X_{i,l,k} \cdot B_{l,t} \cdot Work_{l,t} = A_{i,l} \cdot LOS_i \qquad (17)$$

With $Work_{l,t} = 1$ if bed is operational, 0 otherwise.

To determine if bed is operational or not, system generates randomly 10 values (for 10 beds in our example) following the exponential distribution with a parameter value fixed at $(1/180)$. These values are compared to life duration fixed at 180 and deter- mine if system is still operating or not.

4.3   Model Resolution

The same data and the same initial availability of the third part are used. The result of the simulation of this model gives the schedule shown in the table 5. The cell containing letter O means that the bed is out of order that date and it is being repaired (duration of repair valued at 1 day). Compared to the first planning, patient 30 is refused. This patient was assigned in bed 6 from day 17 to 20. Due to maintenance actions, bed 6 is out of order in this period and system did not find another assignment to this patient. Many patients change their initial assignment compared to the first planning. For example, patient 15 was assigned in bed 1 from day 11 to 16 but reassigned in bed 7 for the same period.

Table 5. Updated planning with maintenance actions

	Beds	Planning horizon													
		1	2	3	4	5	6	7	8	9	10	11	12	13	14
Room 1	1						W	W	M	M		O	25	25	25
	2			2	2	2	2	2			W	W	19	19	19
Room 2	3		3	3	3				12	12	12	12	24	24	24
	4			4	4		W	W	M	M		14	14	14	14
Room 3	5	9	9	9							W	W	18	18	18
	6		M	M	M	M	M	W	W	W	W	W	W		
Room 4	7		3	3	3	W	W	M	M			15	15	15	15
	8	5	5	5	5	5	5	5	17	17	W	W	20	20	20
Room 5	9			10	10	10	10	10	10	21	21	21	21	26	26
Room 6	10	1	1	1	1		W	W	M	M	11	11	11		32

	Beds	Planning horizon													
		15	16	17	18	19	20	21	22	23	24	25	26	27	28
Room 1	1				W			M	W	W				M	M
	2	23	23								33	33	33	33	33
Room 2	3	24													
	4	14			W			M	W	W				M	M
Room 3	5	18	18	18	27	27	27	27	27	27		O			
	6	W	W				O			M	M	W	W	W	W
Room 4	7	15	15		W			M	W	W				M	M
	8	W	W	W	W							M	M		
Room 5	9	26	26	26	26	26	33	33	33	33	33				
Room 6	10	32	32	32	W			M	W	W				M	M

Table 6. Re-planning of refused patients

| Patient(i) | Emergency(i) | I(i) | D(i) | | T(i) | | S(i) | LOS(i) | J(i) | End(i) | Bed(i) | F(i) |
			AD_i	ND_i	AT_i	NT_i						
8	Scheduled	No	6	5	7	8	M	4	6	9	0	R
13	Scheduled	No	8	7	11	12	W	5	8	12	0	R
16	Scheduled	Yes	12	11	14	15	M	4	12	15	0	R
22	Scheduled	Yes	15	14	16	17	W	5	15	19	0	R
28	Scheduled	No	14	13	18	19	W	8	13	20	2	A
30	Scheduled	No	15	25	19	28	M	4	25	28	9	A
31	Scheduled	Yes	17	16	20	21	W	5	17	21	9	A
34	Scheduled	No	24	23	28	29	W	7	22	28	0	R

4.4 Re-planning Scheduled Patients

The change of dates allowed accepting patients 28, 30 and 31 in the new schedule. The result of re-planning is shown in the table 6. The time resolution of the second planning is 1.7 seconds.

5 Conclusion

Thanks to the development of planning model, hospital can anticipate assignment and cost results regarding the acceptance rate, the occupancy rate, and the desired gain in a finite horizon in order to optimize it.

Resources allocation depends on several parameters and constraints mainly hospitalization and patients do not mix dates. A second stage of re-planning is also recommended to seek the possibility of inserting scheduled patients by modifying their earliest and latest hospitalization dates.

The aim of the second part of this paper is to associate maintenance actions plan in the model. In this case, we tried to anticipate equipment deterioration dates and inte-grate them as constraints.

We plan to test the resolution of this model using other solvers like Cplex or Lingo to compare and evaluate performances of Xpress solver.

References

1. Akcali, E., Côté, M.J., Lin, C.: A network flow approach to optimizing hospital bed capacity decisions. Health Care Management Science 9(4), 391–404 (2006)
2. Bechar, S., Guinet, A.: Planification des lits d'un établissement de soins. In: Proceedings of Conférence GISEH, Luxembourg (2006)
3. Ben Bachouch, R., Guinet, A.: Gestion des lits mutualisés d'un établissement hospitalier. In: Proceedings of 7e Congrés International de Génie Industriel, Trois-Rivières, Québec (2008)
4. Lapierre, S.D., Goldsman, D., Cochran, R., DuBow, J.: Bed allocation techniques based on census data. Socio-Economic Planning Sciences 33(1), 25–38 (1999)

5. Murray, C.J.: A note on Bed allocation technique based on census data. Socio Economic Planning Sciences 30(2), 183–192 (2005)
6. Vissers, J.M.H.: Patients flow-based allocation of hospital resources: A case study. European Journal of Operational Research, 356–370 (1998)
7. Wang, T.: Organisation et pilotage des services sur le trajet des urgences. In: Computer Science. INSA de Lyon (2008)

Solving the Production and Maintenance Optimization Problem by a Global Approach

Vinh Thanh Ho[1], Zied Hajej[2], Hoai An Le Thi[1], and Nidhal Rezg[2]

[1] Laboratoire d'Informatique Théorique et Appliquée
[2] Laboratoire de Génie Industriel, de Production et de Maintenance
Université de Lorraine, Ile du Saulcy, 57045 Metz Cedex 01, France
{vinh-thanh.ho,zied.hajej,hoai-an.le-thi,nidhal.rezg}@univ-lorraine.fr

Abstract. This paper concerns the combined production and maintenance plans for a manufacturing system satisfying a random demand. We propose a new optimization model transformed from the production policy model and a global approach for finding the optimal production/maintenance schedule. The numerical experiments illustrate the efficiency of our proposed model and a global method by comparing with the result of the literature works.

Keywords: mixed integer quadratic programming, production plan, maintenance scheduling.

1 Introduction

Recently, maintenance and production scheduling using stochastic optimal control techniques has drawn much attention among researchers. Due to the complexity of the manufacturing systems, decisions pertaining to marketing, production and maintenance have traditionally been treated separately. Clearly, however, analyzing these decisions simultaneously is more realistic and useful from a practical point of view. Accordingly, this study seeks to find the joint optimal production and maintenance strategy for a randomly failing manufacturing system which must satisfy a random product demand over future periods. This is indeed a complex task due to the various uncertainties caused by exogenous and endogenous factors. Moreover, it is interesting to develop an intelligent optimal maintenance strategy considering the deterioration of the manufacturing system as a function of the production rate.

This area attracts many experts and researchers such as Brown and Proschan (1983), Akella and Kumar (1986), Kijima (1988), Boukas and Haurie (1990), Boukas and Yang (1996), Silva and Wagner (2004), Rezg et. al. (2004, 2008), Barták et al. (2009),... (for details, see Hajej et al. (2011) ([6]) and the references therein). Here, we introduce briefly approaches to maintenance and production models. In reality, the failure rate increases with time and according to the utilization of the equipment, a situation rarely studied in the literature. Many maintenance models assume that the system is maintained under fixed operational and environmental conditions. For example, fixed operational conditions assume that the manufacturing system operates at the maximal production rate (hence ignoring the production rate variation). Schutz et al. (2009) proposed model periodic and sequential preventive maintenance policies for a system that performs various missions over a finite planning horizon. To account for variable environmental conditions, Özekici (1995) proposed to take an intrinsic age of the system instead of the actual age, while Martorell et al. (1999) used models of accelerated life.

By considering the systems failure rate variation according to the production rate change, Hajej et al. (2011) proposed an approach to model an integrated maintenance/production policy where the problem can be formulated as a linear-stochastic optimal control problem. Benbouzid-Sitayeb et al. (2006) proposed a Joint Genetic Algorithm (JGA), for joint production and maintenance scheduling problem in permutation flowshop, in which different genetic joint operators (mutation, selection and crossover) are used and proposed a joint structure to represent an individual in with two fields: the first one for production data and the second one for maintenance data. Belkaid et al. (2013) proposed a meta-heuristics based on genetic algorithm to solve hard combinatorial optimization problems and constituted an important alternative resolution for parallel machine scheduling problems. To evaluate the efficiency of the genetic algorithm and proposed heuristics, some performance measures composed of a set of computational experiments are applied.

Our main contributions are threefold. Firstly, we propose a new optimization model, an equivalent deterministic problem transformed from the stochastic production policy model. Secondly, we present a global method for solving this optimization problem by showing the influence of the variation of production rates on the maintenance strategy. Finally, from the numerical experiments, we show the efficiency of our proposed model and the good method for solving such a stochastic linear programming problem by comparing with the result of the literature works.

2   Problem Statement

Now, we describe briefly a model for jointly planning the production and maintenance activities of a single machine M producing one part-type through a single operation in order to satisfy a random demand. During the horizon H, machine M is subject to random failure. The probability density function of time to failure is $f(t)$, while the failure rate $\lambda(t)$ is increasing in both time and production rate $u(t)$. Failures of machine M can be reduced through preventive maintenance activities. Preventive maintenance (PM), usually scheduled periodically at certain time intervals, is a policy aimed at improving the overall reliability and availability of a system. Ideally, one would like to define a PM policy such that the overall cost of system failure, maintenance, and replacement during its production horizon H is minimized.

Under the constraint that the total time needed to perform both maintenance activities (preventive and corrective) is not greater than the finite horizon H, the

Cox model (Cox, 1972) provides an estimate of the treatment effect on survival after adjustment for other explanatory variables. Thus a parametric relationship between risk factors (related to the operational and environmental conditions of each period) and the hazard rate. The model relies mainly on the assumption of proportional hazards, which implies that each factor affects the life steadily over time.

Our first objective is to establish an economical production plan satisfying the random demand under given service level. Secondly, using this optimal production plan, we establish the optimal preventive maintenance plan. The use of the optimal production plan as an input to the maintenance study is justified by the influence which the production rate at each period exerts on the failure rate of the machine. Since the Cox model is used to define the failure law, each period has its distinct failure rate. Meantime, the operational and environmental conditions will impact the optimal scheduling of maintenance actions through the minimization of the average number of failures. The cost of a PM activity are respectively assumed to be strictly lower than the cost of a corrective maintenance action.

2.1   Notation

The main decision variables, cost coefficients and parameters associated with the stochastic problem at hand are listed below:

H finite production horizon
Δt period length of production
$s(k)$ inventory level at the end of the period k $(k = 1, \dots, H/\Delta t)$
$u(k)$ production cadence at period k $(k = 1, \dots, H/\Delta t)$
$U_{\max}$ maximal production cadence
$d(k)$ demand level at period k $(k = 1, \dots, H/\Delta t)$
$f(t)$ probability density function of time to failure for the machine
$R(t)$ reliability function
C_s holding cost of a product unit during the period k
C_{pr} unit production cost
C_p preventive maintenance action cost
C_c corrective maintenance action cost
C total expected maintenance cost per time unit
α probabilistic index (related to customer satisfaction)
mu   monetary unit

2.2   Problem Formulation and Stochastic Production Policy

It is assumed that the horizon H is partitioned equally into N periods of length Δt. Let $\{f_k, k = 1, \dots, N\}$ represent holding and production costs (they will be formulated below) and $\mathbb{E}\{\}$ denotes the mathematical expectation operator, $\mathbb{P}(A)$ denotes the probability of an event A. The following aggregate sequential stochastic linear programming problem provides an optimal production plan over the planning horizon:

$$\min_{u(k)} \mathbb{E}\left\{ \sum_{k=0}^{N-1} f_k(s(k), u(k)) + f_N(s(N)) \right\}$$

subject to:

$$s(k+1) = s(k) + u(k) - d(k) \qquad k = 0, 1, \cdots, N-1 \qquad (1)$$
$$\mathbb{P}[s(k+1) \geq 0] \geq \alpha \qquad k = 0, 1, \cdots, N-1 \qquad (2)$$
$$0 \leq u(k) \leq U_{\max} \qquad k = 0, 1, \cdots, N-1 \qquad (3)$$

Constraint (1) defines the inventory balance equation for each time period. Constraint (2) imposes the service level requirement for each period as well as a lower bound on inventory variables so as to prevent stockouts. Note that the non-negative lower limit in (2) represents a safety stock. Finally, the last constraint defines an upper bound on the production level during each period k.

The model is described by a hybrid state with continuous component, namely the inventory level as given by equation (1) above, with $s(0) = s_0$, where s_0 is the given initial inventory.

The expected production and holding costs for period k are given by:

$$f_k(s(k), u(k)) = C_s \times \mathbb{E}[s(k)^2] + C_{pr} \times u(k)^2.$$

The total expected cost of production and inventory over the finite horizon H can then be expressed as:

$$F(u) = \sum_{k=0}^{N} \{f_k(u(k), s(k))\} = C_s \times \mathbb{E}[s(N)^2] + \sum_{k=0}^{N-1} \left[C_s \times \mathbb{E}[s(k)^2] + C_{pr} \times u(k)^2 \right].$$

Thus, the problem becomes

$$\min_u F(u) = \min_u \left\{ C_s \times \mathbb{E}[s(N)^2] + \sum_{k=0}^{N-1} \left[C_s \times \mathbb{E}[s(k)^2] + C_{pr} \times u(k)^2 \right] \right\} \qquad (4)$$

subject to constraints (1), (2) and (3).

2.3   Maintenance policy

The maintenance strategy under consideration is the well known preventive maintenance policy with minimal repair at failure (Faulkner, 2005). Perfect preventive maintenance is performed periodically at times $k \times T$, $k = 0, 1, \ldots, N$, following which the unit is as good as new. Whenever a failure occurs between preventive maintenance actions, the system undergoes a minimal repair to allow it to continue operating during the current period and hence the failure rate is undisturbed. It is assumed that the repair and replacement times are negligible.

It has been proved in the literature that the average maintenance total cost per time unit is expressed as follows:

$$C_T = \frac{C_p + C_c \times \int_0^T \lambda(t)dt}{T},$$ (5)

where $\lambda(t)$ is the machine failure rate function.

The existence of an optimal preventive maintenance period T^* has been proved in the case of an increasing failure rate.

We next seek to determine the optimal interval k^* at which the preventive maintenance actions must be carried out considering the production plan previously established for the N periods of the planning horizon. For the case where k^* exceeds $N \times \Delta T$, no preventive maintenance is done. In order to calculate the average total maintenance cost per time unit, the analytical model is developed.

For each period k, we use the production rate $u(k)$ earlier established by the optimal production plan. The machine failure rate in each interval will vary according to the intervals production rate. Following the Cox approach, we define the machine rate as follows

$$\lambda_k(t, u(k)) = \lambda_0(t) \times g(u(k)),$$ (6)

where $\lambda_k(t, u(k))$ represents the instantaneous failure rate function at period k according to the production rate $u(k)$; $\lambda_0(t)$ is the failure rate for nominal conditions which is equivalent to the Failure rate with maximal production over the period H; and $g(u(k)) = \dfrac{u(k)}{U_{\max}}$ is the production function representing the operational condition for each period k.

For Weibull law degradation, the failure rate is

$$\lambda_0(t) = \frac{\delta}{\beta} \times \left(\frac{t}{\beta}\right)^{\delta-1},$$ (7)

where β and δ are the scale and the shape parameters respectively.

For the maintenance policy, we seek to find the cost associated with a given schedule of future PM and replacement activities. The joint optimization strategy considers these costs based on optimal production rates previously found by the production policy in order to optimize the maintenance strategy characterized by the optimal time interval between successive PM or replacement activities, $k^* \times \Delta t$.

The analytic expression of the average cost per unit time of maintenance actions is defined by:

$$C(k) = \frac{C_p + C_c \times A_k}{k \times \Delta t}$$ (8)

where A_k corresponds to the expected number of failure, i.e. the average number of failures that can occur during the horizon H, considering the production rate variation for each production period Δt. We recall that the manufacturing system considered in this study is composed of a machine M which produces a single product at the rate $u(k)$ during each Δt period with the reliability function $R_k(t, u(k))$ $(k = 0, 1, \ldots, N - 1; N \times \Delta t = H)$. Since $u(k)$ varies in each production period Δt, it is complex to formulate directly the analytical expression of A_k, which is why we do so by employing the operational age method. Using the maximal production rate and the failure rate, i.e. the nominal failure rate, we determine the expected failure number as follows:

$$A_k = \sum_{i=1}^{k} \int_{\Gamma_i}^{\Gamma_i + \Delta t} \lambda_i(t, u(i)) dt \tag{9}$$

Assume that the nominal failure rate is the failure rate where the production level is maximal. Also, Γ_i is time at which the reliability at the end of period $i - 1$ is identical to that at the beginning of the next period i.

We now determine an analytical expression for Γ_i by Lemma 1 proved in detail in [6].

Lemma 1

$$\Gamma_1 = 0, \ \Gamma_2 = R_2^{-1}(R_1(\Delta t)), \ \Gamma_j = R_j^{-1}(R_{j-1}(\Gamma_{j-1} + \Delta t)) \text{ for } j \geq 3,$$

$R_{\max}^{-1}$: inverse of the reliability with the nominal (maximum) production.
R_{i+1} : reliability at the production rate $u(i + 1)$.
Γ_{i+1} : time at which the reliability at the end of period i is identical to that at the beginning of period $i + 1$.
Γ_j : time at which the reliability at the end of period $j - 1$ is identical to that at the beginning of the next period j.

Consequently, from (6)-(9), the average cost per unit time of maintenance actions can be computed explicitly as the following way:

$$C(k) = \frac{1}{k \times \Delta t} \left(C_p + C_c \times \sum_{i=1}^{k} \left(\frac{u(i)}{U_{\max}} \times \frac{1}{\beta^\delta} \times \left[(\Gamma_i + \Delta t)^\delta - (\Gamma_i)^\delta \right] \right) \right). \tag{10}$$

Moreover, the existence of a local minimum k^* of the maintenance cost $C(k)$ is also proved in [6].

3 New Optimization Model and Solution Method

The main purpose of this section is to propose a new optimization model for minimizing the total cost over the planning horizon H and present a new method for solving this model. Solving such a sequential stochastic linear programming problem under constraints defined above is generally difficult. Let us proceed by transforming the stochastic problem defined above into an equivalent deterministic problem which will then be easier to solve. In this paper, we focus on reformulating the chain-constrained stochastic production-planning problem by a natural way.

3.1 Transformation to an Equivalent Deterministic Problem: A New Model

The Inventory Balance Equation (1)

As mentioned above, the demand variable $d(k)$ is normally distributed with mean $\mathbb{E}(d(k)) = \hat{d}(k)$ and standard deviation $\sigma_{d,k}$ known for each period k and the inventory variable $s(k)$ is statistically described respectively by its mean $\mathbb{E}(s(k)) = \hat{s}(k)$ and variance $\mathbb{E}[(s(k) - \hat{s}(k))^2] = \text{Var}(s(k))$.

Consequently, (1) can be converted into an equivalent inventory balance equation $\hat{s}(k+1) = \hat{s}(k) + u(k) - \hat{d}(k)$. Since $u(k)$ is constant for each interval Δt, we have $\hat{u}(k) = \mathbb{E}(u(k)) = u(k)$ and $\text{Var}(u(k)) = 0$.

The Objective Function (4)

If we assume that $\text{Var}(s(0)) = 0$, $\sigma_{d,k}$ is constant and equal to σ_d for all k, then we can simplify the expected value of the production/inventory costs of Equation (4) as follows.

$$F(u) = C_s \times \hat{s}(N)^2 + \sum_{k=0}^{N-1} \left[C_s \times \hat{s}(k)^2 + C_{pr} \times u(k)^2 \right] + C_s \times (\sigma_d)^2 \times \frac{N(N+1)}{2},$$
(11)

where $\hat{s}(k)$ represents the mean stock level at the end of period k.

Indeed, from the fact $\text{Var}(s(k)) = \mathbb{E}[s(k)^2] - \hat{s}(k)^2$ and $\text{Var}(s(k)) = k \times (\sigma_d)^2$, shown in detail in [6], we have

$$\mathbb{E}[s(k)^2] = k \times (\sigma_d)^2 + \hat{s}(k)^2.$$
(12)

Hence, the equivalent deterministic function (11) is obtained by substituting (12) in the expected cost (4).

The Service Level Constraint (2)

Recall that the variance of inventory variable over the periods of planning horizon is analytically described by $\text{Var}(s(k)) = k \times (\sigma_d)^2$. The random inventory variable $s(k)$ is completely defined by the following equality:

$$s(k) = \hat{s}(k) + \varepsilon_k \times \sqrt{\text{Var}(s(k))} \iff s(k) = \hat{s}(k) + \varepsilon_k \times \sqrt{k}.\sigma_d, \quad (13)$$

where $\varepsilon_k \propto \mathcal{N}(0,1)$ is a standard normal deviate.

Let Φ be a Gaussian cumulative distribution function with mean $\mu = 0$ and standard deviation $\sigma = 1$, Φ^{-1} the quantile function of the standard normal distribution (probit function). From (13) and the fact $\Phi^{-1}(1 - \alpha) = -\Phi^{-1}(\alpha)$, after transforming, the constraint (2) is written as the following inequality:

$$\hat{s}(k+1) \geq \Phi^{-1}(\alpha) \times \sqrt{k+1} \times \sigma_d, \quad (14)$$

Thus, the equivalent deterministic model can now be formulated as follows.

$$\min_u C_s \times \hat{s}(N)^2 + \sum_{k=0}^{N-1} \left[C_s \times \hat{s}(k)^2 + C_{pr} \times u(k)^2 \right] + C_s \times (\sigma_d)^2 \times \frac{N(N+1)}{2} \quad (15)$$

under constraints:

$$\hat{s}(k+1) = \hat{s}(k) + u(k) - \hat{d}(k) \qquad\qquad k = 0, 1, \cdots, N-1 \qquad (16)$$

$$\hat{s}(k+1) \geq \Phi^{-1}(\alpha) \times \sqrt{k+1} \times \sigma_d \qquad k = 0, 1, \cdots, N-1 \qquad (17)$$

$$0 \leq u(k) \leq U_{\max} \qquad\qquad\qquad k = 0, 1, \cdots, N-1 \qquad (18)$$

3.2 Solution Method

Assume that, in practical problem, the initial inventory level s_0, the unit production cost C_{pr}, the holding cost of a product unit C_s, the mean $\hat{d}$ and the standard deviation σ_d of demand d is given in advance and at the initial period ($k = 0$), the variance $\mathrm{Var}(s(0)) = 0$, which means that $s(0) = \hat{s}(0)$.

Henceforth, the optimization problem (15) is equivalent to the following problem:

$$\min_{u} \left\{ G(u) := \sum_{k=0}^{N-1} C_{pr} \times u(k)^2 + \sum_{k=1}^{N} C_s \times \hat{s}(k)^2 \right\} \qquad (19)$$

subject to constraints (16), (17) and (18).

Now, we reformulate the optimization problem (19) as a convex mixed integer quadratic programming.

Set $x = (u, \hat{s}) = (u(0), u(1), ..., u(N-1), \hat{s}(1), ..., \hat{s}(N)) \in \mathbb{Z}^N \times \mathbb{R}^N$, the problem (19) can be written in the following form:

$$\min_{u} G(u) = \min_{x \in C} \frac{1}{2} \langle Ax, x \rangle \qquad (20)$$

where A is a diagonal $2N \times 2N$-matrix with

$$(A)_{ij} = \begin{cases} 2C_{pr} & \text{if } i = j, i = 1, \ldots, N \\ 2C_s & \text{if } i = j, i = N+1, \ldots, 2N, \\ 0 & \text{otherwise} \end{cases}$$

and

$$C = \left\{ x \in \mathbb{Z}^N \times \mathbb{R}^N \left| \begin{array}{ll} \hat{s}(k+1) = \hat{s}(k) + u(k) - \hat{d}(k) & k = 0, 1, \cdots, N-1 \\ \hat{s}(k+1) \geq \Phi^{-1}(\alpha) \times \sqrt{k+1} \times \sigma_d & k = 0, 1, \cdots, N-1 \\ 0 \leq u(k) \leq U_{\max} & k = 0, 1, \cdots, N-1 \end{array} \right. \right\}.$$

Since A is a diagonal matrix whose elements are positive, the objective function G of problem (20) is convex. In addition, C is a nonempty polyhedral convex set. Thus, the optimization model (20) can be considered as a convex mixed integer quadratic programming problem.

Mixed-integer quadratic programming problems and in particular binary quadratic programming problems are known to be NP-hard. They have a wide range of applications in many areas of science, technology, economics and business,... In the literature, existing algorithms for mixed-integer quadratic programming problems are able to globally solve in the small or medium-scale setting. For solving binary quadratic programming problems, there are different global algorithms based on branch-and-bound methods, among them: Billonnet-Elloumi [4], Pardalos-Rodgers [10], Le Thi-Pham Dinh [9], Pham Dinh et al. [11],... and local approach based on DC programming and DCA is developed in [11]. In fact, the binary quadratic programming problem is difficult to solve and the mixed-integer quadratic programming problem is more challenging.

4 Numerical Experiment

In our experiments, the development of joint production-maintenance plans for a hypothetical company is introduced as an example. We compare our proposed model as well as the method for solving new model with the model proposed by [6] in terms of the optimal production plan and the optimal preventive maintenance cost. In [6], for solving the optimization problem, a numerical procedure consisting of dynamic programming is developed. This problem becomes one of finding a sequence of control $\{u_k^* \in U_\alpha, k = 0, 1, \ldots, N - 1\}$ where U_α is a subspace that according to the observed state and the probability measure at each period k.

It is assumed that this company manufactures one product type whose demand fluctuates periodically. Moreover, a production plan is generated for a planning horizon $H = 18$ months, the failure time of machine M characterized by a Weibull distribution with increasing failure rate, implying the existence of an optimal maintenance schedule. In our experiments, the convex mixed integer quadratic programming is solved by the Branch and Bound method.

The main data of the problem are the monthly mean demands $\hat{d}(k)$ given by the sequence (Table 1); $C_{pr} = 3$ mu, $C_s = 2$ mu; $U_{min} = 2, U_{max} = 10$; $s(0) = 10$ ut; $d(k)$, which is extracted from a historical sales report, is assumed Gaussian with $\sigma_d = 1.42$; the degree of customer satisfaction, associated with the service level constraint (2), is equal to 90% ($\alpha = 0.9$).

The optimal production plan is given in Table 2 and the corresponding optimal production cost is 5233.61 ut, while the optimal production cost in [6] is 11333.61 ut. This shows the efficient of our proposed model and our presented method for solving the production policy problem.

For the maintenance policy, the scale and shape parameters of the Weibull distribution are exhibited, respectively, $\beta = 16.79, \delta = 3$, while $C_c = 3000$ mu, $C_p = 500$ mu and $\Delta t = 1$.

We invoked Lemma 1 using the numerical data, which yielded the following

$$\Gamma_j = \left(\frac{u(j-1)}{u(j)}\right)^{\frac{1}{\delta}} \times (\Gamma_{j-1} + \Delta t), j \geq 3 \text{ and } \Gamma_2 = \left(\frac{u(1)}{u(2)}\right)^{\frac{1}{\delta}} \times \Delta t; \Gamma_1 = 0.$$

Figure 1a , Figure 1b present the curve of the average total maintenance cost per time unit, $C(k)$, as a function of k in our proposed model and the compared model, respectively. Table 3 shows in detail the maintenance costs with respect to Figure 1a. We observe that the optimal PM period is $k^* \times \Delta t = 9 \times \Delta t$ with a corresponding optimal cost of $C^* = 89.72$ mu. Meanwhile, with the production policy model in [6], the optimal PM period is $8 \times \Delta t$ and the optimal cost C^* is 111.114 mu.

Table 1. The mean demands $\hat{d}(k)$

k	1	2	3	4	5	6	7	8	9
$\hat{d}(k)$	8	8	9	8	8	8	7	6	4
k	10	11	12	13	14	15	16	17	18
$\hat{d}(k)$	5	7	8	10	8	9	5	6	6

Table 2. The optimal production plan $\mathbf{u}^*(\mathbf{k})$ in our proposed model and $u^*(k)$ in the compared model

k	1	2	3	4	5	6	7	8	9
$\hat{d}(k)$	8	8	9	8	8	8	7	6	4
$u^*(k)$	10	10	10	9	8	8	5	4	2
$\mathbf{u}^*(\mathbf{k})$	3	7	9	8	9	8	7	7	4
k	10	11	12	13	14	15	16	17	18
$\hat{d}(k)$	5	7	8	10	8	9	5	6	6
$u^*(k)$	5	10	10	10	9	10	2	4	6
$\mathbf{u}^*(\mathbf{k})$	5	8	8	10	8	10	5	6	6

Table 3. The minimum maintenance cost C^* in our proposed model

k	1	2	3	4	5	6	7	8	9
$C(k)$	500.19	251.2	170.06	131.51	110.92	99.46	93.35	91.11	**89.72**
k	10	11	12	13	14	15	16	17	18
$C(k)$	91.04	96.11	102.95	112.79	122.79	135.83	145.39	157.1	169.77

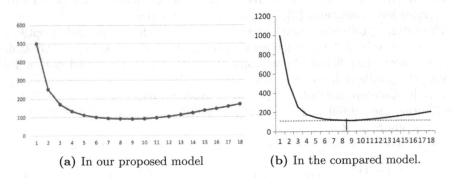

(a) In our proposed model   (b) In the compared model.

Fig. 1. Curve of the average total maintenance cost as a function of k assuming production rate

5   Sensitivity Analysis

In this section, firstly, we study the impact of demand variance and service level on the optimal production and maintenance plans. However, by using the previous optimization procedure, we change the values of standard deviation, and

Table 4. The impact of service level α on the total production/inventory cost F with the fixed demand variance $\sigma_d^2 = 1.42^2$ (std: standard deviation)

α	0.90	0.91	0.92	0.93	0.94	0.95	0.96	0.97	std(α)	std(F)
F	5233.61	5372.61	5466.61	5618.61	5823.61	6105.61	6362.61	6779.61	0.0229	500.15

Table 5. The impact of demand variance σ_d^2 on the total production/inventory cost F with the fixed service level $\alpha = 0.9$ (std: standard deviation)

σ_d	1.42	1.46	1.50	1.54	1.58	1.62	1.66	std(σ_d)	std(F)
F	5233.61	5257.61	5372.61	5466.61	5532.61	5618.61	5757.61	0.08	176.81

service level and then we determine the corresponding optimal plans in order to study the variability of those parameters. In this case, normally, you will remark that the higher value of the total production/inventory cost corresponds to the higher of the service level (Table 4). We can note that, as service level α increases, the principal machine produces more and stores more to meet the demand and to satisfy the higher service level. Another point for production system behavior is that once the variance of demand increases, you will remark that total production/inventory cost increases as a function of the demand variance variability (Table 5). This can be clarified by the reality that, as variance of demand increases, the production rates of principal machines increase, the stock level augments and consequently the total production and inventory cost increases.

Secondly, we can compare between two systems, i.e. comparing our model with another that uses a constant production rate in all horizons (maximal production for each period). In this case, normally you will get as a result of cost reduction engendered by using optimal instead of nominal production rates.

6 Conclusion

In this work, we are interested in a stochastic production planning and maintenance scheduling problems satisfying a random demand. In order to obtain a simultaneous optimal production and maintenance scheduling, we propose a new transformation from a stochastic problem to a deterministic one and a global approach for solving the optimization production problem. The numerical experiments show the efficiency of our proposed approach.

References

1. Barták, R., Salido, M.A., Rossi, F.: Constraint satisfaction techniques in planning and scheduling. Journal of Intelligent Manufacturing 21(1), 5–15 (2010)
2. Belkaid, F., Sari, Z., Souier, M.: A Genetic Algorithm for the Parallel Machine Scheduling Problem with Consumable Resources. International Journal of Applied Metaheuristic Computing (IJAMC) 4(2), 17–30 (2013)
3. Benbouzid-Sitayeb, F., Varnier, C., Zerhouni, N.: Proposition of New Genetic Operator for Solving Joint Production and Maintenance Scheduling: Application to the Flow Shop Problem. In: International Conference Service System and Service Management, ICSSSM 2006, pp. 607–613 (2006)
4. Billionnet, A., Elloumi, S.: Using a mixed integer quadratic programming solver for unconstrained quadratic 0-1 problem. Math. Program. 109(1, Ser.A), 55–68 (2007)
5. Cox, D.R.: Regression models and life-tables. Journal of the Royal Statistical Society 34(2), 187–220 (1972)
6. Hajej, Z., Dellagi, S., Rezg, N.: Optimal Integrated Maintenance/Production Policy for Randomly Failing Systems with Variable Failure Rate. International Journal of Production Research 49(19), 5695–5712 (2011)
7. Hajej, Z., Dellagi, S., Rezg, N.: Joint optimisation of maintenance and production policies with subcontracting and product returns. Journal of Intelligent Manufacturing 25(3), 589–602 (2014)
8. Hajej, Z., Rezg, N., Gharbi, A.: Forecasting and maintenance problem under subcontracting constraint with transportation delay. International Journal of Production Research (2014)
9. Le Thi, H.A., Pham Dinh, T.: A continuous approach for large-scale constrained quadratic zero-one programming. (In Honor of Professor ELSTER, Founder of the Journal Optimization) Optimization 45(3), 1–28 (2001)
10. Pardalos, P.M., Rodgers, G.P.: Computational aspects of a branch and bound algorithm for quadratic zero-one programming. Computing 45, 131–144 (1990)
11. Pham Dinh, T., Nguyen Canh, N., Le Thi, H.A.: An efficient combination of DCA and B&B using DC/SDP relaxation for globally solving binary quadratic programs. J. Glob. Optim. 48(4), 595–632 (2010)
12. Rezg, R., Dellagi, S., Chelbi, A.: Optimal strategy of inventory control and preventive maintenance. International Journal of Production Research 46(19), 5349–5365 (2008)
13. Rezg, N., Xie, X., Mati, Y.: Joint optimization of preventive maintenance and inventory control in a production line using simulation. International Journal of Production Research 44(10), 2029–2046 (2004)
14. Schutz, J., Rezg, N., Leger, J.B.: Periodic and sequential preventive maintenance policies over a finite planning horizon with a dynamic failure law. Journal of Intelligent Manufacturing (2009)
15. Turki, S., Hajej, Z., Rezg, N.: Optimal production planning for a manufacturing system: an approach based on PA. In: IFAC Conference on Manufacturing Modelling Management and Control (IFAC MIM 2013), Saint-Petersburg, Russia, June 19-21 (2013)
16. Turki, S., Hajej, Z., Rezg, N.: Perturbation analysis for optimal production planning of a manufacturing system with influence machine degradation. In: 19th International Federation of Automatic Control (IFAC 2014), Cap Town, South Africa, August 24-29 (2014)

Part V

Scheduling

Exploring a Resolution Method Based on an Evolutionary Game-Theoretical Model for Minimizing the Machines with Limited Workload Capacity and Interval Constraints

Óscar C. Vásquez[1], Luis Osorio-Valenzuela[2], and Franco Quezada[1]

[1] Industrial Engineering Department, University of Santiago of Chile
[2] Electrical Engineering Department, University of Santiago of Chile

Abstract. We present an extension of the machines minimization for scheduling jobs with interval constraints, adding a limited machines workload capacity. We are motivated by the fixed-speed processors minimization problem subject to energy constraints, in which the time resolution is a critical factor for the quality of service in the system. We propose a mixed integer linear programming (MILP) model for an exact solution and explore an alternative resolution method based on a non-cooperative evolutionary theoretical-game model. Our resolution method guarantees a feasible solution to the problem and the computational experiments with a timeout of 3 minutes show that it finds a solution with a number of machines less than or equal to the number of machines for a 97,19% of instances in comparison with the MILP solution over CPLEX 12.6.1.0, in only deciseconds.

Keywords: scheduling, interval constraints, game-theoretical model.

1 Introduction

A classic scheduling problem is the machines minimization for jobs with time interval during which they must be scheduled, such that no two jobs assigned to the same machine overlap in time. This problem has two versions: discrete, where the sets of job intervals are given explicitly and; continuous, where each job has a processing time, which will be contained inside a time window defined by its deadline and its release date.

The complexity of problem is NP-hard [8]. In the approximation algorithms setting, the main results have been developed by [4,5,3,14,9]. Applications of this problem can be found in crew/vehicle scheduling, telecommunications, among others (see [10,11]).

We focus on a problem extension of machines minimization for scheduling jobs with time interval constraint, considering machines with a limited workload capacity, which is not necessary a fixed time interval [12]. This problem is motivated by the fixed-speed processors minimization problem subject to energy

© Springer International Publishing Switzerland 2015
H.A. Le Thi et al. (eds.), *Model. Comput. & Optim. in Inf. Syst. & Manage. Sci.*,
Advances in Intelligent Systems and Computing 360, DOI: 10.1007/978-3-319-18167-7_28

constraints, in which the jobs preemption is disallowed [2,1,4] and the resolution time is a critical factor for the quality of service in the system [6,7].

Our goal is to explore the computation time/solution performance of a resolution method based on a non-cooperative evolutionary theoretical-game model. In a companion paper [13], we introduce a similar method to a particular project scheduling problem from the point the view of natural disasters, obtaining good results in comparison with the optimal solution, specially the resolution time. In this paper, we study a general scheduling problem setting, where an adapted and improved resolution method is applied.

The remainder of this paper is organized as follows: in section two, the statement of the problem is exposed and a mixed integer linear programming (MILP) model is formulated; in section three, the resolution method is described; in section four, the performance analysis of resolution method is carried out by computational experiments; and finally, in section five, the conclusions and directions for further research are given.

2   Statement of the Problem and MILP Model

Formally, our problem is defined as follows. Given a set of n jobs, where each job i is associated with a release date $r_i \in \mathbb{N} \cup \{0\}$, a processing time $p_i \in \mathbb{N}$ and a deadline $d_i \geq r_i + p_i$. Scheduling a job i requires selecting an interval of length p_i between its release date and deadline, and assigning it to a machine, with the restriction that each machine has a limited workload capacity and executes at most one job at any given time. We assume n parallel machines with a workload capacity w greater than or equal to the maximum processing time of activities $\max_i p_i$. Our goal is to schedule all jobs using the minimum possible number of machines. We define a MILP model for our problem as follows:

Decision Variables:

$$x_k = \begin{cases} 1 \text{ if machine } k \text{ is used} \\ 0 \text{ otherwise} \end{cases}$$

$$y_{ik} = \begin{cases} 1 \text{ if job } i \text{ is processed by machine } k \\ 0 \text{ otherwise} \end{cases}$$

$$z_{ij} = \begin{cases} 1 \text{ if job } i \text{ is processed before job } j \neq i \\ 0 \text{ otherwise} \end{cases}$$

$C_i = $ Completion time of job i

Objective Function:

$$\min \sum_{k=1}^{n} x_k \tag{1}$$

Subject to:

$$p_i \leq C_i - r_i \qquad \forall i \tag{2}$$

$$C_i \leq d_i \qquad \forall i \tag{3}$$

$$\sum_{i=1}^{n} y_{ik} p_i \leq w x_k \qquad \forall k \tag{4}$$

$$\sum_{k=1}^{n} y_{ik} = 1 \qquad \forall i \tag{5}$$

$$[(C_j - p_j) - C_i] + \varepsilon \leq N(z_{ij}) \quad \forall i,j \quad j \neq i \tag{6}$$

$$[C_i - (C_j - p_j)] \leq N(1 - z_{ij}) \quad \forall i,j \quad j \neq i \tag{7}$$

$$y_{ik} + y_{jk} \leq 1 + (z_{ij} + z_{ji}) \quad \forall i,j \quad j > i, \forall k \tag{8}$$

$$C_i \in \mathbb{N}, x_k, z_{ij}, y_{ij} \in \{0,1\} \quad \forall i,j,k \tag{9}$$

where ε is a small number and N is a large number. The objective function (1) minimizes the number of machines. Constraints (2) and (3) define the finish-start type of release time-deadline relations of each job. Constraint (4) indicates that if a machine is used, then the machine usage should be less than or equal to its workload capacity. Constraint (5) establishes that a job will be processed by only one machine. Constraints (6), (7) and (8) are of the on-off type and ensure that if a job is processed by a machine, then it does not process another job at the same time. In the case of constraint (6), an ε is incorporated that allows approaching the case in which a job completion time is equal to the start time of another one. We remark two elements: a) the inequality $z_{ij} + z_{ji} \leq 1, \forall i,j; i \neq j$ holds, avoiding duplicity in the processing of the jobs, because if jobs i and j are processed by machine k, they can only be processed in one sequence, i.e. i and j or j and i; b) if $z_{ij} = 0$ then the constraint (8) is inactive and $C_j - C_i \leq p_j$ follows the constraint (7).

3 Resolution Method

Our resolution method is based on a non-cooperative evolutionary game-theoretical model. In this game, a job or a set of them are the players, which maximize the remaining workload of the machine where it is performed, associating or not with another player in order to execute them in a specified schedule on the same machine. The jobs interval and workload constraints are guaranteed by a price mechanism, whereas a Nash equilibrium in pure strategies (PNE) is found. The main idea of the resolution method is to generate a new set of jobs from the PNE in a step, which is incorporated to another new game, in the next step. This process is repeated until new players can not be generated.

In addition, our resolution method adds two new ingredients: a special generation of players in the game at the initial step and a new game as a subroutine for diversification. The goal is to give a better lower bound for our problem and allows to jump the local solutions in order to improve the final result, respectively. We define our game model and describe the algorithms of the resolution method as follows:

Players. Each player $i \in \{1, ..., n\}$ is characterized by α/Ω, where $\alpha = \{1, 2, 3\}$ corresponds to the "type of player" and Ω is the set of jobs. A player type 1 represents a single job. Players of type 2 and 3 are formed by two players, a "predecessor" of features α'/Ω' and "successor" of features α''/Ω''. Both players type 1 and type 2 are differentiated by the time between the completion time of all jobs of the "predecessor" player and the start time of all jobs of the "successor" player. Figure 1 illustrates the difference between a player of type 2 and type 3.

Fig. 1. Difference between players of type 2 and 3

In the case of the type 2 player, it shows that the time of completion of all jobs of the predecessor player with $1/\{1\}$ is equal to the start time of all jobs of the successor player with $1/\{5\}$. Whereas in the case of the player type 3, it has a time equal to $\Delta t > 0$ between each job. Note that the players of type 1 and 2 are determined only by the completion time of all jobs of the predecessor player and the start time of all the jobs of the successor player. Therefore, a player of type 2 could have $\Delta t > 0$ between some of its jobs, because the predecessor and/or successor could be a player of type 3. Figure 2 shows an example of the situation.

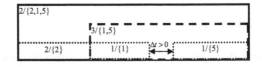

Fig. 2. Player of type 2 conformed by a player of type 3

Strategies. Each player $i \in \{1, ..., n\}$ has a set of strategies $S_i := \{1, ..., i - 1, i, i + 1, ...n, S\}$, where the strategy $s_i \in S_i \setminus \{i, S\}$ means that the player i wants to be after the player s_i on a machine, $s_i = i$ indicates that player i wants to be alone on a machine and $s_i = S$ means that player i wants to be before any player on a machine. In Table 1, we present the definition of the buying and selling price for each of the type of players, which guarantees the job interval constraints [1].

Table 1. Prices definition for types of players

Player	BP (buying price)	SP (selling price)
$1/\{i\}$	$d_i - p_i$	$r_i + p_i$
$2/\{\Omega' \bigcup^* \Omega''\}$	$r_a + \min(BP_{\alpha'/\Omega'} - r_a, BP_{\alpha''/\Omega''} - SP_{\alpha'/\Omega'})$	$SP_{\alpha'/\Omega'} + (SP_{\alpha''/\Omega''} - r_b)$
$3/\{\Omega' \bigcup^* \Omega''\}$	$r_a + \min(BP_{\alpha'/\Omega'} - r_a, BP_{\alpha''/\Omega''} - r_b)$	$SP_{\alpha''/\Omega''}$

Payoffs. The payoff $g_i(s_i, s_{-i})$ of each player i of feature α/Ω is defined by equation (10), where $s_i \in S_i$ and $s_{-i} \in S_{-i} = \bigotimes_{j \neq i} S_j$.

$$
g_i(s_i, s_{-i}) = \begin{cases} \sum_{r \in \Omega'} p_r & \text{if } \forall s_i = i, \; s_{-i} \in S_{-i} \\ \sum_{r \in \Omega' \bigcup^* \Omega''} p_r & \text{if } \sum_{r \in \Omega' \bigcup^* \Omega''} p_r \leq w, \Omega' \cap \Omega'' = \varnothing \text{ and} \\ & \quad \text{a. } s_i = j, s_j = S, BP_i \geq SP_j, \forall s_k \neq j(k \neq i, j) \text{ or} \\ & \quad \text{b. } s_i = S, s_j = i, BP_j \geq SP_i, \forall s_k \neq i(k \neq i, j) \\ & \quad \text{where } \Omega' \text{ is the set of jobs of the player } i \\ & \quad \text{and } \Omega'' \text{ is the set of jobs of the player } j \\ -\infty & \text{otherwise} \end{cases}
$$

$$(10)$$

Solution concept. In this game, the interaction among players is modeled as non-cooperative and the solution concept is a Nash equilibrium in pure strategies (PNE), which always exists since the strategy profile where all players use the strategy "alone" on a machine is a PNE by definition. Therefore, any game has at least one PNE.

Algorithms of resolution method. We define an algorithm with four steps: a) An initial game with only players i of type 1 with $d_i - r_i = p_i$, b) Game with all players, c) Generation of players and d) Game with only "predecessors" players of players of type 3. Here, each "predecessor" player can be bought α times, generating new players if and only if the obtained selling price from the generated player is less than or equal to the start time of the "successor" player associated to the "predecessor" player. The idea of this step aims to generate players for the $\Delta > 0$ of players of type 3. We adopt ALGO 1 and ALGO 2 defined by [13] in order to find a PNE and generate new players as a new set of jobs from generated PNE, respectively. The steps are integrated as follows: step a) is applied at the beginning of the resolution algorithm, whereas steps b), c) and d) are called by a cyclic routine, until new players cannot be generated.

---

[1] $\bigcup^*$ corresponds to the ordered union. For the players type 2 and 3, a and b are the index of the first job in set Ω' and Ω'', respectively.

4   Experimental Study

4.1   Random Instances

We develop a model of random instances in order to evaluate the performance of the proposed resolution method. We chose a release time r_i and a processing time p_i according to the discrete uniform distributions $U(0, N_1)$ and $U(1, N_2)$, respectively; and then we obtain the deadline d_i as a random variable from the discrete uniform distribution $U(r_i + p_i, r_i + p_i + \min(N_1 + N_2 - r_i - p_i, \sigma)), \sigma \geq 0$. This definition of deadline aims to measure the "hardness" of instances by σ subject to a total time of project $N_1 + N_2$, based on the probability to find jobs with overlapped time windows.

The instances of our set test are defined as follows. Fix $N_1 = 100$ and $N_2 = 101$. For each $\sigma \in \{25, 50, 75, 100\}$, we generated 20 instances of $n \in \{25, 50, 75, 100\}$ jobs with a workload capacity $w \in \{\max p_i, \frac{\max p_i + \max d_i}{2}, \max d_i\}$.

4.2   Results

We fix a timeout of 3 minutes in order to analyze the computation time/solution performance of our resolution method and the proposed MILP over CPLEX 12.6.1.0, programming in C++ (Intel Core i7, 2.4 GHz, 8.00 GB RAM, 64 Bit).

We consider $\alpha \in \{0, \lceil n/2 \rceil, n\}$ and use the most favorable value dependent on the value $w \in \{\max p_i, \frac{\max p_i + \max d_i}{2}, \max d_i\}$. For the cases with the same solution, we chose the inferior value of α. Figure 3 shows the obtained results.

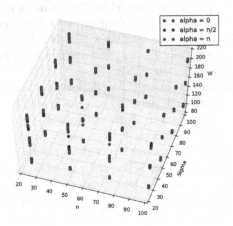

Fig. 3. Performance analysis for α values

In general, the obtained results show that the instances can be efficiently solved by using $\alpha = 0$. In fact, for 889, 66 and 5 instances were more favorable with $\alpha = 0$, $\alpha = \lceil n/2 \rceil$ and $\alpha = n$, respectively.

Later on, we resolve our set test by proposed MILP, obtaining the solution and status (optimal or feasible), the lower bound (LB) and the computation time in seconds. The obtained results are shown in Table 2.

Table 2. Results of MILP

		Solution		Computation time (seconds)				
	Instances	Optimal	Feasible	Ave.	Stand. Dev.	Max.	Min.	Median
25 jobs	240	233	7	8.71	29.51	175.76	0.69	1.87
50 jobs	240	143	97	94.65	66.89	177.66	16.5	49.08
75 jobs	240	0	240	175.44	2.29	178.40	167.80	176.26
100 jobs	240	0	240	168.69	7.39	180.42	140.10	170.92
Total	960	376	584	111.87	78.83	180.42	0.69	168.3

Finally, we compare the behavior of our resolution method with the performance of MILP in terms of solution (number of machines) and computation time (seconds) as shown in Figures 4 and 5.

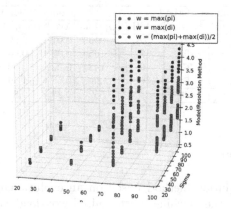

Fig. 4. Performance in solution

The obtained results show that our resolution method computes a feasible solution in only deciseconds, finding a solution with a number of machines less than or equal to the number of machines for a 97,19% of instances (933 instances) in comparison with the MILP solution over CPLEX 12.6.1.0, where a 57,50% of instances (552 instances) are optimal solutions. Table 3 shows the obtained results.

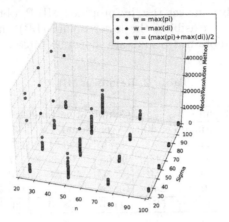

Fig. 5. Performance in time

Table 3. Results of resolution method

	Instances	Resolution Method				Computation time (seconds)				
		< MILP	= MILP	> MILP	Optimal	Ave.	Stand. Dev.	Max.	Min.	Median
25 jobs	240	1	221	18	216	0.01	0.00	0.04	0.00	0.00
50 jobs	240	31	200	9	165	0.05	0.06	0.56	0.01	0.03
75 jobs	240	213	27	0	143	0.22	0.50	6.53	0.03	0.12
100 jobs	240	240	0	0	28	0.66	1.46	11.37	0.12	0.30
Total	960	485	448	27	552	0.23	0.82	11.37	0.00	0.06

5   Final Remark

We developed a resolution method based on an evolutionary game-theoretical
model for minimizing the machines with limited workload capacity and interval
constraints, which ensures feasibility of the solution, convergence and an efficient time performance in comparison with the value obtained by the MILP
solution over CPLEX 12.6.1.0, considering a timeout of 3 minutes. We leave
open the question about the performance of our method in comparison with the
other heuristic or metaheuristic resolution methods, and the incorporation of
our method as a initial point to the optimal resolution of proposed MILP.

Acknowledgments. We are grateful to the anonymous referees who spotted
errors in previous versions of this paper. This work is supported by FONDECYT
grant 11140566.

References

1. Antoniadis, A., Huang, C.-C.: Non-preemptive speed scaling. Journal of Scheduling 16(4), 385–394 (2013)
2. Bampis, E., Letsios, D., Lucarelli, G.: Speed-scaling with no preemptions. In: Ahn, H.-K., Shin, C.-S. (eds.) ISAAC 2014. LNCS, vol. 8889, pp. 257–267. Springer, Heidelberg (2014)
3. Chuzhoy, J., Codenotti, P.: Resource minimization job scheduling. In: Dinur, I., Jansen, K., Naor, J., Rolim, J. (eds.) APPROX and RANDOM 2009. LNCS, vol. 5687, pp. 70–83. Springer, Heidelberg (2009)
4. Chuzhoy, J., Guha, S., Khanna, S., Naor, J.S.: Machine minimization for scheduling jobs with interval constraints. In: Proceedings of the 45th Annual IEEE Symposium on Foundations of Computer Science, pp. 81–90. IEEE (2004)
5. Cieliebak, M., Erlebach, T., Hennecke, F., Weber, B., Widmayer, P.: Scheduling with release times and deadlines on a minimum number of machines. In: Levy, J.-J., Mayr, E.W., Mitchell, J.C. (eds.) Exploring New Frontiers of Theoretical Informatics. IFIP, vol. 155, pp. 209–222. Springer, Boston (2004)
6. Dürr, C., Jeż, L., Vásquez, Ó.C.: Mechanism design for aggregating energy consumption and quality of service in speed scaling scheduling. In: Chen, Y., Immorlica, N. (eds.) WINE 2013. LNCS, vol. 8289, pp. 134–145. Springer, Heidelberg (2013)
7. Dürr, C., Jeż, L., Vásquez, O.C.: Scheduling under dynamic speed-scaling for minimizing weighted completion time and energy consumption. Discrete Applied Mathematics (2014)
8. Garey, M.R., Johnson, D.S.: Computers and intractability, vol. 174. Freeman, New York (1979)
9. Kao, M.-J., Chen, J.-J., Rutter, I., Wagner, D.: Competitive design and analysis for machine-minimizing job scheduling problem. In: Chao, K.-M., Hsu, T.-S., Lee, D.-T. (eds.) ISAAC 2012. LNCS, vol. 7676, pp. 75–84. Springer, Heidelberg (2012)
10. Kolen, A.W.J., Lenstra, J.K., Papadimitriou, C.H., Spieksma, F.C.R.: Interval scheduling: A survey. Naval Research Logistics (NRL) 54(5), 530–543 (2007)
11. Lee, S., Turner, J., Daskin, M.S., Homem-de Mello, T., Smilowitz, K.: Improving fleet utilization for carriers by interval scheduling. European Journal of Operational Research 218(1), 261–269 (2012)
12. Sanlaville, E., Schmidt, G.: Machine scheduling with availability constraints. Acta Informatica 35(9), 795–811 (1998)
13. Vásquez, Ó.C., Sepulveda, J.M., Alfaro, M.D., Osorio-Valenzuela, L.: Disaster response project scheduling problem: A resolution method based on a game-theoretical model. International Journal of Computers Communications & Control 8(2), 334–345 (2013)
14. Yu, G., Zhang, G.: Scheduling with a minimum number of machines. Operations Research Letters 37(2), 97–101 (2009)

Job-Shop Scheduling with Mixed Blocking Constraints between Operations

Christophe Sauvey[1], Nathalie Sauer[1], and Wajdi Trabelsi[2]

[1] LGIPM/Université de Lorraine - UFR MIM, Ile du Saulcy - 57045 METZ CEDEX 01, France
{chistophe.sauvey,nathalie.sauer}@univ-lorraine.fr
[2] LGIPM/ICN Business School - 13, Rue Michel Nay - 54000 NANCY, France
wajdi.trabelsit@icn-groupe.fr

Abstract. Job-shop scheduling problem is a very hard issue in operations research and is studied since a long time, with a wide variety of constraints. This paper addresses the job-shop scheduling problem with a new kind of constraint. The particular issue we have been interested in is about mixed blocking constraints from an operation to its successor in job tasks list. In this paper, a mathematic model is proposed and its results in Mosel Xpress software presented. Then, an evaluation function is proposed and tested with a genetic algorithm on this problem. Results are given and compared to optimal results obtained with the mathematic model on four benchmarks, each composed of twenty instances.

Keywords: Job-shop, scheduling, mixed blocking constraints, mathematic model, meta-heuristics.

1 Introduction

In a context of stressed international markets, production companies try to maximize their productivity, using modern scheduling optimization tools. To achieve this goal, they try to reduce storage capacity, and remove it when possible. In order to cope with always challenging, variable and complex environments and to meet the various requirements of demanding customers, a good scheduling becomes an always increasing point of interest for designers, and the scheduling problematic as well.

Since the early 1950's and the first works of Jackson in 1956 [8], who generalized to the job-shop scheduling problem the algorithm developed by Johnson (1954) for the flow-shop, research on this type of scheduling problem has never stopped. A review paper on this issue presented not less than 260 references in 1999 [7]. Indeed, a lot of approaches have been tested and developed by the research community over the past 60 years. Far from breaking, research intensifies in the 21st century. For example, a famous search engine programmed to find only academic research papers, gives about 4000 references with the keyword "job-shop", with the advanced research "since 2014". Globally, research interests have been put towards complexity studies and heuristics. Thanks to job-shop scheduling obvious fitting with industrial issues, a lot of model extensions have been proposed. In this paper, we propose an extension towards mixed blocking constraints between successive operations of the problem.

© Springer International Publishing Switzerland 2015
H.A. Le Thi et al. (eds.), *Model. Comput. & Optim. in Inf. Syst. & Manage. Sci.*,
Advances in Intelligent Systems and Computing 360, DOI: 10.1007/978-3-319-18167-7_29

Reduction of storage capacity between machines is known in literature as leading to blocking situations. A blocking situation appears when an operation, which can not progress towards its further operation in process, occupies the last machine on which it was processed. In function of the number of machines blocked by a situation and in function of the dates at which blocking situation occurs, it is possible to define different blocking constraints. All the blocking constraints taken into account in this paper are presented in the second part of this paper, after a short consideration for complexity issues. Mathematical models of job-shop scheduling problem are available in the literature. To be able to guarantee an optimal solution to a mixed blocking jobs-shop problem, we developed a mathematical model, based on what had been published in [11] and first adapted to a unique blocking constraint between all operations of a same problem [5]. Benchmark problems are defined and a comparison between exact method and developed genetic algorithm are given. In the last section, we conclude this paper and give some perspectives to our work.

2   Problem Description

Job-shop (JS) scheduling problem is a well known problem of operations research. It is a generalisation of the flow-shop also well known optimization problem. In this paper, we consider a JS with m machines and n jobs. Each job J_i is constituted of n_i operations O_{ij}. Each operation can be executed only by a single machine M_{ij} and each machine cannot execute more than one operation at a time. The problem is to determine the best sequence of jobs minimizing the completion time of last job, also called the makespan.

Blocking constraints are a concept developed and widely studied in the flow-shop (FS) field. Indeed, buffer space capacity between two machines can be considered as limited, or null. Blocking constraints are then considered between machines. A lot of references exist about FS problem with blocking constraints. In [12], we have proposed heuristics and meta-heuristics to approach FS problems with mixed blocking constraints. In JS problems, blocking constraints can be considered between successive operations. Few references have interested to JS problems with blocking constraints. An approach has been proposed to schedule trains as a blocking parallel-machine JS scheduling problem [10]. D'ariano et al. [3] have used blocking constraints to schedule transportation devices traffic, such as aircrafts, metros or also trains. As we did for the classical FS problem, we propose to address the JS problem with mixed blocking constraints between operations.

To make easily understand each blocking case, we decide to present them, in the following figures, in the JS case situation. In Fig. 1, an example of JS with four jobs and three machines are considered. The routings of the jobs are the following ones: J_1 is executed on M_1 and M_3, J_2 is executed on M_2, M_1 and M_3, job J_3 is executed on M_3 and J_4 is executed on M_2, M_1, and M_3. In the Without Blocking (*Wb*) case (Fig. 1a), a machine is immediately available to execute its next operation after its in-course operation is finished.

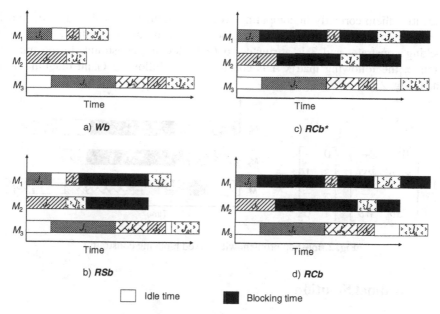

Fig. 1. JS with different blocking constraints

Then, when a machine is available to execute its next operation after its in-process operation is finished only when this operation starts on its next machine in its tasks list, the considered operation sets the considered machine in the Release when Starting blocking (*RSb*) case. In Fig. 1b, the job J_4 can be processed on the machine M_1 only when J_2 has started its execution on the machine M_3. When a machine is available to execute its next operation after its in-process operation is finished only when this operation ends on its next machine in its tasks list, the considered operation sets the considered machine in the special Release when Completing blocking (*RCb**) case. In the example presented in Fig. 1c, specific blocking *RCb** differs from classical blocking *RSb* by the fact that machine M_2 remains blocked by job J_2 until its operation on machine M_1 is finished.

In the *RCb* blocking case (Release when Completing blocking), it is not sufficient for the job to have finished its operation on its next machine in its tasks list, but it should leave it for the further one in its tasks list to release the first considered machine. This blocking constraint links together three machines around a same job. In the example presented in Fig. 1d, machine M_2 is obliged to wait, before treating job J_4, until job J_2 has started on machine M_3. With a *RCb** blocking constraint, it would have been sufficient to wait only until J_2 had finished its operation on M_1. *RCb* blocking constraint has been discovered and treated for the first time in [4].

To describe a JS problem with ***mixed blocking constraints***, we introduce a matrix A which contains blocking constraints between operations. Then, A_{ij} is the blocking constraint between operation j and operation $j+1$ of job i. A is then an n by m-1 matrix (n rows, m-1 columns). Different values are attributed to blocking constraints, in order to identify them correctly in computing programs. Value 0 is attributed to *Wb* blocking constraint, 1 is attributed to *RSb* blocking constraint, 2 is attributed to *RCb** blocking constraint and 3 is attributed to *RCb* blocking constraint. In the previous example, the following matrix A is considered and following Gantt diagram is obtained (Fig. 2):

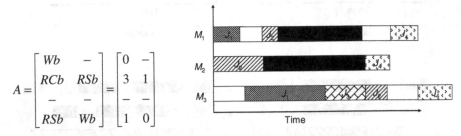

$$A = \begin{bmatrix} Wb & - \\ RCb & RSb \\ - & - \\ RSb & Wb \end{bmatrix} = \begin{bmatrix} 0 & - \\ 3 & 1 \\ - & - \\ 1 & 0 \end{bmatrix}$$

Fig. 2. Jobshop problem with mixed blocking constraints

3 Optimal Solution

Even if we know NP-hardness of JS problems, it is worth developing mathematical models to be able to obtain optimal results on benchmarks, to assess approximate methods such as heuristics or meta-heuristics. We give in this paragraph the mathematical model developed to solve JS problem with mixed blocking constraints. This model is inspired from the model developed in [11], for the case of job shop problem without constraint between operations. A first extension of this model for a unique *RCb* blocking constraint between all operations of the problem had been done and presented in [5]. For the version proposed in this paper, main modification is proposed in equations (3) to (12), where four different blocking constraints (*Wb*, *RSb*, *RCb** and *RCb*) can be taken into account from an operation to another.

3.1 Mathematical Model

Model variables are defined as follows:

- n: number of jobs.
- m: number of machines. In this model, the number of machines is also the maximal number of operations treated by job.
- ni: number of operations of job J_i.
- O_{ij}: j^{th} opération of job J_i.
- $O_i = \{O_{i1}, O_{i2}, \ldots, O_{ini}\}$: Set of operations to be executed for job J_i.
- p_{ij}: Processing time of operation O_{ij}.
- M_{ij}: Machine needed to process operation O_{ij}.

- $\lambda = \sum_{i=1}^{n} \sum_{j=1}^{n_i} P_{ij}$. It is a high value constant.

- $LO_{ij} = \begin{cases} 1 & if \ j \le ni\text{-}2 \\ 0 & else \end{cases}$, $\Leftrightarrow O_{ij}$ is neither the last, nor the penultimate of job J_i.

- $L1_{ij} = \begin{cases} 1 & if \ j = n_i\text{-}1 \\ 0 & else \end{cases}$, $\Leftrightarrow O_{ij}$ is the penultimate operation of job J_i.

- $L2_{ij} = \begin{cases} 1 & if \ j = n_i \\ 0 & else \end{cases}$, $\Leftrightarrow O_{ij}$ is the last operation of job J_i.

- $Bk_{ij} = \begin{cases} 1 & if \ after \ O_{ij} \ blocking \ constraint \ k \ is \ valid \\ 0 & else \end{cases}$,

  with k = 0(*WB*), 1(*RSb*), 2(*RCb**), 3(*RCb*)

Decision variables are defined as follows:

- $Y_{i,j,i1,j1}$ is equal to 1 if operation O_{ij} uses the same machine as operation O_{i1j1}.
- $S_{i,j}$: Starting time of operation O_{ij}.
- C_{max}: Makespan or maximal completing time of the scheduling problem.

The mathematical model is as follows:

$$min \quad C_{max}$$

Subject to the following constraints:

$$S_{ij} \ge S_{i(j-1)} + P_{i(j-1)}, \forall i \in \left[1, \rightleftharpoons, n\right], \forall j \in \left[1, \rightleftharpoons, n_i\right] \tag{1}$$

$$C_{max} \ge S_{in_i} + P_{in_i}, \forall i \in \left[1, \rightleftharpoons, n\right] \tag{2}$$

$$\begin{aligned}
S_{ij} \ge [S_{i_1(j_1+2)}.LO_{i_1j_1} &+ (S_{i_1(j_1+1)} + P_{i_1(j_1+1)}).L1_{i_1j_1} + (S_{i_1j_1} + P_{i_1j_1}).L2_{i_1j_1} - \lambda.Y_{iji_1j_1}].B3_{i_1j_1} \\
+ [(S_{i_1(j_1+1)} + P_{i_1(j_1+1)}).(L1_{i_1j_1} &+ LO_{i_1j_1}) + (S_{i_1j_1} + P_{i_1j_1}).L2_{i_1j_1} - \lambda.Y_{iji_1j_1}].B2_{i_1j_1} \\
+ [S_{i_1(j_1+1)}.(L1_{i_1j_1} + LO_{i_1j_1}) &+ (S_{i_1j_1} + P_{i_1j_1}).L2_{i_1j_1} - \lambda.Y_{iji_1j_1}].B1_{i_1j_1} \\
+ [(S_{i_1j_1} + P_{i_1j_1}).L2_{i_1j_1} &- \lambda.Y_{iji_1j_1}].B0_{i_1j_1},
\end{aligned} \tag{3}$$

$$\forall i,i_1 \in \left[1, \rightleftharpoons, n\right], \forall j \in \left[1, \rightleftharpoons, n_i\right], \forall j_1 \in \left[1, \rightleftharpoons, n_{i_1} - 2\right]$$

$$\begin{aligned}
S_{i_1j_1} \ge [S_{i(j+2)}.LO_{ij} &+ (S_{i(j+1)} + P_{i(j+1)}).L1_{ij} + (S_{ij} + P_{ij}).L2_{ij} - \lambda.(1-Y_{iji_1j_1})].B3_{ij} \\
+ [(S_{i(j+1)} + P_{i(j+1)}).(L1_{ij} &+ LO_{ij}) + (S_{ij} + P_{ij}).L2_{ij} - \lambda.(1-Y_{iji_1j_1})].B2_{ij} \\
+ [S_{i(j+1)}.(L1_{ij} + LO_{ij}) &+ (S_{ij} + P_{ij}).L2_{ij} - \lambda.(1-Y_{iji_1j_1})].B1_{ij} \\
+ [(S_{ij} + P_{ij}).L2_{ij} &- \lambda.(1-Y_{iji_1j_1})].B0_{ij},
\end{aligned} \tag{4}$$

$$\forall i,i_1 \in \left[1, \rightleftharpoons, n\right], \forall j \in \left[1, \rightleftharpoons, n_i - 2\right], \forall j_1 \in \left[1, \rightleftharpoons, n_{i_1}\right]$$

$$S_{ij} \geq [(S_{i_1(j_1+1)} + P_{i_1(j_1+1)}).L1_{i_1j_1} + (S_{i_1j_1} + P_{i_1j_1}).L2_{i_1j_1} - \lambda.Y_{iji_1j_1}].B3_{i_1j_1}$$
$$+[(S_{i_1(j_1+1)} + P_{i_1(j_1+1)}).L1_{i_1j_1} + (S_{i_1j_1} + P_{i_1j_1}).L2_{i_1j_1} - \lambda.Y_{iji_1j_1}].B2_{i_1j_1}$$
$$+[(S_{i_1(j_1+1)}.L1_{i_1j_1} + (S_{i_1j_1} + P_{i_1j_1}).L2_{i_1j_1} - \lambda.Y_{iji_1j_1}].B1_{i_1j_1} \tag{5}$$
$$+[(S_{i_1j_1} + P_{i_1j_1}) - \lambda.Y_{iji_1j_1}].B0_{i_1j_1},$$

$$\forall i, i_1 \in \left[1, \leftrightarrows, n\right], \forall j \in \left[1, \leftrightarrows n_i\right], \forall j_1 \in \left[1, \leftrightarrows, n_{i_1} - 1\right]$$

$$S_{i_1j_1} \geq [(S_{i(j+1)} + P_{i(j+1)}).L1_{ij} + (S_{ij} + P_{ij}).L2_{ij} - \lambda.(1 - Y_{iji_1j_1})].B3_{ij}$$
$$+[(S_{i(j+1)} + P_{i(j+1)}).L1_{ij} + (S_{ij} + P_{ij}).L2_{ij} - \lambda.(1 - Y_{iji_1j_1})].B2_{ij}$$
$$+[(S_{i(j+1)}.L1_{ij} + (S_{ij} + P_{ij}).L2_{ij} - \lambda.(1 - Y_{iji_1j_1})].B1_{ij} \tag{6}$$
$$+[(S_{ij} + P_{ij}) - \lambda.(1 - Y_{iji_1j_1})].B0_{ij},$$

$$\forall i, i \in \left[1, \leftrightarrows, n\right], \forall j \in \left[1, \leftrightarrows n_i - 1\right], \forall j_1 \in \left[1, \leftrightarrows, n_{i_1}\right]$$

$$S_{ij} \geq [(S_{i_1j_1} + P_{i_1j_1}).L2_{i_1j_1} - \lambda.Y_{iji_1j_1}], \forall i, i_1 \in \left[1, \leftrightarrows, n\right], \forall j \in \left[1, \leftrightarrows n_i\right], \forall j_1 \in \left[1, \leftrightarrows, n_{i_1}\right] \tag{7}$$

$$S_{i_1j_1} \geq [(S_{ij} + P_{ij}).L2_{ij} - \lambda.(1 - Y_{iji_1j_1})] \forall i, i_1 \in \left[1, \leftrightarrows, n\right], \forall j \in \left[1, \leftrightarrows n_i\right], \forall j_1 \in \left[1, \leftrightarrows, n_{i_1}\right] \tag{8}$$

$$S_{ij} \geq 0, \forall i \in \left[1, \leftrightarrows, n\right], \forall j \in \left[1, \leftrightarrows, n_i\right] \tag{9}$$

$$Y_{iji_1j_1} \in \{0,1\}, \forall (O_{ij}, O_{i_1j_1}) / M_{ij} = M_{i_1j_1} \tag{10}$$

The meaning of each model constraint is as follows:

- Equation (1) represents the precedence constraint for all operations of a same job.
- Equation (2) defines C_{max} as the completion time of the longest job of the problem.
- Equations (3) and (4) represent precedence constraints of different operations O_{ij} and $O_{i_lj_l}$ on a machine, independently on the knowledge of which of i and i_l is scheduled at first. But, on those two equations, it is assumed that both operations are neither last, nor penultimate operation of their respective jobs. The precedence constraints (3) and (4) are represented by equations (5) and (6) (respectively (7) and (8)), when an operation is the penultimate (respectively the last) of its job
- Equation (9) insures that all operations start at a positive time.
- Equation (10) defines $Y_{iji_lj_l}$ as a binary variable.

3.2   Optimal Solutions for Uniform Blocking Constraint Job-Shop Problems

We have remarked that sometimes the Gantt diagram drawing takes a too long time in front of what is really necessary to understand a mistake in in-course programming. We propose four benchmarks for small sized problem to help researchers to remedy to this issue. We propose 20 problem instances for each problem sizes 5 jobs/3 machines and 5 jobs/5 machines. We give in [13] the values of the benchmarks used in this paper for the four uniform blocking constraint cases *Wb*, *RSb*, *RCb** and *RCb*. A uniform blocking constraint case is when the same blocking is considered after all the operations of the problem. The model has been computed in Mosel-Xpress software on a PC, Intel Pentium 3GHz with 1Go RAM. It can be noticed that, for each problem of any benchmark, optimal values are rising in function of blocking constraint evolution (Table 1).

Table 1. Optimal values for JS problems with uniform mixed blocking constraints for 5 jobs

	5 jobs - 3 machines					5 jobs - 5 machines			
Problem	Wb	RSb	RCb*	RCb	Problem	Wb	RSb	RCb*	RCb
0	141	141	191	191	0	330	345	497	497
1	214	214	248	294	1	356	356	464	511
2	231	231	305	305	2	236	236	307	307
3	158	159	215	215	3	278	278	346	346
4	215	215	320	320	4	425	425	519	563
5	240	255	406	406	5	324	340	517	517
6	242	248	434	456	6	291	291	462	525
7	332	365	616	695	7	296	296	439	537
8	215	215	251	251	8	435	435	461	461
9	293	293	428	428	9	276	276	366	366
10	263	263	383	480	10	301	301	524	609
11	349	349	481	481	11	355	365	558	646
12	203	203	371	371	12	375	375	526	526
13	224	224	320	320	13	249	267	476	476
14	255	278	434	519	14	340	340	498	537
15	311	311	415	459	15	410	418	653	754
16	324	324	463	487	16	340	340	498	537
17	351	390	642	642	17	250	266	419	460
18	351	395	601	628	18	391	391	474	493
19	263	283	457	457	19	368	390	633	671

We did not report computation times in these tables because they are negligible for the considered problem sizes. We have given interest to different (n, m) combinations. In order to give an idea of how computation time increases in function of problem sizes, we reported in Table 2 computation time evolution in function of n and m for JS problems where *RCb* blocking constraint blocks all the operations. We can see that exact method computation times are increasing quickly after 10 jobs /10 machines problem size.

Table 2. Mean computation time (in seconds) of mathematical model with *RCb* constraint

	Machines							
Jobs	5	6	7	10	15	20	50	100
5	0,06	0,15	0,17	0,39	0,64	0,9	2,22	9,15
6	0,14	0,27	0,3	0,94	2,4	7,3	30	76
7	0,24	0,49	0,57	3,2	29	132	1085	7331
8	0,49	1,07	1,4	19,31	347	2707	35667	27007
9	1,07	3,24	4,05	102	4532	24345	88249	>24h
10	5,39	12,92	16,85	1160	58149	>24h	>24h	>24h

The evolution of exact method computation time in function of both number of jobs and number of machines leads us to develop approximate methods. Indeed, with such methods, we are able to give rather good solutions, and sometimes optimal ones, but without any proof, to hard problems. This is the reason why we developed an evaluation function to make this problem fit on the genetic algorithm. To give an idea of the computation time of our approximate method, it solves on Lawrence 15*15 instances [9] in a 110 seconds average time.

4   Evaluation Function for Meta-heuristics

Meta heuristics are good algorithms, developed to explore a solutions space, with the intelligence developed inside. For example, genetic algorithms generate a population of individuates and make them evolve in the solutions space, by crossing two individuates and mutations on individuate, thus imitating nature genetic evolution [6]. It becomes possible to ask for a solution to a problem with a meta-heuristic at the mere condition to develop an evaluation function, compatible with the meta-heuristic mechanism. When such a function is developed, the meta-heuristic is able to evolve in the solutions space described by this function, looking conventionally after the minimum value. The evaluation function we developed is based on Bierwirth vector adaptation presented in the next part.

4.1   Bierwirth Vector

In order to be able to solve the classical job-shop problem with a genetic algorithm he had initially developed for flow-shop problems, Bierwirth proposed a special vector, representing the order in which jobs operations are set on schedule [1]. This vector is composed of the jobs numbers, repeated as many times as the number of operations of the considered job. Jobs numbers are ranged in Bierwirth vector the way the operations must be considered by the meta-heuristic evaluation function. In the original work, the meta-heuristic was a genetic algorithm, but this vector can obviously work with any other meta-heuristic. In order to create an evaluation function adapted to the JS problem, we use chromosomes with $n*m$ length. Then, we use the function "modulo $(n) + 1$". Thus, the function which makes any chromosome corresponding to a Bierwirth vector is injective, as is presented in the following example.

For a 3 jobs / 4 machines problem, we use a chromosome of 12 genes. Let us take for example the chromosome: 3-4-1-12-2-8-7-5-9-11-10-6. This chromosome is directly transformed with the operation "modulo $(3) + 1$" into the following vector: 1-2-2-1-3-3-2-3-1-3-2-1. This last vector is the Bierwirth vector, corresponding to the first considered chromosome. This vector represents the order in which we will take jobs operations to set them on schedule. As we consider m operations per job, we can see that a job number is repeated m times in the vector. The relation defined between the two vectors is obviously injective, and not surjective. For example, the chromosome 3-1-4-6-2-5-7-8-9-11-10-12 would lead to the same Bierwirth vector as the first proposed. To find a solution to the job-shop problem, bijection is not necessary.

4.2   Evaluation Function

Once obtained, on classical JS problems with no blocking constraint, it only remains to set operations, one by one, in the order given by Bierwirth vector, to obtain the corresponding schedule. But, in the case of mixed blocking constraints between operations, the time when a machine becomes available depends on further operations scheduling. In order to avoid conflict situations where no operation is able to be operated any more (characterized by cycles on the conjunctive graph), we set coherent groups of operations, one after the other, when we build the schedule.

In order to know the first time at which a machine treating an operation on which there is either a *RSb* or a *RCb** blocking constraint will be available to treat another operation, one only have to know how is placed the further operation in job tasks list. Then, if it is a *RCb* blocking constraint, we have to know how are placed the two further operations in job tasks list. For an operation on which there is no blocking constraint, we can just place it without preoccupation of its further operations. And, if a group of operations ends on an operation followed with a no blocking constraint, this operation becomes the last of considered group. To be able to propose an evaluation function which copes with such a problem, we propose to group the operations of job in function of their successive blocking constraints. Then, when an operation is encountered in the Bierwirth vector, a grouping of its following operations is operated in function of what has just been described.

To illustrate how the operations are grouped in the evaluation function, we can take the example of following mixed blocking constraints matrix:

$$A3 = \begin{bmatrix} 3 & 1 & 2 & 0 \\ 0 & 0 & 0 & 0 \\ 1 & 0 & 0 & 2 \\ 0 & 3 & 1 & 0 \\ 3 & 0 & 2 & 1 \end{bmatrix}$$

Operations will then be grouped as follows:
Job J_1 : $[O_{11}, O_{12}, O_{13}]$; $[O_{14}, O_{15}]$
Job J_2 : $[O_{21}]$; $[O_{22}]$; $[O_{23}]$; $[O_{24}]$; $[O_{25}]$
Job J_3 : $[O_{31}, O_{32}]$; $[O_{33}]$; $[O_{34}, O_{35}]$
Job J_4 : $[O_{41}]$; $[O_{42}, O_{43}, O_{44}]$; $[O_{45}]$
Job J_5 : $[O_{51}, O_{52}, O_{53}, O_{54}, O_{55}]$

Once explained Bierwirth vector construction and the grouping of operations in function of mixed blocking constraints, evaluation function construction is rather easy to describe. We give in the following lines the pseudo-code of this function.

```
double Makespan_evaluation_function (Bierwirth vector, Blocking matrix)
    k = 1 // indice to travel on Bierwirth vector
    do
        identification of job, operation and machine in process
        grouping of operations with operation in course
        reformulation of Bierwirth vector in function of grouped operations
        place the group of operations in function of blocking constraints
        incrementation of n in function of the number of grouped operations
    while (k < n*m)
    return (maximum of operations completing times)
```

To illustrate how the evaluation function works when it is passed a vector during a meta-heuristic computing, we present here the schedule constructed when the canonical vector 1-2-3- ... -24-25 and the A_3 matrix are passed as input data to the evaluation function we describe just before. The groups of operations are then placed in the following order: $[O_{21}]$ $[O_{31}, O_{32}]$ $[O_{41}]$ $[O_{51}, O_{52}, O_{53}, O_{54}, O_{55}]$ $[O_{11}, O_{12}, O_{13}]$ $[O_{22}]$ $[O_{33}]$ $[O_{42}, O_{43}, O_{44}]$ $[O_{14}, O_{15}]$ $[O_{23}]$ $[O_{34}, O_{35}]$ $[O_{45}]$ $[O_{24}]$ $[O_{25}]$.

5   Benchmarks and Computing Results

In this part, we present the results found with our evaluation function on the proposed benchmarks and the four uniform blocking constraints between all the operations. Then, in a second paragraph, we propose four mixed blocking constraints matrixes to evaluate both our mathematical model and the evaluation function.

The genetic algorithm we develop in our laboratory is able to treat classic chromosomes, each composed of integer numbers strings, as presented in the paragraph dealing with evaluation function. Usually, crossing operation is performed with a one cut crossover, and mutation is done with a one random inversion process at each time mutation procedure is called. With this algorithm, we can adjust the number of people in the population. Stopping criteria which is used is the number of successive populations tested without objective function improvement. Concerning the selection procedure, we can adjust the percentage of best individuates kept from a population to the next, the percentage of individuates issued from crossing procedure and the percentage of random generated new people inserted to the new population. The difference to the total population is filled with the mutation procedure, operated on best individuates of the population.

5.1   Results Obtained with Uniform Blocking Constraints

Results presented in Table 3 have been obtained with the genetic algorithm we develop in the laboratory. For these problems, even if we can see that global performance is satisfying for a first approach of this kind of problems, we can also note that, for some instances, performance is clearly bad. This phenomenon is especially visible on problems concerned with *RSb* blocking, and in a least measure for *RCb** blocking. Performance is very good for *Wb* constraint and correct for *RCb* blocking constraint.

Table 3. Evaluation function error percentages and number of times where $S_{opt} = C_{max}$

		Wb	RSb	RCb*	RCb
j3m3	Sopt = Cmax	20	11	17	20
	Global error	0,0%	11,4%	1,8%	0,0%
j3m4	Sopt = Cmax	20	13	15	17
	Global error	0,0%	17,9%	6,1%	1,8%
j5m3	Sopt = Cmax	19	3	12	19
	Global error	1,5%	24,2%	5,5%	0,3%
j5m5	Sopt = Cmax	20	1	3	6
	Global error	0,0%	45,8%	16,1%	8,8%

Our explanation of this phenomenon is that *RSb* and *RCb** blocking constraints are blocking in-process machine only until respectively starting and completing time of further operation in process. This constraint is optimally taken into account in the mathematical model and not optimally at all in our evaluation function. This penalty is less effective for *RCb* blocking constraint, as it constrains far more strongly machines and operations. This is a point on which we will try to improve the evaluation function in further research works. The very good results obtained with *Wb* blocking constraints validate all the method we propose in this paper, from the Bierwirth vector to its use in the evaluation function and the quality of the genetic algorithm to solve problems up to 25 (5 jobs * 5 machines) genes problems. Indeed, when we are not imposed any supplementary difficulty, we succeed in finding optimal solutions. Moreover, execution time is short, about 2 seconds for j5m5 problems.

5.2 Genetic Algorithm Results for (*n*=5, *m*=5) Mixed Blocking Constraints

In order to prove the ability of our evaluation function combined with the genetic algorithm to find solutions to mixed blocking constraint problems, we propose to test it on four different blocking constraints matrixes. As it is obviously impossible to treat all the possible constraints combinations in such problems, we have arbitrarily defined four blocking matrixes:

$$A1 = \begin{bmatrix} 1 & 2 & 3 & 0 \\ 0 & 1 & 0 & 2 \\ 0 & 2 & 0 & 3 \\ 1 & 1 & 2 & 2 \\ 3 & 2 & 1 & 0 \end{bmatrix} \quad A2 = \begin{bmatrix} 0 & 1 & 2 & 3 \\ 2 & 0 & 1 & 0 \\ 3 & 2 & 0 & 1 \\ 1 & 0 & 0 & 0 \\ 0 & 3 & 1 & 2 \end{bmatrix} \quad A3 = \begin{bmatrix} 3 & 1 & 0 & 2 \\ 0 & 0 & 0 & 0 \\ 1 & 0 & 0 & 2 \\ 0 & 3 & 1 & 0 \\ 3 & 0 & 2 & 1 \end{bmatrix} \quad A4 = \begin{bmatrix} 1 & 0 & 1 & 0 \\ 0 & 2 & 1 & 0 \\ 2 & 1 & 3 & 0 \\ 1 & 3 & 0 & 2 \\ 0 & 0 & 0 & 0 \end{bmatrix}$$

In Table 6, we present results obtained with our genetic algorithm with the evaluation function presented in this paper, when optimal solutions are presented in [13]. We note again that even if global performance is satisfying, the results can strongly vary from a problem instance to another. More, the genetic algorithm gives a global performance of respectively 2.4% and 2.3% on *A2* and *A3* blocking matrixes, instead of 12.6% and 12.3% for *A1* and *A4*. The number of times when it obtains the optimal value is also variable from a blocking matrix to another (Table 4). We think that this may be due to the granularity of the groups of operations induced by the matrixes, but this hypothesis has to be verified in future works. Execution times are also about 2 seconds for each problem.

Table 4. Results on 5 jobs / 5 machines benchmark, with 4 mixed blocking constraints

	A1	*A2*	*A3*	*A4*
Sopt = Cmax	6	12	15	5
Global error	12,6%	2,4%	2,3%	12,3%

6 Conclusion and Perspectives

In this paper, we have proposed to solve job-shop scheduling problems with mixed blocking constraints between operations. As job-shop problems are widely NP-hard, after the presentation of the model we developed and the results obtained, we have proposed a special evaluation function, in order to be able to solve bigger problems with meta-heuristics. This evaluation function is based on Bierwirth vector and a grouping of operations in function of blocking constraints sequence. Solutions obtained with a genetic algorithm have been presented on four benchmarks, each composed of 20 problem instances, given in reference [13] of this paper.

In a future work, we intend to improve the evaluation function to fit at best possible to small problem instances optimal values. Then, we will address Lawrence instances to compare different meta-heuristics on harder instances, with a same evaluation function. The granularity notion about groups of operations has to be investigated. Another research direction in which we are interested is the flexible job-shop scheduling problem. We intend to propose the mixed blocking constraints extension, discussed in this paper, to the case of flexible job-shop scheduling.

References

1. Bierwirth, C.: A generalized permutation approach to job-shop scheduling with genetic algorithms. OR Spektrum 17, 87–92 (1995)
2. Brucker, P.: An efficient algorithm for the job-shop problem with two jobs. Computing 40(4), 353–359 (1988)
3. D'Ariano, A., Samà, M., D'Ariano, P., Pacciarelli, D.: Evaluating the applicability of advanced techniques for practical real-time train scheduling. Transportation Research Procedia 3, 279–288 (2014)
4. Dauzère-Pérès, S., Pavageau, C., Sauer, N.: Modélisation et résolution par PLNE d'un problème réel d'ordonnancement avec contraintes de blocage, pp. 216–217. ROADEF, Nantes (2000)
5. Gorine, A., Sauvey, C., Sauer, N.: Mathematical Model and Lower Bounds for Multi Stage Job-shop Scheduling Problem with Special Blocking Constraints. Information Control Problems in Manufacturing 14(pt. 1), 98–104 (2012)
6. Holland, J.H.: Outline for logical theory of adaptive systems. Journal of the Association of Computing Machinery 3, 297–314 (1962)
7. Jain, A.S., Meeran, S.: Deterministic Job-shop scheduling: Past, present and future. European Journal of Operational Research 113, 393–434 (1999)
8. Johnson, S.M.: Optimal Two and Three Stage Production Schedules with Setup Times Included. Naval Research Logistics Quarterly 1, 61–68 (1954)
9. Lawrence, S.: Supplement to resource constrained project scheduling: an experimental investigation of heuristic scheduling techniques. GSIA, Carnegie Mellon University, Pittsburgh, PA (1984)
10. Liu, S.Q., Kozan, E.: Scheduling trains as a blocking parallel-machine job shop scheduling problem. Computers & Operations Research 36(10), 2840–2852 (2009)
11. Mati, Y.: Les problèmes d'ordonnancent dans les systèmes de production Automatisés: modèle, complexité. PhD Thesis. University of Metz, France (2002) (in French)
12. Trabelsi, W., Sauvey, C., Sauer, N.: Heuristics and metaheuristics for mixed blocking constraints flowshop scheduling problems. Computers and Operations Research 39(11), 2520–2527 (2012)
13. website: http://cs-benchmarks.e-monsite.com

Towards a Robust Scheduling on Unrelated Parallel Machines: A Scenarios Based Approach

Widad Naji, Marie-Laure Espinouse, and Van-Dat Cung

Univ. Grenoble Alpes, G-SCOP, F-38000 Grenoble, France
CNRS, G-SCOP, F-38000 Grenoble, France
46, avenue Félix Viallet - 38031 Grenoble Cedex 1, France
{Widad.Naji,Marie-Laure.Espinouse,Van-Dat.Cung}@grenoble-inp.fr

Abstract. The article deals with scheduling on unrelated parallel machines under uncertainty of processing times. The performance measure is to minimize the makespan C_{max} when the splitting is tolerated. We show through an example that an optimal schedule of nominal data instance prior to its execution may be suboptimal due to minor or strong data uncertainty. Thus, we propose an artificial scenarios based approach to construct and identify a robust schedule to hedge against the uncertainty of processing times. To achieve this, we structure the uncertainty of processing time by mean of discrete scenarios, and evaluate the robustness according to the worst case strategy. We generate a family of artificial scenarios from the list of the potential realizations (processing time scenarios) to provide a family of feasible solutions using Lawler and Labetoulle's linear formulation [10]. Then, we identify a robust solution by evaluating the maximal cost or the maximal regret of each solution when applied to the potential realizations. Extensive empirical tests have been carried out. We have determined that the artificial scenario of maximal processing times *Smax* and the artificial scenario average processing times *Savg* are able to generate the robust schedules regarding the potential realizations.

Keywords: Robust scheduling, processing time uncertainty, makespan, Scenario approach, min-max, min-max regret.

1   Introduction

Scheduling is a decision-making process that occurs in all manufacturing and service systems. It aims to allocate resources to tasks over given time periods with respect to one or several objectives and constraints [1]. A large variety of mathematical models and algorithms (exact or approximate) are designed to solve scheduling problems [2, 3, 4]. The majority of these methods is based on the assumption of predictable universe where all data are known in advance. They consider an instance of values supposed to be representative of the real system, and provide a predictive schedule. However, uncertainty is inevitable in a production environment due to both internal factors (as machine breakdown or lack of staff) and external ones (as market fluctuations, customer requirements, unreliability of suppliers). Indeed, uncertainty is

© Springer International Publishing Switzerland 2015
H.A. Le Thi et al. (eds.), *Model. Comput. & Optim. in Inf. Syst. & Manage. Sci.*,
Advances in Intelligent Systems and Computing 360, DOI: 10.1007/978-3-319-18167-7_30

"an indeterminacy that commonly encounter all decisions whose outcome cannot be predicted in advance" [5]. The uncertainty causes poor precision of numerical values which creates a gap between the deterministic model and the real problem. This gap can lead to suboptimal and/or infeasible solutions.

This paper addresses the problem of scheduling jobs on unrelated parallel machines so as to minimize the makespan C_{max} under uncertainty of processing times and the possibility of jobs splitting. Scheduling on unrelated parallel machines, which is the general case of scheduling on parallel machines, is a very common problem in many production systems: in the textile industry [6], in semi-conductors manufacturing [7, 8, 9, 10], in Printed Wiring Board (PWB) manufacturing line [11, 12], in multiprocessor computer applications [13]. Unrelated parallel machines arise also in the aircraft maintenance process at airports [14], and in the problem of scheduling aircraft arrivals and departures on multiple runways at airports [15], etc.. Among all scheduling criteria, C_{max} is the most considered in the literature. The uncertainty of processing times is arguably one of the most prevalent; it can be justified by many factors such as the demand certainty degree, the conditions of the tools, the mastering of the technology, the operators experience and availability, condition of auxiliary devices for holding the job at the appropriate position on the machine, etc. Despite its theoretical and practical importance, this problem has been little addressed in the literature.

The remainder of this paper is organized as follows. Section 2 describes the problem ($Rm/Split/C_{max}$; p_{ij} uncertain) and shows the limits of the deterministic resolution when uncertainty of processing times arises. Section 3 reviews the main developed robust approaches to deal with uncertainty in scheduling problems. The focus will be on scheduling applications where the uncertainty of processing times is structured by mean of scenarios. Section 4 describes the approach constructed to identify a robust solution to our scheduling problem with scenario uncertainty representation. The proposed framework is of proactive nature; the uncertainties are anticipated and taken into account in the off-line phase. The purpose is to identify a solution that deals with processing times uncertainty over many scenarios. Section 5 describes the computational experiments and results for the approach elaborated. Section 6 summarizes the important results and provides some guidelines for future research.

2   Problem Description

We consider n independent jobs allowed to be processed with any of the m unrelated parallel machines. In this series of same function's machines, each machine i has a job dependent speed v_{ij} to accomplish the processing requirement p_j of job j. As consequence processing times p_{ij} depends on job j and machine i: $p_{ij}=p_j/v_{ij}$. They are arbitrary for all i and j i.e. no particular relation exists between processing times of different couples (i, j). The processing requirement of a job j is considered as the total demand of a product reference in production planning. The minimization of C_{max} implies a high utilization of the machines.

The problem $Rm//C_{max}$ is demonstrated to be NP-hard in the strong sense since it is NP-hard, even if reduced to 2 machines [16]. Under the preemptive assumption, the problem $Rm/pmtn/C_{max}$ can be solved in polynomial time, according to the algorithm proposed by Lawler and Labetoulle [13]. Indeed, preemption implies that it is not necessary to keep a job on a machine until completion; it is allowed to interrupt the processing of a job at any time and put a different job on the machine. When a preempted job is put back on a machine, it only needs the machine for its remaining processing time. But, a job cannot be processed on two machines at the same time.

Lawler's algorithm allows solving the problem $Rm/pmtn/C_{max}$ in two steps [13]:

The first step algorithm is to solve the linear program (see LP1) that computes the optimal values of C^*_{max} and t_{ij}^* the total amount of time spent by jobs j on the machine i.

$$min \ C_{max} \tag{1}$$

s.t.

$$\sum_{i=1}^{m} \frac{t_{ij}}{p_{ij}} = 1 \qquad \forall \ j \in \{1,..,n\} \tag{2}$$

$$\sum_{j=1}^{n} t_{ij} \leq C_{max} \qquad \forall \ i \in \{1,..,m\} \tag{3}$$

$$\sum_{i=1}^{m} t_{ij} \leq C_{max} \qquad \forall \ j \in \{1,..,n\} \tag{4}$$

$$t_{ij} \geq 0 \qquad \forall \ (i,j) \in \{1,..,m\} x \{1,..,n\} \tag{5}$$

LP1: Lawler and Labetoulle's LP

The objective function (1) minimizes the makespan. Constraints (2) ensure that a job j is completely worked with the set of machines, while constraints (3) ensure that the total time that a machine i is working never exceed the makespan C_{max}. Constraints (4) prohibit the splitting, i.e., a job j is proceeded by one and only one machine i at a given time, although it can be divided into a finite number of partial jobs. Lastly, the constraints (5) set the non-negativity of temporal variables t_{ij}.

The second step algorithm is solved via "open shop" theory.

Our scheduling problem namely $Rm/Split/C_{max}$ can also be solved by Lawler and Labetoulle's LP. We circumvent the hypothesis for preemptive scheduling. Alternatively, we regard the problem as a lot sizing problem in production planning. Therefore, n jobs (lots) can be split into continuous sub-jobs (sub-lots) and processed independently on m machines so as to finish the processing of all demands as soon as possible. A job can be processed on different machines at the same time, each sub-job on one machine. As a consequence, constraints (4) are relaxed and we refer to this linear program as LP2. The deterministic problem is reduced to a simple allocation problem where no order is needed to be predetermined.

When we consider the uncertainty of processing times, the deterministic model does not withstand the variance over scenarios. As an illustration let us consider the following example to demonstrate the limits of the deterministic resolution. We suppose that the decision-maker has an instance of nominal values p_{ij} that he regards as certain. This instance is given in Table 1.a.

Table 1. Processing time data according to the nominal instance $Inst_f$ and the real instance

1.a Nominal instance data

p_j/v_{ij}.	$j=1$	$j=2$	$j=3$	$j=4$
$i=1$	30/3	50/2	30/4	100/1
$i=2$	30/3	50/1	30/3	100/2
$i=3$	30/3	50/4	30/4	100/4

1.b Real instance data

p_j/v_{ij}.	$j=1$	$j=2$	$j=3$	$j=4$
$i=1$	50/3	30/2	80/4	50/1
$i=2$	50/3	30/1	80/3	50/2
$i=3$	50/3	30/4	80/4	50/4

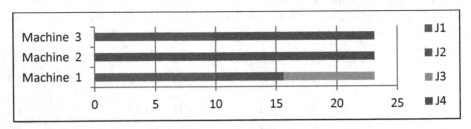

Fig. 1. Optimal schedule of the nominal instance $Inst_f$

The optimal schedule of the nominal instance $Inst_f$, as shown in Fig. 1, can handle all jobs with a makespan equal to 23.12 time units according to *LP2* resolution. The optimal solution of the nominal instance Sol_f* will determine the allocation of the jobs to the machines and the ratio of participation of each machine i to perform a job j.

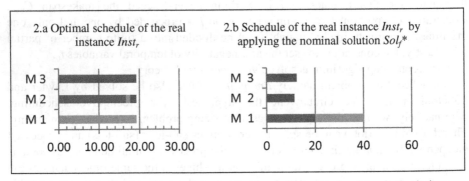

Fig. 2. Schedule of the real instance $Inst_f$ according to the optimal and nominal solutions

If the customer demand requirements change, which is often the case, the processing times will likewise vary and this will lead to a new dataset of p'_{ij} called real instance $Inst_r$ (see Table 1.b). The optimal schedule corresponding to $Inst_r$ permits to complete the tasks on 19.23 time units according to Fig. 2. However, by applying the solution Sol_f* to the real instance $Inst_r$, we obtain a schedule that requires a makespan equal to 40.04 time units (see Fig. 2). The generated decision is then suboptimal:

- The C_{max} increased over 108% the time units compared to the optimal makespan of the real scenario C_{max}^* (if we had known we would save 20 time units).
- Load balancing in the workshop is altered as the machine $M1$ is overused while machines $M2$ and $M3$ are idle in more than half schedule time.

This example demonstrates the weakness of the deterministic scheduling approach to the problem ($Rm/Split/C_{max};p_{ij}$ uncertain). A schedule that is determined to be optimal with regard to the total completion time criterion prior to its execution may be very vulnerable to minor or strong data uncertainty. This observation confirms the necessity to setup an approach that can build robust schedule to deal with the presence of plausible data instances other than the nominal one.

3   Literature Review on Robust Scheduling

The Robust Optimization (RO) approach has been proposed as an alternative to stochastic approach [17, 18, 19], because producing probability distributions is not trivial for many decision makers due to deficient data and interdependencies between uncertainty factors. Moreover, the subsequent problems are not usually tractable. It aims to develop models whose solutions are less sensitive to the uncertainty of the data regardless of a specific distribution of the uncertain parameters. The paradigm of RO dates back to Soyster who was the first to consider what is now called the robust linear programming [20]. Under the robust approach, the uncertainty is modelled by mean of continuous sets [21, 22, 23] or discrete sets called scenarios [24, 25]. The goal of robust optimization is to find decisions that are optimal for the worst-case realization of the uncertainty within the given set. Most authors use the min-max or the min-max regret criteria to evaluate the solution robustness.

Applied to scheduling, a robust solution is a schedule with the best performance in the worst case compared to the corresponding optimal solution to any potential processing time realization [26], and it is able to absorb some level of unexpected events without rescheduling [27]. The robust scheduling approaches could be classified into three families according to [26]: proactive approach, reactive approach and proactive-reactive approach.

The proactive approach is known to be the robust approach. Its use could be motivated by many justifications. It does not ignore uncertainty when constructing the solution and takes a proactive step to anticipate the variability that may occur in the processing times during the execution. And it evaluates a priory the risk that occurs when applying a solution to the potential realizations of uncertainty at expense to the optimal ones. Modeling uncertainty is the basic step in the proactive approach. One of the most practical tools to structure processing times uncertainty is the discrete scenario representation. It consists on defining a finite set of instances that capture the range of futures (potential realizations) without ranking to prepare the robust decision to cope satisfactorily with any one of them [25]. After uncertainty representation, to evaluate the risk that occurs from a specific solution and to identify robust one, the min max and the min max regret strategies are the most used over literature. Kouvelis and Yu consider the following notations to define the min max and min max regret solutions under discrete scenario approach:

Ω: the set of problem scenarios, s: a scenario (potential realization) of the problem, π a feasible solution to the problem and $Z^s(\pi)$ the optimization criterion when π is applied to s.

$Z_A(\pi) = max_{s\in\Omega} Z^s(\pi)$ is the worst case measures the maximal cost of the solution π over all the potential realizations. The min-max solution is the solution that minimizes the maximal cost Z_A. This criterion is very useful for non-repetitive decisions, in competitive situations, or in environments where protection measures are needed .The min-max decisions are established on the expectation that the worst might happen that why they are of conservative nature.

$Z_D(\pi) = max_{s\in\Omega}(Z^s(\pi) - Z^{s^*})$ is the maximal deviation measures the worst loss in performance when we use the solution π at the expense of the optimal solutions. The min-max regret solution is the solution that minimizes the maximal deviation Z_D. This criterion allows the benchmarking of solutions compared to the best possible under any realizable scenario, and the bounding of the magnitude of missed chances.

From a computational effort perspective, the min-max criterion is the easiest, since it does not require the calculation of the optimal decision in each scenario. The min-max regret criterion requires to computing the best solution under any scenario. For many polynomial solvable scheduling problems, their corresponding robust versions (min max or min max regret) are weakly or strongly *NP*-hard when taking into account the uncertainty of processing times represented by mean of discrete scenarios [26, 27, 28, 29], even in the case of single machine [30].

4   Artificial Scenarios Based Approach for Robust Scheduling (*Rm/Split/C$_{max}$;pij uncertain*)

To deal with the problem (*Rm/Split/C$_{max}$;p$_{ij}$ uncertain*), we propose an artificial solutions approach divided into four successive steps as shown in the figure (fig. 4):

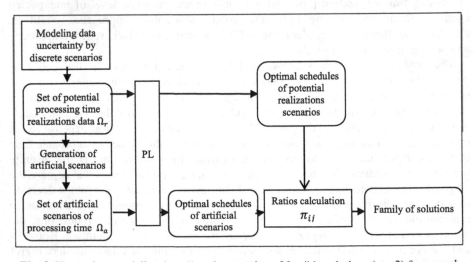

Fig. 3. Uncertainty modeling (step1) and generation of feasible solutions (step2) framework

Step 1 aims to model uncertainty of processing times and resolve the static problems. (Fig. 3)

We structure the uncertainty of processing times by mean of discrete scenarios. Each scenario s reflects a potential realization of the uncertain parameters p_{ij}. We denote by t_{ij}^s, p_{ij}^s, and C_{max}^s, the total amount of time spent by jobs j on the machine i when j is assigned to i, the processing time, and the value of total time completion under the scenario s. For the sake of simplification, we suppose that the uncertainty of processing times is due to demand uncertainty, i.e. the processing requirements p_j^s are uncertain, while the machines speed v_{ij} are known with certainty. Consequently, we generate the family of potential demand scenarios. The corresponding family of potential processing time realizations is then called Ω_r. In order to evaluate the regret of each solution, we will calculate the optimal makespan of each scenario s from Ω_r that we denote it : C_{max}^{s*}.

Step 2 aims to the construction of a set of feasible solutions. (Fig. 3)

Let π^* be the robust absolute solution. π^* is the best solution over all feasible solutions $\pi \epsilon F$ i.e. π^* has the best worst makespan over all scenarios belonging to Ω_r and π^* respects all the constraints of the deterministic problem for each scenario. The robust problem and the robust regret problem are respectively:

$$min_{\pi \epsilon F} \, max_{s \in S} \, C_{max}^s(\pi) \, and \, min_{\pi \epsilon F} \, max_{s \in S} \, C_{max}^s(\pi) - C_{max}^{s*}$$

As the set of feasible solution F is infinite, we will restrict the set of feasible solutions to a special subset $F' \subset F$. This subset will be constructed from solutions very closed to the solutions of real instances in order to get non optimal completion times that will have the same scale of magnitude of optimal ones. We construct a set of artificial scenarios Ω_a from Ω_r. Some special case scenarios are constructed to be tested:

The artificial worst-case scenario s^w in which processing times take the max values among all the values of the potential realizations. This scenario inherently provides temporal protection. This quality is due to the fact that the processing time are extended to cope with scenarios fluctuating on the one hand, and to provide flexibility to the scheduler on the other hand due the offered margin. In this case $p_{ij}^w = max_{s \in \Omega_r} \, p_{ij}^s$ However, the worst-case scenario is viewed as pessimistic. To remedy this, the average scenario s^{avg} is a good alternative because it increases the processing times under average and reduces the duration of tasks that are over the average and also $p_{ij}^{average} = \sum_{s \in \Omega_r} p_{ij}^s \times Card^{-1}(\Omega_r)$

Nevertheless, the average value does not reflect the general tendency of the distribution. For this, we selected to also try out a median scenario s^{median}. The p_{ij}^{median} is the value that divides the list of processing times among Ω_r into a list of higher half and a list of lower half. It's found by arranging all the values from lowest value to highest value and picking the middle one. To complete the artificial scenarios, we also propose to test the best-case scenario s^{min} that shortens all durations of tasks. Hence, $p_{ij}^{min} = min_{s \in \Omega_r} \, p_{ij}^s$.

And finally, to assess the impact of randomness on solving the problem, we construct random scenarios from picked values on the potential processing time realizations values. We also construct random scenarios from values included in the intervals of on the potential processing time realizations. The objective is to test the solution behavior in comparison with solutions built randomly.

The generation of the family of solutions consists on:

(1) Finding the optimal schedule of each scenario $T^* = (t_{ij}^*)$;

(2) Calculating the percentage of participation of each machine i to the realization of a job j under the considered scenario s:

$$\pi_{ij}^{s\prime} = \left(\frac{t_{ij}^{s\prime*}}{p_{ij}^{s\prime}} \right) \qquad \forall s' \in \Omega_r \cup \Omega_a$$

These solutions determine the jobs percentages (rates) affectation to machines with in respect to all the constraints of the deterministic problem which makes them feasible for all potential realizations scenarios. Moreover, these matrix solutions are optimal for scenarios with values around the real values.

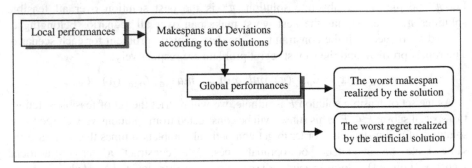

Fig. 4. Local (steps 3) and global performances calculation framework (step 4)

Step 3 aims to calculate the local performances of each solution. (Fig. 4)
The potential realizations are scheduled according to the constructed solutions and the non-optimal makespans resulting from these non-optimal schedules are measured.

$T^s (\pi^{s\prime})$ *is the non-optimal schedule of the scenario s scheduled according to the* constructed solution $\pi^{s\prime}$, then:

$$T^s (\pi^{s\prime}) = (t_{ij}^s)_{i,j} = (\pi_{ij}^{s\prime} \times p_{ij}^s)_{i,j} \qquad \forall s' \in \Omega_r \cup \Omega_a \quad \forall s \in \Omega_r$$

The local performances of the solution $\pi^{s\prime}$ are calculated as:

(1) The makespan $C_{max}^s(\pi^{s\prime})$ of the real scenario s when $\pi^{s\prime}$ is applied to:

$$C_{max}^s(\pi^{s\prime}) = max_{i \in \{1,..,m\}} \sum_{j=1}^n t_{ij}^s \qquad \forall s \in \Omega_r$$

(2) The deviation of $C_{max}^s(\pi^{s\prime})$ from the optimal makespan C_{max}^{s*}:

$$D^s(\pi^{s\prime}) = C_{max}^s(\pi^{s\prime}) - C_{max}^{s*} \qquad \forall s \in \Omega_r$$

Step 4 aims to calculate the global performances, and to identify a robust solution. (Fig. 4)
In this decision-making phase of the process, the objective is to determine the global performance of the constructed solutions and to compare them. The criteria for

measuring the robustness of the solution $\pi^{s'}$ are the worst case completion time Z_A $(\pi^{s'})$ and the worse deviation from the optimal completion times $Z_D (\pi^{s'})$, it is the worst local performance of the solutions applied to all potential realizations

$$Z_A (\pi^{s'}) = max_{s\in \Omega_r} C_{max}^s(\pi^{s'}) \qquad\qquad \forall s' \in \Omega_r \cup \Omega_a$$
$$Z_D (\pi^{s'}) = max_{s\in \Omega_r} C_{max}^s(\pi^{s'}) - C_{max}^{s*} \qquad\qquad \forall s' \in \Omega_r \cup \Omega_a$$

The robust solution is the solution that has the best global performance: $min_{\pi^{s'}\in F'} Z_A$ $(\pi^{s'})$ or $min_{\pi^{s'}\in F'} Z_D(\pi^{s'})$.

Example

Steps (1 and 2): Let's consider the example introduced earlier in section 2. We model processing times uncertainties by mean of scenarios: 3 potential processing time realizations (matrix of data). Scenario 1 and scenario 2 are given in the example section 2. To complete the list, we add scenario 3 presented in table 3.

Table 2. Processing time data according to scenario 3

$p/v_{ij}.$	$j=1$	$j=2$	$j=3$	$j=4$
$i=1$	25/3	90/2	100/4	80/1
$i=2$	25/3	90/1	100/3	80/2
$i=3$	25/3	90/4	100/4	80/4

Table 3. The worst-case artificial scenario and its corresponding solution

3.a The worst-case artificial scenario

$p/v_{ij}.$	$j=1$	$j=2$	$j=3$	$j=4$
$i=1$	50/3	90/2	100/4	100/1
$i=2$	50/3	90 1	100/3	100/2
$i=3$	50/3	90/4	100/4	100/4

3.b The worst-case scenario's solution π^w

$p/v_{ij}.$	$j=1$	$j=2$	$j=3$	$j=4$
$i=1$	0.55	0.0	1.0	0.0
$i=2$	0.45	0.0	0.0	0.53
$i=3$	0.0	1.0	0.0	0.47

We construct the set of artificial scenarios based on the potential realizations using artificial scenarios algorithms generation. Using *Cplex*, we solve all the instances belonging to the set of scenarios. We calculate the solution of each $s \in \Omega_a \cup \Omega_r$. The set of feasible solutions is then constructed. For example, the worst case scenario data and solution are presented in Table 3.

The optimal completion times: $C_{max}^{s1*}= 23.125$; $C_{max}^{s2*}= 19.23$; $C_{max}^{s3*}=29.58$

Steps (3) and (4): We calculate the local and the global performances. For this, we multiply term by term the matrix solutions elements π_{ij} which are the rates participation of the machine i to perform a quantity of products from reference j with the matrix of processing times belonging to the potential realizations. And then, we calculate the makespan under the considered solution. Some computational results are summarized in Table 4.

Table 4. The min-max and min-max regret calculation for some solutions

$C_{max}^{s1}(\pi_1) = 23.12$ $C_{max}^{s2}(\pi_1) = 40.04$ $C_{max}^{s3}(\pi_1) = 43.48$	$Z_A(\pi_1) = 43.48$	$C_{max}^{s1}(\pi_1) - C_{max}^{s1*} = 0$ $C_{max}^{s2}(\pi_1) - C_{max}^{s2*} = 20.81$ $C_{max}^{s3}(\pi_1) - C_{max}^{s3*} = 13.7$	$Z_D(\pi_1) = 20.81$
$C_{max}^{s1}(\pi_w) = 31.16$ $C_{max}^{s2}(\pi_w) = 29.16$ $C_{max}^{s3}(\pi_w) = 31.83$	$Z_A(\pi_w) = 31.83$	$C_{max}^{s1}(\pi_w) - C_{max}^{s1*} = 8.24$ $C_{max}^{s2}(\pi_w) - C_{max}^{s2*} = 9.93$ $C_{max}^{s3}(\pi_w) - C_{max}^{s3*} = 2.25$	$Z_D(\pi_w) = 9.93$

According to the criterion of robustness the solution π_w has the best worse maximal cost over all the real scenarios while the solution π_{median} has the best worse deviation over all the real scenarios.

5   Experiments and Results

In this section, we describe the computational experiments for the approach elaborated to identify a robust solution for ($Rm/Split/C_{max}$; p_{ij} *uncertain*). Our algorithm is coded in Java. Linear programs algorithms are solved by CPLEX. To limit the number of factors that influence the quality of the decision and to standardize the basis of tests, we generated and analyzed instances of tests according to 2 axes of variability: the Size of the Workshop and the Variety of products.

Table 5. The size workshops generation

	3 machines	7 machines	15 machines
10 jobs	Small with low variety	Medium with low variety	High with low variety
50 jobs	Small with medium variety	Medium with medium variety	High with medium variety
200 jobs	Small with high variety	Medium with high variety	High with high variety

Table 6. Average robustness frequency of solution according to the min-max criterion

	π_w	π_{avg}	π_{median}	π_{min}	π_{rand}	π_1, π_2, π_3
M3 _J10	40%	32%	18%	5%	3%	2%
M_3J_50	33%	35%	21%	6%	3%	2%
M_3J_200	32%	35%	21%	6%	4%	2%
M_7J_10	89%	2%	2%	4%	2%	1%
M_7J_50	31%	47%	18%	2%	0%	2%
M_7J_200	30%	45%	20%	3%	1%	1%
M_15_j50	31%	43%	20%	2%	0%	0%
M_15_j200	30%	45%	20%	3%	1%	1%

We can distinguish six kinds of production workshops presented in Table 5. Given all possible combinations of the parameters mentioned above. We count 900 different instances. In order to rank the solutions, we calculate from results the average frequency of robustness of each solution over the 900 tests.

As shown in Table 6, in small workshops with limited variety of products, π_w is the most robust with an average robustness frequency equal to 40%. π_w is followed by π_{avg} and π_{median} which have respectively an average robustness frequency equal to 32% and 16% . π_1, π_2, π_3, π_{min} and π_{rand} are very bad in terms of robustness. In small workshops with medium and high variety of products, the worst-case scenario's solution π_w is no longer the most robust; we remark a decrease in its average robustness frequency is around 32%. However, the worst-case scenario's solution π_w and the average scenario's solution π_{avg} are both more or less at the same level (35%, 32%). In medium size workshops with a small variety of products, the worst case scenario's solution π_w is better than the other solutions; its average robustness frequency is equal to 89%. However, when the variety of products increases again, the worst case scenario's solution π_w is no longer more robust than the average scenario's solution π_{avg}. In medium and high size workshops with medium and high variety of products. We remark that average robustness frequency of π_{avg} exceeds the average robustness frequency of π_w. We can notice that the most influent parameter when analyzing the robustness of the worst case scenario's solution in an unrelated parallel machine problem with splitting is the ratio between the workshop size and the variety of products: More it tends to 1, the more is π_w more robust and vice versa.

Table 7. Average robustness frequency of solution according to the min-max regret

	π_w	π_{avg}	π_{median}	π_{min}	π_{rand}	π_1, π_2, π_3
M3 _J10	24%	50%	16%	7%	2%	1%
M_3J_50	25%	40%	24%	8%	2%	1%
M_3J_200	33%	43%	15%	8%	1%	0%
M_7J_10	60%	30%	10%	0%	0%	0%
M_7J_50	34%	47%	13%	2%	4%	0%
M_7J_200	32%	45%	13%	5%	4%	1%
M_15_j50	35%	44%	14%	2%	3%	2%
M_15_j200	33%	45%	12%	3%	4%	2%

Table 7 shows that for the robust deviation criterion π_{avg} is better than any other solution, except for the case of low size workshop with low production where the average robust deviation frequency of π_w is two times better than the average robustness frequency of π_{avg}.

6   Conclusion and Prospects

This paper has focused on the scheduling problem on unrelated parallel machines with splitting under uncertainty of processing times ($Rm/Split/C_{max}$; p_{ij} *uncertain*). The problem has been little addressed in the literature despite its theoretical and practical importance. We have investigated a new approach to build a robust solution. We structured the uncertainty of processing times by mean of discrete scenarios. Then, we generated a family of artificial scenarios from the potential processing time realizations to provide a family of feasible solutions. To generate robust solutions we evaluated the maximal cost and the maximal regret of each solution when applied to the potential realizations. According to the computational results, the robustness frequency of the worst case scenario's solution π_w where all the processing times take their maximal values, and the average scenario's solution π_{avg} are very close for the min-max criterion except for a special. However for the min-max regret criterion, the average scenario's solution π_{avg} is more robust than π_w. In the future works, we will be interested in studying the theoretical complexity of the robust problem. We propose also to test the elaborated approach to other uncertain scheduling problems.

References

1. Pinedo, M.L.: Scheduling: theory, algorithms, and systems. Springer (2012)
2. Graham, R.L., Lawler, E.L., Lenstra, J.K., Kan, A.R.: Optimization and approximation in deterministic sequencing and scheduling: a survey. Annals of Discrete Mathematics 5, 287–326 (1979)
3. Blazewicz, J., Dror, M., Weglarz, J.: Mathematical programming formulations for machine scheduling: a survey. European Journal of Operational Research 51(3), 283–300 (1991)
4. Brucker, P.: Scheduling algorithms, vol. 3. Springer, Berlin (2007)
5. Liu, B.: Uncertainty theory. Springer, Heidelberg (2010)
6. Silva, C., Magalhaes, J.M.: Heuristic lot size scheduling on unrelated parallel machines with applications in the textile industry. Computers & Industrial Engineering 50(1), 76–89 (2006)
7. Kim, D.W., Na, D.G., Chen, F.F.: Unrelated Parallel Machine Scheduling with Setup times and a Total Weighted Tardiness Objective. Robotics and Computer-Integrated Manufacturing 19, 179–181 (2003)
8. Aubry, A.: Optimisation pour la configuration robuste de systèmes de production de biens et de services. Thèse de doctorat. Institut National Polytechnique de Grenoble-INPG (2007)
9. Zhang, Z., Zheng, L., Weng, M.X.: Dynamic Parallel Machine Scheduling with Mean Weighted Tardiness Objective by Q-Learning. International Journal of Advanced Manufacturing Technology 34(9-10), 968–980 (2007)
10. Rossi, A.: A robustness measure of the configuration of multi-purpose machines. International Journal of Production Research 48(4), 1013–1033 (2010)
11. Yu, L., Shih, H.M., Pfund, M.: Scheduling of unrelated parallel machines: an application to PWB manufacturing. IIE Transactions 34(11), 921–931 (2002)

12. Bilyk, A., Mönch, L.: A variable neighborhood search approach for planning and scheduling of jobs on unrelated parallel machines. Journal of Intelligent Manufacturing 23(5), 1621–1635 (2012)
13. Lawler, E.L., Labetoulle, J.: On preemptive scheduling of unrelated parallel processors by linear programming. Journal of the ACM (JACM) 25(4), 612–619 (1978)
14. Kolen, A.W.J., Kroon, L.G.: On the computational complexity of (maximum) shift class scheduling. European Journal of Operational Research 64(1), 138–151 (1993)
15. Hancerliogullari, G., Rabadi, G., Al-Salem, A.H.: Greedy algorithms and metaheuristics for a multiple runway combined arrival-departure aircraft sequencing problem. Journal of Air Transport Management 32, 39–48 (2013)
16. Lenstra, J.K., Rinnooy Kan, A.H.G., Brucker, P.: Complexity of machine scheduling problems. Annals of Discrete Mathematics 1, 343–362 (1977)
17. Kall, P., Wallace, S.W.: Stochastic programming. Wiley, New York (1994)
18. Prékopa, A.: Stochastic programming. Kluwer Academic Publishers, Dordrecht (1995)
19. Birge, J.R., Louveaux, F.: Introduction to Stochastic Programming. Springer (1997)
20. Soyster, A.L.: Convex programming with set-inclusive constraints and applications to inexact linear programming. Operations Research 21, 1154–1157 (1973)
21. Ben-Tal, A., El Ghaoui, L., NemirowskiI, A.: Robust optimization. Princeton University Press (2009)
22. Bertsimas, D., Pachamanova, D., Sim, M.: Robust linear optimization under general norms. Operations Research Letters 32(6), 510–516 (2004)
23. Bertsimas, D., Sim, M.: The price of robustness. Operations Research 52(1), 35–53 (2004)
24. Mulvey, J.M., Vanderbei, R.J., Zenios, S.A.: Robust optimization of large-scale systems. Operations Research 43(2), 264–281 (1995)
25. Kouvelis, P., Yu, G.: Robust discrete optimization and its applications. Springer (1997)
26. Billaut, J.C., Moukrim, A., Sanlaville, E.: Flexibility and robustness in scheduling. ISTE/John Wiley & Sons (2008)
27. Kouvelis, P., Daniel, R.L., Vairaktarakis, G.: Robust scheduling of a two-machine flow shop with uncertain processing times. IIE Transactions 32(5), 421–432 (2000)
28. Averbakh, I.: On the complexity of a class of combinatorial optimization problems with uncertainty. Mathematical Programming 90(2), 263–272 (2001)
29. Kasperski, A., Kurpisz, A.: Approximating a two-machine flow shop scheduling under discrete scenario uncertainty. European Journal of Operational Research 217(1), 36–43 (2012)
30. Aloulou, M.A., Della Croce, F.: Complexity of single machine scheduling problems under scenario-based uncertainty. Operations Research Letters 36(3), 338–342 (2008)

Part VI

Post Crises Banking and Eco-finance Modelling

Initial Model Selection for the Baum-Welch Algorithm Applied to Credit Scoring

Badreddine Benyacoub[1], Ismail ElMoudden[1], Souad ElBernoussi[1],
Abdelhak Zoglat[2], and Mohamed Ouzineb[1]

[1] Laboratory of Mathematics, Computing and Applications, Faculty of Sciences,
University Mohammed-V, Rabat, Morocco
[2] Laboratory of Applied Mathematics, Faculty of Sciences, University Mohammed-V,
Rabat, Morocco

Abstract. Credit scoring become an important task to evaluate an applicant by a banker. Many models and tools are available for making initial lending decisions. This paper presents an Hidden Markov Models (HMMs) for modeling credit scoring problems. Baum-Welch algorithm an iterative process for estimating HMM parameters are often used to developed such models and improve the pattern recognition for many problems. We introduce HMM/Baum-Welch initial model selection: a tool developed to test the impact of choosing initial model to train Baum-Welch process. Experiments results show that the performance of learned models depend on different way of generating the initial models used in credit scoring.

Keywords: Credit scoring, Hidden markov model, Baum-Welch, Initial model, classification.

1 Introduction

Credit scoring has gained more and more attentions both in academic world and the business community today. Credit scoring is a system creditors use to assign credit applicants to either a "good credit" one that is likely to repay financial obligation or a "bad credit" one who has high possibility of defaulting on financial obligation. Since credit scoring was first developed in the 1950s and 1960s, both statistics and non-statistics based methods are used to research in building credit scoring models to protect from credit fraud. Statistics based methods to build credit scorecards include discriminant analysis, logistic regression, non-linear regression, classification trees, nearest-neighbor approach, multiple-type discrimination and the rest, while non-statistics based methods for scorecard development includes linear programming, integer programming, neural network, genetic algorithm, expert system, and so forth [2],[4],[5],[8].

The Hidden Markov Model (HMM) is a statistical model widely used to represent an observation sequence data. HMM are composed of states, and a sequence data is viewed as a series of observations emitted by the states. Formally, parameters of an HMM are characterized by three stochastic matrices, called the

© Springer International Publishing Switzerland 2015

H.A. Le Thi et al. (eds.), *Model. Comput. & Optim. in Inf. Syst. & Manage. Sci.*,
Advances in Intelligent Systems and Computing 360, DOI: 10.1007/978-3-319-18167-7_31

initial, transition and observation matrices [6]. For credit scoring modeling, two state are considered good and bad borrowers, which are commonly used in the literature. Our model is based on two assumptions, one of which is that the probability of transitioning from state i to state j are not considered, therefore this matrix will be not calculated. Other assumption is concerned the traditional single observation sequence model which is extended to a multiple observation sequences. In our study, a sequence observation is related to a observed variables which characterize information of applicant. Once a model has been developed, it can be used to determine whether a new borrowers belongs to the set [9],[11].

The Baum-Welch method was applied in problem of Hidden Markov Model (HMM). Baum-Welch algorithm is used to find the unknown parameters of an hidden Markov model (HMM), It makes use of the forward-backward algorithm. The BaumWelch algorithm is a particular case of a generalized expectation-maximization (GEM) algorithm. It can compute maximum likelihood estimates and posterior mode estimates for the parameters (transition and emission probabilities) of an HMM, when given only emissions as training data. Baum-Welch is an iterative procedure, which maximize locally the probability of observation sequence [10]. Baum-Welch can be used to learned HMM parameters from a training sequence data. The algorithm starts with an initial model and iteratively updates it until convergence. The convergence is obtained when the estimation technique is guaranteed to adjust the model parameters that maximize the probability of the observation sequence. The local optimum model provided by Baum-Welch and the required iterations for credit scoring are the main discussed work in this paper. We also study the effect of initial HMM chosen as starting points for the Baum-Welch algorithm. The experiment shown that resulting HMM and the number of required iterations depend heavily on the initial model.

2   HMM Background

Baum-Welch algorithm is recognized to solve the third problem of HMM. It based on a combination of two iterative algorithm : forward and backward procedure [10]. In this section, we introduce the form of an Hidden Markov Model, and the estimation procedure of Baum-welch.

2.1   Elements of an HMM

A hidden Markov model (HMM) is a bivariate discrete time process $\{X_k, Y_k, k \geq 0\}$, where $\{X_k\}$ is a Markov chain and, conditional on $\{X_k\}$, $\{Y_k\}$ is a sequence of independent random variables such that the conditional distribution of Y_k only depends on X_k. An HMM has a finite set of states governed by a set of transition probabilities. In a particular state, an outcome or observation can be generated according to an associated probability distribution. It is only the outcome and not the state that is visible to an external observer [5].

Generally, an HMM is characterized by the following assumptions: N, the number of states in the model. M, the number of distinct observations symbols per state. The state transition probability distribution $A = \{a_{ij}\}$. The observation symbol probability distribution in state j, $B = \{b_j(k)\}$. The initial state distribution $\pi = \{\pi_i\}$. A complete specification of an HMM requires the determination of two model parameters, N and M, and estimation of three probability distributions A, B, and π. We use the notation $\lambda = (A, B, \pi)$ to indicate the complete set of parameters of the model.

Given an observation sequence $O = (O_1, O_2, \ldots, O_T)$ and an HMM $\lambda = (A, B, \pi)$. The iterative algorithm proposed by Baum-Welch for reestimation (iterative update and improvement) of HMM parameters, is a mixture of the two iterative procedure, named as Forward-Backward procedure. It used here to compute model parameters λ.

2.2   Baum-Welch Algorithm: Parameter Estimation

The most difficult problem about HMM is that of the parameter estimation. Given an observation sequence $O = (O_1, O_2, \ldots, O_T)$, we want to find the model parameters $\lambda = (A, B, \pi)$ that best explains the observation sequence. The problem can be reformulated as find the parameters that maximize the following probability:

$$argmax_\lambda \mathbb{P}(O|\lambda) \tag{1}$$

Where $\mathbb{P}(O|\lambda)$ is the probability of the observation sequence given the model λ. There is no known analytic method to choose λ to maximize $\mathbb{P}(O|\lambda)$ but we can use a local maximization algorithm to find the highest probability. This algorithm is called the Baum-Welch. It works iteratively to improve the likelihood of $\mathbb{P}(O|\lambda)$. The Baum-Welch algorithm is numerically stable with the likelihood nondecreasing of each iteration. It has linear convergence to a local optima.

Given the model $\lambda = (A, B, \pi)$ and the observation sequence $O = (O_1, O_2, \ldots, O_T)$, we can start training the Baum-Welch method. The training algorithm has the following steps : 1) Initialization of HMM parameters : A, B and π., 2) forward procedure, 3) backward procedure, and 4) reestimation procedure of HMM parameters. Details of these steps can be found in [10]. The Baum-Welch reestimation formulas, based on the concept of counting event occurrences, are described and given by the following expressions:

$$\hat{\pi}_i = expected\ frequency\ (number\ of\ times)\ in\ state\ i \tag{2}$$

$$\hat{a_{ij}} = \frac{expected\ number\ of\ transitions\ from\ state\ i\ to\ state\ j}{expected\ number\ of transitions\ from\ state\ i} \tag{3}$$

$$\hat{b_j}(k) = \frac{expected\ number\ of\ transitions\ from\ state\ j\ and\ observing\ symbol\ k}{expected\ number\ of\ transitions\ from\ state\ j} \tag{4}$$

3   Baum-Welch Algorithm for Credit Scoring

With the rapid development of consumption credit market, and finance industry, the importance of personal credit scoring has become more and more significant nowadays. credit scoring is used to evaluate and reflect a number of practical attribute indicators (e.g. age, account, residents, job, income, etc) of consumers or credit appliers. Some characteristics will be used in the model to estimate the probability to be a good or a bad borrows. In this section, we explain how Baum-Welch procedure can be used for credit scoring problems. In the following, we describe briefly the HMM-model as has been developed by Benyacoub et al. [3], and we show how it applied in credit scoring.

3.1   HMM Model Estimation

In credit scoring, the information provided by a borrower is used to classify him as a good customer or as a bad customer. This information is characterized by a explanatory variables $Y = \{Y^k, k = 1, \ldots, l\}$. Each applicant is evaluated by a risk level X. The risk X is indicated by a set $S = \{e_1, e_2\}$; where e_1 denotes the level of good customers, e_2 denotes the level of bad customers. The probability $p_i = \mathbb{P}(X = e_i)$, $i = 1, 2$, to be in one class estimates the risk level e_1 and e_2. We present a method to model the two probabilities p_1 and p_2 by observed characteristics given by an expert or a banker.

We suppose that the two levels are an hidden states for the banker. However, we use the vector Y as an observed process to evaluate the situation of each member. We obtain a system of linear equations, where the two probabilities p_1, p_2, are related to a set of potential predictor variables $Y = \{Y^k, k = 1, \ldots, l\}$ in the form:

$$\mathbb{P}(X = e_i) = u_{11}^i \mathbb{P}(Y^1 = f_{11}) + \ldots + u_{jk}^i \mathbb{P}(Y^l = f_{jk}) \tag{5}$$
$$+ \ldots + u_{m(l)M}^i \mathbb{P}(Y^l = f_{m(l)l}). \qquad 1 \leq i \leq 2$$

where $Y^l \in \{f_{1k}, f_{2k}, \ldots, f_{m(k)k}\}$, for $1 \leq k \leq l$, and u_{jk}^i, $1 \leq j \leq m(k)$ represents the coefficients associated with the corresponding probability to have the categorical f_{jk} of variable $Y(k)$. The coefficients of corresponding parameters $\{u_{jk}^i\}$ have obtained from the observation equation [3]:

$$X = K^{-1}C^t Y - K^{-1}C^t W \tag{6}$$

where C matrix of probabilities $c_{ji}(k) = \mathbb{P}(Y^k = f_{jk}|X = e_i)$ represent the observation symbol probability distribution. W is vector of error estimation. The matrix K is defined as $K = C^t C$, (t indicate transpose matrix). The matrix C is estimated from dataset by using the formula (4) given above, and the matrix $U = K^{-1}C^t$ gives the model parameters.

X and Y are defined as following :

$$Y = \begin{pmatrix} \{Y^1 = f_{11}\} \\ \cdots \\ \{Y^k = f_{jk}\} \\ \cdots \\ \{Y^l = f_{m(l)l}\} \end{pmatrix} \quad X = \begin{pmatrix} \{X = e_1\} \\ \{X = e_2\} \end{pmatrix}$$

So, the equation (5) can be expressed as follows :

$$\{X = e_i\} = \sum_{q=1}^{p} \sum_{j=1}^{m(q)} u_{jq}^i \{Y^k = f_{jk}\}, \quad i = 1, 2 \tag{7}$$

if we calculate the expectation of each equation presented in expression (6), and as we have $\mathbb{E}[K^{-1}C^t W] = K^{-1}C^t \mathbb{E}[W] = 0$. We obtain probability estimation equation correspond to each risk level as exposed before in the presented system (5).

3.2 Model Parameter Estimation and Training

Given a credit scoring dataset $D_i = \{y_{i1}, y_{i2}, \ldots, y_{il}\}, 1 \leq i \leq n$, each individual i has p attributes such age, account, job, income, residents and so on. The risk level X uses 1 or 0 to denote the type of customers good or bad. It is the target that the model would like to estimate and predict. Here, the paper uses the basic principles of hidden markov model, since the iterative method of Baum-Welch can help to predict good or bad customers.

The Proposed Scoring Model. Our work is to estimate the tow probabilities p_1, p_2, from the data; p_1 (resp p_2) represent the probability to be a good borrow (to be a bad borrow). The state space as defined in section 2 is limited to $S = \{e_1, e_2\}$. Further, each continuous characteristic Y^k will be converted to discrete categorical values. Here, the matrix C is a $m \times 2$ matrix. The two column of matrix C are linearly independent. Therefore, K is an inverse matrix.

As mentioned in the above section, the probability to be a good solvent or bad solvent will be evaluated by a linear relation using observed characteristics. The sum of associated coefficients, correspond to the categories that characterize a borrower in each equation, estimates the two probabilities p_1, p_2. Then we classify it to be a good or bad applicant according to greater probability [3]. In practices, we use a historical credit data to train and test the model.

Our problem is defined in credit scoring union. Each observation symbol $O_t, t = 1, 2, \ldots, T$ from a sequence of observation correspond to a vector of observed data $Y_i = (Y_i^1, Y_i^2, \ldots, Y_i^l), i = 1, 2, \ldots, n$, where n is size of a sample for training model. Datasets are organized such that all the data are represented in discrete categorical values, and to be suitable in HMM form. If each attribute $Y(k), 1 \leq k \leq p$ has m_k categorical classes, then the total number of output symbols for our model is the sum $m_1, m_2, \ldots, m_p$.

Here, we describe the scheduled values of the variants and use the tool of MAT-LAB 7.5 to compute model's results. In general, parameters of hidden markov model are presented by three element: $\lambda = (A, B, \pi)$. After the HMM parameters of initial model $\lambda^0 = (A^0, B^0, \pi^0)$ are learned from instances of trained dataset, we can use iteratively the Baum-Welch procedure to obtain a sequence of HMM models $\lambda^1, \lambda^2, \ldots, \lambda^l$.

In this paper, each element X of the sequence $\{X_t, \ t = 1, 2, \ldots, n\}$ is independent via the presence of other observations. Then, the matrix A which represent state transition probability is not involved in this work. Also, the initial state distribution vector doesn't considered here. Consequently, the HMM-model contains just the matrix B, which represented in our study by the matrix C indicated in before section.

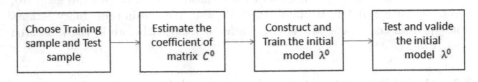

Fig. 1. Initialization

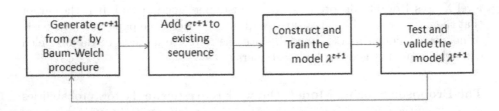

Fig. 2. Iterative procedure

Iterative Generation of Models. An HMM is initially trained with a sequence of observation $O = (O_1, O_2, \ldots, O_T)$. The sequence is obtained from a process of observed characteristics. We use this sequence to estimates the matrix C. Then we train the first model λ^0. Figure.1 shows the complete process of initialization Baum-Welch procedure. As shown in the figure.1, the data is divided in two parts, one for training, and the other for testing. The initialization step consist to construct the first model λ^0 and test it.

Baum-Welch is an iterative algorithm to compute HMM parameter. The whole iterative phase can be depicted as shown in figure.2. Once the process runs, we obtain a new model and we test it accuracy. By repeating this process, we hope to converge to the optimal parameters values for our problem.

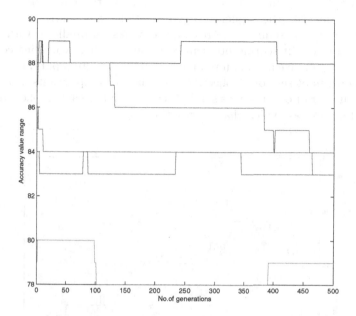

Fig. 3. Accuracy values of models in 1000 generations for Australian data

4   Experimental Results, Analysis and Discussions

The model described in this study were applied in Credit scoring problem. The proposed model are trained with two datasets and used to evaluate the performance of the building HMM/Baum-Welch model: German credit dataset and Australian credit dataset. In this section we present how the developed model can be used in credit scoring environment union.

4.1   Credit Datasets Description

The model developed in this study were performed by two popular datasets in credit scoring: German credit dataset [7] and Australian credit dataset [1]. The two datasets are from UCI machine learning repository. The German dataset contains 1000 applicant of a German bank, where 700 of them are considered as good borrowers while 307 are bad borrowers. Number of Attributes are fixed in 20, which 7 numerical, and 13 categorical. The Australian dataset consist of a set of 690 applicants, where 307 of them are qualified as creditworthy and 383 are not creditworthy. In this set, there are 6 numerical and 8 categorical attributes. Data preprocess is the first step for creating a model for credit scoring. The data are organized to be in suitable form of our HMM/Baum-Welch model.

Our process began to discretize all continuous variables into categorical observations before building the models. Given, a dataset including l attributes, the first attribute contains m_1 classes, the second attribute contains m_2, and so on. Consequently, the sequence of observations was a string of alphabets. Each alphabet in the sequence correspond to a class of predictor variable. Each dataset used in credit scoring problem is a matrix, where each line correspond to an applicant and each column to an attribute. Thus, an input data to the our problem consists of sets of a collection information of applicants arranged in a row. The number of observations symbols correspond to each applicants was the number of attributes, which characterize the dataset.

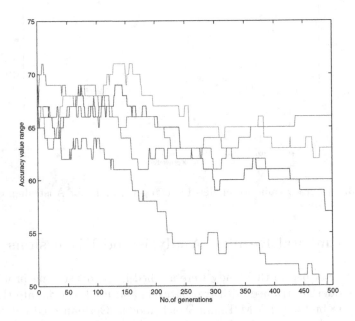

Fig. 4. Accuracy values of models in 1000 generations for German data

4.2   Results and Analysis

In credit scoring problem, the performance of each model developed was measured in the term of average accuracy. Accuracy was the number of correctly classifying cases under the total number cases in a testing set. The dataset is divided into a training set and testing set. For German dataset, we choose 900 instances randomly to train model and the 100 remainder instances is used to test the model. For Australian dataset, 500 instances are selected randomly for training phase and 100 instances for testing. The program used for our experiments was developed in Matlab language. At each iteration, the program provides the accuracy of obtained HMM/Baum-Welch model. To provide a robust model, we generate 1000 models iteratively.

Figure.3 and Figure.4 present the experimental results of the HMM/Baum-Welch model after 1000 generations. The results is obtained for 5 repetitions.

As be shown from the two figures, the accuracy of models began to decrease after 200 iterations for Australian dataset, and after 100 iterations for German dataset. We can conclude from results of 5 repetitions presented in Figure.3 and Figure.4, that the best model can be obtained in a few generations models after initial model.

As mentioned below, The Baum-Welch algorithm starts from an initial model. As shown in in the Figure.3 and Figure.4, the resulting HMM/Baum-Welch and the number of required iterations depend on the chosen initial model. Figure.3 describes the Accuracy variation of the German data and Figure.4 for Australian data. The implementation of iterative procedure defined in section.3 provides several important assumptions, according the experimental results plotted in Figure.3 and Figure.4. We can conclude that, the HMM/Baum-Welch model tend to stabilize in a certain ranges at different generations. As shown, the application of HMM/Baum-Welch algorithm for two dateset is numerically stable with the nondecreasing of accuracy value in a successive iterations, and characterize a linear convergence to a local optima. We have also a convergence to the same value of a local optimum at different times. Therefore, we can suppose that each range represent a set of models give the same value of accuracy, so we can define a class of models as solution of our problem instead one model. Numerically, the set of matrix correspond to the set of models define one class, have approximatively the same value of the determinant.

Since the Baum-Welch algorithm is a local iterative method, the resulting HMM and the number of required iterations depend heavily on the initial model. The effect of the initial model on the result of the Baum-Welch algorithm is

Table 1. Comparison of average accuracy of initial models over German dataset

DATASET	Best Value	Mean Value	Worst Value
$\lambda^o=92$	94	91	88
$\lambda^o=90$	92	91	90
$\lambda^o=89$	91	90	89
$\lambda^o=88$	89	88	87
$\lambda^o=87$	88	87	86
$\lambda^o=86$	89	86.5	84

Table 2. Comparison of average accuracy of initial models over Australian dataset

DATASET	Best Value	Mean Value	Worst Value
$\lambda^o=77$	77	68.05	60
$\lambda^o=73$	76	68	60
$\lambda^o=72$	73	70	68
$\lambda^o=71$	71	65.09	60
$\lambda^o=70$	71	61.25	51
$\lambda^o=69$	71	61.85	52

shown in the Table.1 and Table.2 respectively. From the 5 curves represented in the two figures, we can draw the following conclusions : First, the best model obtained for each repetition depend to the initial model chosen. Second, at each time the choice of the initial model has a strong influence on the number of iterations I_{max} taken to generate models for obtain the good model.

5   Conclusion

We have tested the estimation performance of HMM/Baum-Welch applied in credit scoring environment trained from different initials models on two datasets: German and Australian. From the results, we found that the designing the initial model has an impact of reaching the global optimum. In fact, Baum-Welch procedure performs much better using a good initial model model. It converges faster, better fits the training data and further improve performance of credit scoring models. We can assume that Baum-Welch algorithm searches for a locally-optimal HMM according the result of initial model learned from data. It would also be interesting to experiment with other possible initialization strategies.

References

1. Australian Credit Database, http://ftp.ics.uci.edu/pub/machinelearning-databases/statlog/australian
2. Baesens, B.: Developing intelligent systems for credit scoring using machine learning techniques, Ph.D Thesis, Katholieke Universiteit Leuven (2003)
3. Badreddine, B., Souad, B., Abdelhak, Z., Ismail, E.: Classification with Hidden Markov Model. Applied Mathematical Sciences 8(50), 2483–2496 (2014)
4. Cox, D.R., Snell, E.J.: Analysis of binary data. Chapman & Hall, London (1989)
5. Crook, J.N., Hamilton, R., Thomas, L.C.: A comparison of discriminators under alternative definitions of credit default. In: Thomas, L.C., Crook, J.N., Edelman, D.B. (eds.) Credit Scoring and Credit Control, pp. 217–245. Clarendon, Oxford (1992)
6. Elliot, R.J., Aggoum, L., Moore, J.B.: Hidden Markov Models: Estimation and control. Springer, New York (1995)
7. German Credit Database, http://ftp.ics.uci.edu/pub/machinelearning-databases/statlog/german
8. Hand, J., Henley, W.: Statistical Classification Methods in Consumer Credit Scoring. Computer Journal of the Royal Statistical Society Series a Statistics in Society 160(3), 523–541 (1997)
9. Oguz, H.T., Gurgen, F.S.: Credit Risk Analysis Using Hidden Markov Model. In: IEEE Conferences: 23rd International Symposium on Computer and Information Sciences, ISCIS 2008 (2008), doi:10.1109/ISCIS.2008.4717932
10. Rabiner, L.R.: A tutorial on hidden markov models and selected applications in speech recognition. Proceedings of the IEEE 77, 257–286 (1989)
11. Elliot, R.J., Filinkov, A.: A self tuning model for risk estimation. Expert Systems with Applications 34, 1692–1697 (2008)

Macroeconomic Reevaluation of CNY/USD Exchange Rate: Quantitative Impact of EUR/USD Exchange Rate

Raphael Henry[*], Holy Andriamboavonjy, Jean-Baptiste Paulin, Sacha Drahy, and Robin Gourichon

ECE Paris School of Engineering, France
rhenry@ece.fr

Abstract. During past decade, Chinese monetary policy has been to maintain stability of exchange rate CNY/USD by creating parity between the two currencies. This policy, against market equilibrium, impacts the exchange rate in having low Yuan currency, and keeping attractiveness of Chinese industries. Using macroeconomic and statistic approach, the impact of such policy onto CNY/USD exchange rate is quantitatively determined. We also point out, in this paper, the influence of the EUR/USD exchange rate over the CNY/USD.

Keywords: Macroeconomics Models, Statistical approach, Yuan Floating Exchange Rate, China Banking System, Chinese Monetary Policy, Yuan/US Dollar, Hall & Taylor Theory, Market Equilibrium.

1 Introduction

During more than ten years, the Chinese monetary policy was to maintain the exchange rate Yuan/ Dollars US stable by creating parity between the two currencies. This policy was against the market equilibrium and has caused some political trouble. This strategy permits to keep the Yuan low and thanks to that to keep the attractiveness of the Chinese industries.

This year China began to change its strategy and has a long term plan to compete US Dollars as an international currency [1]. However the Chinese market and the Chinese bank are not ready for abrupt change of policy.

As one of the greatest emerging economies, China is fully committed to follow the reforms of global regulation in order to make its financial system healthier and more resistant [2]. Since January 2013, China follows an approach of the Basel III regulation [3] concerning the credit risk and the market risk, and an elementary approach on the operational risk [4]. So far no-one has investigated the consequence of the disparity of the Yuan and the Dollars US on the Chinese banking system if it wants to respect the Basel III agreements [5].

Thus, we need to build a system which permits us to determine a real exchange rate Yuan/ US Dollars without the influence of the Chinese monetary policy. The macroeconomic exchange rate model has interested many people in the previous years.

---

[*] Corresponding author.

© Springer International Publishing Switzerland 2015

H.A. Le Thi et al. (eds.), *Model. Comput. & Optim. in Inf. Syst. & Manage. Sci.*,
Advances in Intelligent Systems and Computing 360, DOI: 10.1007/978-3-319-18167-7_32

A work has already been done for the Chinese exchange rate model by using their equations issued by an analysis in detail of the economic strategy of the country. "The Yuan Value: the paradoxes of the exchange rate equilibrium" is an example of an issue of the subject. In the model used in this work paper, the equilibrium exchange rate must allow a sufficient growth to reduce the unemployment in China. This model defines explicitly the link between the exchange rate and the strategy of the Chinese economic policy. It also takes into account the under-evaluation of the Yuan. It provides us with an example of how a model can be done from an analysis of the situation of China. Indeed, many models can be adopted when modeling the exchange rate. But as China is a particular country with particular strategies, the model must suit the economy policy as best as possible, in order to apply it for the issue of the flexibility of the Yuan and its consequences, especially on the Chinese banks. This model includes many hypotheses: no purchasing power parity, strategy of China of maintaining a low RMB to increase its economy, and so on.

We chose to adopt a Keynesian equilibrium model: the model of Hall & Taylor [6]. Using this model, the exchange rate is determined independently of the evolvement of the Chinese Central Bank on the Yuan. This market equilibrium exchange rate comes from the equilibrium between the real interest rate (of goods and services) and the monetary exchange rate.

In parallel, we need to determine the future value of the exchange rate Yuan/US Dollars to have future variation entirely correlated with the past fluctuation of the exchange rate. A statistical approach is therefore used, based on the random law followed by the yield of the exchange rate day by day.

2   Macroeconomic and Statistical Model

2.1   Computation of EUR/USD Exchange Rate

Euro currency is used here as a pivot to analyze the influence of Chinese monetary policy onto Yuan/USD exchange rate. To determine the impact of this policy a macroeconomic approach is developed. As one should determine a future exchange rate representing a continuation of this policy, a statistical approach will also be used. Taking Euro currency as a pivot forces to examine Euro/USD exchange rate with the two approaches. In the following, the statistical approach will be called the S model and the macroeconomic approach the M model. The S approach is used with the method of generalized least squares yields applied on Gauss Newton law [7]. It is used here, to predict the future exchange rate modeled with a Monte Carlo method. For the macroeconomic model, an approach using Hall & Taylor (HT) model [8] is used to see the different factors mostly affecting exchange rate changes. It is a linear statistical model, taking into account external constraints and economic policy. Unlike other models, HT model includes deeper structural parameters [9] describing the preferences of inflation, productivity and other variables of constraints [10]. The model is based on the theory of general equilibrium [11] which describes the balance between supply and demand on the market. HT model contains 12 equations with several parameters and economic factors. The macroeconomic HT model is based on 2014 exchange rate historical data to calibrate some parameters, and it is extended for years 2015 and 2016. The model is a monthly model.

Identifying variables determining the exchange rate is an indispensable preliminary work. Then, the endogenous variable that adjusts to the conditions must be well chosen. As the time scale is the month (short-term modeling), major variables such as investment, GDP and savings can't play this role. Thus, there is the interest rate, the exchange rate, the rate of tax. The choice of an "instrumental variable" which we play on is equally important.

Concerning the adjustment variable, the interest rate often plays this role in many countries where the credit system is not really developed and the economy less indebted than the United States or France. After handling equations Hall & Taylor, these equations are obtained:

$$E.P/P_w = P$$
$$\pi^e = \alpha.\pi_{-1} + \beta.\pi_{-2}$$
$$\pi = \pi^e + f.((Y_{-1} - Y_N) / Y_N)$$
$$P = P_{-1}(1+\pi)$$
$$Y = Y_0$$
$$X-M = g - mY - n(q+vr_1)$$
$$M = M_0 - X-M/P_w$$
$$G = G_{annual} / 12$$
$$r_1 = (e+b+g-nq-Y(1-a(1-t)+m)+G) / (d+nv)$$
$$r_2 = (kY - M_{adjusted} / P) / h$$

Endogenous variables	
C_t	Consumption
Y_t	Production
π_t	Inflation index
r_t	Nominal interest rate
I	Investment
E.P/Pw	Expected value of the interest rate (average)
Exogenous variables	
X_t	Exports
M_t	Imports
M	Money
G	Budgetary expenditures
π^e	Expected inflation
π_{-1}	Inflation in the previous quarter
π_{-2}	Inflation in the quarter t-2
Y_{-1}	Production of the previous quarter
Y_N	Production in the quarter
P_t	Price level

Fig. 1. Equations representing the relationship between each macroeconomic variable after processing Hall & Taylor model and above the table introducing the endogenous and exogenous variables

To keep a good macroeconomic model, the following points had to be carefully checked for the model of Hal & Taylor before and during the study: assumptions and simplifications must be reasonable given the particular economy that China is; the model meets the rules of internal logic; it can be used to study the problem of forecasting the exchange rate; forecasts and the observed facts are comparable; Finally, there is no major contradiction when comparing the implications of the model to the facts.

We need to constantly think back and forth between the theoretical model and the facts to calibrate the parameters and analyze if the results are consistent. It took most of the time for computing the macroeconomic model.

The calibration of the model parameters was indeed difficult at first. Analysis of historical and current data and a review of the equations to be used as the work progresses have achieved a satisfactory final calibration.

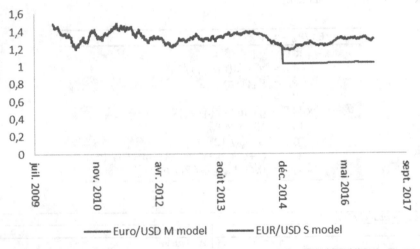

Fig. 2. Comparison of S and M Models Computation of Euro/USD Exchange Rate vs. Time

The largest differences between the two curves are indicated on Tables I and II below:

Table 1. Comparison of Models M and S Computation of Euro/USD Exchange Rate by Value

Largest Difference (Value)	Model M
Model S	0,295595119

Table 2. Comparison of Models M and S Computation of Euro/USD Exchange Rate by percentage

Largest Difference (%)	Model M
Model S	22,236%

It is important to notice that a difference of about 0.3 points exists between models S and M. It is seen that model M shows that there is an equal parity between Euro and US Dollar currencies. It can be deduced from this result that if equilibrium of prices level between Euro zone and USA is respected, it will imply a stable EUR/USD exchange rate around 1.

2.2   Computation of CNY/USD Exchange Rate

From S and M models to compute CNY/EUR and EUR/USD exchange rates, resulting CNY/USD exchange rate can be simulated. CNY/USD exchange rate simulated with both EUR/USD and CNY/USD calculated with S model, and CNY/USD exchange rate simulated with both EUR/USD and CNY/USD calculated with M model are displayed on Figure2.

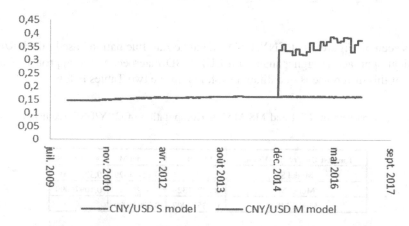

Fig. 3. Comparison of S and M Models Computation of CNY/USD Exchange Rate vs. Time

From the two curves, differences between the two models are ranging between 0,227 and 0,155. In the M model curve, the maximum value is 0.389 from 2015 simulation, and the minimum value is 0.319. Compared to S model maximum value of 0.165, the macroeconomic CNY/USD exchange rate is more than twice the statistical value computed in the future.

2.3   Impact of EUR/USD Exchange Rate onto CNY/USD Exchange Rate

Because of the difference between Euro/USD exchange rate computations, a CNY/USD exchange rate has been created with a Euro/CNY exchange rate computed with M model, Euro/USD exchange rate being computed with S model. This CNY/USD exchange rate will be called the MS model for the rest of the paper.

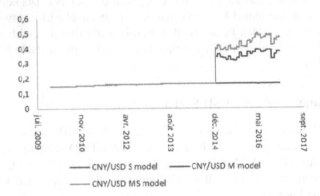

Fig. 4. Comparison of S, M and MS Models Computation of CNY/USD Exchange Rate vs. Time

It is seen on Figure 3that CNY/USD exchange rate fluctuation based on EUR/USD statistical approach is higher than with EUR/USD macroeconomic approach. An example of this difference is quantitatively shown in the two Tables below.

Table 3. Comparison of S, M and MS Models Computation of CNY/USD Exchange Rate by Value

Largest Difference (Value)	Model S	Model M	Model MS
Model S	X	0,22751329	0,33577907
Model M	0,22751329	X	0,10976099
Model MS	0,33577907	0,10976099	X

Table 4. Comparison of S, M and MS Models Computation of CNY/USD Exchange Rate by Percentage

Largest difference by %	Model S	Model M	Model MS
Model S	X	141,00%	206,96%
Model M	141,00%	X	28,59%
Model MS	206,96%	28,59%	X

Here it can be observed on Table IV that MS simulation is 28.59% higher than M simulation. It can be deduced from this analysis that the model used to simulate EUR/USD exchange rate impacts CNY/USD exchange rate simulation over about 30%. Knowing that EUR/USD exchange rate should tend to an equal parity, the model used to simulate CNY/USD dollar exchange rate should be the M model, see Figure 1. The macroeconomic equilibrium between Chinese and US markets should make CNY/USD exchange rate tend to 141% its actual value.

3   Conclusion

Chinese policy on CNY/USD exchange rate to maintain low Yuan currency is nowadays under question. Calculated macroeconomic exchange rate indicates that Yuan currency will suffer a reevaluation of more than twice its actual cost. These results show how much Chinese monetary policy controls actual value of their currency. The impact of EUR/USD exchange rate onto CNY/USD exchange rate is also around 30%. Nowadays, if Chinese monetary policy is to stop pegging Yuan to US Dollar exchange rate, it will also have to take into account the possible parity between Euro and US Dollar. In fact, ECB will probably bring an equal parity EUR/USD with a quantitative easing strategy. This piece of information converges to present macroeconomic equilibrium result. The theoretical CNY/USD exchange rate calculated without intervention of Chinese policy should be determined with an equal parity EUR/USD.

Acknowledgments. The authors are very much indebted to ECE Paris School of Engineering for having provided the environment where the work has been developed, to Pr Pham-Hi for valuable advices and to Pr M. Cotsaftis for help in preparation of the manuscript.

References

1. Bénassy-Quéré, A., Forouheshfar, Y.: The Impact of Yuan Internationalization on the Euro-Dollar Exchange Rate, CESIFO Working Paper Series No. 4149 (March 13, 2013)
2. Kwan, C.H.: China's Unfinished Yuan Reform - Further Appreciation Needed to Redress External Imbalances. Nomura Capital Market Review 9(3) (2006)
3. BCBS: Basel III:A Global Regulatory Framework for more Resilient Banks and Banking Systems, Bank for International Settlements, Basel, Switzerland (2011)
4. RCAP, Assessment of Basel III Regulations – China, Bank for International Settlements, Basel, Switzerland (September 2013)
5. BCBS: Capital Requirements for Bank Exposures to Central Counterparties, Bank for International Settlements, Basel, Switzerland (2012)
6. Macdonald, R., Taylor, M.P.: The Monetary Approach to the Exchange Rate: Rational Expectations, Long-Run Equilibrium, and Forecasting. Staff Papers (International Monetary Fund) 40(1), 89–107 (1993)
7. Hartley, H.O.: The modified Gauss-Newton method for the fitting of non-linear regression functions by least squares. Technometrics 3(2), 269–280 (1961)
8. Macdonald, R., Taylor, M.P.: The Monetary Model of the Exchange Rate: Long-Run Relationships, Short-Run Dynamics and How to Beat a Random Walk. J. International Money and Finance 13(3), 276–290 (1994)
9. Trading Economics, Chinese Monetary Supply,
   http://fr.tradingeconomics.com/china/money-supply-m2
10. Fischer, S., Hall, R.E., et al.: Relative Shocks, Relative Price Variability, and Inflation. Brookings Papers on Economic Activity 1981(2), 381–441 (1981)
11. Frenkel, R., Taylor, L.: Real Exchange Rate, Monetary Policy and Employment: Economic Development in a Garden of Forking Paths, Political Economy Research Institute (September 2006)

Optimal Discrete Hedging in Garman-Kohlhagen Model with Liquidity Risk

Thanh Duong[1], Quyen Ho[2], An Tran[3], and Minh Tran[4]

[1] Quantitative and Computational Finance Lab, John von Neumann, HCMC, Vietnam
thanh.duong@jvn.edu.vn
[2] Department of Mathematics, University of Architecture, HCMC, Vietnam
thucquyen911@gmail.com
[3] Department of Mathematics, University of Science, HCMC, Vietnam
tranvqan@gmail.com
[4] Quantitative and Computational Finance, John von Neumann, HCMC, Vietnam
minh.tran.qcf@jvn.edu.vn

Abstract. In this paper, we study a discrete time hedging and pricing problem using Garman-Kohlhagen model in a market with liquidity costs. We prove that delta hedging is an unique optimal strategy. In particular, the hedging strategy will have expected hedging error is the infinitesimal of the length of the revision interval with order of 3/2. An implicit finite difference method is presented and showed to be stable for solving the PDE required to obtain the option price. Finally, some experiments illustrate the efficiency of our method.

Keywords: discrete time, Garman-Kohlhagen model, liquidity cost, delta hedging.

1 Introduction

Nowadays, liquidity risk is the additional significant risk in the market due to the timing and size a trade. The price process may depend on the activities of traders, especially the trading volume. In the past decade or two, the literature on liquidity risk has been growing rapidly (see [1–10]). These authors developed rigorous models incorporating liquidity risk into the arbitrage pricing theory. They established a mathematical formulation of liquidity costs, admissible strategies, self-financing strategies, and an approximately complete market. They showed the two (approximately modified) fundamental theorems of finance hold under the existence of liquidity risk. They also studied an extension of the Black-Scholes economy incorporating liquidity risk as an illustration of the theory.

In this paper, we study how the classical hedging strategies should be modified and how the prices of derivatives should be changed in a financial market with liquidity costs. We consider a discrete time version of the Garman-Kohlhagen model and a multiplicative supply curve without assuming that the interest rate is zero (see [10], [11], [12]). We prove that delta hedging is still the optimal strategy in the presence of liquidity risks. Using the Leland approximation scheme (see [13]), we obtain a nonlinear partial differential equation which requires the expected hedging error converging to

© Springer International Publishing Switzerland 2015
H.A. Le Thi et al. (eds.), *Model. Comput. & Optim. in Inf. Syst. & Manage. Sci.*,
Advances in Intelligent Systems and Computing 360, DOI: 10.1007/978-3-319-18167-7_33

zero at number of hedging time goes to infinity. We provide an approximate method for solving this equation using a series solution. For an introduction to the theory of Probability and finite difference methods for PDE the reader is preferred to [14–16].

The remaining of the paper is organized as follows. Section 2 introduces the Garman-Kohlhagen model. Section 3 presents the optimal hedging strategy and a pricing PDE. Section 4 studies an analytic solution to the pricing PDE and numerical results are also provided.

2 Model

Consider a situation where we have two currencies: the domestic currency (say rubles), and the foreign currency (say US dollars). The spot exchange rate at time t is denoted by $S(t)$, and by definition it is quoted as

$$\frac{\text{units of the domestic currency}}{\text{unit of the foreign currency}},$$

i.e. in our example it is quoted as rubles per dollar. We assume that the domestic short rate r_D, as well as the foreign short rate r_F, are deterministic constants, and we denote the corresponding riskless asset prices by B_D and B_F respectively. Furthermore, we assume that the exchange rate is modelled by geometric Brownian motion. We can summarize this as follows.

Assumption 1. *We take as given the following dynamics (under the objective probability measure P):*

$$dS = S\alpha_S dt + S\sigma_S d\bar{W},$$
$$dB_D = r_D B_D dt,$$
$$dB_F = r_F B_F dt,$$

where α_S, σ_S are deterministic constant, and $\bar{W}$ is a scalar Wiener process.

Since $B_F(t)$ units of the foreign currency are worth $B_F(t) \cdot S(t)$ in the domestic currency, we immediately have the following result (see [12], pp. 248). Trading in the foreign currency is equivalent to trading in a domestic market with the domestic price process $\tilde{B}_F$, where

$$\tilde{B}_F(t) = B_F(t) \cdot S(t).$$

The dynamics of $\tilde{B}_F$ are given by

$$d\tilde{B}_F = \tilde{B}_F(\alpha_S + r_F)dt + \tilde{B}_F \sigma_S d\bar{W}. \tag{1}$$

We see that our currency model is equivalent to a model of a domestic market consisting of the assets B_D and $\tilde{B}_F$. It now follows directly from the general results that the martingale measure Q has the property that the Q-dynamics of $\tilde{B}_F$ are given by

$$d\tilde{B}_F = r_D \tilde{B}_F dt + \tilde{B}_F \sigma_S dW, \tag{2}$$

where W is a Q-Wiener process. Since by definition we have

$$S(t) = \frac{\tilde{B}_F}{B_F(t)},$$

From (1), (2), we can use Itô's formula to obtain the Q-dynamics of S as

$$dS_t = r_D S_t dt - r_F S_t dt + \sigma S_t dW_t.$$

3 Optimal Hedging Strategy

Let $S(t,x)$ represent the spot price of the deliverable currency at time $t \in [0,T]$ that the trader pays/receives for an order of size $x \in \mathbb{R}$. A positive order $x > 0$ represents a buy, a negative order $x < 0$ represents a sale, and the order $x = 0$ corresponds to the marginal trade. This means that $S(t,0) = S_t$ follows a geometric Brownian motion.

$$dS_t = r_D S_t dt - r_F S_t dt + \sigma S_t dW_t, \quad 0 \le t \le T, \tag{3}$$

where W is a standard Brownian motion, and T is the terminal time of an European contingent claim C $(C = g(S_T))$ for some function g of interest. In this study, we are concerned with discrete time hedging and pricing. Let us consider equally spaced times $0 = t_0 \le t_1 \le t_2 \le ... \le t_n = T$. Set $\Delta t = t_i - t_{i-1}$ for $i = 1, ..., n$. We consider the following discrete time version of (3)

$$\frac{\Delta S}{S} = r_D \Delta t - r_F \Delta t + \sigma Z \sqrt{\Delta t}, \tag{4}$$

where Z is a standard normal variable, and assume a multiplicative supply curve

$$S(t,x) = f(x)S(t,0),$$

where f is a smooth and increasing function with $f(0) = 1$.

Because we consider discrete time trading, the total liquidity costs up to time T is

$$L_T = \sum_{0 \le u \le T} \Delta X_u [S(u, \Delta X_u) - S(u,0)], \tag{5}$$

where $L_{0-} = 0$, and $L_0 = X_0 [S(0, X_0) - S(0,0)]$ (see [2]). Note that the liquidity cost is always non-negative, since $S(t,x)$ is an increasing function of x. Then, the total liquidity costs up to time T could rewrite as

$$L_T = \sum_{i=1}^{n} \Delta X_i [S(t_i, \Delta X_i) - S(t_i,0)] + X_0 [S(0, X_0) - S(0,0)], \tag{6}$$

where $\Delta X_i = X_{t_i} - X_{t_{i-1}}$. X_t represents the trader's aggregate stock holding at time t (unit of money market account). Here, X_t is predictable and optional processes with $X_{0-} \equiv 0$.

When we buy dollars, these are put into a bank account, giving the interest r_F. Buying a currency is like buying a dividend-paying stock with a dividend yield $q = r_F$. Thus, XS units of the foreign currency is worth $r_F XS$ in the domestic currency. We let C_0 denote the value at time 0 of contingent claim C so that the hedging error inclusive of liquidity costs is

$$H = \sum_{i=0}^{n-1} X_{t_i}(S_{t_{i+1}} - S_{t_i}) + r_F X S \Delta t - C + C_0 - L_T - \Delta B, \tag{7}$$

where B is money market account with

$$\Delta B = r_D B \Delta t = r_D (XS - C) \Delta t.$$

Let us consider a European call option C expiring at T with strike price K, and a hedging strategy X. Let $0 = t_0 \leq t_1 \leq \ldots \leq t_n = T$ be equally spaced trading times and $\Delta t = t_i - t_{i-1}$ for $i = 1, \ldots, n$. We now aim to find a hedging strategy to minimize the hedging error. The hedging error over each revision interval is:

$$\Delta H = X \Delta S + r_F X S \Delta t - \Delta B - \Delta C - \Delta X [S(t, \Delta X) - S(t, 0)].$$

Recall that

$$\Delta S = r_D S \Delta t - r_F S \Delta t + \sigma S Z \sqrt{\Delta t},$$

$$(\Delta S)^2 = \sigma^2 S^2 Z^2 \Delta t + \mathscr{O}\left(\Delta t^{3/2}\right),$$

$$(\Delta S)^k = \mathscr{O}\left(\Delta t^{3/2}\right), \quad k = 3, 4, 5, \ldots.$$

Over the small time interval $[t_{i-1}, t_i], i = \overline{1, n}$ we consider the change in the call option value is:

$$\Delta C = C(S + \Delta S, t + \Delta t) - C(S, t)$$

$$= C_S \Delta S + C_t \Delta t + \frac{1}{2} C_{SS} \sigma^2 S^2 Z^2 \Delta t + \mathscr{O}\left(\Delta t^{3/2}\right).$$

The liquidity cost at each interval is:

$$\Delta X (S(t, \Delta X) - S(t, 0)) = \Delta X (f(\Delta X) - 1) S(t, 0).$$

By Taylor expansion with $f(0) = 1$ we have

$$f(\Delta X) - 1 = f(\Delta X) - f(0) = f'(0)\Delta X + \frac{f''(0)}{2}(\Delta X)^2 + \mathscr{O}\left((\Delta X)^3\right).$$

The liquidity cost becomes

$$\Delta X (S(t, \Delta X) - S(t, 0)) = \Delta X \left(f'(0)\Delta X + \frac{f''(0)}{2}(\Delta X)^2\right) S + \mathscr{O}\left(\Delta t^{3/2}\right)$$

$$= f'(0) S (\Delta X)^2 + \mathscr{O}\left(\Delta t^{3/2}\right).$$

Therefore,

$$\Delta H = \Delta S(X - C_S) + \Delta t \left(-C_t + r_F X S - \frac{1}{2}\sigma^2 S^2 Z^2 C_{SS} - r_D(XS - C)\right)$$

$$- f'(0)S(\Delta X)^2 + \mathscr{O}\left(\Delta t^{3/2}\right).$$

A perfect hedging strategy will produce a zero hedging error with probability 1, see [17], [18]. But, considering discrete trading and liquidity costs, it is not possible to produce a strategy whose hedging error equals 0. We now define a strategy which is perfect in discrete time with liquidity risk.

A *perfect discrete time hedging strategy* is a strategy which has expected hedging error is the infinitesimal of the length of the revision interval with order of 3/2, which means $E(\Delta H) \sim \mathcal{O}(\Delta t^{3/2})$. The following theorem will give us the perfect discrete time hedging strategy of our model and then derives a partial differential equation for pricing option.

Theorem 1. *If* $X = C_{S|S=S_{t_i}}$ *and* $C(S,t)$ *is the solution of the following partial differential equation*

$$C_t + \frac{1}{2}\sigma^2 S^2 C_{SS} + (r_D - r_F)SC_S - r_D C + f'(0)S_i^3\sigma^2 C_{SS}^2 = 0, \tag{8}$$

for all $t \in [0,T), S \geq 0$, *with the terminal condition* $C(S,T) = (S-K)^+$. *Then,* X *is the only perfect discrete time hedging strategy with liquidity costs.*

Proof. Part 1. Assuming that X is the perfect discrete time hedging strategy with liquidity costs, which means $E(\Delta H) \sim \mathcal{O}(\Delta t^{3/2})$, we now prove $X = C_{S|S=S_{t_i}}$ and $C(S,t)$ is the solution of (8). Taking expectation, we have

$$E(\Delta H) = \Delta S(X - C_S) + \Delta t \left(-C_t + r_F X S - \frac{1}{2}\sigma^2 S_t^2 C_{SS} - r_D(XS - C)\right)$$
$$- E\left[(\Delta X)^2\right]f'(0)S + \mathcal{O}\left(\Delta t^{3/2}\right). \tag{9}$$

Let $X = g(C_S, t)$, by Taylor expansion, we get

$$g(C_S + \Delta C_S, t + \Delta t) = g(C_S, t) + \frac{\partial g}{\partial C_S}\Delta C_S + \frac{\partial g}{\partial t}\Delta t$$
$$+ \frac{\partial^2 g}{2\partial C_S^2}(\Delta C_S)^2 + \frac{\partial^2 g}{2\partial t^2}(\Delta t)^2 + \frac{\partial^2 g}{\partial C_S \partial t}\Delta C_S \Delta t + \mathcal{O}\left(\Delta t^{3/2}\right),$$

which implies that

$$\Delta X = \frac{\partial g}{\partial C_S}\Delta C_S + \frac{\partial g}{\partial t}\Delta t + \frac{\partial^2 g}{2\partial C_S^2}(\Delta C_S)^2 + \frac{\partial^2 g}{2\partial t^2}(\Delta t)^2 + \frac{\partial^2 g}{\partial C_S \partial t}\Delta C_S \Delta t + \mathcal{O}\left(\Delta t^{3/2}\right).$$

Therefore,

$$(\Delta X)^2 = \left(\frac{\partial g}{\partial C_S}\Delta C_S\right)^2 + \left(\frac{\partial^2 g}{2\partial C_S^2}(\Delta C_S)^2\right)^2 + \mathcal{O}\left(\Delta t^{3/2}\right). \tag{10}$$

However,

$$\Delta C_S = C_{SS}\Delta S + C_{St}\Delta t + \frac{1}{2}C_{SSS}(\Delta S)^2 + \mathcal{O}\left(\Delta t^{3/2}\right),$$
$$(\Delta C_S)^2 = C_{SS}^2(\Delta S)^2 + \mathcal{O}\left(\Delta t^{3/2}\right) = C_{SS}^2\sigma^2 S_t^2 Z^2 \Delta t + \mathcal{O}\left(\Delta t^{3/2}\right), \tag{11}$$
$$(\Delta C_S)^k = \mathcal{O}\left(\Delta t^{3/2}\right), \quad k = 3,4,5,\ldots.$$

Substituting (11) to (10) and taking expectation, we obtain

$$E\left[(\Delta X)^2\right] = \left(\frac{\partial g}{\partial C_S}\right)^2 C_{SS}^2 \sigma^2 S_t^2 \Delta t + \mathcal{O}\left(\Delta t^{3/2}\right).$$

Thus, (9) becomes

$$
\begin{aligned}
E(\Delta H) &= \Delta S(X - C_S) + \Delta t\left(-C_t + r_F X S - \frac{1}{2}\sigma^2 S_t^2 C_{SS}-\right.\\
&\quad \left. - r_D(XS - C) - \left(\frac{\partial g}{\partial C_S}\right)^2 f'(0)\sigma^2 S_t^3 C_{SS}^2\right) + \mathcal{O}\left(\Delta t^{3/2}\right).
\end{aligned}
$$

Since $E(\Delta H) = \mathcal{O}(\Delta t^{3/2})$, we have

$$\Delta S(X - C_S) = 0, \tag{12}$$

and

$$\Delta t\left(-C_t + r_F X S - \frac{1}{2}\sigma^2 S_t^2 C_{SS} - r_D(XS - C) - \left(\frac{\partial g}{\partial C_S}\right)^2 f'(0)\sigma^2 S_t^3 C_{SS}^2\right) = 0.$$
$$\tag{13}$$

From (12),

$$X = C_{S|S=S_{t_i}}. \tag{14}$$

Taking differential, we have

$$\frac{\partial g}{\partial C_S} = 1.$$

Substituting to (13), we have $C(S,t)$ is the solution of the following partial differential equation

$$C_t + \frac{1}{2}\sigma^2 S^2 C_{SS} + (r_D - r_F)SC_S - r_D C + f'(0)S_t^3 \sigma^2 C_{SS}^2 = 0,$$

for all $t \in [0,T), S \geq 0$, with the terminal condition $C(S,T) = (S - K)^+$.

Part 2. Assuming $X = C_{S|S=S_{t_i}}$ and $C(S,t)$ is the solution of (8), we prove that $E(\Delta H) \sim \mathcal{O}(\Delta t^{3/2})$. Then we have X is the perfect discrete time hedging strategy with liquidity costs.

Considering the change of the hedging strategy

$$
\begin{aligned}
\Delta X &= C_S(S + \Delta S, t + \Delta t) - C(S,t)\\
&= C_{SS} S\left((r_D - r_F)\Delta t + \sigma Z\sqrt{\Delta t}\right) + C_{St}\Delta t +\\
&\qquad\qquad + \frac{1}{2}C_{SSS}\sigma^2 S^2 Z^2 \Delta t + \mathcal{O}\left(\Delta t^{3/2}\right),\\
(\Delta X^2) &= C_{SS}^2 \sigma^2 S^2 Z^2 \Delta t + \mathcal{O}\left(\Delta t^{3/2}\right),\\
(\Delta X^k) &= \mathcal{O}\left(\Delta t^{3/2}\right), \quad k = 3,4,5,\ldots.
\end{aligned}
$$

The hedging error over each revision interval is

$$
\begin{aligned}
\Delta H &= X\Delta S + r_F X S \Delta t - \Delta B - \Delta C - \Delta X (S(t,\Delta X) - S(t,0)) \\
&= \Delta S (X - C_S) - C_t \Delta t + r_F X S \Delta t - \frac{1}{2} C_{SS}(\Delta S)^2 - r_D(XS - C)\Delta t - \\
&\qquad\qquad - f'(0)C_{SS}^2 \sigma^2 S^3 Z^2 \Delta t + \mathscr{O}\left(\Delta t^{3/2}\right),
\end{aligned}
$$

Since $X = C_{S|S=S_{t_i}}$, we have

$$
\Delta H = \Delta t \left(-C_t + r_F X S - \frac{1}{2}\sigma^2 S^2 Z^2 C_{SS} - r_D(XS - C) - f'(0)C_{SS}^2 \sigma^2 S^3 Z^2 \right) +
$$
$$
+ \mathscr{O}\left(\Delta t^{3/2}\right).
$$

If C satisfies

$$
C_t + \frac{1}{2}\sigma^2 S^2 C_{SS} + (r_D - r_F)SC_S - r_D C + f'(0)S_t^3 \sigma^2 C_{SS}^2 = 0,
$$

then, by taking expectations, the expected hedging error over each revision interval is

$$
E(\Delta H) \sim \mathscr{O}(\Delta t^{3/2})
$$

$\square$

Corollary 1. *The value of the discrete time delta-hedging strategy* $(X = C_S, Y = C - C_S)$ *where C is the solution of the PDE*

$$
C_t + \frac{1}{2}\sigma^2 S^2 C_{SS} + (r_D - r_F)SC_S - r_D C + f'(0)S_t^3 \sigma^2 C_{SS}^2 = 0,
$$

converge almost surely to the payoff of the option $(S - K)^+$ *including liquidity costs, as* $\Delta t \to 0$.

Remark 1. The expected hedging error over each revision interval is expressed similarly to the one studied by our previous paper (see [10]). However, our approach to the problem is different. We defined a hedging strategy which is perfect in discrete time trading with liquidity risk and proved that the delta hedging is the only perfect discrete time hedging strategy in G-K model with liquidity risks.

4 Analytic Solution of PDE

In this section, we discuss a possible analytic solution for the pricing equation derived in Section 3. We explain briefly how to approximate the option value, including liquidity costs using a series solution.

Let $\alpha = f'(0)$. Now (8) is written as

$$
C_t + \frac{1}{2}\sigma^2 S^2 C_{SS} + \alpha \sigma^2 S^3 C_{SS}^2 + S(r_D - r_F)C_S - r_D C = 0, \tag{15}
$$

for all $t \in [0,T), S \geq 0$, with the boundary condition $C(S,T) = (S-K)^+$.

For sufficiently small $\alpha > 0$, we seek a solution in the form of

$$
\begin{aligned}
C(x,t) &= C_0(x,t) + \alpha C_1(x,t) + \alpha^2 C_2(x,t) + \cdots \\
&= C_0(x,t) + \alpha C_1(x,t) + \mathcal{O}(\alpha^2) .
\end{aligned}
\tag{16}
$$

Inserting (16) into (15), we obtain the following equations for $C_0(S,t)$ and $C_1(S,t)$:

$$
\frac{\partial C_0}{\partial t} + \frac{1}{2}\sigma^2 S^2 \frac{\partial^2 C_0}{\partial S^2} + S(r_D - r_F)\frac{\partial C_0}{\partial S} - r_D C_0 = 0 ,
$$
$$
\text{for all } t \in [0,T), S \geq 0 ,
\tag{17}
$$

with the condition $C_0(S,T) = (S-K)^+$, and

$$
\frac{\partial C_1}{\partial t} + \frac{1}{2}\sigma^2 S^2 \frac{\partial^2 C_1}{\partial S^2} + \sigma^2 S^3 \left(\frac{\partial^2 C_0}{\partial S^2}\right)^2 + S(r_D - r_F)\frac{\partial C_1}{\partial S} - r_D C_1 = 0 ,
\tag{18}
$$

for all $t \in [0,T), S \geq 0$, with the condition $C_1(S,T) = 0$.

The solution to the G-K partial differential (11) is well-known as

$$
C_0(S,t) = Se^{-r_F(T-t)}N(d+) - Ke^{-r_D(T-t)}N(d-) ,
$$

where

$$
N(y) = \frac{1}{\sqrt{2\pi}} \int_{-\infty}^{y} e^{-\frac{z^2}{2}} dz ,
$$

$$
d+ = \frac{1}{\sigma\sqrt{T-t}} \left[ln\frac{S}{K} + \left(r_D - r_F + \frac{\sigma^2}{2}\right)(T-t) \right] ,
$$

$$
d- = \frac{1}{\sigma\sqrt{T-t}} \left[ln\frac{S}{K} + \left(r_D - r_F - \frac{\sigma^2}{2}\right)(T-t) \right] .
$$

Also, it is known that

$$
\frac{\partial^2 C_0}{\partial S^2} = \frac{e^{-r_F(T-t)}}{\sigma S\sqrt{T-t}}N'(d+) = \frac{1}{\sigma S\sqrt{T-t}} \frac{e^{-\frac{1}{2}(d+)^2 - r_F(T-t)}}{\sqrt{2\pi}} .
\tag{19}
$$

Substituting (19) into (18), we obtain

$$
\frac{\partial C_1}{\partial t} + \frac{1}{2}\sigma^2 S^2 \frac{\partial^2 C_1}{\partial S^2} + S(r_D - r_F)\frac{\partial C_1}{\partial S} - r_D C_1 +
$$
$$
+ \frac{S}{2\pi(T-t)}e^{-\frac{\left[ln\frac{S}{K} + \left(r_D - r_F + \frac{\sigma^2}{2}\right)(T-t)\right]^2}{\sigma^2(T-t)} - 2r_F(T-t)} = 0 ,
\tag{20}
$$

with the condition $C_1(S,T) = 0$.

The result of this PDE can be solved by using the Implicit Finite Difference Method. The time-space grid of points is created over which the approximate solution is computed. Time domain $t \in [0,T]$ is discretized by $(N+1)$ points evenly spaced with time step h:

$$
0 = t_0 < t_1 = h < \ldots < t_N = Nh = T .
$$

The natural domain is $(-\infty, +\infty)$ is truncated to $[0, S_{\max}]$, and discretized by an $(M+1)$ point uniform grid with spacial step δS,

$$0 = S_0 < S_1 = \delta < ... < S_M = M\delta = S_{\max} \,.$$

Now we introduce some notations. We build a mesh consisting $(N+1) \times (M+1)$ points (t_i, x_j), and denote the solution of the PDE at these mesh points

$$f_{i,j} = C_1(t_i, S_j),$$

for $0 \le i \le N, 0 \le j \le M$. Taking into account the boundary conditions, we define the following row vectors

$$f_{\bar{\Omega}}^i \equiv [f_{i,j}, j = 0, ..., M]^T \,,$$
$$f_{\Omega}^i \equiv [f_{i,j}, j = 1, ..., M-1]^T \,,$$
$$f_{\partial\Omega}^i \equiv [f_{i,j}, j = 0, M]^T \,.$$

For the implicit method, (20) is discretized using the following formulas at the interior of the domain: for all $1 \le i \le N, 1 \le j \le M-1$,

$$\frac{\partial C_1}{\partial t}(t_i, S_j) = \frac{f_{i+1,j} - f_{i,j}}{h} \,,$$
$$\frac{\partial C_1}{\partial S}(t_i, S_j) = \frac{f_{i,j+1} - f_{i,j-1}}{2\delta} \,, \qquad (21)$$
$$\frac{\partial^2 C_1}{\partial S^2}(t_i, S_j) = \frac{f_{i,j+1} - 2f_{i,j} + f_{i,j-1}}{\delta^2} \,.$$

At the boundary of the domain: $\{(i,j) : (i = N) \text{ or } j = 0, M\}$,

$$C_1(t_N, S_j) = f_{N,j} = 0 \,,$$
$$C_1(t_i, S_0) = f_{i,0} = 0 \,,$$
$$C_1(t_i, S_M) = S_j - Ke^{-r(T-t_i)} \,.$$

For a given node $(i, j), 1 \le j \le M-1$, replacing F with $f_{i,j}$ and the derivatives by the above approximations

$$\frac{f_{i+1,j} - f_{i,j}}{h} + (r_D - r_F)S_j\frac{f_{i,j+1} - f_{i,j-1}}{2\delta} +$$
$$+ \frac{1}{2}\sigma^2 S_j^2 \frac{f_{i,j+1} - 2f_{i,j} + f_{i,j-1}}{\delta^2} - r_D f_{i,j} + w_{i,j} = 0 \,.$$

where $w_{i,j} = \dfrac{S_j}{2\pi(T-t_i)} e^{-\frac{\left[\ln\frac{S_j}{K} + \left(r_D - r_F + \frac{\sigma^2}{2}\right)(T-t_i)\right]^2}{\sigma^2(T-t_i)} - 2r_F(T-t_i)}$.

Denoting $a_j = \dfrac{\sigma^2 S_j^2 h}{2\delta^2}$ and $b_j = \dfrac{(r_D - r_F)hS_j}{2\delta}$, then

$$(f_{i+1,j} - f_{i,j}) + b_j(f_{i,j+1} - f_{i,j-1}) + a_j(f_{i,j+1} - 2f_{i,j} + f_{i,j-1}) -$$
$$-r_D h f_{i,j} + w_{i,j} = 0 \,.$$

By factoring identical nodes, we get

$$(f_{i+1,j} - f_{i,j}) + \underbrace{(a_j - b_j)}_{l_j} f_{i,j-1} + \underbrace{(-2a_j - rh)}_{d_j} f_{i,j} + \underbrace{(a_j + b_j)}_{u_j} f_{i,j+1} + w_{i,j} = 0.$$

The $(M-1)$ equations involving the $(M-1)$ nodes $f_{i,j}, j = 1, ..., M-1$ are then

$$
\begin{array}{cccccccccc}
f_{i,1} & - & f_{i-1,1} & + & l_1 f_{i,0} & + & d_1 f_{i,1} & + & u_1 f_{i,2} & + & w_{i,1} & = 0 \\
f_{i,2} & - & f_{i-1,2} & + & l_2 f_{i,1} & + & d_2 f_{i,2} & + & u_2 f_{i,3} & + & w_{i,2} & = 0 \\
\vdots & & \vdots & & \vdots & & \vdots & & \vdots & & \vdots & \\
f_{i,j} & - & f_{i-1,j} & + & l_j f_{i,j-1} & + & d_j f_{i,j} & + & u_j f_{i,j+1} & + & w_{i,j} & = 0 \\
\vdots & & \vdots & & \vdots & & \vdots & & \vdots & & \vdots & \\
f_{i,M-1} & - & f_{i-1,M-1} & + & l_{M-1} f_{i,M-2} & + & d_{M-1} f_{i,M-1} & + & u_{M-1} f_{i,M} & + & w_{i,M-1} & = 0.
\end{array}
$$

Using notations introduced previously, it is able to reformulate those equations as

$$f_{\Omega}^{i+1} - f_{\Omega}^{i} + \mathbf{P} f_{\bar{\Omega}}^{i} + W = 0,$$

where the matrix $\mathbf{P}$ take the following form

$$
\mathbf{P} = \begin{bmatrix}
l_1 & d_1 & u_1 & 0 & \cdots & & \cdots & 0 \\
0 & l_2 & d_2 & u_2 & & & & \vdots \\
\vdots & & l_3 & d_3 & \ddots & & & \vdots \\
\vdots & & & \ddots & \ddots & u_{M-2} & 0 \\
0 & \cdots & & \cdots & 0 & l_{M-1} & d_{M-1} & u_{M-1}
\end{bmatrix} = [B_{.,1} \quad \mathbf{A} \quad B_{.,2}],
$$

and $W = [w_{i,1} \quad w_{i,2} \quad \cdots \quad w_{i,M-1}]^T$.

By introducing two row vectors $B_{.,1}, B_{.,2}$ and the square matrix $\mathbf{A}$, we may go further

$$f_{\Omega}^{i+1} - f_{\Omega}^{i} + (\mathbf{A} f_{\Omega}^{i} + \mathbf{B} f_{\partial\Omega}^{i}) + W = 0. \tag{22}$$

Isolating f_{Ω}^{i} in the previous equation leads to

$$(\mathbf{I} - \mathbf{A}) f_{\Omega}^{i} = f_{\Omega}^{i+1} + \mathbf{B} f_{\partial\Omega}^{i} + W. \tag{23}$$

Remarking that $f_{\Omega}^{N} = 0$ for all $j \in [1, ..., M-1]$, solving (23) for $i = N-1, ..., 0$ to obtain the solution f_{Ω}^{0}.

Since the equations are solved working backwards in time, superficially (23) says that three unknowns must be calculated from only one known value. This is shown pictorially in Fig.1. However, when all the equations at a given time point are written simultaneously there are $M-1$ equations in $M-1$ unknowns. Hence the value for f at each node can be calculated uniquely.

Two important questions one might ask about any numerical algorithm are when is it stable and if it's stable then how fast does it converge. An iterative algorithm that is unstable will lead to the calculation of ever increasing numbers that will at some point approach infinity. On the other hand, a stable algorithm will converge to a finite solution. Typically the faster that finite solution is reached the better the algorithm. The following Theorem will show the stability of our PDE.

Fig. 1. Implicit Finite Difference Viewed as a Pseudo-Trinomial Tree

Theorem 2. *The Implicit Finite Difference Methods for PDE*

$$\frac{\partial C_1}{\partial t} + \frac{1}{2}\sigma^2 S^2 \frac{\partial^2 C_1}{\partial S^2} + S(r_D - r_F)\frac{\partial C_1}{\partial S} - r_D C_1$$

$$+ \frac{S}{2\pi(T-t)} e^{-\frac{\left[\ln\frac{S}{K} + \left(r_D - r_F + \frac{\sigma^2}{2}\right)(T-t)\right]^2}{\sigma^2(T-t)} - 2r_F(T-t)} = 0,$$

with the condition $C_1(S,T) = 0$, converges for all values of $\rho, \sigma, \delta t$.

We next present some numerical simulations to examine the effect of liquidity costs on option prices.

Table 1. Option prices with liquidity costs and r_F is not equal to zero

Initial spot	Liquidity cost $f'(0)$				
	0 (G-K)	0.0001	0.0005	0.001	0.002
80	1.7056	1.8119	2.2370	2.7684	3.8313
85	2.9786	3.1306	3.7389	4.4992	6.0198
90	4.7624	4.9542	5.7214	6.6803	8.5981
95	7.0829	7.2996	8.1667	9.2506	11.4183
100	9.9251	10.1480	11.0395	12.1540	14.3830
105	13.2423	13.4537	14.2992	15.3561	17.4699
110	16.9687	17.1558	17.9044	18.8401	20.7114
115	21.0305	21.1869	21.8123	22.5940	24.1575

If the foreign short rate r_F equal to zero, the G-K model will be similar to the Black-Scholes model, which we showed in our previous paper.

Table 1 presents the option prices inclusive of liquidity costs with varying $f'(0)$ and initial stock price S_0. The options prices are obtained by solving the PDE has given in Theorem 1 numerically. The parameter values that we used are striking price $K = 100, r_D = 0.04, r_F = 0.02$ and $T = 1$ year. In case $r_F = 0$, we have the result in Table 2. Consistent with intuition, we observe that the option prices increase slightly when the parameter of liquidity costs $f'(0)$ (the slope at 0 of supply curve) increases.

Table 2. Option prices with liquidity costs and r_F is equal to zero

| Initial spot | Liquidity cost $f'(0)$ | | | | |
	0 (Black-Scholes)	0.0001	0.0005	0.001	0.002
80	1.5617	1.5711	1.6089	1.6562	1.7507
85	2.7561	2.7698	2.8246	2.8932	3.0303
90	4.4479	4.4652	4.5343	4.6207	4.7935
95	6.6696	6.6887	6.7650	6.8604	7.0513
100	9.4134	9.4320	9.5063	9.5992	9.7849
105	12.6388	12.6546	12.7181	12.7975	12.9562
110	16.2837	16.2953	16.3414	16.3990	16.5143
115	20.2769	20.2832	20.3080	20.3391	20.4013

References

1. Back, K.: Asymmetric Information and options. The Review of Financial Studies 6, 435–472 (1993)
2. Çetin, U., Jarrow, R., Protter, P.: Liquidity risk and arbitrage pricing theory. Finance and Stochastic 8, 311–341 (2004)
3. Cvitanic, J., Ma, J.: Hedging options for a large investor and forward-backward SDE's. Annals of Applied Probability 6, 370–398 (1996)
4. Çetin, U., Jarrow, R., Protter, P., Warachka, M.: Pricing options in an extended Black Scholes economy with illiquidity: theory and empirical evidence. Review of Financial Studies 19, 493–529 (2006)
5. Duffie, D., Ziegler, A.: Liquidity risk. Financial Analysts Journal 59, 42–51 (2003)
6. Jarrow, R.: Market manipulation, bubbles, corners and short squeezes. Journal of Financial and Quantitative Analysis 27, 311–336 (1992)
7. Frey, R.: Perfect option hedging for a large trader. Finance and Stochastics 2, 115–141 (1998)
8. Frey, R., Stremme, A.: Market volatility and feedback effects from dynamic hedging. Mathematical Finance 7, 351–374 (1997)
9. Subramanian, A., Jarrow, R.: The liquidity discount. Mathematical Finance 11, 447–474 (2001)
10. Tran, M., Duong, T., Ho, Q.: Discrete time hedging with liquidity risk. Southeast-Asian J. of Sciences 2(2), 193–203 (2013)
11. Bank, P., Baum, D.: Hedging and portfolio optimization in financial markets with a large trader. Mathematical Finance 14, 1–18 (2004)
12. Björk, T.: Arbitrage Theory in Continuous Time, 3rd edn. Oxford University Press Inc., New York (2009)
13. Leland, H.E.: Option pricing and replication with transactions costs. Journal of Finance, 1283–1301 (1985)
14. Feller, W.: An Introduction to Probability Theory and Its Applications, 2nd edn., vol. 2. John Wiley & Sons (1970)
15. Duffy, D.J.: Finite Difference Methods in Financial Engineering: A Partial Differential Equation Approach. Wiley (2006)
16. Forsyth, G.E., Wasow, W.R.: Finite Difference Methods for Partial Differential Equations. Wiley (1960)
17. Shreve, S.E.: Stochastic Calculus for Finance I. The Binomial Asset Pricing Model. Springer, New York (2004)
18. Shreve, S.E.: Stochastic Calculus for Finance II. Continuous–Time Model. Springer, New York (2004)

Scientific Methodology to Model Liquidity Risk in UCITS Funds with an Asset Liability Approach: A Global Response to Financial and Prudential Requirements

Pascal Damel[1,3] and Nadège Ribau-Peltre[2]

[1] L.I.T.A. Université de Lorraine, France
[2] C.E.R.E.F.I.G.E. Université de Lorraine, France
[3] C.N.A.M. Paris, France
{pascal.damel,nadege.peltre}@univ-lorraine.fr

Abstract. The financial crisis began with the collapse of Lehman Brothers and the subprime asset backed securities debacle. Credit risk was turned into liquidity risk, resulting from a lack of confidence among financial institutions. To overcome this problem, Basel III agreements for banks and 2009/65/EC UCITS IV Directive (UCITS IV) for regulated mutual funds, enhanced the recognition of liquidity risk. In this article, we focus on the evaluation of liquidity risk for UCITS funds. We will propose a quantified way to apprehend this risk through an asset liability approach. For assets, we will show that liquidity risk is a new type of risk and the current way to deal with is based solely on observed variables without any theoretical link. We propose a heuristic approach to combine the numerous liquidity risk indicators. To model and forecast the liability side we provide several parameters to create a Kalman filter process. This paper brings a comprehensive response to the regulatory requirements.

Keywords: financial crisis, liquidity risk, liquidity scoring, Kalman fiter.

1   Introduction

Liquidity risk is now, after the credit, market and operational risks, the one that has the most mobilized financial regulators since the 2008 financial crisis. Indeed, due to liquidity shortfalls during the financial markets crisis some funds were forced to suspend the redemption of fund units, including UCITS. Regulatory measures have consequently been introduced at an European level. The 2009/65/EC UCITS IV Directive (transposed in Luxembourg in both the Regulation CSSF 10-4 and the CSSF circular 11/512) for the first time contained the requirements to monitor the liquidity of funds.

In UCITS funds we have two dimensions: the asset dimension, where we can find securities (equities, bonds…, and little cash) and the liability dimension where we can find UCITS funds shares subscriptions. The liquidity risk follows this process: the asset manager has to manage at any time the subscriptions and the redemptions of fund's shares. If at any time there are lot of redemptions, those one are at the beginning realized with cash settlement of the funds. After the cash the asset manager has to sell quickly a part of the securities in the portfolio. At this stage, the liquidity level of the securities allows or not the redemption.

The aim of this paper is to propose a methodology to model these two dimensions asset with an original and new liquidity scoring, and liability dimension with a Kalman filter.

2   The Lack of Theoretical Approach toward Asset Liquidity Risk

Most articles dealing with liquidity risk study it under a single dimension, which can be width, depth or immediacy. There are only few multidimensional theoretical approach to our knowledge. Dick-Nielsen et al. used a heuristic and multidimensional approach (using PCA method). But their variables are focused on volumes and prices.

The PCA isn't the best methodology to create a scoring. The limit is the linearity and the Euclidean assumptions. The feature of our research is also to integrate the structure of illiquidity measured by the number of contributors and the Bid Ask spread of each contributor. This accurate database is an added value and is now available with financial providers.

Our approach can be used to assess liquidity for all classical securities (shares, bonds, mutual fund). It is thereby an operational approach because it can be used to measure liquidity risk for a mixed portfolio. In this research we use a logistic regression as a new methodology using the software SAS.

2.1   Liquidity Risk in Financial Publications

Liquidity can be defined as the property of a financial asset. It is its ability to be exchanged against a general equivalent such as money with low transaction costs and little variation in price over a short period. Liquidity risk is the consequence of market imperfections. These imperfections have become pervasively present since most securities exchange is carried out in organized markets, or what we call the exchange of mutual agreement. Moreover, the increasing complexity of products (such as derivatives) tends to compartmentalize markets, so that the exchange price is based on mathematical valuation models (price indicator models). There are several indicators of liquidity risk, which can be put into several categories. Some indicators directly measure transaction costs, while others reflect the market's ability to absorb large trade volumes without a significant change in price.

2.2   Indicators of Liquidity Risk in Financial Publications

Indicators of width

The "Bid Ask" Spread

The Bid-Ask spread plays a significant role in liquidity. The bid price is the highest price investors are willing to pay for a share, while the ask price is the lowest price at which investors are willing to sell a share.

The spread between bid and ask price must cover costs incurred by the market maker (Copeland and Galai, 1983 ; Easley and O'Hara, 1987 ; Roll, 1984 ; Amihud and Mendelson, 1980). A large number of studies showed the link between bid-ask spread and liquidity (Kraus and Stoll, 1972; Keim and Madhavan, 1996). This indicator measures the width of the market (Amihud, Mendelson, 1986 ; Chen, Lesmond, Wei, 2007)

The Stressed Bid-Ask Measure

This Bid-Ask spread is a Bid Ask measure with stochastic stress around the mean.

Indicators of Depth and Resilience

The λ (lambda) Kyle and γ (gamma) of Amihud

The Kyle's lambda coefficient (1985) is used to estimate the market's ability to absorb a high amount of transactions. A high lambda means a low absorption capacity and therefore a lower liquidity. The Kyle lambda is calculated on intra-daily data, which is particularly demanding in terms of data availability, but allows the depth of the market to be measured. To overcome this problem of data availability, Amihud's gamma (2002) calculates an average liquidity ratio from the price and volume of daily transactions.

The Number of Contributors

The number of contributors is often used as one of many indicators to measure the liquidity of a bond or a share (Ericsson and Renault (2001), Gehr and Martell (1992), Jankowitsch and Pichler (2002). Dimson and Hanke (2001) refer to a small number of contributors as a sign of a dealer's market power and thus as a measure of liquidity.

The LOT Indicator, in Reference to the L.O.T (Lesmond, Ogden, Trzcinka, 1999) Model

The LOT model is another way to calculate liquidity risk. LOT is a two-factor evaluation model. The two explanatory factors are the interest rate and return on equity market (in theory, the market portfolio). The authors consider that corporate bonds are hybrid securities between the risk-free sovereign bond market and the riskier stock market.

Indicator of Immediacy: Market Position

Market Position here is an indicator we computed to assess the level of concentration of a fund for a specific bond. As the "Days to Liquidate" indicator, which is usually used for stocks[1], was not available for bonds in the database we used (Reuters), we created an indicator called "Market Position" based on a similar computational method as the "Days to Liquidate" indicator that is available for stocks:

2.3   An Heuristic Method Using Logistic Regression to Create a Liquidity Risk Measure

The Logistic Regression

Logistic regression estimates the probability with a likelihood function of an event occurring or a state of the variable that we have to explain. The logistic regression try to predict the probability (p) that it is 1 (event occurring or weakness of liquidity) rather than 0 (event not occurring or high liquidity). In linear regression, the relationship between the dependent and the independent variables is linear, this assumption is no use in logistic regression. This can be written as:

$$P = \frac{e^{\alpha+\beta x}}{1 + e^{\alpha+\beta x}}$$

Where P is the probability of liquidity level, e is the base of the natural logarithm and are the parameters of the model (as in normal linear regression). The value of P when x is zero indicates how the probability of a 1 changes when x changes by a single unit (the odds ratio). The relation between x and P is nonlinear and so is the liquidity problem that we have to solve. To maximize the Likelihood of the regression logistic, our laboratory also uses the D.C.A. approach developed by Pr. Lethi and Pham Dinh Tao.

2.4   Database and Result: Numerical Tests with Logistic Regression (with SAS Software)

Our database was provided by a Management company of Funds based in Luxembourg.

It includes 15 mutual fund portfolios, each containing between one and eleven compartments (for a total of 83 compartments). 4,817 securities totaling more than 16 billion Euros were held in these 83 compartments.

Graph 1. liquidity rating of bonds
Liquidity rating distribution of the bonds in our database (1 the highest level to 5 the lowest level of liquidity)

---

[1]   The Days-to-Liquidate indicator is defined as the number of days it would take for the portfolio to liquidate the entire position of stock based on the Average Daily Trading Volume. This is calculated by taking the number of shares held in a portfolio and dividing it by the Average Daily Trading Volume over the last 3 months. This is based on the changing Average Trading Volume and is updated daily.

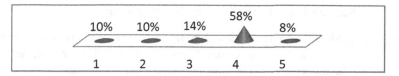

This graph shows that bonds have a low overall level of liquidity: the vast majority of the bonds of our database have a liquidity level of 4 (58%).

Graph 2. liquidity rating of equities
Liquidity rating distribution of the stocks in our database (1 the highest level to 5 the lowest level of liquidity)

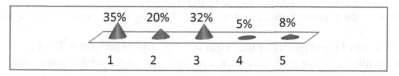

This graph shows that stocks have a higher level of liquidity than bonds. On the other hand, the distribution of shares in terms of liquidity is more heterogeneous than that of bonds.

This multidimensional measure of liquidity offers research opportunities to better understand the link between liquidity risk and market risk as well as the link between the risk of liquidity and the cash settlement period.

3   Liabilities Liquidity Risk

The aim of this third part is to model the liability side i.e. the subscription or redemption of the number of mutual Fund shares. Model the series of net subscription is the challenge. We propose to model the problem with a Kalman filter. The Kalman filter allows forecast and stress scenario to respond to the financial regulatory (CSSF 10-4 and 11/512).

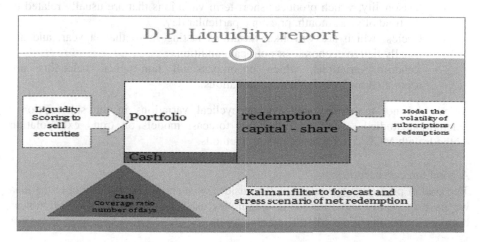

Fig. 1. The Asset liability approach

3.1   Using Kalman Filter to Forecast the Number of Share

In his famous publication "a new approach to linear filtering and prediction problem", R Kalman proposes a forecast model based on probability theory, and more specifically on the properties of conditional Gaussian random variables.

The Kalman filter model assumes a link between the state of the system at time t and the state of the system at time t-1, according the equation (1)

$$S_t F_t\, S_{t-1} + B_t TMS_t + \in_t \tag{1}$$

St :   is the state vector for the system (the number of share in the fund) at time t
TMSt :   is the vector containing any control input
Ft :   is the state transition matrix or the effect of each system state at time t-1 on the state system t.
Bt: is the control input matrix or the effect of each control on vector TMS.
$\in_t$: is a white noise for each parameters in the state vector with a mean multivariate normal distribution.

3.2   The Parameters of Kalman Filter: Equation (1)

- **F_t the state of transition matrix with deterministic trend**

Quantitative forecasts can be time-series forecasts (i.e., a projection of the past into the future) or forecasts based on associative models (i.e., based on one or more explanatory variables). Time-series data may have underlying behaviors that need to be identified by the forecaster. In addition, the forecast may need to identify the causes of the behavior. Some of these behaviors may be patterns or simply random variations. Among the patterns are:

- Trends, which are long-term movements (up or down) in the data.
- Seasonality, which produces short-term variations that are usually related to the time of year, month, or even a particular day.
- Cycles, which are wavelike variations lasting more than a year, and are usually tied to economic or political condition.
- Random variations, which encompass all non-typical behaviors not accounted for by the other classifications.

As we did not noted any seasonal or cyclical variations in the subscription / redemption studied series, we based our forecast models on trend extrapolation methods with MCO criteria or smoothing methods.

A good forecast accuracy:
The goal of predictive techniques is to consider what should be, according to the idea we have of reality and forecasting techniques used. But there are still differences between the results of a prediction and reality itself.

The mean absolute percentage error (MAPE), which is the average absolute deviations compared to observed values, is a good indicator to take account of the quality of forecast.

The average MAPE value of all our series' forecasts is 2,41 %, the lowest value being 0,11 % and the highest 5,29 %. In general a MAPE of 10% is considered very good

The deterministic trend can be used to model F_t the state of transition matrix.

- **TMS$_t$:   is the vector containing any control input**

The most used methodology to control the volatility (ie control input) is in Finance. the Time series model (ARMA – ARIMA – ARCH-GARCH model).

To design the volatility, times series methods were used to model the historical series, in order to compute the standard error of residuals (i.e. random variations). Unlike many time series of financial returns, subscription/redemption series of UCITS funds show only little heteroscedasticity and therefore require little use of ARCH-GARCH models. Moreover, if none of the series showed any seasonal component, the vast majority were not stationary, so stationnarisation was required before modeling.

Here below is a table (table 1) summarizing the different models with the same database seen in the second part.

Table 1. Types of time series models used to model UCITS funds' subscription/redemption series included in our database

Type of modelisation for estimating error	Type of stationnarisation's procedure	Stationnarisation level	Frequency
ARIMA			*81%*
	with differentiation		71%
		Level 0	3%
		Level 1	4%
		Level 2	64%
	with trend		10%
		Level 0	1%
		Level 1	9%
ARCH/GARCH			*6%*
	with differentiation		6%
		Level 2	6%
Others			*13%*

- **B$_t$: is the control input matrix or the effect of each control on vector TMS**

The prediction equation with a time square root about the white noise of TMS$_t$

In Financial risk Management if fluctuations in a stochastic process from one period to the next are independent (i.e., there are no serial correlations or other dependencies) volatility increases with the square root of the unit of time. It is the case with an ARIMA or GARCH white noise.

3.3   The Proposition of Global Kalman Filter

The Kalman-filter is a set of recursive linear estimation steps with a combinaison between the forescast (2), the measurement update (3) and the measurement vector (4) represent the measurement syste

$$\hat{S}_{t/_{t-1}} = F_t\,\hat{S}_{t-1} + B_t TMS_t \tag{2}$$

$$V_{t/_{t-1}} = F_t\,V_{t-1}F_t^T + BQ_t \tag{3}$$

$$M_t = H_t S_t + \alpha_t \tag{4}$$

With
Qt: the process noise covariance matrix associated with the noisy control inputs
Mt: the measurement vector
Ht : the process of measurement
α_t : the white noise of the process measurement (i.e. in Finance the operational risk) with the covariance Rt

We can obtain (3) as follow
  The variance between the forecast and the unknown true value $\hat{S}_{t/_{t-1}}$ and St is

$$V_{t/_{t-1}} = E\left[\left(S_t - \hat{S}_{t/_{t-1}}\right)\left(S_t - \hat{S}_{t/_{t-1}}\right)^{T}\right]$$
$$S_t - \hat{S}_{t/_{t-1}} = F_t(S_{t-1} - \hat{S}_{t-1}) + \epsilon_t$$
$$\Rightarrow V_{t/_{t-1}} = E\left[\left(F_t(S_{t-1} - \hat{S}_{t-1}) + \epsilon_t\right) \times \left(F_t(S_{t-1} - \hat{S}_{t-1}) + \epsilon_t\right)^{T}\right]$$

$$= FE\left[(S_{t-1} - \hat{S}_{t-1}) \times (S_{t-1} - \hat{S}_{t-1})^{T}\right] \times F^T + FE\left[(S_{t-1} - \hat{S}_{t-1})\,\epsilon_t^T\right]$$
$$+ E\left[(\epsilon_t\,S_{t-1} - \hat{S}_{t-1}^{\ T})\right]F^T + E[\epsilon_t\epsilon_t^T]$$

The state error are not correlated

$$E\left[(S_{t-1} - \hat{S}_{t-1})\,\epsilon_t^T\right] = E\left[\left(\epsilon_t\,(S_{t-1} - \hat{S}_{t-1})^{T}\right)\right] = 0$$
$$\Rightarrow V_{t/_{t-1}} = FE\left[(S_{t-1} - \hat{S}_{t-1}) \times (S_{t-1} - \hat{S}_{t-1})^{T}\right] \times F^T + E[\epsilon_t\epsilon_t^T]$$
$$\Rightarrow V_{t/_{t-1}} = FV_{t-1}F^T + Q_t$$

At the end we have the update measurement and the optimal Kalman gain Kt

$$\hat{S}_{t/_t} = \hat{S}_{t/_{t-1}} + K_t\left(M_t - \hat{S}_{t/_{t-1}}\right) \tag{5}$$

$$V_{t/_t} = V_{t/_{t-1}} - K_t H_t V_{t/_{t-1}} \tag{6}$$

$$K_t = V_{t/_{t-1}}H_t^T\left(H_t\,V_{t/_{t-1}}H_t^T + R_t\right)^{-1} \tag{7}$$

3.4   The Test of Kalman Filter: Equation (1)

To illustrate the Kalman Filter
Ft : is the state transition matrix or the effect of each system state at time t-1 on the state system t. We have the best deterministic trend

TMSt :   is the vector containing any control input   or the white noise with the best fit model (ARMA…). The white noise is a Gaussian distribution. We can use several level of confidence (with an appropriate t of student 86% - 95% -99%)

Bt: is the control input matrix i.e. the time square root about the white noise of TMSt

The Forecast is the state transition matrix and the SVaR (share Value at risk with different level of confidence) is the prediction equation in Kalman Filter. We have below an illustration of Forecast and SVaR for 30 days. This process Is rely on the ARIMA – GARCH process and the state transition matrix.

Graph 3. Illustration of equation

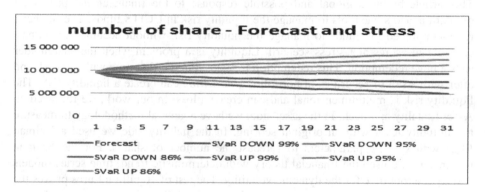

3.5   The Back Testing Test of Kalman Filter Parameter for the First Step: The Forecast

In this section, we have tested the relevance of the Kalman parameters that we have proposed. We have compared the SVaR 30 days with the reality observed between 30 days during two years. We have an illustration with the GARCH noise. The global test shows what the limit of dispersion is relevant with the Gaussian dispersion.

Graph 4. Illustration of equation – GARCH noise

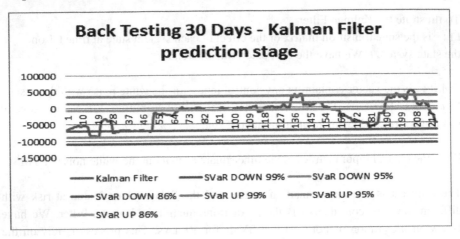

4   Conclusion

This article brings a global and possible response to the financial and prudential regulation, and some tools to manage the liquidity risk in UCITS funds. The financial regulator forced the funds to manage the liquidity risk without accurate guidelines, concrete measures and stress scenarii. Liquidity is a problem when the shareholder asks a huge redemption. This redemption is realized in the first time with the cash and after with a sell of securities in portfolio. This selling can create a liquidity risk. The liquidity risk is multidimensional and can create a loss. In our work, we followed an Asset Liability approach. In the asset side, we have a general methodology to measure the liquidity level with an original scoring. In the liability side we used a Kalman filter with specific parameters to forecast the number of shares. For the Kalman parameters, we used the financial theory with deterministic trend, time series models and square root time for the dynamic volatility. The test of 16 billion Euros proves the relevancy of this approach and confirms the choice of Gaussian assumptions for the Kalman filter.

References

1. Amihud, Y., Mendelson, H.: Dealership markets: market making with inventory. Journal of Financial Economics 15(1), 31–53 (1980)
2. Chen, L., Lesmond, D., Wei, J.: Corporate yield spreads and bond liquidity. The Journal of Finance 62(1), 119–149 (2007)
3. Copeland, T., Galai, D.: Information Effects on the Bid-Ask Spread. Journal of Finance 21(5), 1453–1469 (1983)
4. Dick-Nielsen, J., Feldhutter, P., Lando, D.: Corporate bond liquidity before and after the onset of the subprime crisis. Journal of Financial Economics 103, 471–492 (2012)

5. Dimson, E., Hanke, B.: The Expected Illiquidity Premium: Evidence from Equity Index-linked Bonds, Working Paper, London Business School (2001)
6. Easley, D., O'Hara, M.: Price, Trade Size, and Information in Securities Markets. Journal of Financial Economics 19(1), 69–90 (1987)
7. Ericsson, J., Renault, O.: Liquidity and Credit Risk. Journal of Finance 61(5), 2219–2250 (2006)
8. Jankowitsch, R., Pichler, S.: Estimating Zero-Coupon Yield Curves for EMU Government Bonds, Working Paper, Vienna University of Technology (2002)
9. Keim, D.B., Madhavan, A.: The upstairs market for large-block transactions: Analysis and measurement of price effects. Review of Financial Studies 9(1), 1–36 (1996)
10. Kyle, A.S.: Continuous Auctions and Insider Trading. Econometrica 53(6), 1315–1336 (1985)
11. Kraus, A., Stoll, H.R.: Price Impacts of Block Trading on the New York Stock Exchange. Journal of Finance 27(3), 569–588 (1972)
12. Lesmond, D., Ogden, J., Trzcinka, C.: A new estimate of transaction costs. Review of Financial Studies 12(5), 1113–1141 (1999)
13. Pham Dinh, T., Le Thi, H.A.: Convex analysis approach to d.c. programming: Theory, Algorithms and Applications. Acta Mathematica Vietnamica 2(1), 289–355 (1997)
14. Roll, R.: A Simple Implicit Measure of the Effective Bid-Ask Spread in an Efficient Market. Journal of Finance 39(4), 1127–1139 (1984)

Vietnamese Bank Liquidity Risk Study Using the Risk Assessment Model of Systemic Institutions

Thanh Duong[1], Duc Pham-Hi[2,3], and Phuong Phan[3]

[1] Quantitative and Computational Finance, John von Neumann, HCMC, Vietnam
thanh.duong@jvn.edu.vn
[2] Financial Engineering, ECE Paris Graduate School of Engineering, Paris, France
phamhi@ece.fr
[3] John von Neumann Institute, HCMC, Vietnam
{hi-duc.pham.qcf,phuong.phan.qcf}@jvn.edu.vn

Abstract. This paper presents a liquidity risk management model allows to assess the impact of stress scenarios on a banking system within a top-down approach. The impact of stress scenarios on a banking system includes: (i) individual bank reactions to the shock, (ii) the shock transmission across banks, through interbank networks and financial market channels and (iii) the recover rate, the proportion of the debt a creditor receives in an event of a default. The macro economic model is estimated and simulated quarterly and the data in balance sheet is yearly for the Vietnamese banking system. The results show a high vulnerability of the trading portfolios and interbank market.

Keywords: Liquidity risk, stress testing, contagion, default, risk management.

1 Introduction

In the banking sector, liquidity and liquidity risk are very important factors and become a macroprudential issue in the macro stress tests (see [1]). Liquidity risk occurs when the liquidity of financial institutions is uncertain, the credit rating falls, customer sudden withdrawal and events causing loss of confidence to the lender or the markets in which it depends lost liquidity. If financial institutions do not internalize the externalities of network membership, banks liquidity choices will be suboptimal. As a consequence, liquidity and capital requirements need to be imposed externally and should be set in relation to a banks contribution to systemic risk, rather than on the basis of the banks idiosyncratic risk (see [2]).

In Vietnam, liquidity risk has not been well managed. When banks have problem with liquidity, the central bank solved by funding them, making moral hazard in the State and banks are not interested in risk management strategies. Most of previous studies of Vietnam are to assess the status of the liquidity risk by qualitative methods (see [3,4,5]), there are little research to quantify risks as well as building or using a specific model to calculate liquidity risk and evaluate the

© Springer International Publishing Switzerland 2015
H.A. Le Thi et al. (eds.), *Model. Comput. & Optim. in Inf. Syst. & Manage. Sci.*,
Advances in Intelligent Systems and Computing 360, DOI: 10.1007/978-3-319-18167-7_35

tolerance of bank liquidity through shocks. So, our paper refers to a stress test model that successfully used in the Bank of England and in the Financial Sector Assessment Program of IMF to quantify the liquidity risk and assess the impact of the Vietnamese banks before the financial shocks, see [14].

Macro stress tests, the topic of this paper, are part of the macropruden-tial toolkit that authorities may use to detect system-wide liquidity risks (see [6,7,8,9,10,11,12]). Macro stress tests can be conducted either bottom-up or top-down. We used only top-down stress tests to find vulnerabilities to the entire system by which individual banks can see the risk of a network of other banks to build a reserve fund at risk. The central bank can manage activities of banks to avoid affecting the overall system. In practice, most authorities conduct both top-down and bottom-up stress tests, as these are complementary and allow for valuable cross-checks (see [13]).

We will research about the Risk Assessment Model of Systemic Institutions (RAMSI), a stress-testing model developed at the Bank of England (see [15]). The component models include: a Bayesian vector autoregression model to sim-ulate macroeconomic scenarios (see [16,17]), satellite models for credit and mar-ket risk and net interest income, an interbank network model and an asset price function to simulate fire sales of assets (market liquidity risk) (see [18]). The structure of this paper is presented as follows. Section 2 reviews an overview of RAMSI model. The dataset and the empirical results are given in section 3.

2   An Overview of RAMSI Model

ACRONYMS: AFS (Available for sale asset), BVAR (Bayesian vector autore-gression), Ca (Capital), B (Invest Security), GDP (Gross domestic product), IB (Interbank asset), IMF (International Monetary Fund), H (Mortgage loans), N (Non trading Asset), RAMSI (Risk Assessment Model of Systemic Institutions), SRM (Standard Reference Method), TA (Total Asset), VAR (Vector autoregres-sion), % oTA (Percent of Total asset).

Inputs of RAMSI model include income statement and balance sheet of banks, these data are publicly available. We use the top-down stress tests model, so we do not attend to the small details as the bottom-up model. Second input is a set of macroeconomic variables and domestic finance, forecasts the future development of each income statement and balance sheet of the bank.

According to [15], RAMSI includes a set of econometric equations based on the balance sheet accounting and forecasting of macroeconomic variables to describe the components of the largest UK banks' income. Further, banks responses to exogenous shocks are dictated by simple behavioral rules rather than by the solution to an explicit forward-looking optimisation problem.

In statistics, Bayesian vector autoregression (BVAR) uses Bayesian methods to estimate a vector autoregression (VAR). In that respect, the difference with standard VAR models lies in the fact that the model parameters are treated as random variables, and prior probabilities are assigned to them. Recent research have shown that vector auto regression with Bayesian shrinkage is an appropri-ate tool for large scale dynamic models (see [19]).

In practice, information on the precise classification of bank loans is internal. In such a situation, the data is assumed. The total loss is then distributed over creditors in proportion to the amount of their respective contributions. To clear the network following the default of one or more institutions, in this paper, we will use algorithm in [20].

The values of centrality measures of these networks are node degree centrality, betweenness centrality, and closeness centrality, denoted C_D, C_B and C_C, respectively. In all cases, n represents the total number of banks in the network.

The node degree centrality is calculated by this formula,

$$C_D(i) = \frac{\sum_{i=1}^{g} C_D(n^*) - C_D(i)}{(n-1)}, \tag{1}$$

$C_D(n*)$ is maximum value in the network. The more directed link is, the higher the degree centrality becomes.

The betweenness centrality of a bank,

$$C_B(i) = \frac{\sum_{j<z} \text{of shortest paths between j and z through bank i}}{\text{of shortest paths between j and z}} , \tag{2}$$

reflects the amount of control that this bank exerts over the interactions of other banks in the network.

Closeness centrality,

$$C_C(i) = (n-1)\frac{1}{\sum_{j \in U} d(i,j)} , \tag{3}$$

where d(i, j) is the network theoretic distance between banks i and j, is a measure of how fast information spreads from a given node to other reachable nodes in the network.

In this paper, we will use Balance Sheet equation,

$$A_{i,t} = H_{i,t} + B_{i,t} + N_{i,t} + \sum_{j \in N_i(g)} IB_{i,t}^j , \tag{4}$$

and Capital dynamics equation,

$$C_{i,t+1} = C_{i,t} - L_{i,t} - \sum_{j \in N_i(g)|C_j \leq 0} \left[(1 - RR_j)IB_{i,t}^j\right] , \tag{5}$$

$IB_{i,t}^j$ presents the exposure of a bank j to the bank i in time t. RR is recovery rate. A bank goes to default if the value of its capital becomes negative. When a bank is in distress, it may sell assets, opening up the possibility of an important feedback channel operating via asset prices. In the current version of RAMSI, such fire sales only occur after a bank defaults, and not as a defensive action to stave off failure. A failing bank is assumed to liquidate all its AFS assets. The fire sale discount lasts for one quarter, and the resulting fall in asset prices may lead other banks to incur mark-to-market losses; hence in extreme circumstances these banks may then also fail (see [18]).

3   Data and Empirical Results

Data. Banking system in Vietnam includes 112 banks, among those are 5 state-owned banks, 37 commercial banks, 66 foreign banks and 4 Joint-Venture banks. They are under the supervisoring and control of Central bank. The main businesses of the banks include: deposits, loan and financial investment. In figure 1, we can see that nearly one a half of the banks have the interbank assets over 20% of total asset. It means that the interbank market is very important to the system banks. In this paper we consider 9 listed banks in Table 1.

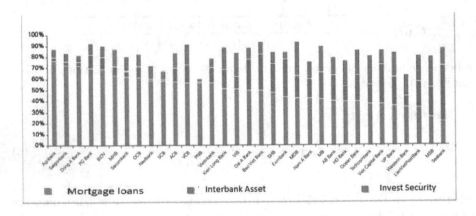

Fig. 1. The main component of asset

Sources: Vietnam Banking Survey 2013, see [22].

The Macroeconomic Model. We will test the stationary of macro variables: the quarterly GDP growth, the level of the three-month T-bill rate, the quarterly inflation of VietNam. To illustrate Bayesian VAR analysis using this data, we work with an unrestricted VAR with an intercept and four lags of all variables included in every equation. Figures 3 presents impulse responses of all three of our variables to all the shocks for two of the priors to see the effects of an impulse which leads to positive long-run effects.

Table 1. Balance sheet structures of 9 banks

BANK	BIDV	VCB	ACB	CTG	EIB	SHB	MBB	STB	NVB
TA	548511	467459	166737	576383	169913	143740	180432	161378	29225
B	69630	65089	34919	83660	14655	18684	49874	22545	3786
% oTA	13%	14%	21%	15%	9%	13%	28%	14%	13%
H	384837	267854	105642	329682	82643	75184	85807	109214	13266
% oTA	70%	57%	63%	57%	49%	52%	48%	68%	45%
IB	47658	89779	7216	57708	57874	30312	26789	7470	4959
% oTA	9%	19%	4%	10%	34%	21%	15%	5%	17%
N	46386	44737	18960	105333	14741	19560	17962	22149	7214
% oTA	8%	10%	11%	18%	9%	14%	10%	14%	25%
Ca	32069	43077	12502	54076	14680	10327	15141	17063	3217
% oTA	6%	9%	7%	9%	9%	7%	8%	11%	11%

Sources: 2013 Financial Statement from Cafef - Unit: billions VND, see [11].

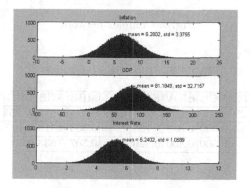

Fig. 2. Data Used In Empirical Illustration

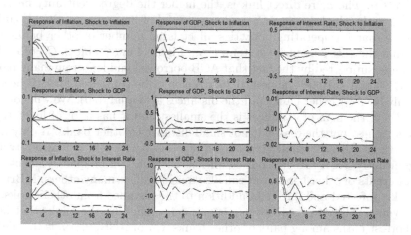

Fig. 3. Posterior of impulse responses

Interbank Network Model. We simulate the interbank assets and liabilities to construct a complex network, in which each bank connects with a large number of other banks in the network.

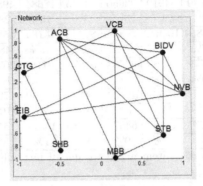

Fig. 4. Complex network

Table 2. Properties of networks

	CTG	VCB	BIDV	STB	EIB	MBB	ACB	SHB	NVB
C_D	0.25	0.5	0.5	0.5	0.25	0.375	0.625	0.25	0.5
C_B	0.261	0.300	0.333	0.253	0.210	0.333	0.419	0.199	0.249
C_C	0.348	0.500	0.400	0.320	-	0.307	0.533	0.500	8

Table 2 reports the values of centrality measures of Complex network, that is node degree centrality, betweenness centrality, and closeness centrality, denoted C_D, C_B, C_C. The more direct link is, the higher the degree centrality becomes. So, bank with a high level of C_D had a potential of becoming a critical bank in the sense that its operating activities affect large number of other banks, and through it the entire financial system. From Figure 4 and Table 2, C_D of ACB is the highest supported by the fact that ACB borrowed and lent from many banks. SHB, CTG and EIB are smallest, 0.25. C_C cannot be computed for EIB because EIB only borrows other banks (all the distances are equal to 0). We will consider C_C if $d > 0$, the sum of distances is the smallest and it has a lot of links. C_C of NVB is highest but the reason it not small distance to other banks. It only lends EIB, so these other distances $= 0$. Therefore, other banks default may not trouble it significantly. ACB satisfied our conditions so ACB can be quickly infected by default events of other banks and can also quickly infect many others. MBB may be weakly infected and infect. In addition to the distance from other banks, the banks vulnerability to credit events is also determined by the banks that lie on the shortest paths among pairs of other banks. C_B of ACB, 0.419, is the highest and C_B of SHB is the smallest. For idiosyncratic shock, which is represented by a default of individual banks, we employ 3 different scenarios: defaults of

Table 3. The recover rates

Bank	BIDV	VCB	ACB	CTG	EIB	SHB	MBB	STB	NVB
RR	0.3169	0.5564	0.4153	0.4198	0.3976	0.4666	0.3886	0.4098	0.4047

individual banks (9 observations), simultaneous defaults of two random banks (36 observations). Default of a bank will affect banks that lend money by the manner loss to interbank asset, then directly write this loss to capital of the affected bank.

For systemic shock, which is represented by a sudden drop in the value of mortgage loans, we start within a shock (drop in housing price) of 5%, then we repeat the experiment more 6 times, so that default rate ranges from 5% to 30% with an increment of 5%. We also apply a shock in which each day the house price drop 5% and the drop continue in 8 days to see the shock development, as well as impact of contagion. In this scenario, there are two forces: direct effects of a shock which reduce banks mortgage loan assets, and the contagion effects which reduce banks interbank assets. The combination of 2 effects makes a systemic shock more damaging.

For every scenario, we will observe the number of defaulted banks, suffered banks, the value of assets losses, capital losses to compare impacts of shocks:

- **Normal scenario:** When no bank goes default, we ran 60 days operation of bank without shocks. It turns out that due to change in stock market, there are no bank goes bankrupt, however their capital lost 21.9% its value, which is a pretty large proportion of capital. The reason is that the proportion of equity in banks total asset is pretty high comparing capital proportion: 9-28% range versus 6-11%. Particularly, MBB have 28% of its asset is equity, while its 8% capital/ total asset. We also try to extend to 100 days, and it lead to MBBs default in 83rd day. However, except MBB other banks are still function well, and their capitals do not remarkably affected by changes of stocks.

- **Scenario with initial default of 1 bank:** Simulations show a significant difference in the number of suffered and defaulted banks. 1 bank default lead to 1.7 banks went default and erode more than 40% total capital in average. A default of one bank can end in loss of 65.36% of initial banking capital, with the lost capital of defaulted banks reaching as much as 36.8 %. It's the case of ACB default which leads to default of more 3 banks including: BIDV, EIB and MBB, the largest number of defaulted banks. In this worst case, 7 banks were suffered. There are also examples of bank defaults does not or barely resulted in any effect. That's the case of Bank 5 (EIB) default (have no impact at all), and Bank 6 (SHB) default. Bank 2 (VCB) default is an example of risk-sharing that does not induce contagion, although its default affects 3 other banks in the system and 52.79 % of total capital was destroyed rank the second position.

It also turn out that EIB is the most sensitive bank to contagion exposure, as it appears in all scenarios which lead to default of other banks. Whenever any of its creditors go bankrupt also lead to its bankruptcy. It due to its capital (9% of total asset) is pretty low comparing to its interbank asset (34 %).

- **Scenario with initial default of 2 banks:** 2 initial defaults led to 3.2 banks
went default and eroded 58.5 % total capital and nearly 36 % total assets in
average. Maximum effect was the same time default of ACB and CTG that led
to more 4 banks went default and all system suffered. This event eroded 74.7 %
of total asset and 88.33 % of total capital.

In contrary, defaults of EIB and NVB did not affect other banks in the system
at all. Only 9.77 % of total asset and 28.4 % of capital is lost.

Table 4 reports selected statistics when different individual banks default. The
meanings of the columns are the following:

   - Def relates to the number of defaulted banks.
   - Suffered relates to the number of suffered banks.
   - DA reports the value of defaulted assets, in billions.
   - DA (%) reports the proportion of defaulted assets of the financial system.
   - ADB reports the value of defaulted assets of defaulted banks, in billions.
   - ADB (%) reports the proportion of defaulted assets of defaulted banks of
the financial system.
   - DC reports the value of defaulted banking capital, in billions.
   - DC (%) reports the proportion of defaulted capital of the financial system.
   - CDB reports the value of defaulted banking capital of defaulted banks, in
billions.
   - CDB (%) reports the proportion of defaulted capital of defaulted banks of
the financial system.

Table 4. Scenario with initial default of banks

	Def	Suf	DA	DA(%)	ADB	ADB(%)	DC	DC(%)	CDB	CDB(%)
Scenerio (Bank default)										
1 bank default										
Mean	1.7	3.2	476,555	19.50	425,133	17.40	85,169	42.13	33,748	16.69
Max	4.0	7.0	1,123,335	45.97	1,065,593	43.60	132,134	65.36	74,392	36.80
Min	1.0	1.0	204,829	8.38	143,740	5.88	55,030	27.22	10,327	5.11
SD	1.0	1.7	333,309	13.64	337,242	13.80	24,075	11.91	24,072	11.91
2 bank default										
Mean	3.2	5.6	877,544	35.91	825,300	33.77	118,266	58.50	66,022	32.66
Max	7.0	9.0	1,825,479	74.70	1,785,716	73.07	178,558	88.33	138,795	68.66
Min	2.0	2.0	238,711	9.77	199,138	8.15	57,470	28.43	17,897	8.85
SD	1.3	1.6	400,830	16.40	408,225	16.70	26,296	13.01	30,971	15.32

- **Housing price drop scenario:** turning to effect of system shocks, its easy to
see that system shock lead to an immediate drop the value of assets. While in
the case of idiosyncratic shock, only one scenario (simultaneous bankruptcy of
3 banks: ACB, CTG and STB lead to complete collapse of the system), in the
case of system shocks, we need to increase the shock to 30% drop in the housing
price lead to default of whole system.

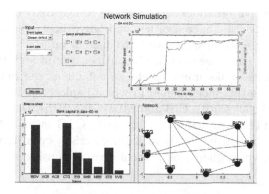

Fig. 5. Scenario with initial default of 2 banks

Table 5. Defaulted asset and defaulted capital when housing price drop from 5-30%

Drop	Def	Suf	DA	DA(%)	ADB	ADB(%)	DC	DC(%)	CDB	CDB(%)
5%	0	0	66,109	2.7	0	0	66,109	33	0	0
10%	0	0	87,922	3.6	0	0	87,922	43.5	0	0
15%	0	0	109,733	4.5	0	0	109,733	54	0	0
20%	6	8	1,632,144	67	1,562,277	64	190,553	94	120,686	60
25%	7	8	1,784,741	73	1,723,655	70.53	198,835	98.35	137,749	68
30%	9	9	2,443,778	100	2,443,778	100	202,152	100	202,152	100

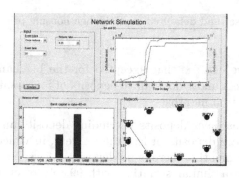

Fig. 6. Defaulted asset and defaulted capital when housing price drop 25%

To be able to see these forces, we should look details in each scenarios:

- When housing price drop from 5 to 15%, no bank goes default, but housing drop shock hugely damage banks capital, from 33 to 54%.

- For the case of drop rate to be 20%, 6 of 9 banks go default, and 2 banks (CTG, STB) are suffered with 94% total capital eroded. The only resistant bank is SHB.

- When drop rate increase to 25%, the whole system collapse with 7/9 went default from 20th (ACB, BIDV), 21th (NVB), 22th (MBB, VCB, EIB), 32th

(STB) right in the period shock were applied. CTG, SHB didnt go default as the cooperative forces that increase their mortgage assets and interbank assets, but their capital were erode significantly to a small value.

- When drop rate increase to 30%, all banks go default mainly by direct effect on mortgage loan as banks immediately go default as the magnitude of drop increase. We can see the phenomenon in which banks go default day after day: BIDV, ACB, NVB (day 20th), EIB and MBB (day 21th). Day 22th experiences default of VCB. STB goes default in day 28th, CTG in day 29th and SHB in day 30th. A drop in housing leads to a severe crisis, even if the magnitude is small. No bank survives when a 50 percent drop in housing.

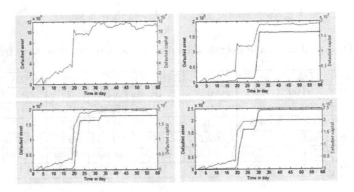

Fig. 7. Defaulted asset and defaulted capital when housing price drop from 15-30%

From Figure 8 we can see that excepted for the big banks have long-term deposits, small banks almost don't have deposits of 5 years, the majority of the deposits is 1 month to 3 months:

- 99.8% NVB customer deposits (excluding deposits and loans from other credit institutions) are deposits of less than 1 year, in which 72.4% is under 1 month, 22.4% is 1-3 months, there is no deposit on 5 years.

- SHB is also in a similar situation with 99.85% of customer deposits are deposits of less than 1 year, in which deposit accounts less than 1 month are 71.7%.

- CTG, STB, EIB medium-term deposits is relatively high comparing with other banks. CTG deposit rate over 5 years is quite high comparing to other banks.

The commercial banks' loans is medium and long term while deposits from customers to focus only on the short term of 1-3 months will make banks face liquidity risks.

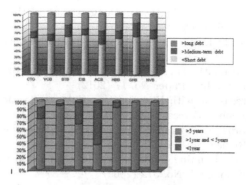

Fig. 8. Structure of debt / deposits of commercial banks

Sources: Vietnam Banking Survey 2013, see [22].l

References

1. Alfaro, R., Drehmann, M.: Macro stress tests and crises: what can we learn? BIS Quarterly Review 3, 1–14 (2009)
2. Cifuentes, R., et al.: Liquidity risk and contagion. Journal of the European Economic Association 3(23), 556–566 (2005)
3. Ngoc, T.B.L., et al.: Getting bank financing: A study of Vietnamese private firms. Asia Pacific Journal of Management 23(2), 209–227 (2006)
4. Nguyen, T.D.K., et al.: Capital structure in small and medium-sized enterprises: the case of Vietnam. ASEAN Economic Bulletin 23(2), 192–211 (2006)
5. Phan, T.T.M., et al.: Effects of Risks on Cost Efficiency: The Case of Vietnamese Banking Industry. Available at SSRN 2312169 (2013)
6. Basel Committee on Banking Supervision: Principles for Sound Liquidity Risk Management and Supervision. Bank for International Settlements (2008)
7. Basel Committee on Banking Supervision: Basel III: International framework for liquidity risk measurement, standards and monitoring. Bank for International Settlements (2010)
8. Basel Committee on Banking Supervision: Liquidity stress-testing: a survey of theory, empirics and current industry and supervisory practices. Bank for International Settlements, Working paper No. 24 (2013)
9. De Haan, L., et al.: Banks' responses to funding liquidity shocks: lending adjustment, liquidity hoarding and fire sales. Dutch Central Bank, Working Paper No. 293 (2011)
10. Van Den End, J.W.: Liquidity Stress-tester, A model for stress-testing banks' liquidity risk. DNB Working papers (2009)
11. Wong, T.C., et al.: A liquidity risk stress-testing framework with interaction between market and credit risks. HongKong Monetary Authority Working Paper (2009)
12. Liliana, S., et al.: Modelling Correlated Systemic Liquidity and Solvency Risks in a Financial Environment with Incomplete Information. International Monetary Fund (2011)

13. Quagliariello, M.: Stress-testing the banking system: methodologies and applications. Cambridge University Press (2009)
14. Schmieder, C., et al.: IMF wide system stress-test. International Monetary Fund (2011)
15. Burrows, O., Learmonth, D., McKeown, J.: RAMSI: a top-down stress-testing model. Financial Stability, Paper No. 17 (2012)
16. Karlsson, S.: Forecasting with Bayesian Vector Autoregressions. Prepared for the Handbook of Economic Forecasting 2 (2012)
17. Todd, R.M.: Improving Economic Forecasting With Bayesian Vector Autoregression. Quarterly Review 8(4) (1984)
18. Aikman, D., et al.: Funding liquidity risk in a quantitative model of systemic stability. Bank of England, No. 372 (2009)
19. Nbura, M., et al.: Large Bayesian vector auto regressions. J. Appl. Econ. 25, 71–92 (2010)
20. Steinbacher, M., et al.: Credit Contagion in Financial Markets: A Network-Based Approach. Available at SSRN 2068716 (2013)
21. Vietnamese financial-economical data source, http://cafef.vn
22. Vietnam Banking Survey, https://www.kpmg.com/Advisory-FS-Vietnam-Banking-Survey-2013.pdf

Part VII
Transportation

Optimal Nodal Capacity Procurement

Marina Dolmatova

Faculty of Computational Mathematics and Cybernetics,
Lomonosov Moscow State University, 119991 Moscow, Russia
ms.marina.dolmatova@gmail.com

Abstract. The paper demonstrates the application of mathematical optimization (non-linear, mixed integer problem) to solve the transportation problem for a multi-node capacity market. Proposed the model that assures adequate cost minimizing generating resources selection and prices capacity within every node on the basis of price bids of market participants and power system structure, and suggest electric power flow distribution. The proposed method of deficiency nodes allocation and pricing can be used for estimation of additional (deficient) capacity value.

Keywords: nodal market, capacity market, optimization, mixed integer program.

1 Introduction

Different optimization mechanisms are used to improve short-term and long-term energy system performance. Bohn et al. [2] presented analysis of optimization model of electricity spot pricing for real power flows over space and time, and offered ex-post transmission pricing given loads and location of binding constraints. The profound research by Schweppe et al. [3] develops optimal pricing for electricity spot markets that respects the particular conditions of electric power transmission systems ignoring some physical parameters of real power dispatch problem. The authors show that optimal electricity prices can be obtained as a byproduct of economic dispatch. Another direction of research is the concept of a dispatchable network proposed by O'Neil [4] and further developed in [5] and [6]. Ruiz et al. [6] propose a lossless linear mixed integer formulation of transmission control. In terms of long-term efficiency, an important role plays generating capacities profile of the energy system that can be created or assessed by means of capacity procurement procedures.

One of the ways to organize power market is to divide it into electricity and capacity markets. The electricity market is intended to cover mainly the variable production costs, while the capacity market is intended for covering capital costs of new capacity. The term *capacity* represents the commitment of an energy producer to maintain the sold volume and be ready to supply it whenever it is needed during the corresponding delivery period. Capacity market models were initially designed as an instrument to increase investment incentives for electricity suppliers. The main idea of these markets is to force and make it profitable

© Springer International Publishing Switzerland 2015   415
H.A. Le Thi et al. (eds.), *Model. Comput. & Optim. in Inf. Syst. & Manage. Sci.*,
Advances in Intelligent Systems and Computing 360, DOI: 10.1007/978-3-319-18167-7_36

for firms (suppliers) to build as many capacities of the needed type and quality on the defined territories as the whole power system requires. Pricing in such markets is effected within local areas (nodes) thus allowing to establish local price signals for consumers and suppliers of capacity. Not all available capacity is selected and paid for through capacity auction, thus increasing competition between capacity suppliers. The return for the selected capacity is guaranteed, whilst capacity not selected at the capacity auction is not paid for. New capacity is a key driver for solid earnings growth in the market since it receives considerably higher capacity payments than the old capacity and gives additional profit from sale of new volumes. The specificity of the capacity auctions is similar to the electricity auctions. The capacity procurement uses the static network, however, the network structure, redistribution of flows rules given lossless transmission assumptions and nonlinearity of the problem make the formulation of these models a subject of advanced study.

Rudkevich et al. [7] formulated the nodal capacity expansion problem in the form of a capacity auction based on the stochastic engines. The method focuses mainly on resource adequacy and provides combined resource adequacy assessment and capacity selection. However, its computation complexity is rather high and implementation appears feasible only for simple networks. The present paper considers a capacity procurement for a network system and proposes a solution of two main market tasks: determination of nodal prices and selection of units (generators) supplying electric capacity. Ahuja et al. [1] provides a detailed treatment of static network flow theory. The procurement model addresses the flow distribution issues and deficient nodes allocation, which may have significant impact in large network implementation. The model specification is given in the second section. The third section offers a variant of solution where the optimization process is divided in three stages to assure explainable results and adequate power flow distribution and to ease the computation. In the first stage the nonlinear mixed integer problem is formulated. Its solution provides allocation of deficient nodes. The second stage presents solution of the linear modification of the initial problem that respects the first stage results. Since the strong duality theorem holds for linear problems, this allows joint optimization of capacity and pricing. The inability to allocate electrical flows directly among the lines causes that for a node connected to two and more lines, one cannot control how much of the out-flow goes along the each line. There is a need to propose a power flow distribution algorithm for lossless transmission, which is done in the third stage. The fourth section provides a sample calculation for the 15 nodes test system.

2   Capacity Procurement Model and Constraints

Consider a stylized optimization model of a multi-nodal capacity market. Omit some technical peculiarities which are redundant in our case. A node can be seen as a territory, inside which constraints on power flows can be disregarded. The model computation should provide the selection of the optimal capacity suppliers profile based on their price bids, and determine nodal prices. Consider a network $M = (N, L)$ with $i \in N$ nodes, lines $(i, j) \in L$, characterized by flows along each line given by the variables matrix $F = [f_{i,j}]$ of order $N \times N$, where $f_{i,j} = 0$ for $(i, j) \notin L$. The flow from node i to node j must be the opposite of the flow from j to i:

$$f_{i,j} = -f_{j,i} \tag{1}$$

The transmission is lossless and a flow in each line must not exceed maximum safe capacity $\overline{F} = [\overline{f}_{i,j}]$. Simultaneous full load of capacity between some pairs of nodes can be unfeasible as these flows affect the same power system's technical restrictions (for instance, use the same network resources). Thus, a flow $f_{i,j}$ along line (i, j) is affected by flows along other lines (k, l). They form a group of lines $gr \in G$ such that each flow $f_{i,j}$, $(i, j) \in gr$ of the group contributes to the constraint $\overline{f}_{gr}$ with weight $w_{i,j}^{gr}$. Constraints on flows between nodes and groups of nodes are given as:

$$f_{i,j} \leq \overline{f}_{i,j},$$
$$\sum_{(i,j) \in gr} w_{ij}^{gr} \cdot f_{i,j} \leq \overline{f}_{gr} \text{ for } \forall(i,j) \in L, \forall gr \in G. \tag{2}$$

In each node $i \in N$ there is a set of generating units (generators) $u \in U_i$ located in that particular node. $\bigcup_{i \in N} U_i = U$ is a set of all the generators in the power system. Generators $u \in U_i$ in node i are characterized by price (cost) bid vector $P_i = [p_{u,i}]$ and maximum unit output vector $\overline{Q}_i = [\overline{Q}_{u,i}]$ of order U_i. Vector of variables $Q_i = [Q_{u,i}]$ represents selected (loaded by optimization process) capacity volumes and is non-negative and constrained: $0 \leq Q_i \leq \overline{Q}_i$, for each $i \in N$. Nodal demand D_i is fixed, specific for each node $i \in N$. Each node can be a source and a sink as well.

The problem must be stable to existence of deficient nodes. Deficient node is a node where the sum of aggregate capacity supply in the node and maximal capacity inflow to the node are lower than the nodal demand. For such a case an implied generator variable $\tilde{Q}_i$ is defined. Non-zero value $\tilde{Q}_i$ determines volume of unsupplied demand in node i and indicates that the node is deficient. Capacity of generators located in a deficient node i covers demand only in that node. The deficient node can only consume, hence the outflow from the node is non-positive:

$$\tilde{Q}_i > 0 \Rightarrow F^i \leq 0 \tag{3}$$

There is no charge for capacity transmission. Thus, free loop flows may occur, replace the generation and cover the demand, affecting the optimal flow values. Denote h_i the volume of inflow in a node $i \in N$. Variables h_i, $i \in N$ are defined from the system (4), minimize volumes of flows along lines and prevent loop flows.

$$f_{ij} = h_i + h_j = 0, \forall(i,j) \in L, i < j. \tag{4}$$

The system must meet the balance condition. For each node i the aggregate capacity flowed out of the node and demand in the node must be equal to selected capacity volumes of generators located in the node taking into account possibility of deficiency:

$$(F1_N)^i + D_i = 1_{U_i}{}^T Q_i + \tilde{Q}_i, \ \forall i \in N \tag{5}$$

Fig. (1, left) shows the 4-node example of flow distribution with generation allocated in nodes 1, 3 and consumption in nodes 2, 4. The balance condition (5) and flow-minimizing system (4), for this network are written as follows:

$$\text{balance} \begin{cases} -f_{12} - f_{13} = -120 \\ f_{12} - f_{23} - f_{24} = 170 \\ f_{13} - f_{23} = -180 \\ f_{24} = 30 \end{cases} \quad \text{flow distribution} \begin{cases} f_{12} = -f_{13} - f_{23} - f_{24} \\ f_{13} = -f_{12} + f_{23} \\ f_{23} = f_{12} + f_{13} - f_{24} \\ f_{24} = f_{12} - f_{23} \end{cases}$$

The solution gives the flows values: $f_{12} = -f_{21} = 50$, $f_{13} = -f_{31} = 70$, $f_{23} = -f_{32} = -150$, $f_{24} = -f_{42} = -30$ (see Fig. (1, right)).

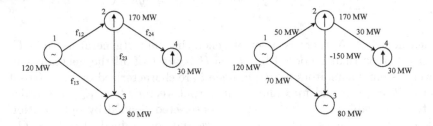

Fig. 1. 4-node flow distribution example

The cost minimization criterion is used to find the capacities values. A non-linear modification of the mixed integer problem of centralized one-sided auction with fixed demand is used. The objective function is constructed as:

$$\min_{Q_i, \tilde{Q}_i, h_i, I_i} \sum_{i \in N} P_i^T Q_i + \frac{K}{D_i} \cdot \tilde{Q}_i + \epsilon \cdot h_i^2 +$$
$$+ I_i \cdot \sum_{j \in N} P_j \cdot \overline{Q}_j, \tag{6}$$
$$K = \max_{u \in U, i \in N} p_{u,i} \cdot \max_{i \in N} D_i$$

The first term is the cost of the selected capacity units. The second term provides that implied generators are loaded least of all. Their weight in the objective function is the penalty coefficient K which is guaranteed to be greater than cost of any other fully loaded generator. In case there might be several deficient nodes which are adjacent to the node with excess of its own supply (supply of generators allocated in this node), then division by D_i in this term ranks the deficient nodes load in order of descending nodal demand. The third term $\epsilon \cdot h_i^2$ provides minimization of flows either positive (inflows) or negative (outflow). The value of the penalty coefficient ϵ is sufficiently small, depends on the computation accuracy and contributes slight impact on the objective function (6). The last term provides mitigation of the number of implied generators, and thus deficient nodes, by making existence of such a generator extremely expensive and equal to total cost of capacity in the system. This follows from the assumption that having one big deficient node is better then several small ones. Integer variables I_i, $i \in N$ identify whether the node is deficient:

$$\forall i \in N \; I_i = \max\{0, \mathrm{sgn}\tilde{Q}_i\} \tag{7}$$

3 Deficient Nodes Location, Capacity Selection and Pricing

The objective function (6) and all the constraints $(1-5)$, (7) form the mixed integer non-linear problem of **the first stage**. The complexity is caused by the need to set deficient nodes and regulate the flows. The form of the problem allows to use CPLEX Solver for computation. The solution gives optimal values of variables $I_i, \tilde{Q}_i, f_{ij}, h_i$, $i \in N, (i,j) \in L$. Thus the directions, values of flows between nodes of the system are known and the locations of implied generators are identified as well as their capacity, which is the volume of the imbalance in the corresponding node (if such exist).

Pricing and units profile determination in **the second stage** requires modification of the problem. Denote the obtained in the first stage values as $\tilde{Q}'_i$, $i \in N$, $F' = [f'_{ij}], (i,j) \in L$. Reduce the number of the flow variables f_{ij} redefining them:

$$0 \le f_{ij} = f_{i,j} \cdot \max\{0, \mathrm{sgn}f'_{i,j}\}, \; \forall(i,j) \in L \tag{8}$$

The linear modification of the problem is formulated as:

$$\min_{Q_i, \tilde{Q}_i} \sum_{i \in N} P_i^T Q_i + \max_{i \in N}(P_i) \cdot \tilde{Q}_i$$

$$\text{subject to } (\mathbf{1_N^T}F)_i - (F\mathbf{1_N})^i + \mathbf{1_{U_i}^T}Q_i + \tilde{Q}_i = D_i, \; \forall i \in N, \; (9')$$

$$f_{ij} \le \overline{f}_{ij}, \forall(i,j) \in L,$$

$$\sum_{(i,j) \in s} w_{ij}^{gr} \cdot f_{ij} \le \overline{f}_{gr}, \forall gr \in G,$$

$$Q_{u,i} \le \overline{Q}_{u,i}, g \in U, \tag{9}$$

$$-f_{ij} \le -f'_{ij} + r, \forall(i,j) \in L,$$

$$-\tilde{Q}_i \le -\tilde{Q}'_i + r, \forall i \in N,$$

$$\tilde{Q}_i \le \tilde{Q}'_i + r, \forall i \in N,$$

$$Q_{u,i}, \; \tilde{Q}_i, \; f_{i,j} \ge 0, \; g \in U, i \in N, (i,j) \in L.$$

This linearization provides that, according to the well known theorem ([8]), dual variables to constraints (9') determine the optimal prices in corresponding nodes. In general case when the flow constraints are saturated, the nodal prices are different. From the complementary slackness condition follows that the dual variable to unsaturated constraint and thus the difference between nodal prices for unsaturated constraints is zero.

The short form of the linear problem (9) can be written as:

$$\max_{Q_j} \sum_{j=1}^{n-L} (-p_j Q_j)$$

$$\text{subject to } \sum_{j=1}^{n} a_{ij} Q_j = b_i, \ i \in \{1, N\}, \qquad (\mu_i)$$

$$\sum_{j=1}^{n} a_{ij} Q_j \leq b_i, \ i \in \{N+1, m\}, \qquad (\mu_i)$$

$$Q_j \geq 0, \ i \in \{1, n\}.$$

(10)

The following notation is used in (10). $Q_{n \times 1} = \{(Q_u)_U, (f_{ij})_L, (\tilde{Q}_i)_N\}$ is a vector of length $n := U + L + N$, represent capacity variables and includes volumes of selected capacity, flows in lines and load of implied generators. $p_{1 \times n} = (p_j)_n$ is a set of the objective function coefficients; $A_{m \times n} = (a_{ij})_{m \times n}$, where $m := 2N + 2L + G + U$, is a m-by-n matrix representing equality and inequality constraints. $b_{m \times 1} = (b_i)_{m \times 1}$ is a column vector of the right side of the constraints. Denote μ_i, $i = \overline{1, m}$ the dual variables. The values of dual variables $\mu_i, i = \overline{1, N}$ to the nodal demand and supply balance equalities (9') are interpreted as nodal prices. If there is no supply deficit in the market, this value represents the cost of supply of the last 1 MW of covered demand with respect to inflows. By the last MW meant the volume of capacity that is supplied by the most expensive generator among the selected generators (Fig. 2). The solution to this problem gives the optimal values of volume variables vector: $\vec{Q} = (Q_{u,i})_{u \in U_i, i \in N}, (f_{i,j})_{(i,j) \in L}, (\tilde{Q}_i)_{i \in N}$ and nodal prices μ_i.

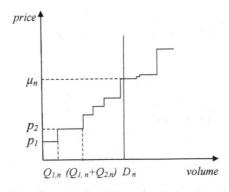

Fig. 2. Example of nodal the supply and demand curves

The third stage aims to better approximate the found flow distribution. Matpower ([9], [10]) power flow problem solution is adapted for the model. Enumerate all the nodes and lines in the system and construct connection matrix $C_{L \times N}$ for lines and nodes of the power system. The entry of the matrix is defined as

$$c_{i,j} = \begin{cases} 1, & \text{if node } j \text{ is the starting node for line } i, \\ -1, & \text{if node } j \text{ is the end node for line } i, \\ 0, & \text{else.} \end{cases}$$

Also admittance matrix $A_{N \times N} = C^T C$ is needed. The balance vector $B_{N \times 1}$, where $b_j = \sum_{u \in U_j} Q_{u,j} - D_j$ is obtained from the second stage results. Define a slack node j_s as the first element of $B_{N \times 1}$ with a positive balance (excess generation). Compute the minor MB_{j_s} of the balance matrix B and the minor $MA_{j_s j_s}$ of the admittance matrix A thereby removing the slack node. Construct the voltage vector $MV_{(N-1) \times 1} = MA^{-1}MB$ for all but slack nodes. Obtain the voltage vector V_N by adding the j_s entry: $v_j = \begin{cases} 0 & \text{for } j = j_s \\ mv_j & \text{for } j \neq j_s \end{cases}$. The flows $f_{i,j}$ in lines are the results of connection and voltage matrices multiplication: $F = CV$. This completes the solution of capacity procurement problem addressing also deficient nodes allocation, pricing and power flow distribution aspects.

4 Sample Calculation

This section tests the capacity procurement using 15 node network. The flow limits between lines are listed in Fig. 3. The system is not short on resources, but has a deficient node because of binding capacity constraint. Fig. 4 shows the summary of system parameters and calculation results. As the result of the optimization process, node 1 is deficient and flow limits constraints are binding for seven lines (Fig. 3), which causes differences in nodal prices.

Source node	1	2	3	4	5	6	7	8	9	10	11	12	13	14	15
1	0	15	0	0	0	0	0	0	0	0	0	0	0	0	0
2	23	0	30	0	45	0	0	0	0	0	0	0	0	0	0
3	0	45	0	17	0	0	0	0	0	0	0	0	0	38	0
4	0	0	15	0	23	0	0	0	0	0	0	0	0	0	0
5	0	30	0	15	0	45	0	0	0	0	0	0	0	0	0
6	0	0	0	0	23	0	15	0	0	0	0	45	0	0	0
7	0	0	0	0	0	23	0	45	0	23	0	0	0	0	0
8	0	0	0	0	0	0	15	0	15	30	0	0	0	0	0
9	0	0	0	0	0	0	0	23	0	0	0	0	0	0	0
10	0	0	0	0	0	0	15	30	0	0	45	0	0	0	0
11	0	0	0	0	0	0	0	0	0	23	0	15	45	0	0
12	0	0	0	0	0	0	0	0	0	0	15	0	0	0	0
13	0	0	0	0	0	45	0	0	0	0	23	0	0	30	0
14	0	0	30	0	0	0	0	0	0	0	0	0	30	0	10
15	0	0	0	0	0	0	0	0	0	0	0	0	0	10	0

Fig. 3. Maximum transmission capacities of the test system

Node	Demand, MW	Number of units	Offer, MW	Cleared, MW	Min bid	Max bid	Nodal Price	Deficiency volume, MW
1	300	9	200	200	42	201	201	78
2	500	14	600	525	28	222	186	-
3	200	11	300	160	54	241	186	-
4	50	8	100	60	49	240	186	-
5	150	7	200	150	32	212	186	-
6	100	8	250	183	30	179	96	-
7	90	9	140	140	53	232	240	-
8	230	9	150	150	58	193	240	-
9	80	8	100	93	113	240	240	-
10	120	9	100	100	50	238	240	-
11	50	9	150	105	33	242	222	-
12	70	8	120	70	25	228	228	-
13	270	9	180	163	20	231	222	-
14	100	10	200	125	47	243	186	-
15	50	4	80	60	24	228	149	-

Fig. 4. Input data and results of test system calculation

5 Conclusion and Further Research

The paper proposed the optimization model of capacity procurement and the description of the solution process. The focus is on the main challenges of network market often overlooked in one-node models and needed for the adequate optimization solution. The results satisfy real power system constraints such as flow limits on transmission lines and groups of lines, suggest electric power flow distribution, and joint pricing and capacity selection. The lossless transmission system assumption can be relaxed and the model modified with linearized power flow assumptions to address losses. However, the specificity of capacity as a product and capacity auctions, which are often conducted for 'node' territories larger than the ones in electricity markets and with respect to significant reserves, make the consideration of AC network rules questionable for capacity markets. The model in the paper can be applied to energy system testing. The proposed method of deficiency nodes allocation and pricing can be used for estimation of additional (deficient) capacity value. This assessment is also important for analysis of possibility and profitability of investment in network and/ or power capacity development. The future work is the study of simultaneous electric capacity and transmission network optimization. Also of interest is the effect of penalty coefficients approach choice on results for large networks.

References

1. Ahuja, R., Magnanti, T., Orlin, J.: Network Flows: Theory, Algorithms, and Applications. Prentice-Hall, Inc., Upper Saddle River (1993)
2. Bohn, R.E., Caramanis, M.C., Schweppe, F.C.: Optimal Pricing in Electrical Networks Over Space and Time. Rand Journal of Economics 15, 360–376 (1984)
3. Schweppe, F.C., Caramanis, M.C., Tabors, R.D., Bohn, R.E.: Spot Pricing of Electricity. Kluwer Academic Publishers, Norwell (1988)

4. O'Neill, R.P., Baldick, R., Helman, U.: Dispatchable transmission in RTO markets. IEEE Transactions on Power Systems 20(1), 171–179 (2005)
5. Fisher, E., O'Neill, R., Ferris, M.: Optimal transmission switching. IEEE Transactions on Power Systems 23(3), 1346–1355 (2008)
6. Ruiz, P.A., Rudkevich, A., Caramanis, M.C., Goldis, E., Ntakou, E., Philbrick, C.R.: Reduced MIP formulation for transmission topology control. In: Proc. 50th Allerton Conf. on Communications, Control and Computing, Monticello, IL, pp. 1073–1079 (2012)
7. Rudkevich, A.M., Lazebnik, A.I., Sorokin, I.S.: A Nodal Capacity Market to Assure Resource Adequacy. In: IEEE Proceedings of 45th Hawaii International Conference on System Sciences, pp. 1876–1887 (2012)
8. Germeyer, Y.B.: Introduction to the Theory of Operations Research, Nauka, Moscow (1971)
9. Zimmerman, R.D.: Matpower 4.0b4 User's Manual, Power Systems Engineering Research Center (Pserc) (2010)
10. Zimmerman, R.D., Murillo-Sánchez, C.E., Thomas, R.J.: MATPOWER's Extensible Optimal Power Flow Architecture. In: 2009 IEEE Power and Energy Society General Meeting, pp. 1–7 (2009)

Real-Time Ride-Sharing Substitution Service in Multi-modal Public Transport Using Buckets

Kamel Aissat[1] and Sacha Varone[2]

[1] University of Lorraine - LORIA, Nancy, France
kamel.aissat@loria.fr
[2] University of Applied Sciences and Arts Western Switzerland (HES-SO),
HEG Genève, Switzerland
sacha.varone@hesge.ch

Abstract. We consider a mix transportation problem, which allows to combine a multi-modal public and a ride-sharing transports, in a dynamic environment. The main idea of our approach consists in labelling interesting nodes of a geographical map with information about either riders or drivers, in so-called buckets. Based on the information contained in these buckets, we compute admissible ride-sharing possibilities. To restrict the needed amount of memory, among the different stops along a public transportation path, we only consider the transshipment nodes, where travellers have to make a change between two modes. Each of those stops are potential pick-up or drop-off stops for ride-sharing. We consider a drivers' maximal waiting time, as well as the maximal driving detour time depending on the actual drive. Each new drive activates a search for new ride-sharing of existing riders. Each new ride activates another process which searches for potential drivers. Among all admissible ride-sharing possibilities, only those which best improve the earliest arrival time are selected. We provide numerical results using real road network of the Lorraine region (FR) and real data provided by a local company. Our numerical experiment shows a running time of a few seconds, suitable for a new real-time transportation application.

Keywords: Multi-modal, Ride-sharing, Public transportation, Real-time.

1 Introduction

Mobility within a region can be done using different transportation modes, either public or private, which might be a function of the preferences or of the possibilities of the commuter. We propose an algorithmic approach to the real-time multi-modal earliest-arrival problem (EAP) in urban transit network, using public transportation and ride-sharing, so that the trip duration is minimized. We start from using public transportation, included walking sub-paths and then, considering (quasi) real-time ride-sharing opportunities, try to transfer sub-paths from public modality to ride-sharing modality. For that purpose, we consider riders, who provide requests to go from one location to another one, and drivers who offer ride-sharing. Riders' goal is to find a quickest trip in a dynamic context that effectively combines the use of several modes of transportation.

© Springer International Publishing Switzerland 2015   425
H.A. Le Thi et al. (eds.), *Model. Comput. & Optim. in Inf. Syst. & Manage. Sci.*,
Advances in Intelligent Systems and Computing 360, DOI: 10.1007/978-3-319-18167-7_37

The search for a quickest or shortest path has been well studied in the literature, and has now very efficient algorithms to solve it, in the order of a millisecond for a continental trip: the authors in [2] survey recent advances in algorithms for route planning. Computation of shortest paths are key components in route planning, but not sufficient to deal with multi-modal problems, since they involve multi-criteria paths, transition or waiting times, etc. which is usually not taken into account in shortest paths on pure road networks. For example, the authors in [1] include several features in their objective function, so that they focus on the modal change node, and propose a two-step algorithm for computing multi-modal routes. An exact algorithm considering arrival and departure time-windows has been proposed in [5]. Multi-criteria search by computing the Pareto set is done in [4].

In public transportation problems, solving earliest-arrival problems (EAP) knowing the departure and arrival station, departure time and timetable information is reviewed in [6,7]. Timetabling information and EAP solving is nowadays often available on-line for users. Real-time journey using ride-sharing opportunities faces the commuting point problem: it has to be decided where the pick-up and drop-off locations have to be located. [3] define this as the 2 synchronization points shortest path problem (2SPSPP). The time complexity of their approach prevents its use in real-time ride-sharing. In our approach, we first find a quickest path using public transportation and consider its different transit stops. We then only allow pick-up and drop-off around those stops to reduce the search space. We then compute a potential drivers' list and further check for admissible drivers. Finally, if a ride-sharing is possible, we update the routes for both rider and driver.

2   Problem Description

The problem to be solved is presented as follows: a user u wishes to go from an origin point O_u to a destination point D_u. He might use either public transportation, ride-sharing or a combination of both. All public transport timetabling are assumed to be known or at least accessible easily. In France, one might use the Navitia API[1]. The main idea of our approach consists in storing for each driver and for each rider information about his travel if passing trough a pick-up or drop-off stop. Each new ride arriving in the system launches a request for a public transportation path to determine the different stops. Each of those stops are considered as potential stops for ride-sharing. A search for drivers around those stops is then run, so that ride-sharing would improve the earliest arrival time to destination, under time-constraints: driver's waiting time, and detour time depending on the actual drive. Another process is run for each new drive, so that information about potential pick-up or drop-off stops are stored.

The network is represented as a directed graph model $G(V, E)$, in which V represents a set of nodes and E the set of arcs. Nodes model intersections and arcs depict street segments. A non negative weight is associated to each arc: its cost.

---

[1] http://www.navitia.io

A path that minimizes the sum of its cost is called a quickest/shortest path. We define a *stop* to be the location for which a transit or road node exists. Stops correspond to bus stops, subway stations, parking, etc. In this multi-modal context, G is modeled as the union of multiple networks, representing different travel modes (bike, car, public transport, ...). Our model uses public transport network and road network. TABLE 1 lists the notations used throughout this paper.

Table 1. A List of Notations

Notation	Definition
$s \rightsquigarrow e$	Driving quickest path between s and e.
$\delta(s,e)$	Duration of a quickest path between s and e.
$d(s,e)$	Distance as the crow flies between s and e.
$\hat{\delta}(s,e)$	Estimated smallest duration from s to e, such that $\hat{\delta}(s,e) = \frac{d(s,e)}{v_{\max}}$, where $v_{\max}$ is the maximal speed.
λ_k	Detour coefficient, $\lambda_k \geq 1$.
$\tau(x)_{ab}$	Time required for moving from modality a to b at x.
$t_a^u(x,m)$	Arrival time at x using the m transport modality for user u.
$t_d^u(x,m)$	Departure time at x using the m transport modality for user u.
$w_{\max}^k$	Maximum waiting time for driver k at pick-up point.
L_x	List of potential drivers at x as pick-up stop.

We will further note as p the public transportation modality, and as c the car modality. The ride-sharing process is the following: a user (rider/driver) generates a request through his smartphone. The request is sent to a central server. Potential drivers/riders are then contacted. Those might accept or reject the request. If several drivers/riders have accepted the request, then the rider/driver might choose which one he prefers.

A driver k is characterized by his position O_k at current time t_0^k and his destination D_k. A driving route from its origin to its destination through a stop x_i has to respect a maximum waiting time at x_i, and a maximum detour time, the latter being a multiple of a shortest duration between the driver's origin and destination. An admissible ride-sharing route for the driver and the rider is defined as follows:

Definition 1. (Admissible ride-sharing)
We say that a path $x_i \rightsquigarrow x_j$ is an admissible ride-sharing for a driver k and a rider u if and only if a maximal driver's waiting time $w_{\max}^k$ at a pick-up location (1), a maximum driver's detour (2) and a rider's latest arrival time (3) constraints are satisfied, i.e,

$$t_a^u(x_i,p) + \tau(x_i)_{pc} - t_a^k(x_i,c) \leq w_{\max}^k \qquad \text{waiting} \quad (1)$$

$$\delta(O_k,x_i) + \max\left\{t_a^u(x_i,p) + \tau(x_i)_{pc} - t_a^k(x_i,c), 0\right\} + \delta(x_i,x_j)$$
$$+\delta(x_j,D_k) \leq \lambda_k \delta(O_k,D_k) \qquad \text{detour} \quad (2)$$

$$\max\left\{t_a^u(x_i,p) + \tau(x_i)_{pc}, t_a^k(x_i,c)\right\} + \delta(x_i,x_j) \leq t_a^u(x_j,p) \qquad \text{arrival} \quad (3)$$

Constraint (1) is composed of the arrival time $t_a^u(x_i,p)$ at x_i by the traveller using public transportation plus the transshipment time $t_a^k(x_i,c)$ to walk from the stop towards the meeting point with the driver, and the arrival time $t_a^k(x_i, c)$ at x_i for the driver. If their difference is positive, then the driver have to wait. Else, the rider has to wait. Constraint (2) is the sum of the duration $\delta(O_k, x_i)$ for a driver k to reach stop x_i, plus a potential waiting time, plus the ride duration $\delta(x_i, x_j)$, plus the duration $\delta(x_j, D_k)$ to reach its final destination. Constraint (3) expresses that the total duration for the rider to reach stop x_j using a ride-sharing mode between x_i and x_j should not exceed its arrival time at x_j using only public transportation.

Each constraint (1), (2) and (3) takes into account $\tau(x_i)_{pc}$ the time required for moving from the public transportation modality to the ride-sharing modality.

For each new offer $O_k D_k$ in the system, related information about time, driver k are stored at each potential stop where driver k might meet a rider, while respecting its *detour* constraint.

3   Methodology

In a dynamic environment, announcements of riders and drivers are arriving continuously at arbitrary times. Hence, matches between riders and drivers have to be computed many times during the day which implies that future requests are unknown during each planning cycle. So, the decision to wait or not for other requests depends on the choice of the user. The underlying application assumes that GPS coordinates of drivers can be known, or at least estimated, continuously. We hereby describe the process for each new driver and each new rider entering the system, and how we remove outdated information. We use a double list of so-called buckets which contain information about riders and drivers: the system maintains for each potential pick-up or drop-off node, information about which driver and/or which rider is/are potential ride-sharing members. In order to reduce the amount of memory, considered riders' journey are only those belonging to a public transportation mode.

Definition 2. *(Drivers bucket $B_d(v)$ $v \in V$)*
The drivers bucket $B_d(v)$ *associated with a node v is the set of triplets, composed of a driver k, his earliest arrival time at v and his latest departure time from v, such that the drivers might pick-up riders at node v without violating the detour constraint (2).*

$$B_d(v) := \{(k, t_a^k(v, c), t_d^k(v, c))| \ constraints \ (2) \ is \ satisfied \ \}. \tag{4}$$

A driver k is contained in $B_d(v)$ if and only if the duration of his trip via v does not exceed a maximal detour: $\delta(O_k, v) + \delta(v, D_k) \leq \lambda_k \delta(O_k, D_k)$. The entries of this driver k is added in the drivers bucket $B_d(v)$ by descending order according to his latest departure time $t_d^k(v, c) = t_o^k + \lambda_k \delta(O_k, D_k) - \delta(v, D_k)$ (the latest time is the first in the heap $B_d(v)$).

Definition 3. *(Riders bucket $B_r(v)$ $v \in V$)*
The riders bucket $B_r(v)$ *associated with a node v is the set of triplets, composed of a rider u, his arrival time at v and his departure time from v, using public transportation.*

$$B_r(v) := \{(u, t_a^u(v, p), t_d^u(v, p))\}. \tag{5}$$

In other words, $B_r(v)$ contains the list of riders and their pick-up time windows. Entries in $B_r(v)$ are sorted by ascending order according to arrival times.

We also maintain the OD paths with the time related information for each driver and each rider until she leaves the system.

3.1 Adding a Driver's Offer

Each new driver entering into the system generates new ride-sharing opportunities. Let's denote by k the driver, O_k his origin and D_k his destination. We compute the integration of those new ride-sharing offers by determining all possible pick-up/drop-off locations. For this purpose we use a modified bidirectional A^* algorithm (BA). The bidirectional search algorithm works with two simultaneous searches of the shortest path, one from the source node O_k and the second reversely from the target node D_k simultaneously, until "the two search frontiers meet". As in A^*, BA search is guided by a heuristic function that estimates the remaining duration to the target (in the forward search) and another from the source (in the backward search). We maintain two priority queues, one, as in simple A^*, from the source O_k, denoted by Q_1, and another one from the target D_k, denoted by Q_2. The latter is a forward search in the reversed graph G^{-1}, in which each arc (u, v) of G is replaced by (v, u) in G^{-1}.

Thus, for a given node v, the cost function used in the direct search (Q_1) is $\delta(O_k, v) + \hat{\delta}(v, D_k)$, whereas in the reverse search (Q_2), we use the cost function $\hat{\delta}(O_k, v) + \delta(v, D_k)$. (see figure 1)

At each iteration, we select the node with the smallest overall tentative cost, either from the source O_k or to the target D_k. In order to do this, a simple comparison between the minima of the two priority queues Q_1 and Q_2 is sufficient.

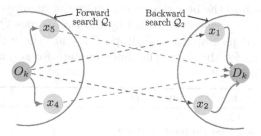

Fig. 1. Situation of BA algorithm before finding the duration $\delta(O_k, D_k)$. The two searches are guided by estimates of the remaining durations (dashed red lines for the first search Q_1 and dashed blue lines for the second search Q_2). Thus, only the left half-disc around O_k is visited for the first search Q_1, and the right half-disc around D_k for the second search Q_2. The solid lines correspond to the exact durations determined after each labelling from either Q_1 or Q_2.

Once we select a node v in one queue that has already been selected in the other queue, we get the first tentative cost of a shortest path from O_k to D_k. Its cost is the cost of the path found by the forward search from O_k to v, plus the cost of the path found by the backward search from v to D_k.

Even when the two parts of the tentative solution of the forward and the backward meet each other, the concatenated solution path is not necessarily optimal. To guarantee optimality, we must continue until the tentative cost of the current minima of the queues is greater than the current tentative shortest path cost (which then corresponds to the cost of the shortest path). Once the shortest path O_k to D_k is found, then $\lambda_k \delta(O_k, D_k)$ is known. At this level, the search doesn't stop but continue until the cost of the current node v is greater than $\lambda_k \delta(O_k, D_k)$. (see figure 2)

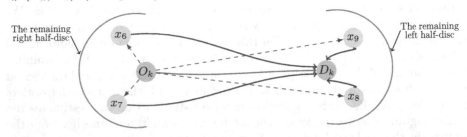

Fig. 2. Duration $\delta(O_k, D_k)$ is found by the BA algorithm. The search space still expands until all nodes with label smaller than $\lambda_k \delta(O_k, D_k)$ are labelled. At this point, the remaining nodes to be labelled are situated in the right half-disc around the target O_k or in the left half-disc around D_k. The red line is the duration $\delta(O_k, D_k)$. Only nodes that satisfy the detour constraint are kept.

In our case, we set $\lambda_k = 1.2$, which means that all customers allow trip duration at most 20% greater than that of a quickest path. For each node v settled by the two queues $\mathcal{Q}_1$ and $\mathcal{Q}_2$, the *detour* constraint (2) is verified. If valid, then the new element is inserted into the *bucket* $B_d(v)$ by decreasing order according to their latest departure times. Then the riders bucket $B_r(v)$ is scanned in order to find the best rider with whom the driver will share his trip. Algorithm 1 is run for each node settled by the two queues $\mathcal{Q}_1$ and $\mathcal{Q}_2$. Rider u^* and its associated pick-up x_{i*} and drop-off x_{j*} are those that generate the greatest time-savings for the rider. The different steps for adding a carpool offer are defined in Algorithm 2.

Algorithm 1. Scan the riders bucket $B_r(v)$

Require: Offer k, $B_r(v)$.
Ensure: Best rider u^*, Best drop-off stop x_{j*}.
    σ^* : Best time-savings generated for rider u^* having v as pick-up stop.
1: Initialization: $u^* \leftarrow -1$, $\sigma^* \leftarrow 0$, $x_{j*} \leftarrow -1$
2: **for all** u in $B_r(v)$ **do**
3:     **for each** stop x_j in path of rider u situated after v **do**
4:         $\sigma = t_a^u(x_j, p) - (\max\{t_a^k(v, c), t_a^u(v, p) + \tau(v)_{pc}\} + \delta(v, x_j))$
5:         **if** $k \in B_d(x_j)$ **and** (v, x_j) forms an *admissible ride-sharing* path for driver k
            and rider u **and** $\sigma > \sigma^*$ **then**
6:             $x_{j*} \leftarrow x_j$, $u^* \leftarrow u$, $\sigma^* \leftarrow \sigma$
7:         **end if**
8:     **end for**
9: **end for**

Algorithm 2. Adding carpool offer k

Require: Offer k, detour coefficient λ_k.
Ensure: Best rider u^*, Best pick-up x_{i^*}, Best drop-off x_{j^*}, Update drivers' list bucket
1: Initialization : $\mathcal{Q}_1.Insert(O_k)$ and $\mathcal{Q}_2.Insert(D_k)$
  $Meeting\_locations\_list \leftarrow \emptyset$, $Search = $ true, $\delta(O_k, D_k) \leftarrow \infty$
2: **while** $Search == True$ **do**
3:   $Current\_node \leftarrow \min(\mathcal{Q}_1.GetFirst(), \mathcal{Q}_2.GetFirst())$
4:   Mark $Current\_node$ as visited and remove it from appropriate queue
5:   **if** $Current\_node$ is marked as visited by both $\mathcal{Q}_1$ and $\mathcal{Q}_2$ **then**
6:     $Meeting\_locations\_list \leftarrow Meeting\_locations\_list \cup \{Current\_node\}$
7:     **if** $\delta(O_k, Current\_node) + \delta(Current\_node, D_k) < \delta(O_k, D_k)$ **then**
8:       Update $\delta(O_k, D_k) \leftarrow \delta(O_k, Current\_node) + \delta(Current\_node, D_k)$
9:     **end if**
10:     **if** $\min\big(\delta(O_k, Current\_node), \delta(Current\_node, D_k)\big) > \lambda_k \delta(O_k, D_k)$ **then**
11:       $Search = $ False
12:     **end if**
13:   **end if**
14:   Update the neighbors' cost of the $Current\_node$ and *insert* them in appropriate queue according to cost function $\big(\delta(O_k, Current\_node) + \hat{\delta}(Current\_node, D_k)$ for $\mathcal{Q}_1$ and $\hat{\delta}(O_k, Current\_node) + \delta(Current\_node, D_k)$ for $\mathcal{Q}_2\big)$.
15: **end while**
16: **for all** v in $Meeting\_locations\_list$ **do**
17:   **if** $\delta(O_k, v) + \delta(v, D_k) \leq \lambda_k \delta(O_k, D_k)$ **then**
18:     $t_a^k(v, c) = t_o^k + \delta(O_k, v)$, $t_d^k(v, c) = t_o^k + \lambda_k \delta(O_k, D_k) - \delta(v, D_k)$
19:     $B_d(v) := B_d(v) \cup \{(k, t_a^k(v, c), t_d^k(v, c))\}$
      {insert operation in descending order according to $t_d^k(v, c)$}
20:     Scan the riders bucket $B_r(v)$ using **Algorithm 1**
21:     Update best rider u^*, best pick-up x_{i^*} and best drop-off x_{j^*}
22:   **end if**
23: **end for**

3.2 Removing Outdated Information in Buckets

Each time a node v is visited, the bucket is updated so that all outdated information are removed. Since a bucket is sorted in decreasing time, the removal of its outdated elements can be done in linear time. The buckets are reinitialized each 24 hours in our application.

3.3 Adding a Rider's Request

Each new ride entering the system corresponds to an origin-destination $O_u D_u$ path, with an initial time t_0. The first action is to provide a public transportation path. A second action is to determine if there are some ride-sharing opportunities which decrease the arrival time to destination. The different steps are described below:

Step 1. The first step gets the multi-modal public transportation path $P = O_u, x_1, \ldots, x_{nbs}, D_u$ starting at time t_{O_u}, using a public transportation API. For each stop x_i in P, we associate arrival time $t_a^u(x_i, p)$ and departure time $t_d^u(x_i, p)$, as illustrated by Figure 3.

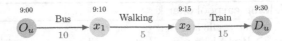

Fig. 3. A rider u start travelling at time $t_u = 9{:}00$ from his origin O_u, which is also the first stop x_0, to his destination D_u. Path P is composed of two transshipment stops $\{x_1, x_2\}$.

Then possible driving substitution sub-paths along P are computed. Durations of these driving paths will then be used to test if there is an improvement for the rider by switching from a public transportation to a ride-sharing. Only driving paths which decrease the earliest arrival time to stops x_j or D_u are kept. A pick-up time-window on each stop x_i is determined by the arrival time and departure time at x_i, as illustrated by Figure 4.

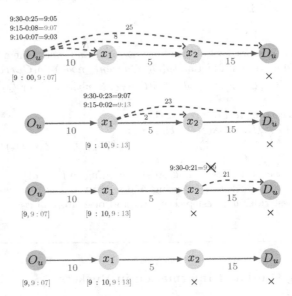

Fig. 4. Determination of pick-up time-window for each potential stop. The red numbers above the nodes are the calculation of the latest departure times, so that the arrival time at x_j is less than or equal to the arrival time using public transportation. The interval below a node is composed of the earliest and the latest departure time. There is no time-window associated with x_2, since it is not possible to decrease any arrival time from this stop using ride-sharing.

Step 2. The goal of this step is to find a set of drivers able to pick-up and drop-off the rider at some stops, within their associated time-windows. Two constraints for the benefit of drivers are considered: a maximum waiting time and a maximum detour. This phase also includes the determination of the drivers' quickest paths towards each stop, computed reversely from each pick-up stop. The information contained in the drivers bucket $B_d(v)$ allows to restrict the computation of the quickest paths. Thus, for each stop x_i a list of potential drivers L_{x_i} is created.

Step 3. This step searches for ride-sharing opportunities and select the best one. For each possible substitution path $x_i \rightsquigarrow x_j$, determined from the set of drivers in $L_{x_i} \cap L_{x_j}$, its *admissibility* is verified. Among all admissible ride-sharing $x_i \rightsquigarrow x_j$, only those allowing maximal savings are chosen. Algorithm 3 explains this process, which is illustrated by Figure 5. The same operation is repeated for each $x_i \in P$, and the best pick-up stop x_{i^*}, the best drop-off stop x_{j^*} as well as the driver k^* that generates the most positive time-savings are selected. Algorithm 4 describes it.

Algorithm 3. Scan the potential drivers' list L_{x_i}

Require: Demand u, potential drivers at x_i i.e. L_{x_i}.
Ensure: k^* : The best driver having x_i as pick-up stop and x_j as drop-off stop
         σ^* : Time-savings generated by sub-path (x_i, x_j).
1: Initialization : $\sigma^* \leftarrow 0$, $k^* \leftarrow -1$
2: **for all** k in L_{x_i} **do**
3:     **for each** $x_j \in P$, such that x_j is situated after x_i **do**
4:        **if** $k \in B_d(x_j)$ **and** (x_i, x_j) form an *admissible ride-sharing* path for the driver
         k and the rider u **then**
5:          **if** $\sigma^* \geq t_a^u(x_j, p) - \big(\max\{t_a^u(x_i, p) + \tau(x_i)_{pc}, t_a^k(x_i, c)\} + \delta(x_i, x_j) \big)$ **then**
6:             $\sigma^* \leftarrow t_a^u(x_j, p) - \big(\max\{t_a^u(x_i, p) + \tau(x_i)_{pc}, t_a^k(x_i, c)\} + \delta(x_i, x_j) \big)$
7:             Update the best driver k^*
8:          **end if**
9:        **end if**
10:    **end for**
11: **end for**

Algorithm 4. Best driver in all drivers buckets

Require: L_{x_i}, Demand u.
Ensure: Best driver k^*, Best pick-up stop x_{i^*}, Best drop-off stop x_{j^*}.
1: Initialization: $k^* \leftarrow -1$, $x_{i^*} \leftarrow -1$, $x_{j^*} \leftarrow -1$
2: **for all** x_i in P **do**
3:     Find the driver k with greater time-savings having x_i as pick-up stop, using
      Algorithm 3
4:     Update the best driver k^*, Update the best pick-up stop x_{i^*} and drop-off stop x_{j^*}
5: **end for**

Step 4. This last step is an updating process that occurs each time a ride-sharing (x_{i^*}, x_{j^*}) with positive time-savings has been found. The rider's route has to be updated to take into account his new arrival time $t_1 = \max\{t_a^{k^*}(x_{i^*}, c), t_a^u(x_{i^*}, p) + \tau(x_{i^*})_{pc}\} + \delta(x_{i^*}, x_{j^*})$ at x_{j^*}. The update also occurs on the driver's side: its route becomes $O_{k^*} \rightsquigarrow x_{i^*} \rightsquigarrow x_{j^*} \rightsquigarrow D_{k^*}$, where O_{k^*} is the current location of the driver. Then *Step 1* is run again to take into account the stop x_{j^*} as the new starting location at time t_1, as well as updating the rider's bucket.

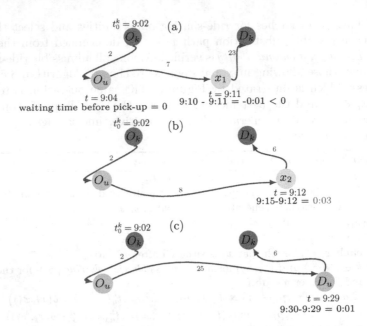

Fig. 5. Substitution sub-path for potential driver k contained in the list L_{O_u}. Driver $k \in L_{O_u}$ is contained in $\{L_{x_1}, L_{x_2}, L_{D_u}\}$. Ride-sharing (O_u, x_1) increases the earliest arrival time at x_1; therefore it is cancelled. Between ride-sharing (O_u, D_u) and (O_u, x_2), the last one better improves the arrival time at x_2; it is therefore the best substitution sub-path.

4   Numerical Results

Computational experiments have been conducted to evaluate the effectiveness of the proposed approach in terms of arrival time. Our practical application is tested on real data and real road network. Our methodology switches between car and public-transportation networks. Stop projections from public transportation to road network and computing shortest paths using car is done using Open-StreetMap[2]. Public transportation timetabling in Lorraine region is provided by the *navitia* API. The proposed methods were implemented in C# Visual Studio 2010. The experiments were done on an Intel Xeon E5620 2.4 Ghz processor, with 8 GB RAM memory.

Offers-Demands Data. Real data is provided by Covivo company[3]. These data concern employees of Lorraine region travelling between their homes and their work places. The real data instance is composed of 1513 participants. Among them, 756 are willing to participate in ridesharing service only as drivers. The rest of participants (i.e. 757) are willing to use other modes in order to reach

---

[2] http://www.openstreetmap.org
[3] http://www.covivo.fr

their destinations. Thus, the data instance is composed of 756 offers and 757 demands. The smallest and the greatest trip's distances are 2 km and 130 km, respectively. The set of offers and demands are filtered such that we can never find a driver's offer and a rider's demand which have either the same starting or ending locations. This way will also allow to test the flexibility of our approach compared to the traditional ride-sharing. The early departure time and the latest departure time for each trip are fixed at 7:00 a.m. and at 8:00 a.m, respectively. The detour time of the driver is fixed to at most 20% of his initial trip duration.

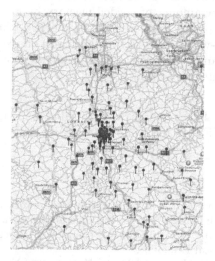

Fig. 6. Map visualization of ride-sharing offers. Red and blue points are the geocoded home and work locations, respectively.

Computational Results. The company which provided the data is still in test process and asked for an embargo of 18 months before the publication of numerical results. Some insight about the efficiency of our approach can nevertheless be given, but the lack of available benchmarks does not allow a direct comparison with previous research. First, this shows that a true real-time application is possible since the whole process running on a personal computer requires only a few seconds. Second, using the alternative hypothesis that the difference between the arrival time at destination using only public transportation and the arrival time at destination partially using ride-sharing, we can show that there is a statistically significant gain for the rider.

5 Conclusion

The problem described in this paper is a first step in designing new dynamic multi-modal transportation process. Multi modalities usually includes pedestrian, cycling, private car or public bus or train transportations. Our approach allows to also include ride-sharing without need changing the process of public-transport. As far as we know, it is the first attempt to solve a mix public-transport and ride-sharing problem in real time under matching constraints.

We have used a simple objective function, which, although appropriate for certain types of users, might be redefined in several terms: the total monetary cost for the rider, the deviation from the origin-destination trip for the driver, the number of transshipment stops, etc. Our approach can integrate those features as a weighted sum in the objective function.

Our approach is sensitive to the number of information to be stored. We have contained the necessary amount of needed memory by means of three features: a restriction of pick-up and drop-off location along a path, a limited detour constraint, and a time-window for pick-up and drop-off. Nevertheless, our practical implementation and tests on real data instances show the viability of our approach for a real-time application.

Acknowledgments. This research was partially funded thanks to the swiss CTI grant 15229-1 PFES-ES received by Sacha Varone.

References

1. Ambrosino, D., Sciomachen, A.: An algorithmic framework for computing shortest routes in urban multimodal networks with different criteria. Procedia - Social and Behavioral Sciences 108, 139–152 (2014), operational Research for Development, Sustainability and Local Economies
2. Bast, H., Delling, D., Goldberg, A., Müller-Hannemann, M., Pajor, T., Sanders, P., Wagner, D., Werneck, R.: Route planning in transportation networks. MSR-TR-2014-4 8, Microsoft Research (January 2014)
3. Bit-Monnot, A., Artigues, C., Huguet, M.J., Killijian, M.O.: Carpooling: the 2 synchronization points shortest paths problem. In: 13th Workshop on Algorithmic Approaches for Transportation Modelling, Optimization, and Systems (ATMOS), Sophia Antipolis, France, vol. 13328, p. 12 (September 2013)
4. Delling, D., Dibbelt, J., Pajor, T., Wagner, D., Werneck, R.F.: Computing multi-modal journeys in practice. In: Bonifaci, V., Demetrescu, C., Marchetti-Spaccamela, A. (eds.) SEA 2013. LNCS, vol. 7933, pp. 260–271. Springer, Heidelberg (2013)
5. Liu, L., Yang, J., Mu, H., Li, X., Wu, F.: Exact algorithms for multi-criteria multi-modal shortest path with transfer delaying and arriving time-window in urban transit network. Applied Mathematical Modelling 38(9-10), 2613–2629 (2014)
6. Müller-Hannemann, M., Schulz, F., Wagner, D., Zaroliagis, C.: Timetable information: Models and algorithms. In: Geraets, F., Kroon, L.G., Schoebel, A., Wagner, D., Zaroliagis, C.D. (eds.) Railway Optimization 2004. LNCS, vol. 4359, pp. 67–90. Springer, Heidelberg (2007)
7. Pyrga, E., Schulz, F., Wagner, D., Zaroliagis, C.: Efficient models for timetable information in public transportation systems. J. Exp. Algorithmics 12, 2.4:1–2.4:39 (2008)

Train Timetable Forming Quality Evaluation: An Approach Based on DEA Method

Feng Jiang[*], Shaoquan Ni, and Daben Yu

School of Transportation and Logistics, Southwest Jiaotong University, Chengdu, China
railwaylife@163.com, nishaoquan@126.com, happy_yx@hotmail.com

Abstract. Train timetable is the key file for train operation organization. China has a complex train network and a unified train operation organization, this makes it difficult to form a periodical timetable and the forming of train timetable is mostly done by manual work. This paper proposes an evaluation index for the quality of train timetable forming and solves the problem with a data envelopment analysis (DEA) method. A real-world train timetable in China is taken as a research case, the efficiency index expectation (REE) is 0.8412, efficiency index variance (REV) is 0.0044, and an efficient distribution function (EDF) with a 0.1interval is given.

Keywords: train timetable evaluation, data envelopment analysis, train path efficiency.

1 Introduction

The state owned China Railway Net has a complex network containing more than 110,000km railway, including more than 15,000km high-speed railway. The extremely busy railway network requires a safe and effective operation. The train operation requires an organization plan formed in advance, in most cases, called *train timetable* or *train diagram*. Generally, the train timetable stipulates the rail sector occupation time and period for every train path and the train timetable forming should consider an extra time added up to the "ideal running times" [1] to make the train timetable more robust [2-5]. Also lots research [1, 6-8] focuses on creating a suitable timetable (most are periodical) for real-world railway net.

In China, the large scale rail net operates by 18 railway bureaus under the unified coordination of China Railways Cooperation (CRC), the train timetable forming has a two-level construction: First, for cross-bureau train path, which means the train path`s start and terminal station not in the same bureau, the CRC will gather all the bureau timetable forming stuffs to negotiation, then decide the train path`s arrival time and departure time for each station, and make a cross-bureau frame timetable called cross-bureau train path frame timetable. Then for within-bureau train path, i.e. start and terminal station in the same bureau, the corresponding railway bureau organizes their timetable forming stuffs to arrange the train path under the cross-bureau train path frame timetable, the processes are shown in Fig.1.

---

[*] Corresponding author.

© Springer International Publishing Switzerland 2015
H.A. Le Thi et al. (eds.), *Model. Comput. & Optim. in Inf. Syst. & Manage. Sci.*,
Advances in Intelligent Systems and Computing 360, DOI: 10.1007/978-3-319-18167-7_38

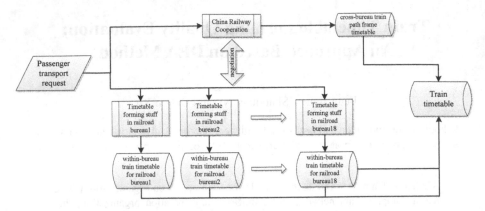

Fig. 1. Forming process of train timetable in China

Because of the unified transportation plan forming organization of CRC and complex transportation situation, it is almost impossible to create a periodic train timetable by computer. Until today, manually forming and adjusting of train path still play the dominant role in CRC`s train timetable forming. The train timetable will be adjusted for many times a year and will affect the railway organization dramatically. Therefore, the timetable forming quality should be evaluated for after it was made.

The forming of train timetable is series project which inquires transport data. Some data for relative departments is shown as Table.1.

Table 1. Transport Data Supplied by Relative Departments

Department	Transport data supposed to supply
Transport plan department	Number of train path needed, stop station for each train path
Track maintenance division	Top speed limit for each sector of rail track
Signal division	Headway for each train in each rail sector and station
Train operation division	Operating procedure and operating time standard

After the collection of transport data, the plan department, including passenger department and freight department, will make the train operation plan include the number and start/terminal station for each train path, known as train operation plan. Then the timetable forming department, called transportation department in China, determines the arriving and leaving time for each train path and forms the practical train timetable then publishes it. Besides, the train timetable is used to determine the working plan for the relative transport department. The train timetable making flow can be seen in Fig.2.

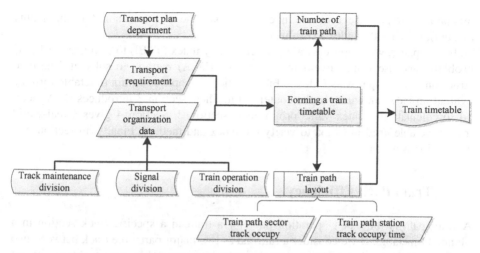

Fig. 2. The Train Timetable Forming Flow

As operation program file, the quality of train timetable will affect transportation significantly. The train timetable is the core file of complex train operation as can been seen in Table.2.

Table 2. Train Operation under Specific Timetable

train operation relative with train timetable		
Running time between each rail sectors and stations	Train stop time Stop time in each station	Maintenance time Train turnover time

However, the train timetable forming is a very complex process and there is much research on the train timetable evaluation which mainly focused on the train timetable evaluation index [9-10], the dynamic performance of the train timetable [3-5, 11]. But almost all the timetable evaluation indexes reflect the timetable quality by considering the average speed and train stop distribution situation. Fig.2 shows once the planning department estimated the market demand and the passenger department as well as freight department decided the start time domain for each train path, the timetable forming plan is fixed. From this point, the train path is the only thing that a timetable maker can operate. So we can realize that the train timetable forming quality is the working level that a timetable maker use transport resources to draw train paths and adjust according to the transport demands. Accordingly, the train path quality is the key to evaluating the forming quality of train timetable.

The track situation varies a lot from region to region, and the transport situation such as the number of the trains, train speed limit, varies as well. The existing index value depends on circumstances directly, so we can define them as *absolute index*. The *absolute index* does not consider the transport circumstance and varies a lot from different tracks. It is priority to establish a *relative index* which is comparable under different transport situation to evaluate the train path quality. The main purpose of this paper is to propose a relative index to evaluate the train timetable forming quality (TTFQ), then use the index to evaluate TTFQ

In this paper, we propose a train path efficiency index (TPEI) to evaluate the TTFQ problem, and use Data Envelopment Analysis (DEA) method to solve it. The construction of the paper is as follows: First, section 2 proposes a train timetable forming evaluation index and analyze the content of it. Then, section 3 introduces DEA method to calculate the index value proposed in section 2. Section 4 gives a real-world train timetable used in China to verify the index and method. Finally, in section 5, a conclusion is given.

2   Train Path Efficiency

A train path is the exact operation path for a train in a specific track section in a planned timetable. If we consider a railroad as two major parts: the track between two stations and the track within a station, while each train path has some technical feature such as various kinds of speed. It is easy to transfer the train path quality into a efficiency evaluation problem, where every train path consume same kind of transport resources such as sector track occupation time and station track occupation time, produce same kind of train path technical indexes such as running speed and travelling speed. To evaluate train path efficiency, a train path efficiency index for train path i is defined as follows:

$$\eta_i = \frac{u\Psi_i}{v\Theta_i} \tag{1}$$

Where: Ψ_i is the technical index set of train path i, Θ_i is the transport resource set that is consumed by the train path i, $0 < i < N$, N is the number of train paths, u and v is the weight vectors of Ψ_i and Θ_i which is used to computer all η_i.

The difference for train organization circumstance determines that Ψ_i and Θ_i is not equal under different situation, but there is a same rule to evaluate a train path quality, which is minimize the transport resource consumption and maximize the technical index output. Thus, η_i is comparable under different timetable forming situation and it is more subjective to evaluate the TTFQ using η_i.

To calculate η_i, we must analyze the type of the transport resource consumption and the technical index feedback of a train path.

2.1   Train Path Transport Resource Consumption

A train path will occupy the sector track during the train running through and occupation the station track while it stops in the station.

Train Path Sector Track Occupation Time

If a train path i runs through n_i sectors, for all $1 \le s \le n_i$, the length for each sector s is d_{is}, the train speed in sector s is v_{is}, then we can get the sector occupation time for train path i is :

$$t_{isec} = \sum_{s=1}^{n} \frac{d_{is}}{v_{is}} \qquad (2)$$

Train Path Station Occupation Time

The station track occupation time for a certain train path start from the signal preparation of the train, end until the station throat is cleared and meet the demands for the operation of the next train. Consider the train path i stopping at station $c_{i1}, c_{i2}, \cdots, c_{ij}, \cdots, c_{im_i}$, and then the station track occupation time for station c_{ij} $(1 \le i \le N, 1 \le j \le m_i)$ is:

$$t_{ista}^{c_{ij}} = t_{approach}^{c_{ij}} + t_{stop}^{c_{ij}} + t_{leave}^{c_{ij}} \qquad (3)$$

In the formula, $t_{approach}^{c_{ij}}$ stands for the running time since the signal preparing for the train until it arrives at the station, the formula is:

$$t_{approach}^{c_{ij}} = \frac{L_{break}^{c_{ij}} + L_{pro}^{c_{ij}} + L_{in}^{c_{ij}}}{v_{st}^{c_{ij}}} + t_{apr}^{c_{ij}} \qquad (4)$$

Where $L_{break}^{c_{ij}}$ is the break distance for train i in station c_{ij}, $L_{pro}^{c_{ij}}$ is the protection distance for station c_{ij}, $L_{in}^{c_{ij}}$ is the home signal to stop point distance of station c_{ij}, $v_{st}^{c_{ij}}$ is the average approaching speed until stop in station c_{ij} for train i, $t_{apr}^{c_{ij}}$ is the approaching time.

$t_{stop}^{c_{ij}}$ is the stop time for train i in station c_{ij}, the formula is:

$$t_{stop}^{c_{ij}} = t_{c_{ij}}b - t_{c_{ij}}a \qquad (5)$$

Where $t_{c_{ij}}a$, $t_{c_{ij}}b$ are the arrive and departure times respectively of train i in station c_{ij}.

$t_{leave}^{c_{ij}}$ is the departure interval for station c_{ij}, the formula is:

$$t_{leave}^{c_{ij}} = \frac{L_{signal}^{c_{ij}} + L_{block}^{c_{ij}} + L_{train}^{i}}{v_{leave}^{c_{ij}}} + t_{lev}^{c_{ij}} \qquad (6)$$

Where $L_{signal}^{c_{ij}}$ is the stop point to starting signal distance for train i in station c_{ij}, $L_{block}^{c_{ij}}$ is the leaving block length of station c_{ij}, L_{train}^{i} is the train length, $v_{leave}^{c_{ij}}$ is the average leaving speed for train i in station c_{ij}, $t_{lev}^{c_{ij}}$ is the starting time. All the symbol's diagram meaning has been explained in detail, see [12].

Then we can get the precise train path station track occupation time as follows:

$$t_{ilsta} = \sum_{j=1}^{m_i} t_{ilsta}^{c_{ij}} = \sum_{j=1}^{m_i} \left(\frac{L_{break}^{c_{ij}} + L_{pro}^{c_{ij}} + L_{in}^{c_{ij}}}{v_{st}^{c_{ij}}} + t_{apr}^{c_{ij}} + t_{c_{ij}} b - t_{c_{ij}} a + \frac{L_{signal}^{c_{ij}} + L_{block}^{c_{ij}} + L_{train}^{i}}{v_{leave}^{c_{ij}}} + t_{lev}^{c_{ij}} \right) \quad (7)$$

2.2   Train Path Feedback Technical Index

The main feedback technical index of a train path is speed. To consider the equipment speed utilization and equilibrium of stop, we propose 3 kinds of speed index, which are average running speed, average travel speed and travel speed.

Average Running Speed

Denoted by rail speed limit v_{lims} in sector s of train path i, the locomotive speed limit v_{limloc} and the vehicle speed limit v_{limvie}. Suppose v_{lims}, v_{limloc}, v_{limvie} stand for the top track speed in sector s, top locomotive and vehicle speed limit, then the average running speed for train path i is :

$$v_{isec\,s}^{l} = \frac{1}{n} \sum_{s=1}^{n} \frac{d_s}{\max\left(\frac{d_s}{v_{lims}}, \frac{d_s}{v_{limloc}}, \frac{d_s}{v_{limvie}} \right)} \quad (8)$$

Average Travel Speed

Denoted by planned running speed v_{is}, starting-stopping extra time t_{ils}^{qt}, extra running time t_{ils}^{e} and additional time t_{ils}^{b}. The starting-stopping extra time t_{ils}^{qt} is:

$$t_{ils}^{qt} = \left(t_{c_{ij}} a - t_{c_{ij-1}} b \right) t_{stand}^{q} + \left(t_{c_{ij+1}} a - t_{c_{ij}} b \right) t_{stand}^{i} \quad (9)$$

where t_{stand}^{q}, t_{stand}^{i} is the stopping extra time and starting extra time, other time standard could be added according to the real require, then the average travel speed for train path i is :

$$v_{isec\,s}^{r} = \frac{1}{n} \sum_{s=1}^{n} \frac{d_{is}}{\frac{d_{is}}{v_{is}} + t_{ils}^{e} + t_{ils}^{b} + t_{ils}^{qt}} = \frac{1}{n} \sum_{s=1}^{n} \frac{d_{is} v_{is}}{d_{is} + v_{is} \left(t_{ils}^{e} + t_{ils}^{b} + t_{ils}^{qt} \right)} \quad (10)$$

Travel Speed

Determined by the total travel distance and total travel time, the travel speed for train path i is:

$$v_i^r = \frac{\sum_{s=1}^{n} d_{is}}{t_{c_{im}} a - t_{c_{i1}} b} \quad (11)$$

(8) reflects the maximum speed level that the train path i could achieve, (10) reflects the stop equilibrium, under a certain number of stops, if a train path stop distribution is concentrated, the average travel speed will be worse than the train path with even distribution.

3   Train Path Efficiency Index Solution and Timetable Forming Quality Evaluation

The train path efficiency is relative with the transportation organization situation; it should all the train paths` co-effectiveness in a specific train timetable. If we There are several ways to analyze the relative efficiency, in this section, a mathematical express of the definition is given and a DEA method is applied to solve it.

3.1   Index Definition and Solve Method Based on DEA

The index is defined as the efficiency of a group of production unit. Each production unit is evaluated individually as well as with regard to its counterparts` efficiency. Obviously, we can suppose that each train path uses p input items expressed as positive numbers φ_i^j (value of item j used by unit i) to deliver m output items valued by θ_i^j (value of item j produced by unit i), consider the relative importance of the items denoted by α_i^j and β_i^j , that the efficiency is defined by:

$$e_i = \frac{\sum_{j=1}^{m} \alpha_i^j \varphi_i^j}{\sum_{j=1}^{p} \beta_i^j \theta_i^j}, \alpha_i^j, \beta_i^j > 0 \tag{12}$$

Data envelopment analysis (DEA) is a technique for assessing and ranking the performance of corporations or other entities where an entire array of indicators of performance is to be evaluated. It was proposed by [13] and the procedure of finding the best virtual producer can be formulated as a linear program. For a group of *producers* that consume same kind of resources and produce same kind of profits, the DEA method allows us to solve the efficiency problem for each *producer* with the model below:

$$e_i = \frac{\sum_{j=1}^{m} \alpha_i^j \varphi_i^j}{\sum_{j=1}^{p} \beta_i^j \theta_i^j} \tag{13}$$

$$s.t. \begin{cases} \dfrac{\alpha_i^j \varphi_i^j}{\beta_i^j \theta_i^j} \leq 1, i = 1, \cdots, n \\ \alpha \geq 0, \beta \geq 0 \end{cases} \tag{14}$$

The method is indeed very effective to solve the efficiency problem by solving n linear programs with the importance values α_i^j, β_i^j depended on index i are solved by the linear program itself. Thus we can ensure that the train path efficiency is comparable among different transportation situations.

3.2   Solution Model for Train Path Efficiency

The problem solved with train path efficiency index is a multi-input-output system efficiency evaluation problem, where each train path consumes a certain amount of transport resources, produces some technical output. So the value of train path efficiency index is evaluating the efficiency in terms of resource utilization versus output performance for each production unit (train path). According to the DEA method described in section 3.1, we can built a train path efficiency solve model as :

$$\theta_{i_0}^{CCR} = \max \frac{u_1 v_{i_0|\text{sec}s}^t + u_2 v_{i_0|\text{sec}s}^r + u_3 v_{i_0}^r}{\mu_1 t_{i_0|\text{sec}} + \mu_2 t_{i_0|\text{sta}}} \tag{15}$$

$$s.t. \begin{cases} \dfrac{u_1 v_{i|\text{sec}s}^t + u_2 v_{i|\text{sec}s}^r + u_3 v_i^r}{\mu_1 t_{i|\text{sec}} + \mu_2 t_{i|\text{sta}}} \leq 1, i = 1, \cdots, N \\ u \geq 0, \mu \geq 0 \end{cases} \tag{16}$$

Where $\theta_{i_0}^{CCR}, (1 \leq i_0 \leq N)$ stands for the efficiency of train path i_0 and is equal to η_{i_0} proposed in (1), u_1, u_2, u_3 and μ_1, μ_2 are isolated parameters used to compute $\theta_{i_0}^{CCR}$ and furthermore, their computation is decided by the whole linear program group, this means their value are not depend on i_0, N is the number of train path.

Set $U = (u_1, u_2, u_3)^T$, $M = (\mu_1, \mu_2)^T$, $X_i = (t_{i|\text{sec}}, t_{i|\text{sta}})^T$, $Y_i = (v_{i|\text{sec}s}^t, v_{i|\text{sec}s}^r, v_i^r)^T$, $i = 1, \cdots, N$, $\alpha = \dfrac{1}{M^T X_{i_0}}$, $\beta = \alpha M$, $\gamma = \alpha u$, use Charnes-Cooper transformation to transform the model into a linear program:

$$\theta_{i_0}' = \max \gamma^T Y_{i_0} \tag{17}$$

$$s.t. \begin{cases} \beta^T X_i - \gamma^T Y_i \geq 0, i = 1, 2, \cdots, N \\ \beta^T X_{i_0} = 1 \\ \forall \beta \geq 0, \forall \gamma \geq 0 \end{cases} \tag{18}$$

By solving N linear programing problems, we can get the value set of train paths efficiency, use the train paths efficiency set, we can analyze the forming quality of the train timetable supposed to be evaluated.

3.3   Train Diagram Forming Quality Evaluation

After getting the train path efficiency value set $H=(\eta_1,\eta_2,\cdots,\eta_N)$, we can evaluate TTFQ by the distribution of train path efficiency, introduce three functions such as:
Efficiency index expectation (REE):

$$E_\eta = \frac{1}{N}\sum_{i=1}^{N}\eta_i \tag{19}$$

Efficiency index variance (REV):

$$D_\eta = \frac{1}{N}\sum_{i=1}^{N}\left(\eta_i - E_\eta\right)^2 \tag{20}$$

Efficient distribution function (EDF) with an interval τ :

$$f\left(\lceil\eta_i\rceil,\lceil\eta_i\rceil-\tau\right) = \frac{\sum_{i=1}^{n}\lceil\eta_i\rceil}{N}\times100\% \tag{21}$$

REE shows the average train path efficiency in a train timetable, which is positive correlation with train timetable forming quality. REV shows the degree of dispersion of the train path efficiency in a train timetable, which is negative correlation with train timetable forming quality. EDF shows the distribution situation of the train path efficiency in a train timetable. Through the three evaluation indices, the working level of train timetable maker could be evaluated.

4   Case Study of a Real-World Train Timetable in China

This section presents a case study of our proposed train timetable forming evaluation. The case is Shibantan station to Xinqiao block post sector timetable of Chengdu to Suining railway. This rail section with length of 119.2km connected two large passenger station which is Chengdu station and Chengdu East station, with two direction which is Dazhou and Chongqing (seen Fig.3). The maximum track speed limit of this section is 200km/h, while the train speed limit is divided into 4 levels, shown in Table 3.

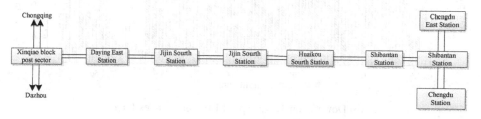

Fig. 3. Railway Section Case for Evaluation

Table 3. Kinds of Train Speed

Train speed limit	Train number title	number of train
200km/h	D	64
160km/h	Z	2
120km/h	T\K/X	43
100km/h	K	2

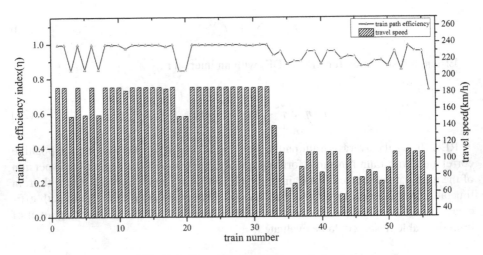

(a) Up Train Travel Speed-Efficiency Index Layout

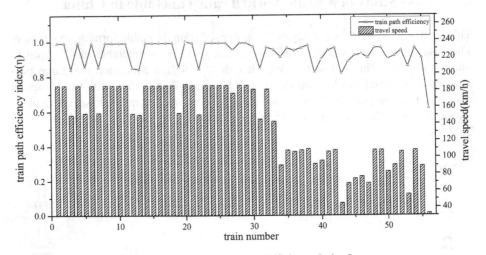

(b) Down Train Travel Speed-Efficiency Index Layout

Fig. 4. Train path travel speed-efficiency index value layout

There are 56 trains operating in this section per day, we use stopping extra time t_{stand}^{q} and starting extra time t_{stand}^{l} represent $t_{approach}^{c_{ij}}$ and $t_{leave}^{c_{ij}}$ for a simplified calculation. Set the parameters as: $t_{stand}^{q} = 2$min, $t_{stand}^{l} = 2$min, $t_{lls}^{e} = t_{lls}^{b} = 0$.

The solution model for train path efficiency is a linear program group contains 112 liner programs; each sub-linear program includes 5 variables, 113 constraints. Solve the model with Lingo11. Fig.4 shows the train path efficiency value and travel speed layout, where the up train runs from Shibantan station to Xinqiao block post sector, down train runs from Xinqiao block post sector to Shibantan station.

In the case, the REE is 0.8412, the REV is 0.0044, and the EDF with an interval of 0.1 is shown in Table.4.

Table 4. Efficient Distribution Function (EDF) with 0.1 interval

EDF	(1,0.9)	(0.9,0.8)	(0.8,0.7)	(0.7,0.6)	(0.6,0)
number	82	28	1	1	0
percentage	73.21%	25%	0.89%	0.89%	-

From the results we can see:

The proposed train path efficiency is relative with the transport situation. In the case, the train path efficiency value does not direct correlation with the train speed limit. So we can compare the train path efficiency value among different kinds of train path without concerning the speed limit or other factors. This is very important for the evaluation of train timetable forming quality, for the train path efficiency is an objective evaluation index; we can evaluation the train timetable forming quality between various rail track sections.

Because of the speed differential, some slow trains (with maximum speed of 120km/h or100km/h) have to stop to let fast train overpass. Eventually, the slow train have to wait for a long time to ensure the fast train`s operation is not influenced. That is the main cause for slow trains suffers low train path efficiency.

The train timetable forming quality is preferable with the high REE and most train path has a relatively high train path efficiency performance (over 98% with an efficiency value more than 0.8).

5 Conclusion

There are few researches on the forming quality of train timetable. In this paper, we described the train timetable forming process in China, and described the key problem to evaluate the forming quality, take the train path efficiency to evaluate the train timetable forming quality, then a relative evaluation index for train timetable forming evaluation is proposed and the solving model is built, then verified using a real-world train timetable in China. The result shows that the proposed evaluation index is relative and practical.

The forming evaluation of train timetable is a complex and systematic project, the evaluation index expansion and method search will be considered in the future work.

Acknowledgements. The author would like to express his thanks for the data and financial support from National Railway Train Diagram Research and Training Center, Southwest Jiaotong University, Chengdu, China. And the support of Chinese Natural Science Foundation under Grant Number61273242, 61403317 as well as technology department of China Railway Cooperation under Grant Number 2013X006-A, 2013X014-G,2013X010-A,2014X004-D.

References

1. Vansteenwegen, P., Van Oudheusden, D.: Developing Railway Timetables which Guarantee a Better Service. European Journal of Operational Research 173, 337–350 (2006)
2. Engelhardt-Funke, O., Kolonko, M.: Analysing Stability and Investments in Railway Networks using Advanced Evolutionary Algorithms. International Transactions in Operation Research 11, 381–394 (2004)
3. Goverde, R.M.P.: Optimal Scheduling of Connections in Railway Systems. Technical report, TRAIL, Delft, The Netherlands (1998)
4. Goverde, R.M.P.: Synchronization Control of Scheduled Train Services to Minimize Passenger Waiting Times. In: Proceedings of the Fourth TRAIL Annual Congress, Part2, TRAIL Research School, Delft, The Netherlands (1998)
5. Goverde, R.M.P.: Transfer Stations and Synchronization. Technical Report, TRAIL, Delft, The Netherlands (1999)
6. Kroon, L., Peeters, L.: A Variable Trip Time Model for Cyclic Railway Timetabling. Transportation Science 37, 198–212 (2003)
7. Liebchen, C., Möhring, R.: A Case Study in Periodic Timetabling. Electronic Notes in Theoretical Computer Science 66(6) (2002)
8. Liebchen, C., Peeters, L.: Some Practical Aspects of Periodic Timetabling. In: Chamoni, P., Leisten, R., Martin, A., Minnemann, J., Stadtler, H. (eds.) Operations Research 2001. Springer, Germany (2002)
9. Peng, Q., Bao, J., Wen, C., et al.: Evaluation Theory and Method of High-speed Train Diagrams. Journal of Southwest Jiaotong University 4, 969–974 (2013) (in Chinese)
10. Zhu, T., Wang, H., Lu, B.: Method for Optimally Selecting the Train Diagram of Passenger Dedicated Line Based on Vague Sets. China Railway Science 32, 122–127 (2011) (in Chinese)
11. Goverde, R.M.P.: Railway Timetable Stability Analysis Using Max-plus System Theory. Transportation Research Part B 41, 179–201 (2007)
12. Zhang, Y., Tian, C., Jiang, X., et al.: Calculation Method for Train Headway of High Speed Railway. China Railway Science 34, 120–125 (2013) (in Chinese)
13. Charnes, A., Cooper, W.W., Rhodes, E.: Measuring the Efficiency of Decision Making Units. European Journal of Operational Research 2, 429–444 (1978)
14. Cooper, W.W., Seiford, L.M., Tone, K.: Data Envelopment Analysis. Kluwer Academic Publishers (2002)

Transit Network Design for Green Vehicles Routing

Victor Zakharov and Alexander Krylatov

Saint-Petersburg State University, Universitetskaya Nab. 7-9, 199034
Saint-Petersburg, Russia

Abstract. Transport of modern worldwide cities generates harmful
emissions that accounts for nearly half of the total pollution in urban
areas. Therefore local governments are forced to stimulate appearance of
new fuels and technological innovations in urban mobility. Nevertheless
there is a lack of methodological tools for supporting decision makers in
spheres of motivation for using green vehicles by drivers and allocation
available green capacity. This paper is devoted to the problem of green
and non-green traffic flow assignment on the network consisting of green
and non-green routes. The approach of defining green routes (green subnetwork) which are fully loaded and provide less travel time for green
vehicles are developed. Conditions of well-balanced green subnetwork are
obtained explicitly for the network of parallel routes.

Keywords: routing, Wardrop equilibrium, green routes allocation, transit network design.

1 Introduction

Transportation sectors of modern worldwide cities consume a large portion of
fossil fuel and significantly contribute to greenhouse gas (GHG) emissions. According to Federal State Statistics Service of Russian Federation since 2000 in
this country emission of road transportation were 41,9% of all greenhouse gas
[1]. In particular, the use of passenger cars contributes to 15% of the overall
carbon dioxide emissions. In Europe passenger cars emitted 12% of air contam-
inants and there exist increasing dynamic of this indicator [2]. In 2011, 27%
of total GHG emissions in the U.S. comes from transportation end-user sector
and passenger vehicles are responsible for 43% of this total share [3]. Developing
countries of South America demonstrate the same state. While industrial sector
of Brazil emitted in 2010 29% of air pollution, road transport contributed 43%
[4]. Certainly, such situation influence on quality of life directly and authorities are interested in decisions that could change situation in the direction of
pollution decreasing.

Emissions from on-road vehicles depend on several attributes that include:
fleet composition, age distribution of the vehicles, fuel type, atmospheric attributes, operational characteristics, congestion, and travel choices made by the
road users. Simultaneously, without a doubt, green vehicles cause the least harm

© Springer International Publishing Switzerland 2015   449
H.A. Le Thi et al. (eds.), *Model. Comput. & Optim. in Inf. Syst. & Manage. Sci.*,
Advances in Intelligent Systems and Computing 360, DOI: 10.1007/978-3-319-18167-7_39

to the atmosphere. Literature offers different definitions of the term green vehicle but the focus is on separating the term Absolutely Green Vehicle (AGV) from the term Low Greenhouse Gas-Emitting Vehicles. The AGV is a zero-emission vehicle, releasing no pollutants. In other words, a vehicle using water or the energy of the Sun is an absolutely green vehicle. Electric vehicles, on the other hand, are considered to be AGV vehicles only if the electricity used for propulsion is obtained from renewable energy sources. If, however, harmful substances are released in the production of electricity, this is not an AGV vehicle. Vehicles running on biodiesel, natural gas and the like are low greenhouse gas-emitting vehicles [5,6].

The government is interested in increasing of green vehicles fleet on the transportation network to resist greenhouse gas emission. To motivate drivers for using green vehicles decision maker could define special green subnetwork consisted of routs for green vehicle transit only. The point is how green vehicles could be provided with sufficiently attractive alternative. Since the government possess information about current quantity of green vehicles the question could be reformulated quantitatively. How many transportation routes should be offered for green vehicles transit only (let us call such routes as green routes as opposed to non-green routes). Indeed, on the one hand, if green routes are partially loaded but non-green routes are overloaded then transportation network will be unbalanced [7]. On the other hand, if green routes could be overloaded by fleet of green vehicles then it will be not so attractive for drivers to use green vehicles. Thereby, it is necessary to find conditions that could guarantee well-balanced using of green and non-green routes of the network. In this study we are going to define such conditions.

The paper is structured as follows: the network of parallel routs is investigated in Section 2. Some generalizations are given in Section 3. Section 4 is devoted to final conclusions.

2   The Network of Parallel Routes

In this section we are going to consider a network with one origin and one destination. Here we assume that there are only parallel alternative routes (without intersections) between origin and destination points. Thereby, one part of these parallel routes could be declared as green routes and another – as non-green routes. We believe that green routes could be used only by green vehicles while non-green routes could be used by both green and non-green vehicles.

Consider transportation network presented by digraph consisted of one origin-destination pair of nodes and n parallel links. Each link is associated with route from origin to destination. We use following notation: $N = \{1, \ldots, n\}$ – set of numbers of all routes; $N_1 = \{1, \ldots, n_1\}$ – set of numbers of green routes; $N_2 = \{n_1 + 1, \ldots, n_2 = n\}$ – set of numbers of non-green routes; F^1 – quantity of green vehicles using the network; F^2 – quantity of non-green vehicles using the network; f_i for $i = \overline{1, n_1}$ – traffic flow of green vehicles through route i; f_i for $i = \overline{n_1 + 1, n_2}$ – traffic flow of green and non-green vehicles through route i; $f =$

$(f_1, \ldots, f_n)$, $f^1 = (f_1, \ldots, f_{n_1})$ and $f^2 = (f_{n_1+1}, \ldots, f_{n_2})$; t_i^0 – free travel time through route i, $i = \overline{1,n}$; c_i – capacity of route i, $i = \overline{1,n}$; $t_i(f_i) = t_i^0 \left(1 + \frac{f_i}{c_i}\right)$ – travel time through congested route i, $i = \overline{1,n}$. We are modeling travel time as linear BPR-delay function [8].

Our task is to define conditions that guarantee: 1) user-equilibrium on the whole network; 2) using of green routes only by green vehicles; 3) lack of inefficiency: all green routes should be used. Actually, user-equilibrium demands all vehicles have absolutely the same travel time from origin to destination. It means that green vehicles go to the green routes until travel time through green routes less or equal then travel time through non-green routes. Mathematically such a problem is expressed by the following optimization program:

$$\min_f z(f) = \min_f \left(\sum_{i=1}^{n_1} \int_0^{f_i} t_i(u)du + \sum_{i=n_1+1}^{n_2} \int_0^{f_i} t_i(u)du \right) , \tag{1}$$

subject to

$$\sum_{i=1}^{n_1} f_i \leq F^1 , \tag{2}$$

$$\sum_{i=n_1+1}^{n_2} f_i \geq F^2 , \tag{3}$$

$$\sum_{i=1}^{n_1} f_i + \sum_{i=n_1+1}^{n_2} f_i = F^1 + F^2 , \tag{4}$$

$$f_i \geq 0 \quad \forall i = \overline{1, n_2} . \tag{5}$$

One can see that for any fixed n_1 unknown variables of formulated problem (1)-(5) are flows through all available routes (both green and non-green). Wherein the decision of this problem is user-equilibrium (because of form of objective function [9]) and constraint (2) guarantees using of green routes only by green vehicles. Generally, solving of this problem demands complex computational procedure. Simultaneously boundary value of n_1 could be estimated directly. Indeed, boundary value of n_1 corresponds to the situation when all green vehicles use only green routes and their travel time is less or equal to travel time of non-green vehicles using non-green routes. Then let us find this boundary value of n_1.

First of all, it should be mentioned that solution of the problem (1)-(5) could contain routes with zero flow (if $\exists j = \overline{1,n}$: $f_j = 0$). Such problem could appear when initial set of possible routes is badly defined. We are going to consider case of well-defined initial set of routes or fully loaded. It is important task to define conditions when initial set of routes is fully loaded to avoid additional unnecessary operations during solving the problem (1)-(5).

Without loss of generality we assume that when n_1 is defined routes renumbering as follows

$$t_1^0 \leq \ldots \leq t_{n_1}^0 \text{ and } t_{n_1+1}^0 \leq \ldots \leq t_{n_2}^0 . \tag{6}$$

Lemma 1. *Assume that all fleet of green vehicles F^1 uses only possible green routes. Each green rout from the set of n_1 routes is used if and only if*

$$F^1 > \sum_{i=1}^{n_1} c_i \left(\frac{t_{n_1}^0}{t_i^0} - 1 \right) \tag{7}$$

and each non-green rout from the set of $|n_2 - n_1|$ routes is used if and only if

$$F^2 > \sum_{i=n_1+1}^{n_2} c_i \left(\frac{t_{n_2}^0}{t_i^0} - 1 \right) . \tag{8}$$

Proof. If all green vehicles use only green routes the optimization problem (1)-(5) can be considered as two independent problems:
1) For green vehicles:

$$\min_{f^1} z^1(f^1) = \min_{f^1} \sum_{i=1}^{n_1} \int_0^{f_i} t_i(u)du , \tag{9}$$

subject to

$$\sum_{i=1}^{n_1} f_i = F^1 , \tag{10}$$

$$f_i \geq 0 \quad \forall i = \overline{1, n_1} . \tag{11}$$

2) For non-green vehicles:

$$\min_{f^2} z^2(f^2) = \min_{f^2} \sum_{i=n_1+1}^{n_2} \int_0^{f_i} t_i(u)du , \tag{12}$$

subject to

$$\sum_{i=n_1+1}^{n_2} f_i = F^2 , \tag{13}$$

$$f_i \geq 0 \quad \forall i = \overline{n_1 + 1, n_2} . \tag{14}$$

Consider Lagrangian of the problem (9)-(11):

$$L^1 = \sum_{i=1}^{n_1} \int_0^{f_i} t_i(u)du + \omega^1 \left(F^1 - \sum_{i=1}^{n_1} f_i \right) + \sum_{i=1}^{n_1} \eta_i f_i$$

and differentiate it

$$\frac{\partial L^1}{\partial f_i} = t_i(f_i) - \omega^1 + \eta_i = 0 \ ,$$

where ω^1 and $\eta_i \geq 0$ for $i = \overline{1, n_1}$ are Lagrange multipliers. Due to Kuhn-Tucker conditions

$$t_i(f_i) \begin{cases} = \omega^1 \text{ if } f_i > 0 \ , \\ \leq \omega^1 \text{ if } f_i = 0 \ . \end{cases} \tag{15}$$

for $i = \overline{1, n_1}$.

When inequality $f_i > 0$ holds for $i = \overline{1, n_1}$ it means that each green rout from the set of n_1 routes is used. According to (15) if $f_i > 0$ then $t_i^0 \left(1 + \frac{f_i}{c_i} \right) = \omega^1$ and $f_i = \left(\frac{\omega^1}{t_i^0} - 1 \right) c_i > 0$ for $i = \overline{1, n_1}$. Consequently $\omega^1 > t_i^0$ for $i = \overline{1, n_1}$ and according to (6) $\omega^1 > t_{n_1}^0$. Thus we obtain (7)

$$F^1 = \sum_{i=1}^{n_1} f_i = \sum_{i=1}^{n_1} \left(\frac{\omega^1}{t_i^0} - 1 \right) c_i > \sum_{i=1}^{n_1} \left(\frac{t_{n_1}^0}{t_i^0} - 1 \right) c_i \ .$$

In the same way we can investigate the problem (12)-(14) and obtain (8). Lemma is proved.

Lemma 1 gives the first validation rule. If inequality (7) holds then all green routs will be used when assumed that whole fleet of green vehicles use only green routes. Simultaneous execution of conditions (7) and (8) means that transportation network is fully loaded – each route is used. At the same time condition (7) could not guarantee that all green vehicles will use only green routes. To make green vehicles use only green routes decision maker has to create such conditions that it would be preferable for drivers to reach destination through green subnetwork.

Theorem 1. *Assume that initial set of routes is fully loaded. The fleet of green vehicles F^1 uses only green routes if and only if*

$$\frac{F^1 + \sum_{i=1}^{n_1} c_i}{\sum_{i=1}^{n_1} \frac{c_i}{t_i^0}} \leq \frac{F^2 + \sum_{i=n_1+1}^{n_2} c_i}{\sum_{i=n_1+1}^{n_2} \frac{c_i}{t_i^0}} \ . \tag{16}$$

Proof. If all green vehicles use only green routes the optimization problem (1)-(5) can be considered as two independent problems (9)-(11) and (12)-(14). Due to (15) we obtain $\sum_{i=1}^{n_1} \left(\frac{\omega^1}{t_i^0} - 1 \right) c_i = F^1$ and consequently

$$\omega^1 = \frac{F^1 + \sum_{i=1}^{n_1} c_i}{\sum_{i=1}^{n_1} \frac{c_i}{t_i^0}} \ .$$

According to (15):

$$t_i(f_i) = \frac{F^1 + \sum_{i=1}^{n_1} c_i}{\sum_{i=1}^{n_1} \frac{c_i}{t_i^0}} \text{ if } f_i > 0 \ , \tag{17}$$

as soon as n_1 is defined in such a way that all green routes are used the equation (17) is true for $i = \overline{1, n_1}$. Thereby (17) is travel time of any car from green vehicles fleet.

We can prove in the same way for (12)-(14) that

$$t_i(f_i) = \frac{F^2 + \sum_{i=n_1+1}^{n_2} c_i}{\sum_{i=n_1+1}^{n_2} \frac{c_i}{t_i^0}} \text{ if } f_i > 0 \ , \tag{18}$$

for $i = \overline{n_1 + 1, n_2}$. Equation (18) defines travel time of any car from non-green vehicle fleet. Hence inequality (16) claims that for any vehicle from green fleet it is more reasonable to use green routes in sense of travel time.

The theorem is proved.

Theorem 1 gives the second rule of defining n_1. Really, if the government provides green vehicle with such amount of routes that condition (16) holds then drivers will identify the main possible advantage of using green vehicles – less travel time.

3 General Network Topology

Consider the network presented by directed graph G that includes a set of consecutively numbered nodes N and a set of consecutively numbered arcs A. Let R denote the set of origin nodes and S – the set of destination nodes ($R \cap S = \emptyset$). We also use following notation: K^{rs} – the set of possible routes between origin r and destination s; K_1^{rs} – the set of green routes, $K_1^{rs} \subset K^{rs}$; K_2^{rs} – the set of non-green routes, $K_2^{rs} \subset K^{rs}$; $K_1^{rs} \cap K_2^{rs} = \emptyset$ and $K_1^{rs} \cup K_2^{rs} = K^{rs}$; F_1^{rs} and F_2^{rs} – quantities of green and non-green vehicles using the network between origin r and destination s respectively; f_k^{rs} for $k \in K_1^{rs}$ – traffic flow of green vehicles through route k; f_k^{rs} for $k \in K_2^{rs}$ – traffic flow of green and non-green vehicles through route k; x_a – traffic flow on the arc $a \in A$, $x = (\dots, x_a, \dots)$; c_a – capacity of link $a \in A$; $t_a(x_a)$ – travel time through congested link $a \in A$; $\delta_{a,k}^{rs}$ – indicator: $\delta_{a,k}^{rs} = 1$ if link a is a part of path k connecting O-D pair r-s, and $\delta_{a,k}^{rs} = 0$ otherwise. Now problem (1)-(5) can be reformulated for the case of general network topology:

$$\min_{K_1^{rs}, x} Z(x) = \min_{K_1^{rs}, x} \sum_{a \in A} \int_0^{x_a} t_a(u) du \ , \tag{19}$$

subject to

$$\sum_{k \in K_1^{rs}} f_k^{rs} \le F_1^{rs} \quad \forall r, s \ , \tag{20}$$

$$\sum_{k \in K_2^{rs}} f_k^{rs} \ge F_2^{rs} \quad \forall r, s \ , \tag{21}$$

$$\sum_{k \in K_1^{rs}} f_k^{rs} + \sum_{k \in K_2^{rs}} f_k^{rs} = F_1^{rs} + F_2^{rs} \quad \forall r, s \ , \tag{22}$$

$$f_k^{rs} \geq 0 \quad \forall k, r, s , \tag{23}$$

with definitional constraints

$$x_a = \sum_r \sum_s \sum_{k \in K_1^{rs}} f_k^{rs} \delta_{a,k}^{rs} + \sum_r \sum_s \sum_{k \in K_2^{rs}} f_k^{rs} \delta_{a,k}^{rs} . \tag{24}$$

Unknown variables of optimization program (19)-(24) are the set of green routes and flows through all available routes (both green and non-green). The most important thing in formulated problem is that its solution gives certain set of green routes – green subnetwork. To find optimal solution of this problem we can consider numbers of green routes as parameters controlled by road authority. Thus we really would have two level control system. Assignment of the green vehicles has to take into account optimal reaction of drivers of green and non-green vehicles based on user-equilibrium concept.

In previous section we defined procedure that could support decision making in sphere of green routes allocation. Indeed, the main criterion for decision maker when she is going to provide part of common routes solely for one kind of transport (for example, green vehicles) is effectiveness. Developed approach allows us to find such amount of green routes that:

– travel time of any amount of green vehicles (from zero to all green vehicles) using green routes is less or equal to travel time of rest traffic flow;
– if all green vehicles use only green routes then there is no one green route to be unused (there are no unclaimed routes).

The first condition guarantees such absolute advantage as less travel time for green vehicles that use green routes. The second condition confirms that transportation network is used effectively. Compliance with these two conditions, on the one hand, motivates drivers use green vehicles and, on the other hand, does not use capacity of transportation network wastefully.

To summarize, investigation of simple transportation network of parallel routes allows us to develop approach that could be written as following procedure:

1. Define the initial set of possible routes as fully loaded.
2. Define the set of green routes so to provide green vehicles with less travel time in contrast with the rest traffic flow.

Let us apply this approach to the network of general topology. First of all, we could estimate boundary state of the set K_1^{rs}. Boundary state of the set K_1^{rs} corresponds to the situation when all green vehicles use only green routes and their travel time is less or equal to travel time of non-green vehicles using non-green routes. In such a case we can consider problem (19)-(24) as two independent problems:

1) For green vehicles:

$$\min_x Z(x) = \min_x \sum_{a \in A} \int_0^{x_a} t_a(u) du , \tag{25}$$

subject to

$$\sum_{k \in K_1^{rs}} f_k^{rs} = F_1^{rs} \quad \forall r, s \ , \tag{26}$$

$$f_k^{rs} \geq 0 \quad \forall k \in K_1^{rs}, r, s \ , \tag{27}$$

with definitional constraints

$$x_a = \sum_r \sum_s \sum_{k \in K_1^{rs}} f_k^{rs} \delta_{a,k}^{rs} \ . \tag{28}$$

2) For non-green vehicles:

$$\min_x Z(x) = \min_x \sum_{a \in A} \int_0^{x_a} t_a(u) du \ , \tag{29}$$

subject to

$$\sum_{k \in K_2^{rs}} f_k^{rs} = F_2^{rs} \quad \forall r, s \ , \tag{30}$$

$$f_k^{rs} \geq 0 \quad \forall k \in K_2^{rs}, r, s \ , \tag{31}$$

with definitional constraints

$$x_a = \sum_r \sum_s \sum_{k \in K_2^{rs}} f_k^{rs} \delta_{a,k}^{rs} \ . \tag{32}$$

Optimization problems (25)-(28) and (29)-(32) are ordinary programs for finding user-equilibrium [9]. Unfortunately it is impossible to obtain analytic conditions as in case of parallel routes. However there are numerous approaches today concerning solving such problems. Firstly Frank-Wolfe algorithm was employed for this goal [10]. Thereby decision maker could find optimal green subnetwork doing following actions:

1. Define initial set of green routes K_1^{rs} for each origin-destination pair.
2. Solve problems (25)-(28) and (29)-(32) employing existing tools or approaches.
3. Check if travel time for green vehicles moving through green subnetwork less or equal travel time of the rest traffic flow.
4. Check if all green routes are used in case when all green vehicles go through green subnetwork.
5. If stages 3 and 4 are false then go to stage 1. Else – optimal green subnetwork is found.

One can see that defining green subnetwork on network of general topology is quite complex problem. To solve problems (25)-(28) and (29)-(32) it is necessary to use special computational technologies with specific software. Simultaneously, sometimes decision maker need solely estimations but not absolutely accurate values. Then obtained in section 2 conditions could be adjuvant tool for decision making on the general network topology. Indeed consider one of the central districts of Saint-Petersburg – Vasileostrovsky district (Fig. 1). Assume that authority decided to define some green routes and the set of possible green routes consist of 4 alternatives: red, yellow, blue and green lines. We believe that these 4 routes could be considered as parallel (with certain assumptions). So for brief estimation of the best decision one could easily use conditions developed in section 2. If such a first estimation shows that set of green routes is not empty one could proceed to the more accurate investigations.

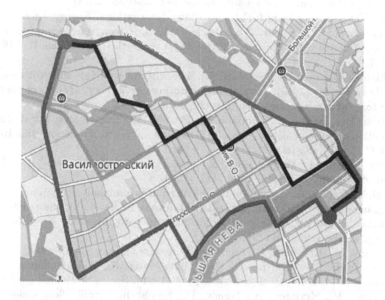

Fig. 1. Vasileostrovsky district of Saint-Petersburg

4   Conclusion

This paper is devoted to the problem of green and non-green traffic flow assignment on the network consisting of green and non-green routes. The approach of defining green routes (green subnetwork) which are fully loaded and provide less travel time for green vehicles are developed. Conditions of well-balanced green subnetwork are obtained explicitly for the network of parallel routes. It is discussed that experts and decision makers could employ both general approach and explicit conditions obtained for the network of parallel routes depending on the particular situation.

Further development would be done in the direction of similar conditions obtaining for the case of competitive routing [11,12]. Moreover additional effort has to be spent in the evaluation of the changes in distribution of GHG emission when green subnetwork is appeared. Authors are going to deal with it in next studies.

References

1. Federal State Statistics Service, http://www.gks.ru
2. Commission of the European Communities, Proposal for a Regulation of the European Parliament and of the Council - Setting Emission Performance Standards for New Passenger Cars as Part of the Community's Integrated Approach to Reduce CO2 Emissions from Light-Duty Vehicles. Dossier COD/2007/0297 (2007)
3. EPA, U., U.S. Transportation Sector Greenhouse Gas Emissions: 1990–2011. Office of Transportation and Air Quality EPA-420-F-13-033a (2013)
4. IEA, CO2 emissions from fuel combustions. International Energy Agency (2012)
5. Jovanovic, A.D., Pamucar, D.S., Pejcic-Tarle, S.: Green vehicle routing in urban zones – A neuro-fuzzy approach. Expert Systems with Applications 41, 3189–3203 (2014)
6. Serbian Department of Energy (SDE), The Alternative Fuels and Advanced Vehicles Data Center (2013), http://www.afdc.energy.gov.rs/afdc/locator/stations/state (accessed February 24, 2013)
7. Beltran, B., Carrese, S., Cipriani, E., Petrelli, M.: Transit network design with allocation of green vehicles: A genetic algorithm approach. Transportation Research Part C 17, 475–483 (2009)
8. Traffic Assignment Manual. In: U.S. Bureau of Public Roads (eds.) U.S. Department of Commerce, Washington, D.C. (1964)
9. Sheffi, Y.: Urban transportation networks: equilibrium analysis with mathematical programming methods, 416 p. Prentice-Hall Inc., Englewood Cliffs (1985)
10. Frank, M., Wolfe, P.: An algorithm for quadratic programming. Naval Res. Logistics Quarterly 3, 95–110 (1956)
11. Zakharov, V., Krylatov, A.: Equilibrium Assignments in Competitive and Cooperative Traffic Flow Routing. In: Camarinha-Matos, L.M., Afsarmanesh, H. (eds.) PRO-VE 2014. IFIP AICT, vol. 434, pp. 641–648. Springer, Heidelberg (2014)
12. Zakharov, V., Krylatov, A., Ivanov, D.: Equilibrium traffic flow assignment in case of two navigation providers. In: Camarinha-Matos, L.M., Scherer, R.J. (eds.) PRO-VE 2013. IFIP AICT, vol. 408, pp. 156–163. Springer, Heidelberg (2013)

Part VIII

Technologies and Methods for Multi-stakeholder Decision Analysis in Public Settings

A Systematic Approach to Reputation Risk Assessment

Anton Talantsev

Stockholm University, Department of Computer and Systems Sciences/DSV,
Postbox 7003, SE-164 07 Kista, Sweden
antontal@dsv.su.se

Abstract. Corporate reputations are intangible assets that provide organizations
with significant market advantages. Nevertheless, too often corporate decisions
and actions, being strategically sound within the traditional management paradigm, meet unexpected backlash by key stakeholders, loss of reputation, and
inevitable dire consequences for business. Drawing conclusions from the reputation measurement literature, this study further conceptualizes and operational-
izes the notion of reputation risk. Having adopted multiple stakeholder view on
corporate reputation, this paper presents a format for stakeholder profile, which
guides systematic reputation risk identification. Furthermore, we utilize a multi-
criteria approach for aggregating reputation risk assessments and prioritizing
identified risks. Finally, we demonstrate application of the approach on a state-
owned company.

Keywords: reputation risk assessment, stakeholder analysis, reputation management, opportunities management.

1   Introduction

Corporate reputations are intangible assets that provide organizations with definite
strategic advantages, including added financial value [1]. While benefits of good corporate reputation are not questionable, majority of organizations fail to properly manage it. As Benjamin Franklin rightfully noted: "It takes many good deeds to build a
good reputation, and only one bad one to lose it". Some of the obstacles for controlling the reputation are its high dynamism, complexity, and causal ambiguity. The
intensity of these features has only boosted in the era of social media and other new
communication technologies [2]. Moreover, organizations nowadays have to be more
exposed to authorities and general public, not least due to the growing practice of
CSR reporting, as well as, the social demand for corporate transparency in developed,
higher-income markets.

While decision and risk analysis is integral part of corporate risk management systems, conventional risk assessment tools and practices have yet to be adapted and
applied to embrace the new phenomena of reputation with its implications and risks
[3]. Therefore, with the purpose of systematic identification of reputation risks, this
study contrasts stakeholders expectations under organizational key performance areas
(KPAs) with a given business decision; in this way highlighting gaps - causes of reputation threats and gains. Further, magnitudes of the gaps are assessed by means of a

linguistic rating scale, and aggregating the assessments through a multiple criteria decision-making approach.

The paper is organized as follows: a literature review on the concept of reputation (Section 2), then a conceptual methodology for reputation risk identification and assessment through stakeholder expectations (Section 3) is followed by a methodology for quantitative assessment of reputation risks (Section 4), and adoption to a real case study (Section 5). Finally conclusions are discussed (Section 6).

2   Background

2.1   What Is Reputation?

The concept of reputation attracts researchers from multitude of domains, including economy, psychology, sociology, marketing and strategic management. While there is a common idea of corporate reputation as the beliefs or opinions that are generally held about the organization [4], Fombrun et al. [5] concluded that there is no consensus on reputation definition neither within practitioners from different realms nor within academia. Moreover, there are several adjacent notions such as identity, image, legitimacy, corporate social responsibility (CSR), which are not seldom used interchangeably. For this reason several authors attempt to conceptually distinct the competing notions [6, 7].

One of such linked phenomena is CSR reporting. Bebbington et al. [6] draw a line between the two arguing that the CSR reporting is both an output and a part of a broader reputation risk management. CSR practices and indirectly reputation are often explained from the perspective of a legitimacy theory (also referred to as strategic legitimacy theory). Kaplan and Ruland [8] define organizational legitimacy as "a process, legitimation, by which an organization seeks approval (or avoidance of sanction) from groups in society". Mathews [9] further explains legitimacy as follows: "Organizations seek to establish congruence between the social values associated with or implied by their activities and the norms of acceptable behavior in the larger social system in which they are a part. In so far as these two value systems are congruent we can speak of organizational legitimacy. When an actual or potential disparity exists between the two value systems there will exist a threat to organizational legitimacy". Tilling [10] compares legitimacy to any resource a business requires in order to operate. "Certain actions and events increase that legitimacy, and others decrease it. Low legitimacy will have particularly dire consequences for an organization, which could ultimately lead to the forfeiture of their right to operate"[10 no page number].

Similarly, the resource-based view (RBV) on a firm considers reputation as one of a firm's most important strategic resources [11, 12]. Indeed, good corporate reputation fulfills the VRIN (or VRIO) criteria: Valuable, Rare, In-imitable, Non-substitutable [see 13]. Finally, along with the RBV of a firm Bergh et al. [12] explain the role and importance of reputation through few other theories including transaction cost economics, signaling theory and social status research.

Therefore, while there is a multitude of empirical evidences of reputation impact on business (including financial performance [14], website commitment [15] , brand extensions [16] and many others), there is still no one theoretical explanation for existence of reputation.

Many authors agree that reputation should be defined as reflections of how well or how badly different groups of interested people – stakeholders – view a company [17]. Examples of primary stakeholder include investors, employees, customers and suppliers; while secondary stakeholders, not having direct impact on organization process, can be trade unions, NGOs, regulators and local or national government. It's worth noting here, that a stakeholder is not an exclusive unit, meaning that a group might perform several stakeholder roles. As instance, in the defense industry the government often plays three distinct roles: policy maker, owner and customer.

In order to go beyond single stakeholder focused approach (such as ubiquitous customers only focused surveys), both academia and practitioners have called for a multiple stakeholder approach to conceptualize the corporate reputation construct. In this light, Bebbington et al. [6] see reputation risk management as enhancement of a stakeholder theory [see 18]. The stakeholder theory suggests that reputation and CSR disclosures are motivated by an aim to manage certain stakeholders, who provide resources to the organization and thus define its operations and even existence. Stakeholder literature seeks to define stakeholders which "count" [19] as well as mapping them against different dimensions, such as "power", "support", "interest" and others [19]. Such stakeholder analysis further guides communication efforts prioritization and development of communication strategies. Thus, organizational endeavors to nurture corporate reputation should be in major extent motivated by the need to attract different resources from different stakeholders. Therefore, the multiple stakeholder paradigm poses organizations a task of managing multiple corporate images and assessments.

2.2   Operationalizing Reputation

Research focusing on corporate reputation assessment and measurement has made major progress in conceptualization and operationalization of the concept. A common approach to assess reputation is to identify its drivers, and then get those evaluated by selected groups of people, although not necessarily stakeholders. Bebbington et al [6] review several approaches to conceptualize reputation and reveal five common elements of reputation: financial performance, quality of management, social and environmental responsibility performance, employee quality, and the quality of the goods/services provided. Schwaiger [20] finds similar categories, while additionally highlighting fair attitude towards competitors, transparency and openness, credibility. As instance, the most famous Reputation Quotient [5], giving an aggregate rank score for each corporation in an annual survey, included 8 categories: familiarity, operational capability, strategic positioning, industry leadership, distinctiveness, credibility, influential and carrying. It has been refined over years with the latest RepTrak Pulse [21] - simplified and shorter-term oriented approach. Although not specifically designed for reputation measurement but as reputation building blocks Brady [22] suggested another seven reputation drivers: Knowledge and skills, Emotional connections, Leadership, vision and desire, Quality, Financial credibility, Social credibility,

Environmental credibility. Nevertheless, the list is in great extent consistent with the Reputation Quotient as well as other predecessors. Thus, approaches for decomposing reputation are mostly consistent amongst themselves.

Nevertheless, while definitely advancing conceptualization of reputation, current reputation measurement approaches have some limitations for direct applications in corporate reputation risk analysis. Foremost, the aim of most reputation measurement surveys is to rank companies among peers. The rankings represent current and past experiences of stakeholders. This makes the measurements useful to link past events to the reputation changes, but it's problematic to derive any future reputation risks for a given firm.

Secondly, usually reputation is defined as an aggregate of general attitudes towards reputation drivers of a firm. In particular some reputation measurements relies on random sampling of public (e.g., [20]), thus, veiling views of particular stakeholders. Alternatively, a restricted set of financially oriented stakeholders (managers and analysts) is taken, which limits generalizability of conclusions and implications.

Finally, even if different stakeholders are specifically considered (as in the Reputation Quotient case [5]), the multiples stakeholders' perceptions are then aggregated to produce an overall score. Provided that perceptions among different stakeholder groups are not necessarily homogenous, if not conflicting [2, 6, 23], an organization might have a high Reputation Quotient score, but still not a good reputation among a key stakeholder group such as an owner or customers. The same issue arises with aggregating perceptions amongst different reputation drivers.

Therefore, aggregated reputation indices conceal particular stakeholders and reputation within particular category. They hinder comprehension of polarity of different stakeholders' perceptions, thus, impede informed risk and decision analysis and following targeted reputation management strategies and actions.

2.3   What Is Reputation Risk?

Reputation risk is conventionally understood as a potential loss of reputation. Reputation risks arise when there is a gap between expectations and reality [24, 25]. Thus, reputation risk magnitude can be simply defined as "Reality minus Expectations". Identification of reputation risks is a gap-analysis exercise, which, in opposite to reputation measurement should be a simpler task.

Traditionally reputation management focuses on improving perceptions of organization achievements, since changing beliefs and perceptions can increase or decrease the gap. However, the gaps are not caused by misperceptions only. Perceptions can correspond to reality, but stakeholder might have expectations exceeding what a company can deliver. In such a case inflated expectations carry even higher risks than being undervalued, effectively requiring making strategic actions [3]. Similarly, Aula [2] argues that in the era of social media strategic reputation management should concentrate on ethics, implying that "an organization cannot just look good; it has to be good" (p. 46).

Overwhelming majority of publications consider reputation risks from a negative perspective only, i.e., reputation can only be damaged but never gained due to the occurred gap. Seeing risks as only threats is not exceptional for reputation – it is still a dominant view in risk management practices despite the calls for embracing opportunities as another counterpart of risk [cf. 26-28]. In the case of reputation risks it is conceptually and practically useful to distinguish between risks caused by inflated expectations and those caused by undervalued performance. Considering threats along with potential gains should provide managers with a holistic and balanced view on reputation risks. This, in turn, should enable comprehensive decision and risk analysis.

The highlighted points create necessary prerequisites for reputation management becoming a central part of strategic decision-making. Liehr-Gobbers and Storck [29] demonstrate how communication can be instrumental for fulfilling strategic changes in the company, arguing that "communication needs to become an integral part of the strategic process" [29]. Not only reputation should be nurtured and exploited for business opportunities, but also the business-driven decisions should be evaluated in terms of their reputational consequences.

3   Conceptual Framework for Reputation Risk Identification

Derived from the reputation risk management literature [24, 25, 30] a set of interlinked definitions are further introduced. The definitions are framed in a conceptual model for reputation risk (**Fig. 1**).

Since reputation is an intangible asset formed by stakeholders' attitudes and expectations towards a company's exposed performance, therefore, reputation risk "involves an organization acting, behaving or performing in a way that falls short of stakeholder expectations" [25]. Therefore, a *reputation risk* can be defined as an uncertain change in the corporate reputation due to a gap between stakeholder expectations and company's exposed performance. A magnitude of the "performance – expectation" gap, i.e., in what extent a particular organization's performance differ from a particular stakeholder's expectation on the performance, we define as *reputation risk impact*. As discussed in the previous section, reputation risk impact can be not only negative (expectations are bigger than performance) but also positive (expectations are lower than performance). Thus, the suggested approach should be able to identify reputational opportunities for exploitation as well.

Any past, current or future business decision, activity, or behavior (business action) induces potential changes in performance. Such an action not yet exposed to stakeholders is a potential *reputation risk source* defined by ISO [31] as "element which alone or in combination has the intrinsic potential to give rise to risk".

Further, stakeholders have different perspectives and concerns as well as corporate performance has multitude aspects. *Performance*, in this context, can be defined as organization's achieved and publicly exposed past results, along with intended and on-going business actions within key reputation areas. *Key reputation areas* (KPAs) are similar to the notion of reputation drivers. They represent different aspects of company's performance, upon which stakeholders have perceptions and expectations. Each KPA can be broke down into subcategories or attributes detailing a focal KPA.

KPAs may reflect organization's values and goals, however they should also reflect what stakeholders believe are important for the organization to achieve.

Stakeholder is defined as "any group or individual who can affect or is affected by the achievement of the organization's objectives" [32]. Each group has a unique relationship with a company, a unique set of values and goals, and distinct expectations. Moreover, stakeholders' attitudes are not static but change over time and with changes to the business environment.

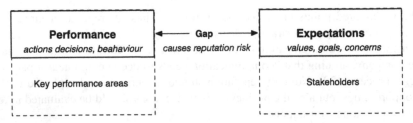

Fig. 1. Reputation risk model

In order to advance analysis of reputation risks stakeholder profiling is suggested. *Stakeholder profiling* is a continuous identification, monitoring and evaluation of stakeholders' attitudes and expectations within a defined set of KPAs. Thus, a stakeholder profile should be formed for each stakeholder and list expectations for each KPA. Different KPAs have different importance for different stakeholders. Indeed, shareholders must have bigger concern about financial performance of a firm then for corporate social responsibility, while different NGOs would do the opposite. Therefore, the same "performance–expectations" gap might have different impact on different stakeholders. Such importance priorities of KPAs ought to be captured and represented in a stakeholder profile. Furthermore, different stakeholders are of different importance for an organization, which also should be reflected in stakeholder profiles. Finally, stakeholder profiles should be regularly updated to catch any possible shifts within the mentioned parameters.

In the suggested framework reputation risk sources are used as inputs for identifying reputation risks. Particularly, the identification is done by contrasting expectations defined in the stakeholder profiles against the risk source being analyzed. Decision makers or experts are to analyze if and how a given risk source contradicts any expectations, thus triggers reputation risks.

Eccles [24] describes a case of American Airlines trying to stave off bankruptcy in 2003, which may illustrate a use-case for the framework. While negotiating a major reduction with unions, the executive board decided to approve retention bonuses for senior managers and paying a huge amount to a trust fund to secure executive pensions. No surprise the company's reputation with unions got ruined once they found out about the decision. As a few of the consequences, the unions revisited the concessions package they already had approved, and the CEO Donald J. Carty lost his job. Did the board analyze unions and/or employers profiles, this should have highlighted obvious contradictions between that board decision (i.e., risk source) with the expectations under such KPAs as corporate responsibility, leadership, and financial performance.

4   Method for Quantitative Reputation Risk Assessment

Having identified reputation risks using the stakeholder profiles, it is possible to further analyze the risks using the same structure implied by the stakeholder profile. In the following section we describe a formal method for reputation risk assessment. The definitions are further introduced as for one reputation risk source of interest.

Let r_i be an impact of a reputation risk induced by the given reputation risk source, w_{jk}^{KPA} - relative importance of the j:th KPA containing the expectation, such that $w_{jk}^{KPA} > 0$ and $\sum w_{jk}^{KPA} = 1$ for a k:th stakeholder, w_k^S - relative importance of the stakeholder, such that $w_k^S > 0$ and $\sum w_k^S = 1$.

Aggregated reputation risk impact of stakeholder.
Let R_k be the set of risks associated with the k:th stakeholder. Partition R_k into $^-R_k = \{r \in R_k, r \leq 0\}$ and $^+R_k = \{r \in R_k, r \geq 0\}$. Then $I_k^S = w_k^S \sum_{r_i \in R_k, j} r_i w_{jk}^{KPA}$, where I_k^S is aggregated reputation impact of a k:th stakeholder, j is the index for j:th KPA.

The aggregated threat of k:th stakeholder $^-I_k^S$ is given by the

$$^-I_k^S = w_k^S \sum_{r_i \in \, ^-R_k, j} r_i w_{jk}^{KPA}, \tag{1}$$

and the k:th stakeholder opportunity ($^+I_k^S$) is given by the

$$^+I_k^S = w_k^S \sum_{r_i \in \, ^+R_k, j} r_i w_{jk}^{KPA}. \tag{2}$$

Aggregated reputation risk impact on KPA.
Let $I_j^{KPA} = \sum_{r_i, k} r_i w_{jk}^{KPA} w_k^S$ be the global impact on a j:th KPA. Then the global threat on a j:th KPA ($^-I_j^{KPA}$) index is given by the

$$^-I_j^{KPA} = \sum_{r \leq 0, k} r w_{jk}^{KPA} w_k^S, \tag{3}$$

and the global favor on a j:th KPA ($^+I_j^{KPA}$) index is given by the

$$^+I_j^{KPA} = \sum_{r > 0, k} r w_{jk}^{KPA} w_k^S. \tag{4}$$

Global reputation risk impact.
Finally, let $I = \sum_k I_k^S$ be the global impact of the given reputation risk source. Then the risk source threat index (^-I) is given by the $\sum_k \, ^-I_k^S$, and the risk source opportunity index (^+I) is given by the $\sum_k \, ^+I_k^S$.

5   Application Case

The framework has been probed in cooperation with a state-owned company operating in Europe. Following the action research strategy, a high level manager dealing with public communication was collaborating in adaptation of the framework.

The following is the case, which was analyzed with the suggested methodology. The company made a recent executive decision to introduce a new resources supplier. The decision was to pursue the supply chain diversification strategy, which should increase competition among the resource suppliers, thus potentially reducing buying prices as well as increasing quality of the supply. Other benefits of the strategy are flexibility and supply risks reduction. One particular supplier was selected as another partner, however none of the selection criteria reflected reputation impact and any potential disagreements among company's stakeholders.

For the analysis initial set of key stakeholder groups was identified, which included employees, customers, owners, suppliers, pressure groups/NGOs, supervisory bodies, and politicians. Particular instances, i.e., organizations and persons, were listed within each stakeholder group.

During one of the sessions an interesting discussion raised around media as stakeholder. It was agreed that media, albeit powerful in communication, was not a stakeholder per se, but rather a communication mediator between the company and its stakeholders. Media's intrinsic interest is deemed to lie in "news-making", i.e., revealing "performance-expectations" gaps and communicating them to stakeholders. In terms of the presented framework media does not appear to have their specific expectations. With this motivation media was excluded from the initial list, though further research and discussion are secured.

Further, an initial set of KPAs with corresponding attributes was derived from the literature on reputation management. The set was revised with the company representatives aligning it to the company's promoted values and CSR reporting formats. The final set defined for the company consists of six KPAs: safety, environment and sustainability, performance, leadership, innovation, governance and citizenship.

Further, selected stakeholders were profiled. Information on the current stakeholder expectations were partially derived from the satisfaction surveys the company regularly performed among key stakeholders, such as employees, customers, and owners. Another set of expectations was derived from analysis of reputation issues cases within the industry. Finally, the collected expectations were classified within the defined KPAs and associated attributes. The Fig. 2 presents a sample of the "government/politicians" expectations framed in a stakeholder profile. Note that some original formulations in the profile were omitted or generalized due to the non-disclosure agreement.

Guided by the evidences from the aforementioned studies on the company's stakeholders the KPAs weights for each stakeholder were assigned with the direct rating method [see 33]. Stakeholders' relative importance weights were elicited using a cardinal pairwise comparison as in the analytical hierarchical process [34]. Table 1 exhibits a sample of the derived weights.

Table 1. Elicited weights for KPAs and Stakeholders

KPAs	Shareholders ($w_1=0,41$)	Suppliers ($w_2=0,33$)	Politicians ($w_3=0,26$)
Citizenship	0,1	0,05	0,29
Governance	0,1	0,16	0,18
Innovation	0,15	0,16	0,12
Leadership	0,2	0,16	0,12
Performance	0,25	0,26	0,06
Sustainability and Safety	0,2	0,21	0,23

Risk identification and assessment with stakeholder profiles.
Having profiled stakeholders, the profiles were analyzed for identifying "decision-expectations" gaps. Particular threats and opportunities were formulated for each gap. The **Fig. 2** depicts the "Government/Politicians" profile as an example (some original formulations were omitted or generalized due to the non-disclosure agreement).

KPA	Attributes	Stakeholder expectations	Description of possible threats/opportunities	Impact
Citizenship	Social responsibility			
Governance	Best practices	Evidence of best practice befitting a global business	The project is an evidence of global co-operation, openness and fairness	Insignificant Opportunity
	Transparency	Evidence of a corporate code and group policy	The supplier has been selected through comprehensive analysis.	Insignificant Opportunity
	Compliance	Follow national, EU guidelines and regulations	Relying on the supplier can be seen controversial	Significant Threat
Innovation	Expertise	Evidence that it has the right skills and technologies to service the industry	The new supplier has world leading technologies	Insignificant Opportunity
Leadership	Strategy	The business model & strategy for execution are both clear and sound	The executive decision is an example of the diversed supply strategy	Minor Opportunity
Performance	Taxes	Company is a sustainable and significant tax-payer	Cutting costs potentially increases amount of tax payments	Moderate Opportunity
Sustainability and Safety	Security	Company provides operations with no harm to environment, people, infrastructure	The supplier provides secure and safe resources	Insignificant Opportunity

Fig. 2. Government/Politicians profile with risks

Further, impact magnitudes of the identified risks (r_i) were subjectively assessed using a bipolar 12-term linguistic scale (Table 2), where each term is associated with an integer in the [-8;8] interval. The idea behind the scale design is to give non-linear progression to the risk impact, in this way boosting prioritization of higher impact risks. Moreover, the extreme linguistic values, namely "catastrophic threat" and "outstanding opportunity", gained relatively high scores to reflect usually stronger confidence when people assign extreme values.

Finally, the assessments were aggregated both for each stakeholder (as defined in 1 and 2) and for each KPA (as defined in 3 and 4). The Fig. 3 and Fig. 4 depict aggregated reputation risk assessments for three of the stakeholders.

Reputation risk assessment.
The analysis indicated that among several stakeholders three were mostly affected. Those are shareholders, current and potential suppliers (excluding the newly selected one), and politicians. Foremost, the owners and shareholders should favor the decision, as it brings evidences of sound strategic decision-making in the company (Leadership), potential profits (Performance), as well as transparent governance practices and global cooperation (Governance).

Table 2. Linguistic gap magnitude scale

Linguistic term	Numeric value
Catastrophic Threat	-8
Significant Threat	-5
Moderate Threat	-3
Minor Threat	-2
Insignificant Threat	-1
No Impact	0
Insignificant Opportunity	1
Minor Opportunity	2
Moderate Opportunity	3
Significant Opportunity	5
Outstanding Opportunity	8

Stakeholder	Favour	Threat
Politicians	0.2483	-0.234
Shareholders	0.83025	0
Suppliers	0.0528	-0.5874
Aggregated favour/threat	1.13135	-0.8214

Fig. 3. Aggregated stakeholder threat and favor

KPA	Favour	Threat
Performance	0.5437	-0.429
Governance	0.1142	-0.234
Innovation	0.0156	-0.1584
Sustainability and Safety	0.0299	0
Leadership	0.3348	0
Citizenship	0	0
Aggregated favour/threat	1.0382	-0.8214

Fig. 4. Aggregated KPA threat and opportunities

However, some potential issues were raised among the suppliers and, surprisingly, the politicians/government stakeholder group. As intended, the extra supplier should increase competition between the current ones and raise the entry barrier for new-comers (Performance). This also raises the bar for technological requirements of the supplied resources (Innovation).

Furthermore, the national government of the selected supplier was deemed to be controversial. Thus, it could be tempting for local and national politicians to raise this as issue for political agenda. Especially given that the company is state-owned, thus, politicians perceive themselves as "owners" along with the profile ministry and other regulators.

Finally, while critical threats were identified the framework application has also revealed opportunities, which can be further used as counter-arguments for communicating with the stakeholders. As instance, the new supply chain diversification strategy should benefit end-customers with lower prices, thus, it should increase profit tax payments to local and national budgets due to the reduction costs — both points are in favor for the governmental expectations. If the "pluses" are properly communicated they should be able to overcome the "minuses" (see the aggregated threat/favor for politicians in the Fig. 3). Moreover, the proposed approach directly indicates the stakeholders most receptive to such arguments.

6   Conclusion

Being a highly valuable but still an intangible asset corporate reputation is hard to measure and manage, as well as, conceptualize and operationalize in general. This is also true for the reputation-associated risks.

In this paper we tackle an issue of corporate reputation risk analysis. First we suggest a conceptual model for identification of reputation risks induced by endogenous factors, such as business decisions, actions, behaviors. Having adopted a multi-stakeholder perspective on reputation, a format for stakeholder profiling is suggested. The format implies eliciting different stakeholders' expectations and classifying the expectations within key performance areas of a company. Reputation risk is further identified based upon a gap between a focal endogenous factor and stakeholders' expectations.

We further offer a formalization for quantitative evaluation of reputation risk impact, utilizing a bipolar linguistic scale. The methodology integrates reputation risk impact through multiple stakeholder perspectives, as well as, multiple key performance areas in the form of additive weighting. While particular weight elicitation methods were selected for an application case, the formalization is deemed to be flexible to the choice of weight elicitation schemes.

The methodology is applicable to any industry and any set of stakeholders. We analyze a case of a state-owned company, which recently decided to introduce a new supplier as execution of their supply chain diversification strategy. The methodology underlines which key performance areas are mostly affected by the decision, both negatively and positively. Furthermore, not only did the applied methodology reveal potential reputation threats, but it also indicates reputation gain opportunities to balance the potential adverse effects (as perceived by individual stakeholders). In fact, the application methodology received high appraisal from the company management, which highlighted its high practical applicability and relevance, ease of use, and insightful analysis.

This study provides both academic and practical insights that could be immediately useful for operational as well as strategic reputation and communication management. The demonstrated reputation risk analysis has shifted the company's focus on issue management and crisis communication to proactive management of reputation risks. The conducted analysis is able to further guide prioritization of reputation risks, development of pre-emptive actions and assessment of their efficiency; and we encourage further development of these applications. Finally, as the application case revealed, the approach can be used as a framework for conveying reputation consequences of decisions to decision-makers.

References

1. Little, P.L., Little, B.L.: Do Perceptions of Corporate Social Responsibility Contribute to Explaining Differences in Corporate Price-Earnings Ratios? A Research Note. Corp. Reputation Rev. 3, 137–142 (2000)
2. Aula, P.: Social media, reputation risk and ambient publicity management. Strategy & Leadership 38, 43–49 (2010)
3. Liehr-Gobbers, K., Storck, C.: Future Trends of Corporate Reputation Management. In: Helm, S., Liehr-Gobbers, K., Storck, C. (eds.) Reputation Management, pp. 235–238. Springer, Heidelberg (2011)
4. Stevenson, A. (ed.): "Reputation". Oxford Dictionary of English. Oxford University Press (2010)
5. Fombrun, C.J., Gardberg, N.A., Sever, J.M.: The Reputation QuotientSM: A multi-stakeholder measure of corporate reputation. J. Brand. Manag. 7, 241–255 (2000)
6. Bebbington, J., Larrinaga, C., Moneva, J.M.: Corporate social reporting and reputation risk management. Acc. Auditing Accountability J. 21, 337–361 (2008)
7. da Camara, N.Z.: Identity, Image and Reputation. In: Helm, S., Liehr-Gobbers, K., Storck, C. (eds.) Reputation Management, pp. 47–58. Springer, Heidelberg (2011)
8. Kaplan, S.E., Ruland, R.G.: Positive theory, rationality and accounting regulation. Critical Perspectives on Accounting 2, 361–374 (1991)
9. Mathews, R.: Socially responsible accounting. Chapman & Hall (1993)
10. Tilling, M.V.: Refinements to legitimacy theory in social and environmental accounting. Commerce Research Paper Series (2004)
11. Flanagan, D.J.: The Effect of Layoffs on Firm Reputation. Journal of Management 31, 445–463 (2005)
12. Bergh, D.D., Ketchen, D.J., Boyd, B.K., Bergh, J.: New Frontiers of the Reputation–Performance Relationship: Insights From Multiple Theories. Journal of Management 36, 620–632 (2010)
13. Barney, J.B., Hesterly, W.S.: VRIO framework. In: Strategic Management and Competitive Advantage, pp. 68–86 (2010)
14. Schnietz, K.E., Epstein, M.J.: Exploring the Financial Value of a Reputation for Corporate Social Responsibility During a Crisis. Corp. Reputation Rev. 7, 327–345 (2005)
15. Casalo, L.V., Flavián, C., Guinalíu, M.: The Influence of Satisfaction, Perceived Reputation and Trust on a Consumer's Commitment to a Website. Journal of Marketing Communications 13, 1–17 (2007)
16. Hem, L.E., de Chernatony, L., Iversen, N.M.: Factors Influencing Successful Brand Extensions. Journal of Marketing Management 19, 781–806 (2003)
17. Larkin, J.: Strategic Reputation Risk Management. Palgrave Macmillan, Basingstoke (2002)
18. Freeman, R.E., Reed, D.L.: Stockholders and Stakeholders: A New Perspective on Corporate Governance. California Management Review (1983)

19. Mitchell, R.K., Agle, B.R., Wood, D.J.: Toward a Theory of Stakeholder Identification and Salience: Defining the Principle of Who and What Really Counts. Acad. Manage. Rev. 22, 853–886 (1997)
20. Schwaiger, M.: Components And Parameters Of Corporate Reputation – An Empirical Study. Schmalenbach Business Review (SBR) 56, 46–71 (2004)
21. Ponzi, L.J., Fombrun, C.J., Gardberg, N.A.: RepTrak™ Pulse: Conceptualizing and Validating a Short-Form Measure of Corporate Reputation. Corp. Reputation Rev. 14, 15–35 (2011)
22. Brady, A.K.O.: The sustainability effect: rethinking corporate reputation in the 21st century. Palgrave Macmillan, Basingstoke (2005)
23. Davies, D.: Risk Management—Protecting Reputation: Reputation Risk Management— The Holistic Approach. Computer Law & Security Review (2002)
24. Eccles, R.G., Newquist, S.C., Schatz, R.: Reputation and Its Risks. Harvard Business Review, 104–114 (2007)
25. Honey, M.G.: A Short Guide to Reputation Risk. Gower Publishing, Ltd. (2012)
26. Hillson, D.: Extending the risk process to manage opportunities. International Journal of Project Management 20, 235–240 (2002)
27. Olsson, R.: In search of opportunity management: Is the risk management process enough?, vol. 25, pp. 745–752 (2007)
28. Lechler, T.G., Edington, B.H., Gao, T.: Challenging Classic Project Management: Turning Project Uncertainties Into Business Opportunities 43, 59–69 (2012)
29. Liehr-Gobbers, K., Storck, C.: How to Manage Reputation. In: Helm, S., Liehr-Gobbers, K., Storck, C. (eds.) Reputation Management, pp. 183–188. Springer, Heidelberg (2011)
30. Fombrun, C.J.: Reputation. Harvard Business Press (1996)
31. ISO/Guide 73:2009, Risk management — Vocabulary. International Organization for Standardization
32. Freeman, R.E.: Strategic Management: A Stakeholder Approach. Pitman (1984)
33. Riabacke, M., Danielson, M., Ekenberg, L.: State-of-the-Art Prescriptive Criteria Weight Elicitation. Advances in Decision Sciences 2012, 1–24 (2012)
34. Saaty, T.L.: The Analytic Hierarchy and Analytic Network Processes for the Measurement of Intangible Criteria and for Decision-Making. In: Multiple Criteria Decision Analysis: State of the Art Surveys, vol. 78, pp. 345–405 (2005)

Properties and Complexity of Some Superposition Choice Procedures

Sergey Shvydun[1,2,*]

[1] National Research University Higher School of Economics, Moscow, Russia
[2] V.A. Trapeznikov Institute of Control Sciences of Russian Academy of Sciences,
Moscow, Russia
shvydun@hse.ru

Abstract. In this paper, the application of two-stage superposition choice procedures to the choice problem when the number of alternatives is too large is studied. Two-stage superposition choice procedures consist in sequential application of two choice procedures where the result of the first choice procedure is the input for the second choice procedures. We focus on the study of properties of such choice procedures and evaluate its computational complexity in order to determine which of two-stage superposition choice procedures can be applied in the case of large amount of alternatives.

Keywords: Superposition, Choice Problem, Choice Procedures, Computational Complexity.

1   Introduction

In recent years, due to the enormous increase of the amount of information, which led to a larger number of alternatives and criteria, there has been more attention given to the choice problem. The choice problem can be defines as follows. Consider a finite set A with $|A| > 2$ alternatives where any subset $X \in 2^A$ may be presented for choice. Denote by $C(\cdot)$ a choice function that performs mapping $2^A \rightarrow 2^A$ with restriction $C(X) \subseteq X$ $C(X) \subseteq X$ for any $X \in 2^A$. The choice consists in selection according to some rule from some presentation X of the non-empty subset of alternatives $Y \subseteq X$ $Y \subseteq X$ (non-empty subset of «best» alternatives).

There are a lot of different choice procedures that allow to choose and rank alternatives from initial set. All choice procedures can be divided into 5 following groups:

1. Scoring rules;
2. Rules, using majority relation;
3. Rules, using value function;
4. Rules, using tournament matrix;
5. q-Paretian rules.

---

[*] Corresponding author.

© Springer International Publishing Switzerland 2015
H.A. Le Thi et al. (eds.), *Model. Comput. & Optim. in Inf. Syst. & Manage. Sci.*,
Advances in Intelligent Systems and Computing 360, DOI: 10.1007/978-3-319-18167-7_41

Unfortunately, most choice procedures cannot be applied when we deal with a large amount of alternatives due to its computational complexity. For instance, choice procedures based on the majority relation require pairwise comparisons of all alternatives and, consequently, if the number of alternatives is more than 10^5, more than 10^{10} comparisons should be performed which is not always possible to do in a sufficient time. Moreover, there are a lot of situations when after applying some choice procedures the remaining set of alternatives is too large. It means that other choice procedures should be applied in order to narrow the initial set of alternatives. Thus, there is a need to consider new approaches how to aggregate individual preferences in case of the large number of alternatives.

In this paper the application of the idea of superposition to the choice problem is studied. Let us remind that by superposition of two choice functions $C_1(\cdot)$ and $C_2(\cdot)$ we mean a binary operation $\odot$, the result of which is a new function $C*(\cdot)=C_2(\cdot)\odot C_1(\cdot)$, having the form $\forall X \in 2^A \; C^*(X)=C_2(C_1(X))$ [1]. In other words, superposition consists in sequential application of choice functions where the result of the previous choice function C_1 is the input for the next choice function C_2.

The interest in superposition of choice procedures can be explained by several reasons. First, choice procedures based on the idea of superposition have a low computational complexity, which is crucial in the cases when the number of alternatives or/and criteria is very large. The use of superposition allows to reduce the complexity by applying choice procedures with a low computational complexity on first stages and more accurate choice procedures on final stages. Thus, the results can be obtained in a reasonable time. Second, the use of superposition allows to avoid situations when the remaining set of alternatives is too large through the use of additional choice procedures.

Thus, we consider the simplest type of superposition of choice procedures – two-stage superposition choice procedures. We focus on the study of properties and computational complexity of such choice procedures. The study of properties gives us information on how the final choice is changed due to changes of preferences, a set of feasible alternatives and a set of criteria while the study of computational complexity of two-stage choice procedures determines the list of choice procedures that can be applied when we deal with a large amount of alternatives.

1.1   Description of Choice Procedures

In this work we study two-stage superposition choice procedures which use scoring rules and rules, using majority relation on the first stage and scoring rules and rules, using majority relation, value function or tournament matrix on the second stage.

In Table 1 we provide a list of such procedures which are studied in this work.

Table 1. A list of studied choice procedures

№	Name of choice procedure	Type of choice procedure
1	Simple majority rule	Scoring rules
2	Plurality rule	
3	Inverse plurality rule	
4	q-Approval rule	
5	Run-off procedure	
6	Hare's rule (Ware's procedure)	
7	Borda's rule	
8	Black's procedure	
9	Inverse Borda's rule	
10	Nanson's rule	
11	Coombs' procedure	
12	Minimal dominant set	Rules, using majority relation
13	Minimal undominated set	
14	Minimal Weakly stable set	
15	Fishburn's rule	
16	Uncovered set I	
17	Uncovered set II	
18	Richelson's rule	
19	Condorcet winner	
20	Core	
21	k-stable set	
22	Threshold rule	Rules, using value function
23	Copeland's rule 1	
24	Copeland's rule 2	
25	Copeland's rule 3	
26	Super-threshold rule	
27	Minimax procedure	Rules, using tournament matrix
28	Simpson's procedure	

The definition of such choice procedures is given in [1-4].

1.2   Normative Conditions

All normative conditions, which characterize different choice procedures, can be divided in the following groups [1, 5-8]

- Rationality conditions;
- Monotonicity conditions;
- Non-compensatory condition.

Let us describe now these conditions.

Rationality Conditions
There are four rationality conditions for choice functions.

1. Heritage condition (**H**)

$$\forall X, X' \in 2^A, X' \subseteq X \Rightarrow C(\vec{P}_{X'}, X') \supseteq C(\vec{P}_X, X) \cap X', \tag{1}$$

where $\vec{P}_X$ is a contruction of a preference profile $\vec{P}$ onto a set $X \subseteq A$, $X \neq \emptyset$, i.e.,
$\vec{P}_X = (P_1 / X, .., P_n / X)$, $P_i / X = P_i \cap (X \times X)$.

If the presented set is narrowed by eliminating some alternatives, those chosen from the initial set and remaining in the narrowed set will be chosen from the narrowed set.

2. Concordance condition (**C**)

$$\forall X', X'' \in 2^A \rightarrow C(\vec{P}_{X' \cup X''}, X' \cup X'') \supseteq C(\vec{P}_{X'}, X') \cap C(\vec{P}_{X''}, X''). \tag{2}$$

Condition **C** requires that all the alternatives chosen simultaneously from X' and X'' be included in the choice when their union $X' \cup X''$ is presented.

3. Independence of Outcast of options (**O**)

$$\forall X, X' \in 2^A, X' \subseteq X \setminus C(\vec{P}_X, X) \Rightarrow C(\vec{P}_{X \setminus X'}, X \setminus X') = C(\vec{P}_X, X). \tag{3}$$

Condition **O** requires that the narrowing of X by rejection of some or all the alternatives not chosen from the initial set X does not change the choice.

4. Arrow's choice axiom (**ACA**)

$$\forall X, X' \in 2^A, X' \subseteq X \Rightarrow \begin{cases} if\ C(\vec{P}_X, X) = \emptyset,\ then\ C(\vec{P}_{X'}, X') = \emptyset, \\ if\ C(\vec{P}_X, X) \cap X' \neq \emptyset,\ then\ C(\vec{P}_{X'}, X') = C(\vec{P}_X, X) \cap X'. \end{cases} \tag{4}$$

For the case where the choice is not empty, Condition **ACA** requires that the alternatives chosen from the initial set X and left in the narrowed one X' and only such alternatives be chosen from X'.

Monotonicity Conditions

1. Monotonicity condition 1

$$\forall X \in 2^A, x \in C(\vec{P}_X, X), \forall \vec{P}_X, \vec{P}_X':$$
$$(\forall a, b \in X, aP_i b \Leftrightarrow aP_i' b \,\&\, \exists y \in X, yP_i x \Rightarrow xP_i' y) \Rightarrow x \in C\left(\vec{P}_X', X\right). \tag{5}$$

Let alternative x be chosen from initial set X. Suppose now that the relative position of alternative x was improved while the relative comparison of any pair of other alternatives remains unchanged. Monotonicity condition is satisfied if alternative x is still in choice.

2. Monotonicity condition 2

$$\forall X \in 2^A, x, y \in C(\vec{P}_X, X), \ X' = X \backslash \{x\}, X'' = X \backslash \{y\}$$
$$\to x \in C(\vec{P}_{X''}, X'') \,\&\, y \in C(\vec{P}_{X'}, X'). \tag{6}$$

Let two alternatives x, y be chosen from initial set X. Monotonicity condition 2 is satisfied if one of the chosen alternative (x or y) is still in choice when the other chosen alternative (y or x) is eliminated.

3. Strict monotonicity condition

$$\forall X \in 2^A, \forall y \in X, y \notin C(\vec{P}_X, X), \forall \vec{P}_X, \vec{P}_X':$$
$$(\forall a, b \in X, aP_i b \Leftrightarrow aP_i' b \,\&\, aP_i y \Rightarrow yP_i' a)$$
$$\to C\left(\vec{P}_X', X\right) = \begin{bmatrix} C(\vec{P}_X, X) \ or \\ \{y\} \ or \\ C(\vec{P}_X, X) \cup \{y\}. \end{bmatrix} \tag{7}$$

The change of relative position of unchosen alternative $y \in X \backslash C(\vec{P}_X, X)$ $y \in XC(\vec{P}_X, X)$ so that its position will be improved while the relative comparison of any pair of other alternatives remains unchanged leads to the choice of alternative y or/and all alternatives that were in initial choice $C(\vec{P}_X, X)$.

Non-compensatory Condition
According to non-compensatory condition, low estimates of one criterion cannot be compensated by high estimates on other criteria.

2   Two-Stage Superposition Choice Procedures

A list of two-stage superposition choice procedures studied in this paper is provided in Table 2.

Table 2. Two-stage choice procedures

№	Stage 1	Stage 2
1-121	Scoring rules (11 procedures)	Scoring rules (11 procedures)
122-231		Rules, using majority relation (10 procedures)
232-286		Rules, using value function (5 procedures)
287-308		Rules, using tournament matrix (2 procedures)
309-418	Rules, using majority relation (10 procedures)	Scoring rules (11 procedures)
419-518		Rules, using majority relation (10 procedures)
519-568		Rules, using value function (5 procedures)
569-588		Rules, using tournament matrix (2 procedures)

Thus, 588 two-stage procedures of 8 different types are studied. Before proceeding to the study of properties of two-stage choice procedures, it is necessary to make some notes.

First, as not all two-stage choice procedures make any sense, there were found 168 two-stage procedures the second stage of which does not change the choice. For instance, two-stage choice procedures which use simple majority rule, run-off procedure, Hare's procedure, Coombs' procedure or Condorcet winner on the first stage do not make any sense as such procedures choose no more than one alternative. Thus, 168 choice procedures were excluded from further consideration.

Second, there were found 25 two-stage choice procedures equivalent to one of existing choice procedures. For instance, two-stage choice procedure "Minimal dominant set-Single majority rule" is equivalent to Condorcet winner. Another example is two-stage choice procedure "Minimal dominant set –Minimal weakly stable set" which is equivalent to minimal weakly stable set. Properties of such procedures fully coincided with properties of such existing procedures. However, these two-stage procedures cannot be excluded from further consideration as the computational complexity of some of them can be lower than the complexity of existing procedures.

Finally, it is necessary to mention that properties of two-stage choice procedures which use Black's procedure on the first stage fully coincide with properties of two-stage choice procedures which use Borda's rule if there is no Condorcet winner.

Thus, it remains to study properties of 395 two-stage choice procedures.

3   A Study of Properties for Choice Procedures

A study of properties for choice procedures can be divided in two steps. First, it is necessary to study properties of 28 existing choice procedures which are used in two-stage choice procedures. Second, properties of 395 two-stage procedures are studied.

A study of properties is the following. If a choice procedure does not satisfy given normative condition, a counter-example is provided. On the country, if a choice procedure satisfy given normative condition a necessary proof is followed. The results of the study of properties are given in Theorem 1.

Theorem 1. Information on which choice procedures satisfy given normative conditions is provided in Table 3 and Table 4.

Table 3. Properties of existing choice procedures («+» - choice procedure satisfies given normative condition, «-» - choice procedure does not satisfy given normative condition, «X» - normative condition does not exist for a given choice procedure)

№	Choice procedure	Rationality conditions				Monotonicity conditions			Non-compensatory condition
		Condition H	Condition C	Condition O	Condition ACA	Monotonicity condition 1	Monotonicity condition 2	Strict monotonicity condition	
1	Simple majority rule	+	-	+	-	+	X	-	-
2	Plurality rule	-	-	-	-	+	-	-	-
3	Inverse plurality rule	-	-	-	-	+	-	-	-
4	q-Approval rule	-	-	-	-	+	-	-	-
5	Run-off procedure	-	-	-	-	-	X	-	-
6	Hare's rule (Ware's procedure)	-	-	-	-	-	X	-	-
7	Borda's rule	-	-	-	-	+	-	-	-
8	Black's procedure	-	-	-	-	+	-	-	-
9	Inverse Borda's rule	-	-	-	-	-	-	-	-
10	Nanson's rule	-	-	-	-	-	-	-	-
11	Coombs' procedure	-	-	-	-	-	X	-	-
12	Minimal dominant set	-	+	+	-	+	-	-	-
13	Minimal undominant set	-	-	-	-	+	-	-	-
14	Minimal weakly stable set	-	-	-	-	+	-	-	-
15	Fishburn's rule	-	-	-	-	+	-	-	-
16	Uncovered set I	-	-	-	-	+	-	-	-
17	Uncovered set II	-	+	-	-	+	-	-	-
18	Richelson's rule	-	-	-	-	+	-	-	-
19	Condorcet winner	+	+	-	-	+	X	-	-
20	Core	+	+	-	-	+	+	-	-
21	k-stable set (k>1)	-	-	-	-	+	-	-	-
22	Threshold rule	-	-	-	-	+	-	-	+
23	Copeland's rule 1	-	-	-	-	+	-	-	-
24	Copeland's rule 2	-	-	-	-	+	-	-	-
25	Copeland's rule 3	-	-	-	-	+	-	-	-
26a	Super-threshold rule (fixed threshold)	+	+	+	+	+	+	+	-
26b	Super-threshold rule (threshold depends on X)	-	-	-	-	+	+	-	-
27	Minimax procedure	-	-	-	-	+	-	-	-
28	Simpson's procedure	-	-	-	-	+	-	-	-

Table 4. Properties of two-stage choice procedures («+» - choice procedure satisfies given normative condition, «-» - choice procedure does not satisfy given normative condition, «X» - normative condition does not exist for a given choice procedure)

Two-stage choice procedure		Rationality conditions				Monotonicity conditions			Non-compensatory condition
		Condition H	Condition C	Condition O	Condition ACA	Monotonicity condition 1	Monotonicity condition 2	Strict monotonicity condition	
Stage 1	**Stage 2**								
Borda's rule Black's procedure	Simple majority rule Run-off procedure Hare's rule (Ware's procedure) Condorcet winner	-	-	-	-	+	X	-	-
Plurality rule Inverse plurality rule q-Approval rule (q>1)	Simple majority rule Condorcet winner	-	-	-	-	+	X	-	-
Minimal undominant set Core	Simple majority rule	+	+	-	-	+	X	-	-
Fishburn's rule Uncovered set I Uncovered set II Richelson's rule	Simple majority rule Condorcet winner	+	+	-	-	+	X	-	-
Borda's rule Black's procedure	Inverse Borda's rule Nanson's rule k-stable set (k>1)	-	-	-	-	+	-	-	-
Fishburn's rule Richelson's rule Uncovered set I	Minimal dominant set Uncovered set I Richelson's rule	-	-	-	-	+	-	-	-
Minimal undominant set	Fishburn's rule Uncovered set II Copeland's rule 1-3	-	-	-	-	+	-	-	-
Plurality rule Inverse plurality rule q-Approval rule (q>1) Borda's rule Black's procedure	Plurality rule Inverse plurality rule q-Approval rule (q>1) Borda's rule Black's procedure Minimal dominant set Minimal undominant set Minimal weakly stable set Fishburn's rule Uncovered set I Uncovered set II Richelson's rule	-	-	-	-	+	-	-	-

Table 4. (*continued*)

Two-stage choice procedure		Rationality conditions				Monotonicity conditions			Non-compensatory condition
		Condition H	Condition C	Condition O	Condition ACA	Monotonicity condition 1	Monotonicity condition 2	Strict monotonicity condition	
Stage 1	**Stage 2**								
	Core Threshold rule Copeland's rule 1-3 Super-threshold rule (threshold depends on X) Minimax procedure Simpson's procedure								
Minimal undominant set Uncovered set II	Uncovered set I Richelson's rule	-	-	-	-	+	-	-	-
Uncovered set I	Core	-	-	-	-	+	-	-	-
Minimal dominant set Minimal undominant set	Super-threshold rule (threshold depends on X) Minimax procedure Simpson's procedure	-	-	-	-	+	-	-	-
Core	Borda's rule Black's procedure Inverse Borda's rule Nanson's rule Minimal dominant set Minimal undominant set Minimal weakly stable set Fishburn's rule Uncovered set I Uncovered set II Richelson's rule Copeland's rule 1-3 Minimax procedure Simpson's procedure	+	+	-	-	+	+	-	-
Uncovered set II	Minimal dominant set	-	+	-	-	+	-	-	-

Two-stage choice procedures not included in Table 4 do not satisfy any given normative condition.

4   Computational Complexity of Two-Stage Choice Procedures

Based on computational complexity of existing choice procedures and number of remaining alternatives we can divide all two-stage procedures in several groups in accordance with their computational complexity (see Table 5).

Table 5. A computational complexity of two-stage superposition choice procedures («...» - any studied choice procedure)

Two-stage choice procedure	
Stage 1	**Stage 2**
Choice procedures with a low computational complexity	
Plurality rule q-Approval rule (q>1)	...
Inverse plurality rule Borda's rule Black's procedure	Simple majority rule Run-off procedure Hare's rule (Ware's procedure) Borda's rule Black's procedure Condorcet winner Plurality rule Threshold rule Inverse plurality rule q-Approval rule (q>1) Super-threshold rule
Computational complexity depends on initial set of alternatives	
Inverse plurality rule Borda's rule Black's procedure	Inverse Borda's rule Nanson's rule Core Copeland's rule 1-3 Minimax procedure Simpson's procedure Coombs' procedure
Choice procedures with average computational complexity	
Inverse Borda's rule Nanson's rule Core Minimax procedure Simpson's procedure	Simple majority rule Run-off procedure Hare's rule (Ware's procedure) Borda's rule Black's procedure Condorcet winner Plurality rule Threshold rule Inverse plurality rule q-Approval rule (q>1) Super-threshold rule Inverse Borda's rule Nanson's rule Core Copeland's rule 1-3 Minimax procedure Simpson's procedure Coombs' procedure
Coombs' procedure	...
Choice procedures with a high computational complexity	
Inverse plurality rule Borda's rule Black's procedure Inverse Borda's rule	Minimal dominant set Minimal undominant set Minimal weakly stable set Fishburn's rule

Table 5. (*continued*)

Two-stage choice procedure	
Stage 1	Stage 2
Nanson's rule Core	Uncovered set I, II Richelson's rule k-stable set (k>1)
Minimal dominant set Minimal undominant set Minimal weakly stable set Fishburn's rule Uncovered set I, II Richelson's rule k-stable set (k>1)	...

To prove the results the run-time complexity was calculated for two-stage choice procedures from Table 5. For the case of 300 thousands of alternatives two-stage choice procedures with a low computational complexity calculated the results in less than 3 seconds. Two-stage choice procedures for which the computational complexity depends on initial set of alternatives calculated the results in less than 0.4-27 minutes. Two-stage choice procedures with average computational complexity calculated the results in less than 5 hours. The run-time for choice procedures with a high computational complexity was not calculated as they do not allow to obtain the results in a reasonable time.

5   Conclusion

In this paper, we studied the properties of 28 existing and 395 two-stage choice procedures, which can be used in various multi-criteria problems. It was defined which choice procedures satisfy given normative conditions, showing how a final choice is changed due to the changes of preferences or a set of feasible alternatives. Such information leads to a better understanding of different choice procedures and how stable and sensible is a set of alternatives obtained after applying some choice procedure.

The results show that only Simple majority rule, Condorcet winner, Core and Threshold rule with fixed threshold level satisfy condition **H**, only Minimal dominant set, Uncovered set II, Condorcet winner, Core and Threshold rule with fixed threshold level satisfy condition **C**, and only Simple majority rule, Minimal dominant set and Threshold rule with fixed threshold level satisfy condition **O**. As for two-stage choice procedures, most of them do not satisfy any normative conditions. Only some of them satisfy monotonicity condition 1. More information is provided in Table 3 and Table 4.

To compute run-time complexity of choice procedures the average computational complexity was used. All choice procedures were divided in different groups (see Table 5). It was shown that two-stage choice procedures, which use choice procedures with a high computational complexity on the first stage, require more time than other procedures. It means that such procedures are not recommended to use when the number of alternatives is too large. Two-stage choice procedures which use on the first stage choice procedures with a low computational complexity and on the second stage - with a high computational complexity can be used in the case of large amount of alternatives, however, its application depends on the number of alternatives remained after the first stage. Two-stage choice procedures, which use on both stages choice procedures with a low computational complexity, can be used in the case of large amount of alternatives with no restrictions.

References

1. Aizerman, M., Aleskerov, F.: Theory of Choice. Elsevier, North-Holland (1995)
2. Aleskerov, F.T., Khabina, E.L., Shvarts, D.A.: Binary relations, graphs and collective decisions. HSE Publishing House, Moscow (2006)
3. Aleskerov, F.T., Kurbanov, E.: On the degree of manipulability of group choice rules. Automation and Remote Control 10, 134–146 (1998)
4. Volsky, V.I.: Voting rules in small groups from ancient times to the XX century. Working paper WP7/2014/02. HSE Publishing House, Moscow (2014)
5. Moulin, H.: Axioms of cooperative decision-making. Cambridge University Press, Cambridge (1987)
6. Aleskerov, F.T., Yuzbashev, D.V., Yakuba, V.I.: Threshold aggregation for three-graded rankings. Automation and Remote Control 1, 147–152 (2007)
7. Aleskerov, F., Chistyakov, V., Kalyagin, V.: Social threshold aggregations. Social Choice and Welfare 35(4), 627–646 (2010)
8. Aleskerov, F.: Arrovian Aggregation Models. Kluwer Academic Publishers, Dordrecht (1999)

A Cross-Efficiency Approach for Evaluating Decision Making Units in Presence of Undesirable Outputs

Mahdi Moeini*, Balal Karimi, and Esmaile Khorram

Chair of Business Information Systems and Operations Research (BISOR),
Technical University of Kaiserslautern, Postfach 3049,
Erwin-Schrödinger-Str., D-67653 Kaiserslautern, Germany
`moeini@wiwi.uni-kl.de`
Department of Mathematics, Azad University (Karaj Branch), Karaj, Iran
`b.karimi1987@gmail.com`
Department of Mathematics and Computer Science,
Amirkabir University of Technology, Hafez Ave., Tehran, Iran
`eskhor@aut.ac.ir`

Abstract. Data Envelopment Analysis (DEA) is a mathematical programming approach for measuring efficiency of Decision Making Units ($DMUs$). In traditional DEA, a ratio of weighted outputs to inputs is examined and, for each DMU, some optimal weights are obtained. The method of *cross-efficiency* is an extension to DEA by which a matrix of scores is computed. The elements of the matrix are computed by means of the weights obtained via usual models of DEA. The cross-efficiency may have some drawbacks, e.g., the cross-efficiency scores may be multiple due to the presence of several optima. To overcome this issue, secondary goals are used. However, this method has never been used for peer evaluation of *DMUs with undesirable outputs*. In this paper, our objective is to bridge this gap. For this end, we introduce a new secondary goal, test it on an empirical example with undesirable outputs, report the results, and finally, we give some concluding remarks.

Keywords: Data Envelopment Analysis, Cross-Efficiency, Secondary Goals, Undesirable Outputs.

1 Introduction

In mathematical programming, Data Envelopment Analysis (DEA) is known as a methodology that is used for evaluating the efficiency of decision making units ($DMUs$). In general, $DMUs$ have multiple inputs and multiple outputs. The pioneer work of Charnes et al. [3] (well known as model CCR) uses the ratio of weighted multiple outputs over weighted multiple inputs for computing the efficiency scores of each DMU. The BCC model of Banker et al. [1] is

---

* Corresponding author.

© Springer International Publishing Switzerland 2015   487
H.A. Le Thi et al. (eds.), *Model. Comput. & Optim. in Inf. Syst. & Manage. Sci.*,
Advances in Intelligent Systems and Computing 360, DOI: 10.1007/978-3-319-18167-7_42

another classical model in DEA. Following these models, different variants of CCR and BCC as well as some new models have been proposed for evaluating DMUs [5]. Despite the efficiency and wide usage of DEA in identifying the most performant $DMUs$, the classical models have some drawbacks. For example, self-evaluation character of DEA models is considered as an important issue [15]. In order to overcome these flaws, other approaches have been proposed [4, 7, 15, 19]. One of them is the technique of *cross-efficiency*. In this approach, a peer evaluation replaces the self-evaluation of $DMUs$ and the efficiency scores are computed through linking each DMU to all $DMUs$ [15]. The performance of cross-efficiency has been approved through numerous applications (see e.g., [2, 15, 19], and more references therein). However, cross-efficiency, at its turn, suffers from a major drawback: in general, the cross-efficiency scores are not unique [7, 15]. In order to overcome this problem, several secondary goals have been proposed and studied [7, 15–17, 19, 21].

Furthermore, among different applications of DEA, we find some applications related to sustainable development and environmental aspects. More precisely, among the usual outputs of a DMU, there may exist some outputs that are known as *undesirable outputs*. As an example, one can cite the excessive carbon emissions (as undesirable outputs) of a manufacturing firm. Due to importance of this class of problems, there have been several studies about undesirable inputs/outputs. The first paper is due to Fare et al. [9] that has been followed by some other researchers (see e.g., [10, 12, 18, 20, 23]).

There were some studies concerning the application of cross-efficiency in peer evaluation of $DMUs$ with undesirable inputs/outputs [24]. However, to the best of our knowledge, in all of such studies there is no emphasize on the drawbacks of the cross-efficiency approach. This paper aims at bridging the gap between cross-efficiency with secondary goal and the DEA models addressing undesirable outputs. In order to attain this objective, we bring together all necessary materials, such as suitable production possibility set (PPS) and a new secondary goal. This procedure produces a multi-objective model for which we use max-min model that focuses on worst-case evaluations [8, 11]. Some experiments have been carried out on a practical data set [22]. A comparative study on the results demonstrates interesting observations about the usefulness of the proposed approach.

The current paper is organized as follows: In Section 2, we present notations, the prerequisites, and the concept of cross-efficiency. Section 3 is an introduction to data envelopment analysis with undesirable outputs. In this section, a self-evaluation model as well as cross-efficiency models for undesirable outputs are presented. Section 4 is devoted to the numerical experiments on a practical data set and some comments on the observations are provided. Section 5 includes the concluding remarks as well as some directions for future research.

2   Technical Materials: Prerequisites, Cross-Efficiency and Secondary Goal

In this section, we present the technical materials and notations that we are going to use through out this paper. Furthermore, we present a new secondary goal that will be used in used in evaluating $DMUs$.

2.1   Prerequisites

Suppose that we have a set of $DMUs$ and each DMU_j (for $j \in \{1, \ldots, n\}$) produces s different outputs from m different inputs. The following optimization model describes the multiple form of the well known CCR model for evaluating DMU_d (where $d \in \{1, \ldots, n\}$):

$$(CCR): \quad E^*_{dd} = \max \quad \frac{\sum_{r=1}^{s} u_{rd} y_{rd}}{\sum_{i=1}^{m} v_{id} x_{id}} \tag{1}$$

$$\text{s.t: } E_{dj} = \frac{\sum_{r=1}^{s} u_{rd} y_{rj}}{\sum_{i=1}^{m} v_{id} x_{ij}} \le 1 : \quad j = 1, \ldots, n, \tag{2}$$

$$u_{rd} \ge 0 : \quad r = 1, \ldots, s, \tag{3}$$

$$v_{id} \ge 0 : \quad i = 1, \ldots, m. \tag{4}$$

The optimal value of this model, i.e., E^*_{dd}, is the optimal relative CCR efficiency score of DMU_d (where $d \in \{1, \ldots, n\}$).

Let us suppose that (v^*_d, u^*_d) is the vector of optimal weights for DMU_d. Using these optimal weights, the cross-efficiency of DMU_j is computed as follows:

$$E^*_{dj} := \frac{\sum_{r=1}^{s} u^*_{rd} y_{rj}}{\sum_{i=1}^{m} v^*_{id} x_{ij}} : j = 1, \ldots, n.$$

We observe that E^*_{dj} reflects a peer evaluation of DMU_j with respect to DMU_d. If we consider DMU_d as a target DMU and solve the CCR model for each $d = 1, \ldots, n$, we will get n CCR scores and their corresponding optimal weights. Using these optimal weights, we can compute $(n - 1)$ cross-efficiency values. If we put all of these CCR and cross-efficiency values in a single $n \times n$ matrix, we get $[E_{ce}]$, that is the *cross-efficiency matrix* of the set of $DMUs$. For each column j of $[E_{ce}]$, we can compute the average of its elements that is considered as the overall performance of DMU_j [15, 21] and is known as the cross-efficiency score of DMU_j. These average scores can then be used for ranking $DMUs$.

It is well-known that the optimal weights obtained through the CCR model are not necessarily unique [7, 15, 17, 21]. Due to this fact, we may have different cross-efficiency scores and, consequently, this drawback may reduce the usefulness of cross-efficiency as a ranking method. In order to overcome this issue, the general approach consists in introducing *secondary goals* [7, 15–17]. More precisely, a complementary model is introduced and solved for the aim of screening out the alternative solutions and getting solutions that satisfy the conditions imposed by the secondary goals. Introducing secondary goals will produce new models to solve and make possible to screen out and select suitable weights. However, this approach is not as perfect as we expect and consequently, different tries have been done to find the best secondary goals [7, 15–17]. In the sequel, we propose a new secondary goal by taking into account some technical points.

2.2   A Linear Model as Secondary Goal

In the sequel, we present a model as *secondary goal*. The objective of introducing a secondary goal consists in screening multiple optima in order to get a suitable efficiency score. First of all, we solve the CCR model for evaluating DMU_d and assume that E_{dd}^* is the CCR-efficiency score of DMU_d. Then, we consider the following linear programming model:

$$(LP - CCR): \quad \max \left(\sum_{r=1}^{s} u_{rd} y_{rd} - E_{dd}^* \sum_{i=1}^{m} v_{id} x_{id} \right) \tag{5}$$

$$\text{s.t:} \ \sum_{r=1}^{s} u_{rd} y_{rj} - E_{jj}^* \sum_{i=1}^{m} v_{id} x_{ij} \leq 0 : \quad j = 1, \ldots, n, \tag{6}$$

$$\sum_{r=1}^{s} u_{rd} + \sum_{i=1}^{m} v_{id} = 1, \tag{7}$$

$$u_{rd} \geq 0 : \quad r = 1, \ldots, s, \tag{8}$$

$$v_{id} \geq 0 : \quad i = 1, \ldots, m. \tag{9}$$

Where E_{jj}^* is the efficiency score for DMU_j (for $j \in \{1, \ldots, n\}$) and the constraint (7) is a normalization constraint. In fact, the contraint (7) excludes the null vector from the set of feasible solutions. Suppose that (v_d^*, u_d^*) is the vector of optimal weights for $(LP - CCR)$. The following theorem states that the optimal value of this model is equal to *zero*.

Theorem 1. *The optimal value of* $(LP - CCR)$ *model is equal to zero.*

Proof: Since E_{dd}^* is the optimal value for CCR with the optimal vector (u^*, v^*) ≥ 0, we have

$$\frac{\sum_{r=1}^{s} u_{rd}^* y_{rd}}{\sum_{i=1}^{m} v_{id}^* x_{id}} = E_{dd}^* \text{ and } \frac{\sum_{r=1}^{s} u_{rd}^* y_{rj}}{\sum_{i=1}^{m} v_{id}^* x_{ij}} \leq 1 : j = 1, \ldots, n; j \neq d.$$

If we define E_{jj}^* as the optimal value of CCR for DMU_j, then

$$\frac{\sum_{r=1}^{s} u_{rd}^* y_{rj}}{\sum_{i=1}^{m} v_{id}^* x_{ij}} \leq E_{jj}^* : j = 1, \ldots, n; j \neq d.$$

In other words

$$\sum_{r=1}^{s} u_{rd}^* y_{rd} - E_{dd}^* \sum_{i=1}^{m} v_{id}^* x_{id} = 0 \text{ and } \sum_{r=1}^{s} u_{rj}^* y_{rd} - E_{jj}^* \sum_{i=1}^{m} v_{id}^* x_{ij} \leq 0 : \forall j \neq d.$$

Let us define $k = \frac{1}{\sum_{r=1}^{s} u_{rd}^* + \sum_{i=1}^{m} v_{id}^*}$ (we know that $\sum_{r=1}^{s} u_{rd}^* + \sum_{i=1}^{m} v_{id}^* \neq 0$). We observe that (ku^*, kv^*) is a feasible solution for $LP - CCR$, with 0 as the objective value. Furthermore, $(LP - CCR)$ in a maximization problem of a non-positive objective function. Hence, we come up with the conclusion that the optimal value of $(LP - CCR)$ is equal to 0. ∎

Corollary: According to Theorem 1, we conclude that the efficiency score obtained by $(LP - CCR)$ is equal to the efficiency score provided by the model CCR.

In order to screen the multiple optima of CCR, we consider the following Multi-Objective model as the *Secondary Goal* (MO-SG).

$(MO - SG)$:

$$\max_{j=1,\ldots,n; j \neq d} \left(\sum_{r=1}^{s} u_{rd} y_{rj} - E_{jj}^* \sum_{i=1}^{m} v_{id} x_{ij} \right).$$

$s.t.$

$$\sum_{r=1}^{s} u_{rd} y_{rj} - E_{jj}^* \sum_{i=1}^{m} v_{id} x_{ij} \leq 0 \; : j = 1, \ldots, n; j \neq d,$$

$$\sum_{r=1}^{s} u_{rd} y_{rd} - E_{dd}^* \sum_{i=1}^{m} v_{id} x_{id} = 0,$$

$$\sum_{r=1}^{s} u_{rd} + \sum_{i=1}^{m} v_{id} = 1,$$

$$u_{rd} \geq 0 \; : r = 1, \ldots, s,$$

$$v_{id} \geq 0 \; : i = 1, \ldots, m.$$

Where E_{dd}^* represents the amount of efficiency score for DMU_d and is obtained by solving CCR. The purpose of *(MO-SG)* model consists in providing optimal weights for DMU_d with fixing the efficiency of this unit and maximizing the efficiency score of other $DMUs$.

The scientific literature in multi-objective optimization includes different approaches for solving $(MO-SG)$ (see e.g., [8]).

3   Undesirable Outputs

In this section, we present an approach for peer evaluation of $DMUs$ in presence of undesirable outputs. Undesirable outputs can be in different forms. Perhaps, some of the most notable examples of undesirable outputs are *excessive carbon emissions* in a manufacturing firm and *tax payments* in business operations. Due to importance of these factors, it is necessary to study the ways of reducing undesirable outputs and ranking the most efficient firms in terms of the least undesirable productions. In this paper, we aim at introducing a procedure for peer evaluation of $DMUs$ (having some undesirable outputs). This becomes possible by means of the technique of cross-efficiency and using the new secondary goal that we presented in Section (2.2).

3.1   Data Envelopment Analysis for Undesirable Outputs

For $j \in \{1, \ldots, n\}$, suppose that DMU_j is a unit with the input vector x_i such that the vectors v_j and w_j are, respectively, the desirable and undesirable output vectors of DMU_j. Each DMU_j uses m inputs x_{ij} $(i = 1, \ldots, m)$ to produce s desirable outputs v_{rj} $(r = 1, \ldots, s)$ and h undesirable outputs w_{tj} $(t = 1, \ldots, h)$. Kousmanen et al. [13, 14] introduced the following linear programming model for evaluating DMU_d (where $d \in \{1, \ldots, n\}$)

$$\min \theta \tag{10}$$

$$s.t. \quad \sum_{j=1}^{n} (\eta_j + \mu_j) x_{ij} \leq \theta x_{id} \quad : \quad i = 1, \ldots, m, \tag{11}$$

$$\sum_{j=1}^{n} \eta_j v_{rj} \geq v_{rd} \quad : \quad r = 1, \ldots, s, \tag{12}$$

$$\sum_{j=1}^{n} \eta_j w_{tj} = w_{td} \quad : \quad t = 1, \ldots, h, \tag{13}$$

$$\sum_{j=1}^{n} (\eta_j + \mu_j) = 1, \tag{14}$$

$$\eta_j \geq 0, \mu_j \geq 0 \quad : \quad j = 1, \ldots, n. \tag{15}$$

Where $\eta_j, \mu_j \geq 0$ (for $j = 1, \ldots, n$) are structural variables for the production possibility set (PPS) (see [13, 14]). This model seeks for decreasing the inputs of DMU_d by rate of θ. The model (10)-(15) is an input oriented model with an optimal value of $\theta^* \in (0, 1]$. If $\theta^* = 1$, we say that DMU_d is technical efficient. If we write the dual of (10)-(15), we get the following model that we name it *Undesirable Multiple Form* (UMF):

$$E_{dd}^* = \max \left(\sum_{r=1}^{s} q_{rd} v_{rd} + \sum_{t=1}^{h} \pi_{td} w_{td} + \alpha_d \right) \tag{16}$$

$$s.t.$$

$$- \sum_{i=1}^{m} p_{id} x_{ij} + \sum_{r=1}^{s} q_{rd} v_{rj} + \sum_{t=1}^{h} \pi_{td} w_{tj} + \alpha_d \leq 0 \quad : \quad j = 1, \ldots, n, \tag{17}$$

$$- \sum_{i=1}^{m} p_{id} x_{ij} + \alpha_d \leq 0 \quad : \quad j = 1, \ldots, n, \tag{18}$$

$$\sum_{i=1}^{m} p_{id}x_{id} = 1, \tag{19}$$

$$p_{id} \geq 0 : \quad i = 1, \ldots, m, \tag{20}$$

$$q_{rd} \geq 0 : \quad r = 1, \ldots, s. \tag{21}$$

Where the dual variables p_{id} (for $i = 1, \ldots, m$) are defined as the weights for the inputs of DMU_d and q_{rd} (for $r = 1, \ldots, s$) (respectively, π_{td} (for $t = 1, \ldots, h$)) are the weights for desirable outputs (respectively, undesirable outputs) of DMU_d.

Let us suppose that $(p_d^*, q_d^*, \pi_d^*, \alpha_d^*)$ is the optimal solution vector of UMF model in evaluating DMU_d (where $d \in \{1, \ldots, n\}$). The cross-efficiency of DMU_j, by using the profile of weights that has been provided by DMU_d, is computed as follows [6]:

$$E_{dj}^* = \frac{\sum_{r=1}^{s} q_{rd}^* v_{rj} + \sum_{r=1}^{h} \pi_{td}^* w_{tj} + \alpha_d^*}{\sum_{i=1}^{m} p_{id}^* x_{ij}} \quad : \quad j = 1, \ldots, n \tag{22}$$

By solving UMF model for each $d \in \{1, \ldots, n\}$ and using (22), we can construct an $n \times n$ matrix $[E_{ce}^*]$ that is the *matrix of cross-efficiency*.

We know that the UMF model may have alternative solutions and consequently, we may have several cross-efficiency matrices. In order to overcome this problem, we can proceed as we explained in Section 2. To this end, we do a *cross-efficiency evaluation with secondary goal* for each DMU_d (for $d = 1, \ldots, n$). First of all, we solve the UMF model for each DMU_j ($j = 1, \ldots, n$) and we get E_{jj}^*, then we consider the (MO-SG) model as the following version in which the undesirable outputs are taken into account:

$(MO-SG)_{undesirable}$:

$$\max_{j=1,\ldots,n;j\neq d} \left(\sum_{r=1}^{s} q_{rd} v_{rj} + \sum_{t=1}^{h} \pi_{td} w_{tj} + \alpha_d - E_{jj}^* \sum_{i=1}^{m} p_{id} x_{ij} \right) \tag{23}$$

s.t.

$$\sum_{r=1}^{s} q_{rd} v_{rd} + \sum_{t-1}^{h} \pi_{td} w_{td} + \alpha_d - E_{dd}^* \sum_{i=1}^{m} p_{id} x_{id} = 0, \tag{24}$$

$$\sum_{r=1}^{s} q_{rd} v_{rj} + \sum_{t=1}^{h} \pi_{td} w_{tj} + \alpha_d - E_{jj}^* \sum_{i=1}^{m} p_{id} x_{ij} \leq 0 : j = 1, \ldots, n, \tag{25}$$

$$-\sum_{i=1}^{m} p_{id} x_{ij} + \alpha_d \leq 0 : j = 1, \ldots, n, \tag{26}$$

$$\sum_{r=1}^{s} q_{rd} + \sum_{t=1}^{h} \pi_{td} + \sum_{i=1}^{m} p_{id} = 1, \tag{27}$$

$$p_{id} \geq 0 \ : i = 1, \ldots, m, \qquad (28)$$

$$q_{rd} \geq 0 \ : r = 1, \ldots, s. \qquad (29)$$

We note that this model is similar to (MO-SG); however, this model takes into account the undesirable factors of $DMUs$. By keeping the score of DMU_d at its own level, the model (23)-(29) seeks for the optimal weights such that the efficiency scores of other $DMUs$ are *maximized*. In order to solve this multi-objective model, many approaches can be chosen from the literature of multi-objective programming. One of the classical approaches consists in using the *max-min method* [8, 11]. In this approach, we introduce a supplementary variable, e.g., ϕ for transforming the multi-objective model (23)-(29) to the following linear programming (LP) model (where $J = \{1, \ldots, n\}$).

$$\max \phi \qquad (30)$$

$$s.t.$$

$$\phi \leq \sum_{r=1}^{s} q_{rd} v_{rj} + \sum_{t=1}^{h} \pi_{td} w_{tj} + \alpha_d - E_{jj}^* \sum_{i=1}^{m} p_{id} x_{ij} \ : j \in J \setminus \{d\} (31)$$

$$\sum_{r=1}^{s} q_{rd} v_{rd} + \sum_{t=1}^{h} \pi_{td} w_{td} + \alpha_d - E_{dd}^* \sum_{i=1}^{m} p_{id} x_{id} = 0, \qquad (32)$$

$$\sum_{r=1}^{s} q_{rd} v_{rj} + \sum_{t=1}^{h} \pi_{td} w_{tj} + \alpha_d - E_{jj}^* \sum_{i=1}^{m} p_{id} x_{ij} \leq 0 \ : j \in J, \qquad (33)$$

$$- \sum_{i=1}^{m} p_{id} x_{ij} + \alpha_d \leq 0 \ : j \in J, \qquad (34)$$

$$\sum_{r=1}^{s} q_{rd} + \sum_{t=1}^{h} \pi_{td} + \sum_{i=1}^{m} p_{id} = 1, \qquad (35)$$

$$p_{id} \geq 0 \ : i = 1, \ldots, m, (36)$$

$$q_{rd} \geq 0 \ : r = 1, \ldots, s. \ (37)$$

To sum up, the following procedure outlines the process of forming the matrix of cross-efficiency by using the proposed secondary goal and for evaluating $DMUs$ having undesirable outputs.

Procedure of Cross-Efficiency with Secondary Goal (PCE-SG) (in presence of undesirable outputs)

Step 1. For each $j = 1, \ldots, n$, solve the model UMF to obtain E_{jj}^*.
Step 2. For each $d = 1, \ldots, n$, solve the *max-min* model (30)-(37) to obtain the solution vector $(p_d^*, q_d^*, \pi_d^*, \alpha_d^*)$ as the profile of weights for DMU_d.

Step 3. For each $d = 1, \ldots, n$, use the vector $(p_d^*, q_d^*, \pi_d^*, \alpha_d^*)$ in (22) to construct the dth row of the matrix of cross-efficiency.

Once the matrix of cross-efficiency is formed, we can compute the *cross-efficiency score (CES)* of each DMU_j (for $j = 1, \ldots, n$). The cross-efficiency score of DMU_j is computed by taking the average of the entries on the jth column of the matrix (see also [6, 15]), i.e.,

$$\overline{E_j^*} := \frac{\sum\limits_{d=1}^{n} E_{dj}^*}{n} \tag{38}$$

4   Numerical Example

In this section, the presented cross-efficiency approach is used to evaluate the efficiency of Chinese provinces in matter of the environmental criteria [22]. In this evaluation, the data of 16 Chinese provinces are considered. Table 1 includes the whole data. In this table, each province is considered as a DMU and it has two inputs (the total energy assumption (x_1) and the total population (x_2)) and four outputs, including one desirable output (GDP, denoted by v) and three undesirable outputs: the total amounts of industrial emissions of waste water (w_1), waste gas (w_2), and waste solids (w_3) [22].

We apply the procedure (PCE-SG) to form the matrix of cross-efficiency. Then, we use (38) to compute the cross-efficiency score (CES) of each DMU.

In order to solve the linear programming (LP) models, we use the standard solver *Lingo 11* on a *Pentium Dual-Core CPU*, with 4GB RAM and 2.10 GHz. The computational time for solving LP models is negligible. The results are presented in Table 2. In this table, we present the meaningful information of the cross-efficiency matrix. More precisely, for each DMU, we present: minimum score (min), maximum score (max), average of scores (CES) (that is the *Cross-Efficiency Score (CES)* as we defined in previous sections), standard deviation of scores (SD), range of scores ($range$), and the $rank$ of the DMU.

The ranking of $DMUs$ is based on the CES values, that is known as the *cross-efficiency score* [6, 15]. We note that the max values correspond to the efficiency scores in conventional CCR model. The CCR model ranks the $DMUs$ by a self-evaluation procedure. However, by using the cross-efficiency, we take its advantage as a weighting procedure that helps us to rank $DMUs$ by a peer evaluation of all $DMUs$. In this context, we use the new secondary goal that lets us to consider the best performance of all DMU.

As we observe in Table 2, through the procedure (PCE-SG), the information provided by the method of cross-efficiency helps us in ranking $DMUs$ in an explicite way and (in this example) without any tie. This fact shows the pure advantage of using a peer evaluation of $DMUs$ through the method of cross-efficiency and, particularly, the proposed secondary goal. More precisely, by means of (PCE-SG) we can take into account more information about all $DMUs$ and the ranking is based on these information.

Table 1. The data corresponding to 16 $DMUs$: inputs (total energy assumption, total population) and desirable outputs (GDP) and undesirable outputs (total industrial emission of waste water, waste gas, waste solids)

DMU	Inputs		Desirable Output	Undesirable Outputs		
	x_1	x_2	v	w_1	w_2	w_3
1	6570	15096	12153.03	8713	4408	1242.4
2	5874	10473	7521.85	19441	5983	1515.7
3	15576	30378	7358.31	39720	23693	14742.9
4	19112	38429	15212.49	75159	25211	17221.4
5	7698	24349	7278.75	37563	7124	3940.5
6	23709	68371	34457.30	256160	27432	8027.8
7	15567	45472	22990.35	203442	18860	3909.7
8	8916	32097	12236.53	142747	10497	6348.9
9	19751	83974	19480.46	140325	22186	10785.8
10	13708	50862	12961.10	91324	12523	5561.5
11	13331	56820	13059.69	96396	10973	5092.8
12	24654	84981	39482.56	188844	22682	4740.9
13	7030	25284	6530.01	65684	12587	2551.8
14	16322	72471	14151.28	105910	13410	8596.9
15	8032	40460	6169.75	32375	9484	8672.8
16	8044	33504	8169.80	49137	11032	5546.7

Table 2. The results of cross-efficiency (CE) method for ranking 16 $DMUs$

DMU	min	max	range	SD	CES	rank
1	0.10	0.94	0.84	0.264	0.48	13
2	0.23	1	0.77	0.267	0.64	7
3	0.12	1	0.88	0.338	0.73	3
4	0.19	1	0.81	0.281	0.71	4
5	0.15	0.88	0.73	0.199	0.54	11
6	0.25	0.92	0.67	0.255	0.61	9
7	0.23	1	0.77	0.298	0.65	6
8	0.19	1	0.81	0.275	0.79	1
9	0.11	0.77	0.66	0.211	0.49	12
10	0.12	0.64	0.52	0.165	0.46	14
11	0.19	1	0.81	0.312	0.77	2
12	0.15	0.68	0.53	0.182	0.44	15
13	0.12	1	0.88	0.328	0.71	5
14	0.09	0.64	0.55	0.168	0.43	16
15	0.07	1	0.93	0.330	0.60	10
16	0.12	0.95	0.83	0.270	0.62	8

As we observe in Table 2, there are 8 efficient $DMUs$ in terms of traditional efficiency evaluation (see the column "max") and this is a drawback of self-evaluation. According to Table 2, there is just one cross-efficient DMU, that is DMU_8. Particularly, DMU_8 is efficient not only in the traditional sense, but also in terms of the *cross-efficiency with secondary goal*. In terms of efficiency, DMU_8 dominates the other $DMUs$. It is important to note that the dominance of DMU_8 with respect to the others has a strong sense, i.e., DMU_8 is efficient by taking into account the *best performance* of all $DMUs$.

Finally, in this example, the procedure (PCE-SG) is used for evaluating $DMUs$ in presence of *undesirable outputs*. This procedure helps to identify the best $DMU(s)$ and produces a ranking approach. By means of these results, we can recognize the necessary improvements that must be done in the inputs and the outputs (desirable and/or undesirable) of any inefficient DMU in order to obtain efficient $DMUs$ with the least undesirable factors.

5 Conclusion

In this paper, we discussed about a cross-efficiency approach for evaluating the performance of $DMUs$ in presence of undesirable outputs. More precisely, we studied a cross-efficiency approach by introducing a new *secondary goal*. This approach consists in constructing a multi-objective model that can be transformed to an LP model in many ways, for example, the classical *max-min method* [8, 11]. The introduced approach has been extended (in a natural way) for evaluation of $DMUs$ with *undesirable outputs*. We applied our approach on a real-world data set [22] involving *undesirable outputs*. According to the results, our approach is useful in peer-evaluation of $DMUs$ and helps decision makers in ranking units by taking into account the best performance of all $DMUs$; particularly, in the case of ties in conventional DEA scoring systems. The current study is in progress by considering alternative secondary goals. The complementary results will be reported in future.

Acknowledgements. Mahdi Moeini acknowledges the chair of Business Information Systems and Operations Research (BISOR) at the TU-Kaiserslautern (Germany) for the financial support, through the research program "CoVaCo".

References

1. Banker, R.D., Charnes, A., Cooper, W.W.: Some models for estimating technical and scale inefficiences in data envelopment analysis. Management Science 30, 1078–1091 (1984)
2. Beasley, J.E.: Allocating fixed costs and resources via data envelopment analysis. European Journal of Operational Research 147, 198–216 (2003)
3. Charnes, A., Cooper, W.W., Rhodes, E.: Measuring the efficiency of decision making units. European Journal of Operational Research 1, 429–444 (1987)
4. Chen, T.Y.: An assessment of technical efficiency and cross-efficiency in Taiwan's electricity distribution sector. European Journal of Operational Research 137, 421–433 (2002)
5. Cook, W.D., Seiford, L.M.: Data envelopment analysis (DEA): Thirty years on. European Journal of Operational Research 2, 1–17 (2009)

6. Cooper, W.W., Ramon, N., Ruiz, J.L., Sirvent, I.: Avoiding Large Differences in Weights in Cross-Efficiency Evaluations: Application to the Ranking of Basketball Players. Journal of CENTRUM Cathedra 4(2), 197–215 (2011)
7. Doyle, J., Green, R.: Efficiency and cross efficiency in DEA: Derivations, meanings and the uses. Journal of the Operational Research Society 54, 567–578 (1994)
8. Ehrgott, M.: Multicriteria Optimization. Springer (2005)
9. Fare, R., Grosskopf, S., Lovel, C.A.K.: Multilateral productivity comparisons when some outputs are undesirable: a nonparametric approach. The Review of Economics and Statistics 66, 90–98 (1989)
10. Fare, R., Grosskopf, S., Lovel, C.A.K., Yaiswarng, S.: Deviation of shadow prices for undesirable outputs: a distance function approach. The Review of Economics and Statistics 218, 374–380 (1993)
11. Gulpinar, N., Le Thi, H.A., Moeini, M.: Robust Investment Strategies with Discrete Asset Choice Constraints Using DC Programming. Optimization 59(1), 45–62 (2010)
12. Hailu, A., Veeman, T.: Non-parametric productivity analysis with undesirable outputs: an application to Canadian pulp and paper industry. American Journal of Agricultural Economics, 605–616 (2001)
13. Kuosmanen, T.: Weak Disposability in Nonparametric Productivity Analysis with Undesirable Outputs. American Journal of Agricultural Economics 37, 1077–1082 (2005)
14. Kuosmanen, T., Poidinovski, V.: A Weak Disposability in Nonparametric Productivity Analysis with Undesirable Outputs: Reply to Fare and Grosskopf. American Journal of Agricultural Economics 18 (2009)
15. Liang, L., Wu, J., Cook, W.D., Zhu, J.I.: Alternative secondary goals in DEA cross efficiency evaluation. International Journal of Production Economics 36, 1025–1030 (2008)
16. Rodder, W., Reucher, E.: A consensual peer-based DEA-model with optimized cross-efficiencies: input allocation instead of radial reduction. European Journal of Operational Research 36, 148–154 (2011)
17. Rodder, W., Reucher, E.: Advanced X-efficiencies for CCR- and BCC-models: towards Peer-based DEA controlling. European Journal of Operational Research 219, 467–476 (2012)
18. Seiford, L.M., Zhu, J.: Modeling undesirable factors in efficiency evaluation. European Journal of Operational Research 48, 16–20 (2002)
19. Sexton, T.R., Silkman, R.H., Hogan, A.J.: Data envelopment analysis: Critique and extensions. In: Silkman, R.H. (ed.) Measuring Efficiency: An Assessment of Data Envelopment Analysis, vol. 3, pp. 73–105. Jossey-Bass, San Francisco (1986)
20. Tone, K.: Dealing with undesirable outputs in DEA: a slacks-based measure (SBM) approach. Presentation at NAPWIII, Toronto (2004)
21. Wang, Y.M., Chin, K.S.: Some alternative models for DEA cross efficiency evaluation. International Journal of Production Economics 36, 332–338 (2010)
22. Wu, C., Li, Y., Liu, Q., Wang, K.: A stochastic DEA model considering undesirable outputs with weak disposability. Mathematical and Computer Modeling (2012), doi:10.1016/j.mcm.2012.09.022
23. Wu, J., An, Q., Xiong, B., Chen, Y.: Congestion measurement for regional industries in China: A data envelopment analysis approach with undesirable outputs. Energy Policy 57, 7–13 (2013)
24. Wu, J., An, Q., Ali, S., Liang, L.: DEA based resource allocation considering environmental factors. Mathematical and Computer Modelling 58(5-6), 1128–1137 (2013)

Author Index

Printed in the United States
By Bookmasters